Sustainability
SCIENCE

Sustainability SCIENCE

Managing Risk and Resilience for Sustainable Development

Second Edition

PER BECKER

Division of Risk Management and Societal Safety,
Lund University, Sweden

Climate, Environment, and Sustainability,
NORCE Norwegian Research Centre, Norway

Unit for Environmental Sciences and Management,
North-West University, South Africa

ELSEVIER

Elsevier
Radarweg 29, PO Box 211, 1000 AE Amsterdam, Netherlands
The Boulevard, Langford Lane, Kidlington, Oxford OX5 1GB, United Kingdom
50 Hampshire Street, 5th Floor, Cambridge, MA 02139, United States

Notices

ISBN: 978-0-323-95640-6

For information on all Elsevier publications visit our website at https://www.elsevier.com/books-and-journals

Publisher: Candice Janco
Acquisitions Editor: Jennette McClain
Editorial Project Manager: Howi De Ramos
Production Project Manager: Paul Prasad Chandramohan
Cover Designer: Christian Bilbow

Typeset by TNQ Technologies

CONTENTS

Acknowledgements

Issues of risk, resilience and sustainability touch all of us on this planet. Throughout my career, I have had the opportunity to work in many different contexts. I have worked in academia and for local and national authorities, international organisations and even a short period for a consultancy firm. For a large part of my career, I have been involved in humanitarian assistance and international development cooperation, which, together with my later academic career, have taken me to all continents except Antarctica. This has provided me with ample opportunity to meet many people who, in different ways, have shared their views, experiences, skills and knowledge concerning various aspects covered in this book. I am eternally grateful for all the time and kind attention they have given me.

I would also like to thank all my colleagues with whom I have had so many exciting dialogues concerning philosophical, theoretical and practical issues captured in this book, especially my friends and colleagues at Lund University, the Swedish Civil Contingency Agency (MSB), UNICEF and within the Red Cross and Red Crescent Movement, with whom I have spent the most time, but also at the various universities, international organisations and NGOs I have had the pleasure to work with during my career. Finally, this book was conceived through a sometimes rather painful birth, with too many evenings and weekends of labour. I would, therefore, like to thank my wife and sons for their patience and support.

CHAPTER 1

Introducing the book

Introduction

The world population has been estimated to have increased almost six times (Maddison, 2001, p. 28), the global economy around 50 times (Maddison, 2001) and the global CO_2 emissions about 500 times (Boden et al., 2011) from the onset of the industrial revolution to the turn of the millennium. These developments continue, with the global population increasing by another third since then—from around six billion to eight billion (United Nations, 2019)—while the world economy tripled (World Bank, 2023) and the global CO_2 emissions from fossil fuels and land use change increased by a third in the same period (Friedlingstein et al., 2022). These developments have always placed growing strains on the world's natural resources and environment (e.g., Dong et al., 2018; Fan & Qi, 2010; Gadda & Gasparatos, 2009; Maja & Ayano, 2021; Syvitski, 2008), while vast inequalities persist and even deepen both between and within states (e.g., Bywaters, 2009; Diffenbaugh & Burke, 2019; Hill & Jorgenson, 2018). Although the last century saw a global increase in life expectancy (Riley, 2001) and a decrease in child mortality (Ahmad et al., 2000, p. 1175) and adult illiteracy (Parris & Kates, 2003, pp. 8070—8071), economic development was highly unequal rendering the same wealth in the final decade of the last century to the richest 1% in the world as to the poorest 57% (Milanovic, 2002, p. 50). However, this already staggering inequality has been seriously exacerbated in the new millennium, with the richest 1% now owning 44% while the poorest half of the global population owns less than 1% of all wealth (Credit Suisse, 2021, p. 25). To reduce poverty while striving towards a more viable use of natural resources, it is vital to make future development more sustainable.

Regardless of whether one focuses on economic growth or more human-centred parameters, most uses of the concept of development have one thing in common. They project some scenarios into the future in which the variables of interest develop over time along a preferred expected course. This scenario is, in modern society, not believed to be predestined or predetermined in any way but dependent on a wide range of human activity, environmental processes, etc. The complexity and dynamic character of the world is, instead, continuously creating a multitude of possible futures (Japp & Kusche, 2008, p. 80), causing uncertainty as to what real development will materialise (Figure 1.1).

Being unable to see into the future, as well as generally incapable of predicting it (Simon, 1990, pp. 7—8; Taleb, 2007), modern individuals, organisations and societies

Sustainability Science, Second Edition
ISBN 978-0-323-95640-6, https://doi.org/10.1016/B978-0-323-95640-6.00005-1

Figure 1.1 What world do we want? *(Source: kwest | Shutterstock.com.)*

resort to the notion of risk to make sense of their uncertain world (Zinn, 2008, pp. 3–10). Risk is a contested concept, but to talk about risk entails some idea of uncertain futures and their potential impacts on what human beings value (Renn, 1998, p. 51; Sörensen et al., 2016, p. 335). This use of risk also entails that risk must be defined in relation to some preferred expected outcome (Kaplan & Garrick, 1981; Luhmann, 1995, pp. 307–310; Zinn, 2008, p. 4; Becker & Tehler, 2013, p. 19). If the risk is related to potential deviations from a preferred expected future, we must endeavour to reduce such risk to safeguard our development objectives.

Many courses of events and their underlying processes may negatively impact development in the short or long term. Abrupt changes in political leadership, global financial crises, algal bloom, epidemic outbreaks, droughts, cyclones and outbreaks of communal violence are just a few examples of initiating events that may set off destructive courses of events. Behind these often dramatic courses of events lay processes of change which are less sensational but may have far-reaching indirect impacts, such as environmental degradation (Boardman, 2006; Ellwanger et al., 2020; Huang et al., 2020; Tan et al., 2022), demographic and socio-economic processes (Satterthwaite et al., 2009, pp. 11–19; Wilkinson, 2005), globalisation (Beck, 1999; Tu et al., 2019), changing antagonistic threats (Kaldor, 1999; von Solms & van Niekerk, 2013) and the increasing complexity of modern society (Berndtsson et al., 2019; Perrow, 2008). In addition, we have the mounting threats of climate change, not only potentially increasing the frequency and intensity of destructive extreme weather events (Cook et al., 2020; Elsner et al., 2008; IPCC, 2012; Nicholls & Cazenave, 2010), but changing everyday life for vast numbers of people.

These courses of events and their underlying processes rarely exist in isolation, neither from each other nor from the development activities and processes they impact. It is,

Figure 1.2 Artificial turf may have benefits today, but it is a significant source of microplastics. *(Source: Koonsiri Boonnak | Dreamstime.com.)*

therefore, not only vital to ensure that development gains are durable in the face of destructive courses of events and their underlying processes but also that the means to reach the development gains do not augment or create new risks that hinder development for future generations (WCED, 1987, p. 43) (Figure 1.2).

Purpose of the book

As I attempt to show in this book, the increasing complexity and dynamic character of our world demand conceptual and practical approaches to sustainable development that help us grasp and manage uncertainty, complexity, ambiguity and dynamic change. I argue that risk is a key concept in this context, as considering sustainability requires us to think ahead into an uncertain future. I also assert that the concept of resilience is central, and take it further by providing a conceptual framework that gives practical guidance for analysing and developing the resilience of societies, communities, organisations, etc. This book is, therefore, necessarily transdisciplinary, drawing upon contributions from a wide range of disciplines (e.g., anthropology, archaeology, design, engineering, geography, public administration, sociology, etc.) and integrating them under the premise of Sustainability Science—the premise of bringing together 'scholarship and practice, global and local perspectives from north and south, and disciplines across' all sciences (Clarke & Dickson, 2003, p. 8060) to address the core challenges of humankind (Clarke & Dickson, 2003; Heinrichs et al., 2016; Kates et al., 2001; Lang et al., 2012).

Sustainability Science asserts that to facilitate the much-needed shift towards sustainable development, we must be able to span the range of spatial and temporal scales of various phenomena, manage complexity, and recognise a wide range of perspectives as useable knowledge from both society and science (Kates et al., 2001, p. 641). This task is formidable, but I intend to contribute by presenting one approach to risk, resilience and sustainability designed to tackle it, leaving you to judge if this approach is useful for your purposes.

The book is both descriptive, in the sense of describing how the world is, and prescriptive, in the sense of prescribing what it ought to be and what we ought to do to get there. However, I attempt to maintain scientific rigour in both, showing how traditional science and design science can complement each other when our needs for explaining and understanding various phenomena give way to solving real-world problems. In other words, when we shift from being mainly concerned with the pursuit of knowledge (e.g., Weber, 1949) to focusing on designing artefacts for satisfying predefined purposes (Simon, 1996, pp. 4–5).

In short, the purpose of the book is to present a coherent framework for grasping and addressing issues of sustainability in our increasingly complex and dynamic world.

Demarcation of the book

Sustainable development is both conceptually and practically a broad and multifaceted issue (Kates et al., 2001; WCED, 1987). It is an issue of paramount importance for the continued existence of the world as we know it. At its core lies the idea that in planning for the future, we must think about what to do and not to do today to bring about that future (Simon, 1990, p. 11). The main part of sustainability must, in other words, be forward-looking, although we must also learn from our past and recognise our present challenges.

The Oxford Dictionary defines *sustainable* as 'able to be upheld or defended' (Oxford English Dictionary, 2020). It indicates a somewhat double meaning, which not only provides a linguistic link between safety and sustainability but also denotes two requisite parts for sustainable development. Safety—'the condition of being protected from or unlikely to cause danger, risk, or injury' (Oxford English Dictionary, 2020)—is, in other words, closely related to the notion of being 'able to be defended'. However, safety often connotes immediate or short time spans, while sustainability typically connotes gazing further into the future. That said, both entail acting now, and Sustainability Science states the necessity to be able to integrate such a range of temporal scales (Kates et al., 2001, p. 641). Safety is, thus, engulfed by sustainability if looking beyond the immediate. If focusing on the potential of future destructive courses of events, at least partly resulting from or related to human activity, we typically assert that such activity or development is not sustainable. While the same situation, but with an immediate focus, would instead evoke notions of an unsafe condition or practice.

The other requisite part of sustainable development, related to the notion of being 'able to be upheld', is equally important and highlights how we exploit our resources to maintain or develop some aspect of society over time. Regardless of how closely related these two parts are, this book will focus mainly on the notion of sustainability in the sense of protecting what human beings value, now and in the future, and not to the same extent on the notion of sustainability in the sense of management of our vital resources—although overusing them may lead to courses of events that do impact something we value and want to protect.

Structure of the book

It can always be debated how to structure the contents of a book best to guide the readers through such a multifaceted topic. I have chosen to divide the book into parts addressing issues of sustainability from three different angles—the descriptive, the conceptual and the transformative. Hence, going from the concrete to the abstract and back to the concrete again. In other words, in addition to the current introductory chapter, this book consists of three parts with five chapters each and a concluding chapter (Figure 1.3).

The first part, *Part I—The State of the World*, sets the stage in Chapter 2 by presenting a broad historical overview of the development and past sustainability problems of humankind, of related social change, and of the invention of risk as a reaction to our increasing appreciation of our own agency. Chapter 3 is devoted to our growing awareness of our sustainability challenges, as depicted in two strings of world conferences—starting in Stockholm in 1972 and, most lately, in Stockholm again 50 years later—and to the vision of a sustainable society they portray. Chapter 4 focuses on our boundaries for sustainability, drawing on the work of influential Earth System scientists and a subsequent social scientific critique. Chapter 5 presents a comprehensive account of a broad range of symptomatic events of the dire state of our sustainability, which over the last decade have finally made their way to be included as core sustainability challenges in their own right. I end the first part with Chapter 6, emphasising the dynamic character of our world by presenting central processes of change that continuously transform it and the risk we live under.

The second part, *Part II—Approaching the World*, provides a coherent framework for grasping and addressing sustainability challenges in our complex and dynamic world. Chapter 7 presents a set of key assumptions and concepts that provide the foundation for my approach to facilitating sustainable development, such as development, sustainability, risk and resilience. Then, in Chapter 8, I suggest an operationalisation of the concept of resilience by connecting the conceptual to the actual and linking this approach

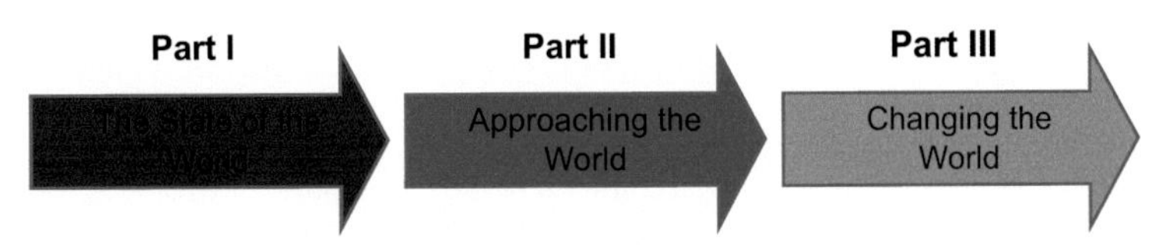

Figure 1.3 The three parts of the book.

to other established approaches to risk, safety and sustainability. Chapter 9 is dedicated to complexity—what it is, how it manifests, and its consequences. In Chapter 10, I elaborate on governing and the governmentalisation of sustainability—the overall process through which sustainability challenges become something governable on the societal level. Finally, in Chapter 11, I elaborate on why it is helpful to approach our world as human-environment systems and how to do it. It includes key aspects of the intrinsically human in our human-environment systems, which we still have to keep in mind when attempting to see parts of our complex world as wholes.

The third part, *Part III—Changing the World*, starts in Chapter 12 by presenting ideas of science and change. Here, I focus on the role of science *for* change and on demonstrating how traditional science and design science complement each other and provide a rigorous way of bridging the divide between 'is' and 'ought'. Chapter 13 elaborates on important and increasingly problematic issues of knowledge resistance and resistance to change. Chapter 14 elaborates on capacity development as an intentional process to strengthen resilience with the assistance of outsiders. I end the last part with Chapter 15, presenting ideas that describe and prescribe social change for a more sustainable world, and Chapter 16, speculating around the possibility that we may already be in a new great revolution.

Finally, the book concludes with my final remarks in Chapter 17, closing the attempted rhetorical loop by tying the three parts together and reflecting on how each contributes to grasping and addressing sustainability issues in our increasingly complex and dynamic world.

Conclusion

Our world is in a dire state, and to steer it towards a more sustainable future, we must be able to recognise and cope with uncertainty, complexity, ambiguity and dynamic change. It is a genuinely transdisciplinary endeavour, requiring the contribution of various disciplines while integrating them under the umbrella of Sustainability Science. Although direct conservation of natural resources also is a requisite for sustainability, this book mainly focuses on sustainability in the sense of protecting what human beings value for our present and future. The concepts of risk and resilience are central to this enterprise and constitute vital frames for all three parts of this book.

References

Ahmad, O. B., Lopez, A. D., & Inoue, M. (2000). The decline in child mortality: A reappraisal. *Bulletin of the World Health Organization, 78*, 1175–1191.

Beck, U. (1999). *World risk society*. Polity.

Becker, P., & Tehler, H. (2013). Constructing a common holistic description of what is valuable and important to protect: A possible requisite for disaster risk management. *International Journal of Disaster Risk Reduction, 6*, 18–27. https://doi.org/10.1016/j.ijdrr.2013.03.005

Berndtsson, R., Becker, P., Persson, A., Aspegren, H., Haghighatafshar, S., Jönsson, K., Larsson, R., Mobini, S., Mottaghi, M., Nilsson, J., Nordström, J., Pilesjö, P., Scholz, M., Sternudd, C., Sörensen, J., & Tussupova, K. (2019). Drivers of changing urban flood risk: A framework for action. *Journal of Environmental Management, 240*, 47–56. https://doi.org/10.1016/j.jenvman.2019.03.094

Boardman, J. (2006). Soil erosion science: Reflections on the limitations of current approaches. *Catena, 68*(2–3), 73–86. https://doi.org/10.1016/j.catena.2006.03.007

Boden, T., Marland, G., & Andres, B. (2011). *Global CO2 emissions from fossil-fuel burning, cement manufacture, and gas flaring: 1751–2008*. Carbon Dioxide Information Analysis Center.

Bywaters, P. (2009). Tackling inequalities in health: A global challenge for social work. *British Journal of Social Work, 39*, 353–367.

Clarke, W. C., & Dickson, N. M. (2003). Sustainability science: The emerging research program. *Proceedings of the National Academy of Sciences, 100*, 8059–8061.

Credit Suisse. (2021). Global wealth report 2021. Credit Suisse.

Cook, B. I., Mankin, J. S., Marvel, K., Williams, A. P., Smerdon, J. E., & Anchukaitis, K. J. (2020). Twenty-first century drought projections in the CMIP6 forcing scenarios. *Earth's Future, 8*(6), 1–20. https://doi.org/10.1029/2019ef001461

Diffenbaugh, N. S., & Burke, M. (2019). Global warming has increased global economic inequality. *Proceedings of the National Academy of Sciences of the U S A, 116*(20), 9808–9813. https://doi.org/10.1073/pnas.1816020116

Dong, K., Hochman, G., Zhang, Y., Sun, R., Li, H., & Liao, H. (2018). CO2 emissions, economic and population growth, and renewable energy: Empirical evidence across regions. *Energy Economics, 75*, 180–192. https://doi.org/10.1016/j.eneco.2018.08.017

Ellwanger, J. H., Kulmann-Leal, B., Kaminski, V. L., Valverde-Villegas, J. M., Veiga, A. B. G. D., Spilki, F. R., Fearnside, P. M., Caesar, L., Giatti, L. L., Wallau, G. L., Almeida, S. E. M., Borba, M. R., Hora, V. P. D., & Chies, J. A. B. (2020). Beyond diversity loss and climate change: Impacts of Amazon deforestation on infectious diseases and public health. *Anais da Academia Brasileira de Ciencias, 92*(1), e20191375. https://doi.org/10.1590/0001-3765202020191375

Elsner, J. B., Kossin, J. P., & Jagger, T. H. (2008). The increasing intensity of the strongest tropical cyclones. *Nature, 455*, 92–95. https://doi.org/10.1038/nature07234

Fan, P., & Qi, J. (2010). Assessing the sustainability of major cities in China. *Sustainability Science, 5*, 51–68. https://doi.org/10.1007/s11625-009-0096-y

Friedlingstein, P., O'Sullivan, M., Jones, M. W., Andrew, R. M., Gregor, L., Hauck, J., Le Quéré, C., Luijkx, I. T., Olsen, A., Peters, G. P., Peters, W., Pongratz, J., Schwingshackl, C., Sitch, S., Canadell, J. G., Ciais, P., Jackson, R. B., Alin, S. R., Alkama, R., … Zheng, B. (2022). Global carbon budget 2022. *Earth System Science Data, 14*(11), 4811–4900. https://doi.org/10.5194/essd-14-4811-2022

Gadda, T., & Gasparatos, A. (2009). Land use and cover change in Japan and Tokyo's appetite for meat. *Sustainability Science, 4*, 165–177. https://doi.org/10.1007/s11625-009-0085-1

Heinrichs, H., Martens, P., Michelsen, G., & Wiek, A. (Eds.). (2016). *Sustainability science: An introduction*. Springer.

Hill, T. D., & Jorgenson, A. (2018). Bring out your dead!: A study of income inequality and life expectancy in the United States, 2000-2010. *Health & Place, 49*, 1–6. https://doi.org/10.1016/j.healthplace.2017.11.001

Huang, J., Zhang, G., Zhang, Y., Guan, X., Wei, Y., & Guo, R. (2020). Global desertification vulnerability to climate change and human activities. *Land Degradation & Development, 31*(11), 1380–1391. https://doi.org/10.1002/ldr.3556

IPCC. (2012). *Managing the risks of extreme events and disasters to advance climate change adaptation (SREX)*.

Japp, K. P., & Kusche, I. (2008). Systems theory and risk. In J. O. Zinn (Ed.), *Social theories of risk and uncertainty: An introduction* (pp. 76–105). Blackwell Publishing.

Kaldor, M. (1999). *New and old wars: Organized violence in a global era*. Polity Press.

Kaplan, S., & Garrick, B. J. (1981). On the quantitative definition of risk. *Risk Analysis, 1*, 11–27. https://doi.org/10.1111/j.1539-6924.1981.tb01350.x

Kates, R. W., Clarke, W. C., Corell, R. W., Hall, J. M., Jaeger, C. C., Lowe, I., McCarthy, J. J., Schellnhuber, H. J., Bolin, B., Dickson, N. M., Faucheux, S., Gallopín, G. C., Huntley, B., Jodha, N. S., Kasperson, R. E., Mabogunje, A., Matson, P. A., Mooney, H., Moore, B., … Svedin, U. (2001). Sustainability science. *Science, 292*, 641—642.

Lang, D. J., Wiek, A., Bergmann, M., Stauffacher, M., Martens, P., Moll, P., Swilling, M., & Thomas, C. J. (2012). Transdisciplinary research in sustainability science: Practice, principles, and challenges. *Sustainability Science, 7*(S1), 25—43. https://doi.org/10.1007/s11625-011-0149-x

Luhmann, N. (1995). *Social systems*. Stanford University Press.

Maddison, A. (2001). *The world economy: A millennial perspective*. Organisation for Economic Cooperation and Development.

Maja, M. M., & Ayano, S. F. (2021). The impact of population growth on natural resources and farmers' capacity to adapt to climate change in low-income countries. *Earth Systems and Environment, 5*, 271—283. https://doi.org/10.1007/s41748-021-00209-6

Milanovic, B. (2002). True World Income Distribution, 1988 and 1993: First calculation based on household surveys alone. *The Economic Journal, 112*, 51—92. https://doi.org/10.1111/1468-0297.0j673

Nicholls, R. J., & Cazenave, A. (2010). Sea-level rise and its impact on coastal zones. *Science, 328*(5985), 1517—1520. https://doi.org/10.1126/science.1185782

Oxford English Dictionary. (2020). *Oxford dictionary of English*. Oxford University Press.

Parris, T. M., & Kates, R. W. (2003). Characterizing a sustainability transition: Goals, targets, trends, and driving forces. *Proceedings of the National Academy of Sciences, 100*, 8068—8073.

Perrow, C. B. (2008). Disasters evermore? Reducing our vulnerabilities to natural, industrial, and terrorist disasters. *Social Research: An International Quarterly of Social Sciences, 75*, 733—752.

Renn, O. (1998). The role of risk perception for risk management. *Reliability Engineering & System Safety, 59*, 49—62. https://doi.org/10.1016/S0951-8320(97)00119-1

Riley, J. C. (2001). *Rising life expectancy: A global history*. Cambridge University Press.

Satterthwaite, D., Huq, S., Reid, H., Pelling, M., & Romero Lankao, P. (2009). In J. Bicknell, D. Dodman, & D. Satterthwaite (Eds.), *Adapting to climate change in urban areas: The possibilities and constraints in low- and middle-income Nations* (pp. 3—47). Earthscan.

Simon, H. A. (1990). Prediction and prescription in systems modeling. *Operations Research, 38*, 7—14.

Simon, H. A. (1996). *The sciences of the artificial*. MIT Press.

Sörensen, J., Persson, A., Sternudd, C., Aspegren, H., Nilsson, J., Nordström, J., Jönsson, K., Mottaghi, M., Becker, P., Pilesjö, P., Larsson, R., Berndtsson, R., & Mobini, S. (2016). Re-thinking urban flood management—time for a regime shift. *Water, 8*, 332—346. https://doi.org/10.3390/w8080332

Syvitski, J. P. M. (2008). Deltas at risk. *Sustainability Science, 3*, 23—32. https://doi.org/10.1007/s11625-008-0043-3

Taleb, N. N. (2007). *The black swan: The impact of the highly improbable*. Random House.

Tan, D., Adedoyin, F. F., Alvarado, R., Ramzan, M., Kayesh, M. S., & Shah, M. I. (2022). The effects of environmental degradation on agriculture: Evidence from European countries. *Gondwana Research, 106*, 92—104. https://doi.org/10.1016/j.gr.2021.12.009

Tu, C., Suweis, S., & D'Odorico, P. (2019). Impact of globalization on the resilience and sustainability of natural resources. *Nature Sustainability, 2*(4), 283—289. https://doi.org/10.1038/s41893-019-0260-z

United Nations. (2019). *World population prospects: The 2019 revision*. https://doi.org/10.1017/CBO9781107415324.004

von Solms, R., & van Niekerk, J. (2013). From information security to cyber security. *Computers & Security, 38*, 97—102. https://doi.org/10.1016/j.cose.2013.04.004

WCED. (1987). *Our common future*. Oxford University Press.

Weber, M. (1949). *The methodology of the social sciences*. The Free Press.

Wilkinson, R. (2005). *The impact of inequality: How to make sick societies healthier*. Routledge.

World Bank. (2023). *World development indicators database*. https://data.worldbank.org.

Zinn, J. O. (2008). Introduction. In J. O. Zinn (Ed.), *Social theories of risk and uncertainty: An introduction* (pp. 1—17). Blackwell Publishing.

PART I

The state of the world

CHAPTER 2

Our past defining our present

Introduction

We increasingly appreciate that we are rapidly approaching crossroads from which there is no return once we fail to choose the right direction (Rockström et al., 2009; Steffen et al., 2015). We may have passed several of them already (Li et al., 2021). However, this is not the only time in history that society has been at such crossroads, although it may be the first time the entire planet is at stake. We have been challenged before, and we have prevailed. The question is if we can learn from our past when understanding our present. I think we need to.

Although there are valid objections against mixing up an idea's origins and validity, often referred to as the 'genetic fallacy' (Cohen & Nagel, 1934, pp. 388–390), there are solid arguments for why history matters. One of the more influential of these comes from the great sociologist Ernest Gellner (1989, p. 12), who does not object to the 'genetic fallacy' in itself but against how it is mistakenly extrapolated to argue that we do not need to be concerned with our past when assessing options for our future. Instead, Gellner advises that we study our past to understand our options for our future without prejudging our potential choices (Gellner, 1989). Social change over time results from a combination of choices made in particular historical contexts that influence what choices are possible in the future. It is, in other words, not one necessary mechanism leading to social change, but instead a complex mix of economic, ideological and political factors (Hall, 1986, pp. 5–6).

This chapter presents an overview of our history concerning sustainability—how we conquered the Earth and how we changed how we understand and interact with our environment and ourselves. Such a task is daunting, and I am aware of the inevitability of crude simplifications that may provoke devoted archaeologists, anthropologists, historians and sociologists. However, I view it as necessary to at least hint at the rich knowledge that these disciplines and others have to offer to understand the core challenges for our present and our future.

Conquering our dynamic world

We live on an extraordinary planet. It has been changing continuously for more than 4.5 billion years— from a burning inferno without an atmosphere to a planet with continents and oceans. Although early life forms had appeared almost four billion years before, it was

Sustainability Science, Second Edition
ISBN 978-0-323-95640-6, https://doi.org/10.1016/B978-0-323-95640-6.00012-9

not until 200,000 years ago that the first anatomically modern humans treaded the African soil (Haywood, 2011). This relatively young species was exceptionally good at adapting to new environments. They spread across Africa before favourable climate conditions facilitated migration into the Arabian Peninsula, starting around 120,000 years ago (Timmermann & Friedrich, 2016). The migration out of Africa continued, and humans spread from continent to continent, replacing the Neanderthals and other earlier human species (Mellars, 2006, p. 9381). Around 10,000 years ago, all continents of the world were colonised, except Antarctica (Haywood, 2011) (Figure 2.1).

The more successful early migrations out of Africa around 60,000 years ago also seem to have coincided with favourable climate conditions (Eriksson et al., 2012), as well as with our ancestors reaching full behavioural modernity (Mellars, 2006)—that is, the appearance of fully articulate speech, intelligence and creativity relative to humans today (Eaton, 2006, p. 2). Regardless if this cognitive change was sudden or more gradual, it represents a fundamental transformation and is referred to as the Upper Paleolithic Revolution (Bar-Yosef, 2002). The resulting boost in the complexity of certain groups' technological, economic, social, and cognitive behaviour gave them a competitive edge over others (Mellars, 2006), who in turn would either learn or lose. Although our prevailing ancestors remained hunters and gatherers for many millennia, systematic exploitation of raw material and tool production became common practice, specialised utensils and hunting tools appeared, symbols, decorations, and jewellery emerged, and long-distance exchange networks for raw material and manufactured products were established (Bar-Yosef, 2002, pp. 365–368).

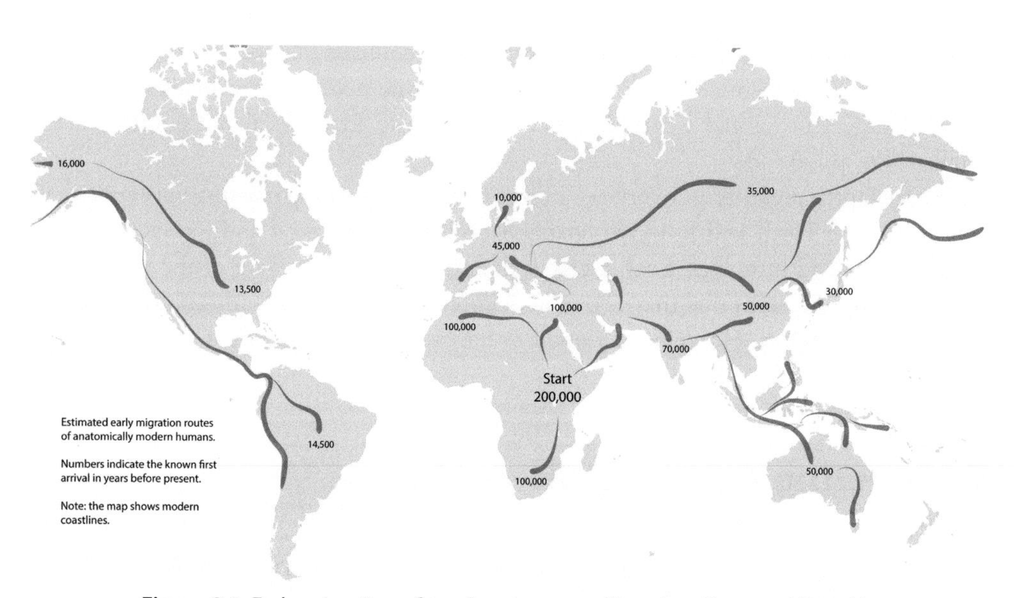

Figure 2.1 Early migration of modern humans. *(Based on Haywood (2011).)*

Although humans had colonised all corners of our planet in 10,000 BCE, the global population remained around 4 million (Kremer, 1993, p. 683; Smith & Archer, 2020, p. 837) and their environmental impact nominal. Despite all the advances of the Upper Paleolithic Revolution, humans lived off their environment without consciously changing it. The challenge for these hunter-gatherer societies was the production of sufficient quantities of food, regulated by the natural carrying capacity of each ecosystem in which they moved. This challenge meant that local overpopulation had only two outcomes: migration or starvation (Fischer-Kowalski & Haberl, 1997, p. 66).

Then, between 9500 BCE and 5000 BCE, groups of humans in several independent locations around the world started cultivating plants and domesticating animals (Gupta, 2004; Shennan, 2018, p. 1; Zeder, 2011). This change marked a second fundamental transformation of society and is referred to as the Neolithic Revolution. The reasons behind this transformation have been heavily debated over the years (Bowles & Choi, 2013; Weisdorf, 2005) and nobody can be entirely sure why such a step was taken (Bar-Yosef, 2017; Hall, 1986, p. 27). However, Shennan (2018) draws on an impressive range of studies and presents a convincing evolutionary perspective, explaining the earliest known emergence and spread of agriculture in the Near East. He argues with painstaking empirical backing how agriculture emerged because hunter-gatherers overhunted their primary game and had to broaden their diet, which under the then favourable climatic and soil conditions in what is referred to as the Fertile Crescent (Figure 2.2) led to growing dependence on agricultural resources that were dense and sustainable. This dependence made people stay put and suddenly had more time to have sex, which led to more children who, in combination with better food security, had better chances of surviving to adulthood (Shennan, 2018). Shennan (2018) then shows how agriculture spread because it enabled people to reproduce and colonise new areas with low-density hunter-gatherer populations, so long as their knowledge, practices, crops and animals were passed on to their children. In any case, it marked the start of humans actively changing their environment to suit their purposes (Haberl et al., 2011, p. 2).

The first farmers in the Fertile Crescent were cultivators of wild cereals (i.e. einkorn, emmer wheat, and barley), pulses (i.e. lentils, peas, chickpeas, and bitter vetch) and flax (Weiss & Zohary, 2011), while rice, millet and soybeans were domesticated in China (Zhao, 2011) and early South American farmers seem to have started with peanuts and squash (Bar-Yosef, 2017, p. 320). Regardless of what they cultivated, growing food may be considered a lucky coincidence as gatherers may accidently have dropped seeds on fertile places in or around their temporary dwellings (Weisdorf, 2005). These early hunter-gatherer farmers were semisedentary, moving around seasonally to secure their livelihood. In addition to farming, humans started to domesticate animals, such as goats, sheep, cattle and pigs (Shennan, 2018; Zeder, 2011), which they had been hunting for millennia and were familiar with (Bar-Yosef, 1998, p. 151). As their primary game dwindled and farming and animal rearing proved viable, they completed the transition to

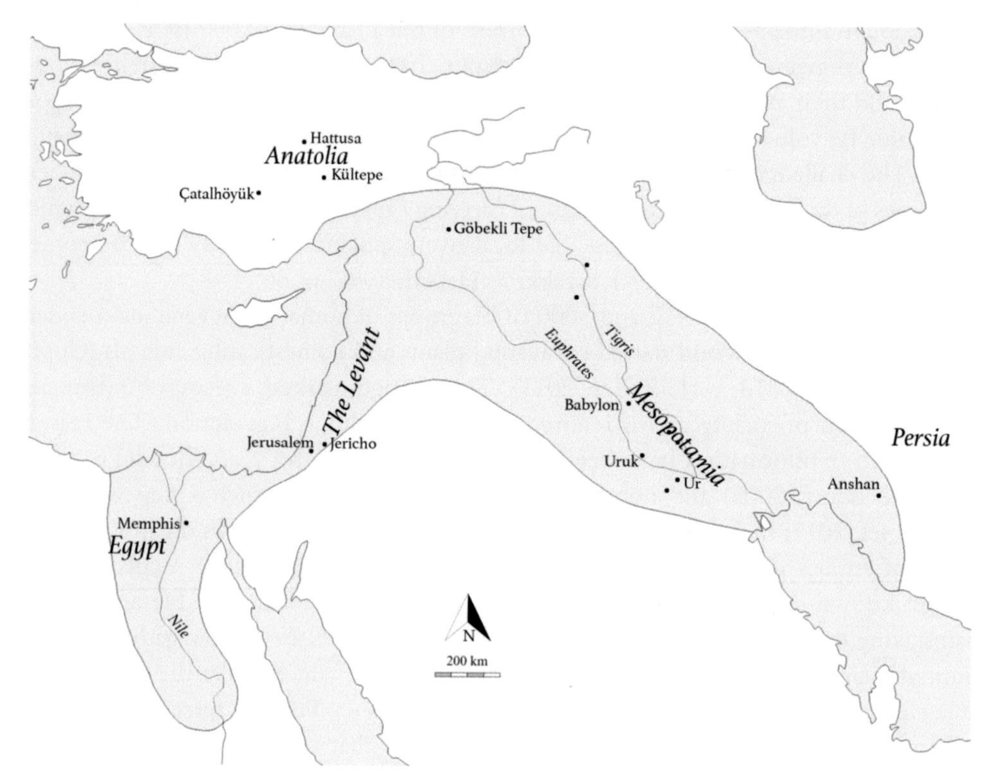

Figure 2.2 Map of the Fertile Crescent and the ancient Near East (current coastlines).

becoming entirely sedentary (Shennan, 2018; Woolf, 2020, pp. 48–54). This transition led to the appearance of the first permanent human settlements in the world, around Göbekli Tepe in the Fertile Crescent (Figure 2.2)—arguably the oldest temple on Earth.

The Neolithic Revolution did not only result in changes in food production practices and movement patterns. Technological advances in farming, such as irrigation, made surplus food production possible, and developments in storage technology, such as pottery, made it possible to store food over time (Goring-Morris & Belfer-Cohen, 2011). However, the introduction of agriculture initially led to declining health and body size (Bowles & Choi, 2019, p. 2194; Cohen & Crane-Kramer, 2007) before the greater reliability of food supplies later increased the fertile age span of women, and the shifts in diet and living conditions resulted in positive impacts on health and body size (Bar-Yosef, 1998, pp. 147–151). Consequently, the population growth increased more than seven times between 10,000 BCE and 5000 BCE and doubled again as the new practices spread over the world in the next 1000 years (Kremer, 1993, p. 683). However, it is important to note that the overall trend of growing populations hides substantial fluctuations—so-called population booms and busts—as the initial success of many of these early agrarian societies could not be sustained due to various reasons, such as climate change and soil overuse (Shennan, 2018, pp. 212–213). Humankind had experienced its first significant sustainability problems.

Growing populations, in combination with permanent settlements and surplus production, resulted in the formation of villages, allowing an emerging division of labour and the accumulation of wealth (Childe & Shennan, 2009, pp. 3–5). When villages were large enough, they became viable gene pools, reducing or removing the necessity to move long distances to find partners (Bar-Yosef, 1998, p. 151). While there are examples of villages from this period large enough to be considered proto-cities—like Çatalhöyük (Figure 2.2) housing between 3500 and 8000 inhabitants (Hodder, 2014, p. 6)—most were much smaller (Woolf, 2020, p. 54). The division of labour did not only emerge within villages but also between villages with different specialities and access to different resources (Kelly, 1979, pp. 39–40; Childe & Shennan, 2009, p. 13), which further developed organised trade.

More complex levels of social alliances emerged, and the importance of territory and ownership increased (Bar-Yosef, 1998, p. 151; Bowles & Choi, 2019). Agrarian societies developed more advanced institutions for inheriting wealth, skills and social networks between generations (Borgerhoff Mulder et al., 2009). The division of labour had, in other words, not only an impact on production but on social status (Borgerhoff Mulder et al., 2009), resulting in increased social stratification based on competition for power within communities (Kuijt, 2000, p. 77). With the potential of accumulating wealth and new notions of territory, ownership, and power came organised armed conflict (Childe & Shennan, 2009, pp. 59–60; Haas, 2001).

Villages started to be protected by palisades (Childe & Shennan, 2009, p. 60) and purpose-made weapons for war appeared (Haas, 2001; Weir, 2005). Pack animals, boats, and wheeled vehicles made it possible to gather food in a few locations (Taylor, 2012, p. 420), which also facilitated protection. The surplus of food in the more advanced communities could support a substantial number of individuals fully released from food production and could focus on other functions (Childe, 1950, p. 8). The global population had in 4000 BCE reached seven million (Kremer, 1993, p. 683), and the first cities emerged (Woolf, 2020, p. 68) (Figure 2.3).

These cities were locations of centralised administration that asserted domination over their surroundings, which at times grew too substantial but initially relatively short-lived empires (Taylor, 2012, pp. 437–439). However, the more refined division of labour in these more advanced societies resulted in plenty of remnants of their existence in architecture, art and other cultural artefacts (Childe, 1950; Woolf, 2020). In these early Mesopotamian cities, such as Uruk, Ur and Babylon (Figure 2.2), the first known true writing appeared (Richardson, 2012). The discovery of copper and the subsequent invention of bronze in the ancient Near East around 3300 BCE made weapons deadlier and tools more efficient, boosting the development and regional domination of the societies

Figure 2.3 The great city of Uruk in Mesopotamia; the first city on Earth.

that mastered it. The world population had in 3000 BCE reached 14 million (Kremer, 1993, p. 683), and cities were spreading into Persia, the Levant, Anatolia and Egypt (Figure 2.2) (Manzanilla, 1997; Woolf, 2020). However, cities not only spread but were invented again and again in different parts of the world (Woolf, 2020, p. 14) (Figure 2.4).

Aggregating large populations in cities turned out not only to be beneficial but also challenging. Higher population density led to new diseases (Emberling, 2003, p. 256). Overirrigation caused agricultural land salinisation and soil overuse led to erosion and desertification of large areas (Desvaux, 2009, p. 224). Insufficient knowledge about maintaining soil fertility caused, in other words, the deterioration of agricultural land at increasing distances from the cities, eventually contributing to their downfall (Taylor, 2012, p. 429)—another early experience of critical sustainability problems. However, the relatively local character of these problems, extending over a few thousand square kilometres at most, limited their effect on the planetary systems as wholes (Desvaux, 2009, p. 224).

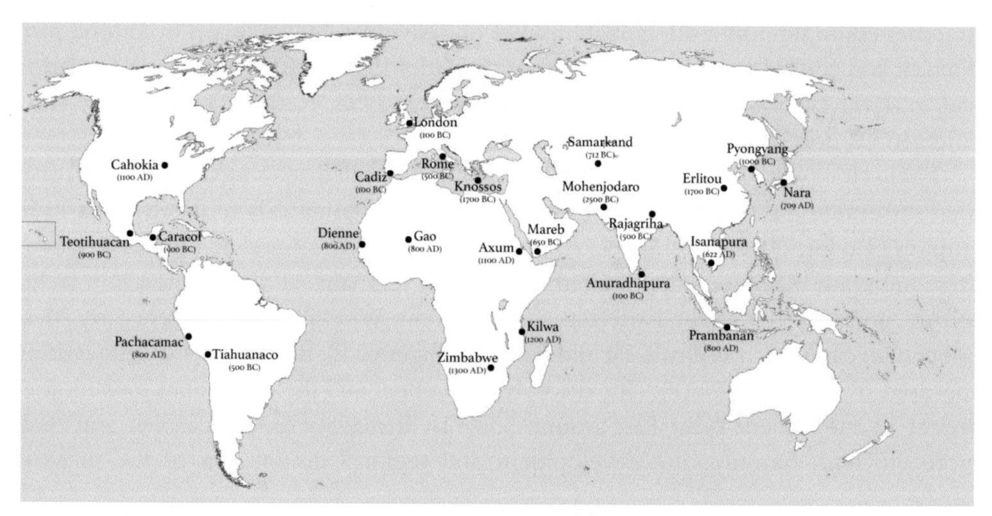

Figure 2.4 Examples of early cities around the world. *(Based on the research by Reba et al. (2016).)*

New technology was developed to sustain the growing populations of cities, such as terracing (Figure 2.5) and river replenishment of fertility (Taylor, 2012, pp. 429—430). Our ancestors were, in other words, not only actively changing their environment by clearing fields and creating monocultures but also changing the flow of rivers and the topography to suit their purposes. The capacity to move soil and rock had improved as the tools developed from bones and antlers to increasingly sophisticated tools of bone, wood, stone and bronze (Hooke, 2000, p. 843). Although the Palaeolithic flint mines were impressive considering the available technology, the pyramids of Egypt and other massive Bronze Age structures worldwide are still mesmerising scholars and visitors alike.

The use of bronze tools was limited by accessibility and cost, prioritising its use for weapons and other purposes of the rich and powerful while restricting its use for ordinary people (Hooke, 2000). Then, around 1300 BCE, again in the Near East, iron was discovered. The new metal turned out to be even more versatile than bronze and more 'democratic' in the sense of being more abundant and accessible for ordinary people (Hooke, 2000). While the human capacity to shape the landscape increased as this discovery spread over the world, resulting in increased productivity and more astonishing artefacts (Hooke, 2000), the way of life stayed more or less the same for the vast majority of people.

It took centuries for the use of iron to spread over the world, reaching Britain around 800 BCE (Cunliffe, 2005, p. 90), China around 600 BCE (Higham, 1996, p. 103) and South Africa around CE 200 (Miller & Van Der Merwe, 1994, p. 12). During this time, empires rose and fell, sometimes due to sustainability problems in balancing resources, population and political ambition (Tainter, 2000, pp. 19—23). Although hard

Figure 2.5 The ancient innovation of terracing. *(Photo by Tine Steiss, shared on the Creative Commons.)*

labour and new inventions repeatedly helped to overcome locally experienced ecological constraints, unexpected side effects emerged in the form of new risks, environmental problems and increasing demands for labour and energy (Haberl et al., 2011, p. 4).

Gradual change over millennia caused the development of notably different agrarian societies worldwide (Haberl et al., 2011). However, regardless of their differences, they all shared one fundamental obstacle for their development that could not be solved by gradual expansion and improvement of their agrarian mode of production. All these societies depended almost entirely on biomass from agriculture and forestry to cover their total energy needs in terms of food and labour, with only marginal contributions from water- and wind power (Haberl et al., 2011). The global population grew from 190 million in CE 200 to 350 million in 1400 (Kremer, 1993, p. 683), and the dependency on biomass made the availability and use of land increasingly crucial for the sustainability of each society (Haberl et al., 2011, p. 4). Although the causes are heavily debated, intense migrations started in Asia and Europe early in this period and rearranged the ethnic map. New geopolitical entities formed, and the population growth fluctuated heavily from century to century as armed conflict (e.g., Mongol Invasions) and disease (e.g., Black Death) recurred (Kremer, 1993, p. 683). There was a constant pressing need to find new solutions to sustainability problems.

One response to these problems was exploring unknown territory, at least partly to acquire additional resources through colonisation or trade. Although there are records of early Phoenician, Greek and Chinese explorers, as well as of legendary explorers like Ahmad Ibn Fadlan, Leif Eriksson, Marco Polo and Ibn Battuta, it was not until the fifteenth century that the intensity of exploration exploded (Gosch & Stearns, 2008). The emerging and fiercely competitive European powers (re)discovered America, found the sea route to India and circumnavigated the world, effectively transforming Europe from being a peripheral and relatively backward part of the world to its hegemon (Wolf, 1997). The subsequent colonisation of 'new' continents provided additional land for producing food and accumulating wealth (Desvaux, 2009, pp. 224—225), to the great detriment of the original populations (Figure 2.6).

Another response was science. Egyptians, Greeks, Chinese and Arabs had endeavoured into scientific inquiry much earlier. Yet, the late Renaissance ideas of Descartes and the following Enlightenment thinking of Locke, Newton, Voltaire, Hume, Smith, Kant and others marked an evident change. Scientific knowledge production intensified significantly (Hassan, 2003), and the printing press, which was invented a couple of centuries earlier, made it possible to make the results available to large groups of people (Eisenstein, 1980)—further increasing the speed of knowledge development and innovation.

The increasing accumulation of knowledge and access to additional resources from the colonies incubated around 1750 in Britain, the start of a third fundamental transformation of society: the Industrial Revolution (Desvaux, 2009, p. 225). After having been

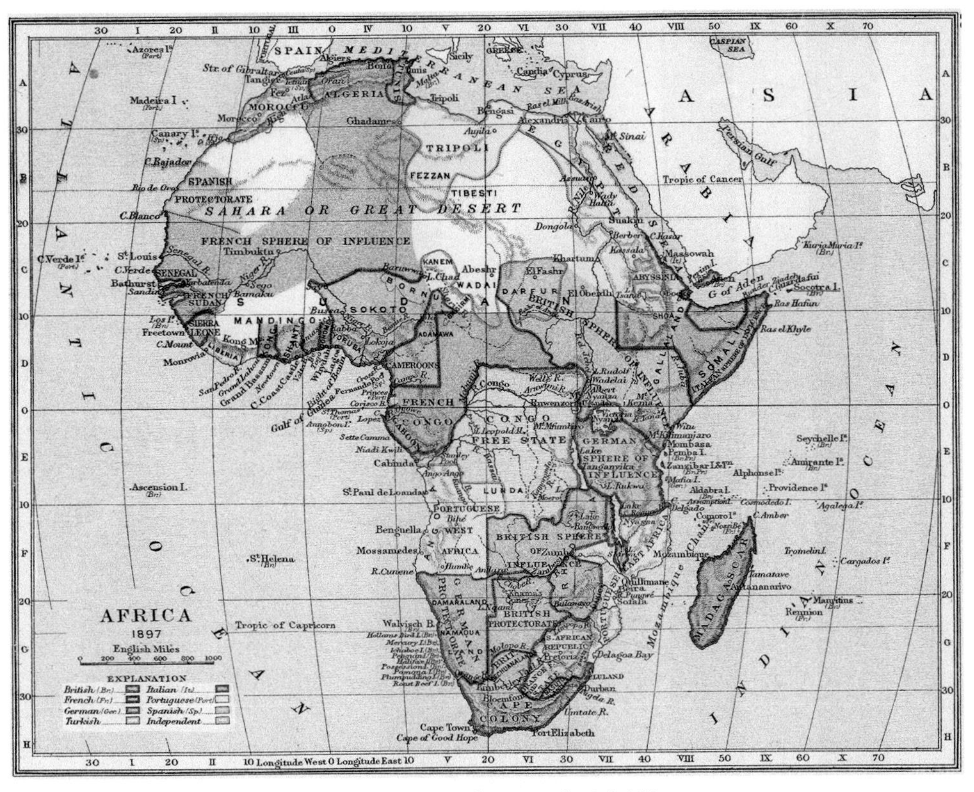

Figure 2.6 Map of 1897 colonial Africa.

more or less entirely dependent on biomass for millennia, fossil fuel emerged and provided a solution to the main sustainability problem of agrarian society (Haberl et al., 2011, p. 4). The steam engine and later combustion engines revolutionised the exploitation of natural resources, manufacturing and transportation (Desvaux, 2009, p. 225). Parallel developments in food production—with agricultural machinery, fertilisers and pesticides—medicine and public health caused population growth to skyrocket in the industrialising societies (Desvaux, 2009). As the Industrial Revolution spread over the world, starting to substantially changing France from 1815 (Dunham, 1955), Germany, the United States and Japan in the 1860s and 1870s (Green, 1939; Smith, 1955, pp. 1868–1880; Veblen, 2003) and China in the 1950s (FitzGerald, 1981), the world population increased exponentially (Figure 2.7), from 720 million in 1750, reaching one billion around 1820, two billion around 1930 and three billion in 1960 (Kremer, 1993, p. 683).

Similar to all earlier solutions to sustainability problems throughout history, the new energy source of fossil fuel, which the industrialised societies have become so fundamentally dependent on, did, unfortunately, also entail adverse side effects. With populations

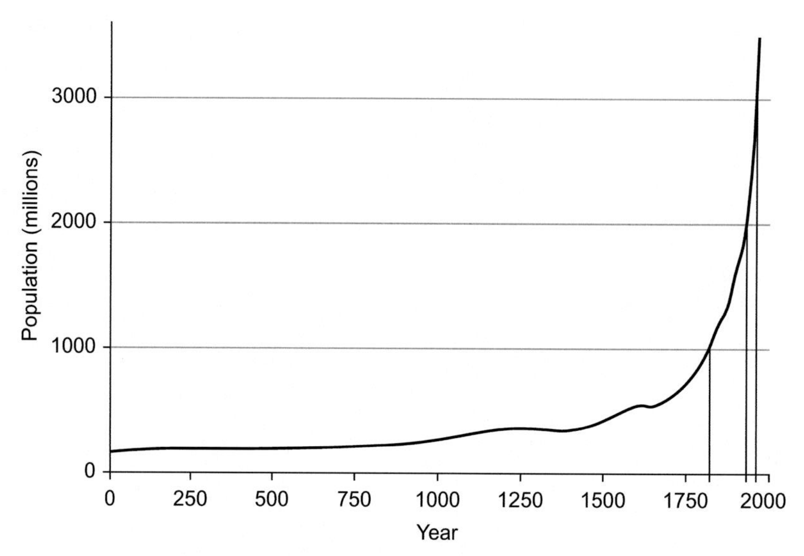

Figure 2.7 Global population over the last two millennia. *(Based on Kremer (1993, p. 683).)*

and economies growing at increasing speed, our world faces global sustainability challenges of complexity and scale never experienced before throughout history.

Social change over millennia

The great transformations of the Neolithic Revolution and the Industrial Revolution did not only transform our relationship with our environment but also our relationships among ourselves. Regardless if these social changes were causes or effects of the transformations described in the previous section, they remain vital aspects for understanding our past, as well as our present and our future.

These two great transformations divide human history into three parts: hunter-gatherer or pre-agrarian, agrarian, and industrial society (Gellner, 1989). Although these categories of societies can be divided into further subcategories (Hall, 1986), they are maintained throughout this section to structure an overall picture of past social change. As Gellner puts it:

> *These three kinds of society differ from each other so radically as to constitute fundamentally different species, notwithstanding the very great and important diversity which also prevails within each of these categories*
>
> ***(Gellner, 1989, p. 16).***

However, the three kinds of societies should not be seen as inevitable developmental steps, as societies can and do stay within a category (Gellner, 1989). There are, for instance, still small hunter-gatherer societies around the world, such as Aché in Paraguay

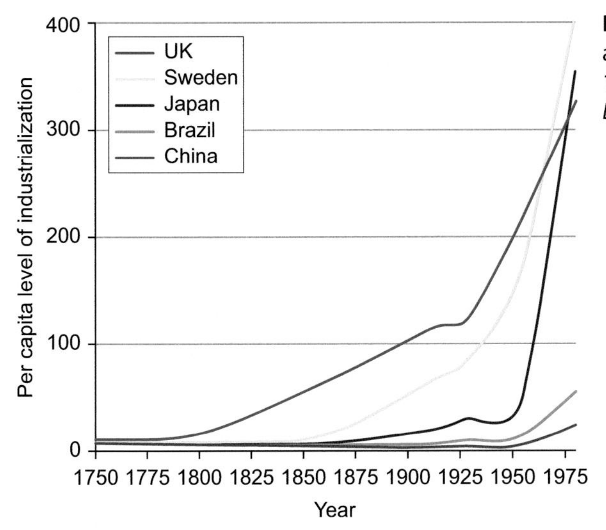

Figure 2.8 Per capita levels of industrialisation 1750–1980 (100 = UK in 1900; 1913 boundaries). *(Based on Bairoch (1982).)*

(910 people), Hadza in Tanzania (1200 people), Jul'hoansi in Southern Africa (45,500 people) and Meriam in Australia (220 people) (Eberhard et al., 2022). What is certain is that only hunter-gatherer societies existed in the first part of our history, that agrarian societies vastly dominated the second part, and that industrial societies have been increasingly dominating the third part as industrialisation spread steadily but unevenly over the world (Figure 2.8).

While not inevitable developmental steps, it is inconceivable for a society to transition directly from hunter-gatherer to industrial society (Gellner, 1989, p. 16). Transitioning back from industrial to agrarian society or from agrarian to hunter-gatherer society is conceivable but extremely unusual (Gellner, 1989). For example, external forces have pushed agrarian people into barren land (Xu et al., 2010, p. 9), or the Moriori of the South Pacific who abandoned agriculture for hunting and gathering as they settled in the lush environment of Chatham Island that, combined with their low population density, provided a stable source of subsistence with less hard labour (Pryor, 2004, pp. 23–24). Generally, pre-agrarian societies were not the most affluent, but certainly the most leisured throughout history, with perhaps only 3 hours of work per day to sustain themselves (Hall, 1986, p. 27). Early farming, on the other hand, was indeed tedious and labour-intensive, and going agrarian was not as commonsensical as initially believed (Weisdorf, 2005, p. 562).

It is important to note that early pre-agrarian societies left us no written records. It is thus problematic to discern whether the few remaining hunter-gatherer societies are representative or atypical just because they have lasted the two great transformations (Gellner, 1989, pp. 35–36). However, what we think we know about hunter-gatherer societies can be summarised into a few main ideas. First, hunter-gatherer societies depend on what can be killed or found and have, as presented in the previous section, little or no means for producing and accumulating wealth (Gellner, 1989, p. 16). They are also generally characterised as exhibiting egalitarian social relations and joint production and ownership of property (Rushforth, 1992, p. 489), causing Marxists to refer to these societies as being governed by 'primitive communism' (Wolf, 1997, p. 75). Moreover, hunter-gatherer societies are small, generally dominated by kinship (Rushforth, 1992, p. 489) and with a low division of labour (Gellner, 1989, p. 16). The exchange between individuals in these societies is characterised by reciprocity (Polanyi, 2001) and emphasises informal leadership, personal autonomy, cooperation, generosity and equality (Rushforth, 1992, p. 489). Considering all these traits, it is difficult to understand why anybody would like to change that in the first place (Figure 2.9).

Agrarian societies, or 'Agraria' as Gellner (1989) so famously named this type of societies, are, per definition, dependent on what can be planted and harvested. They produce and store food and can hoard other forms of wealth necessary for making tools, weapons, symbols and other artefacts for enhancing the quality of life (Gellner, 1989, p. 16). Compared to hunter-gatherer societies, agrarian societies are capable of great size and complexity, with large populations and elaborate social stratification and division of labour (Gellner, 1989, pp. 16—17). It has been suggested that most agrarian societies require over 90% of the population for agricultural production to support a small elite (Hall, 1986, p. 149). It has even been questioned if these societies could be called societies in any modern sense since the vast majority of their population live in 'laterally insulated communities of agricultural producers', while only a minority live in 'stratified, horizontally segregated layers of military, administrative, clerical and sometimes commercial ruling class' (Gellner, 1989, p. 9) (Figure 2.10).

Figure 2.9 Hadza, one of the few remaining hunter-gatherer societies in the world. *(Photo by Woodlouse, shared on the Creative Commons.)*

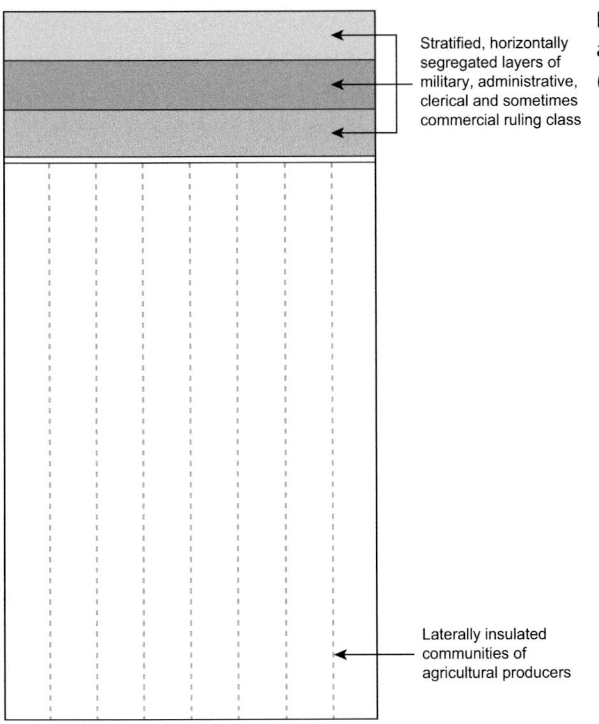

Figure 2.10 The social structure of agrarian society. *(Based on Gellner (1983, p. 9).)*

A central aspect of agrarian societies is the emergence of a specialised ruling class and a specialised clerisy, most often as two separate groups (Gellner, 1989, p. 17). Hence, it is not only production that differentiates agrarian societies from hunter-gatherer societies, but also the scale and importance of coercion in and legitimisation of the social order (Gellner, 1989). It is, in other words, not a coincidence that the first temple emerged in Göbekli Tepe at the same time as agriculture. The exchange between individuals in these societies is characterised by redistribution (Polanyi, 2001), generally enforced by a more or less hierarchical system of predatory ruling classes. In exchange for protection, the lower levels have to provide the level above some stipulated quantity of labour, goods and services (North & Thomas, 1971, p. 780): 'Sheep are fleeced, but safe from wolves' (Buckley & Rasmusen, 2000, p. 314). However, as the lower levels enter into this system, they are subject to the coercive authority of the level above. They cannot dissolve the bond even though the initial benefits may have ceased long ago (North & Thomas, 1971, p. 8).

Agrarian societies are, by definition, based on the innovation of agricultural food production. Additional innovation occurs in these societies, but not as a continuous, cumulative and exponential process (Gellner, 1989, p. 17). Instead, the focus is on stability, with some agrarian societies at least seemingly organised to avoid potentially disruptive innovations and maintain the status quo (Gellner, 1989). An example is the superimposition of military and religious power in the Ottoman Empire that impeded innovation and hindered many European discoveries from being adapted and utilised (Hall, 1986, p. 109). In short, agrarian societies are generally 'stagnant, oppressive, and dominated by the sword and superstition' (Gellner, 1989, p. 238).

The final category of societies, industrial society, or 'Industria' in Gellner's vocabulary (Gellner, 1989), is one in which the production of food is a minority activity and where production, in general, is based on powerful and continuously developing technology (Gellner, 1989, p. 17). In contrast to agrarian societies, the foundation for industrial societies is not resting on any specific innovation but rather on the discovery that science and the application of its findings to address real-world problems are feasible (Gellner, 1989, pp. 17–18). Although the ancient Greeks had emphasised reason much earlier, post-Enlightenment industrial society builds on the belief that applying reason, not faith in ancient authority, brings social benefits (Withers, 2007, p. 2).

Innovations in the agricultural sector reduce its need for labour, opening up human resources for other sectors (Hall, 1986, p. 149). The old agrarian division of labour between those who fight, pray and work is replaced by a more homogeneous but mobile population of functional specialists (Gellner, 1989, pp. 277–278). Productivity is exceptional in relation to agrarian society (Hall, 1986, p. 150) and the continuous and cumulative process of innovation is not only marked 'by a complex division of labour, but also by a perpetually changing occupational structure' (Gellner, 1989, p. 18). The rise in productivity tilts, in other words, the balance in favour of instrumentally efficient specialists and towards an increasingly more extensive and ultimately all-embracing market (Gellner, 1989, p. 277). This shift also changes the balance of power away from the predatory ruling classes to the producers (Gellner, 1989, pp. 238–242), and exchange between individuals is characterised by market principles (Polanyi, 2001).

Industrial society has replaced the agrarian patron-client system with a centralised and hierarchical bureaucracy (Hall, 1986, p. 151). Direct coercion is much less common and occurs only in certain exceptional circumstances, and the state has monopolised the legitimate use of violence. In other words, industrial society citizens are 'relatively seldom obliged to do things by weapons held at their throats' (Gellner, 1989, p. 232). Notable exceptions include internal armed conflict or dissidence, one-sided political violence and organised crime, which, regardless of their horrors, are relatively rare in relation to the total population of industrial society. Legitimacy derives no longer from notions of divine rule or religious rituals, at least not in the modern liberal version of industrial society, but through ritualised competition (Hearn, 2021). Be it a general election, house bidding or recruitment process (Figure 2.11).

When contrasting hunter-gatherer societies with agrarian societies, it is clear that the former seems much nicer for most of the population than the latter. How come, then,

Figure 2.11 Ritualised competition in the workplace. *(Source: alphaspirit/Shutterstock.com.)*

that most of the world's societies committed to this fundamental transformation? The answers are many and varied. For instance, if a hunter-gatherer society depended on having a small population, perhaps most failed to control the size of their population, and the adoption of agriculture became necessary to sustain it (Hall, 1986, p. 27; Shennan, 2018). That is plausible, especially for the emergence and initial spread of the first agrarian societies. However, considering the predatory feature of agrarian societies and their much greater potential for ideological-, economic-, military- and political power (Mann, 1986), it is more plausible to conceive the subsequent expansion of this great transformation as hunter-gatherer societies being swallowed or forced to adapt in the face of expanding neighbours to whom violence came more naturally (Gellner, 1989, pp. 154–155).

Similarly, and with much more unambiguous empirical support, the second great transformation from agrarian to industrial society spread because the transformed societies were so technically superior that staying agrarian meant a grave risk of being subjugated and exploited. However, it is more complicated to explain the transformation of the first society: Great Britain, which it was called after the Acts of Union in 1707. This unique first transformation resulted from a complex combination of economic, ideological and political factors (Hall, 1986, pp. 5–6), which aligned what Gellner (Gellner, 1989, p. 277) calls the three spheres of human activity—production, coercion, cognition—in a favourable way. It is important to note that such transformation is not only challenging and changing relationships between countries but also inevitably causing massive social disruption within them (Hall, 1986).

Considering how much we think we know about the Neolithic Revolution and what we do know about the Industrial Revolution, it is more or less obvious that they cannot plausibly be attributed to any purposeful human design (Gellner, 1989, p. 20). As Gellner puts it:

> *In each case, the new social order, due to be ushered in by history, was so radically discontinuous and different from its predecessor, within which its gestation had taken place, that it simply could not be properly anticipated or planned or willed. Those who sowed knew not what they would reap, nor did those who abandoned the plough and sword for trade, production and innovation. This point in no way applies, of course, to the subsequent diffusion of a new social order, once established and successful in one location. On the contrary: once a new and visibly more powerful order is in existence, it can be, and commonly is, consciously and deliberately emulated. Those who emulate may also end up with more than they intended and bargained for, but that is another story*
>
> ***(Gellner, 1989, p. 20).***

However, we can learn from these great transformations that fundamental social changes are possible. Considering the immense differences between hunter-gatherer society, agrarian society and industrial society in production, coercion, cognition, exchange and so on, a future society may be equally different. We can also learn that a new great transformation may require each society to pass through industrial society. It will likely start in one place due to a unique and complex combination of various factors. The new type of society will only spread if it gives advantages in competition with others. It is also important to note that such a new great transformation is unlikely to be designed by purposeful human activity. At least not with a clear picture of the end result until we get there.

Another critical aspect of great transformations is that nobody could anticipate them beforehand, while they appear entirely self-evident in hindsight. Gellner describes this human disinclination for thinking outside the dominant discourse, prevailing paradigm or episteme well in his now classic quote:

> *Men and societies frequently treat the institutions and assumptions by which they live as absolute, self-evident, and given. They may treat them as such without question, or they may endeavour to fortify them by some kind of proof*
>
> ***(Gellner, 1989, p. 11).***

Gellner's words resonate disturbingly well with our current global dialogue and lack of concrete action on key sustainability challenges, which are elaborated on in Chapters 3 and 4 below. For instance, the trade of carbon offsets applies contemporary market principles to address the challenge of climate change (Figure 2.12). However, recent history has seen more rapid changes in the transactions of humankind than ever before. Even if the prevailing paradigm, dominated by the rules of market capitalism, is rarely questioned in the mainstream, there are signs indicating at least a vision of a new and fundamentally different society—more on this vision of 'Sustainia' in Part III (Chapter 16). So, perhaps we are in the middle of a new great transformation. It is difficult to tell, but if we continue only to think within the frames of our current paradigm, I am afraid that the world as we

Figure 2.12 Carbon offsets—changing the world by not changing the world?

know it is lost in a not-too-distant future. It is, in other words, more crucial now than ever to understand past social change to facilitate social change towards a more sustainable future.

The invention of risk

As we started the initially slow transition towards modernity in sixteenth and seventeenth-century Europe, new ideas began to form concerning the contingent nature of our world (Giddens, 2002, pp. 21–23). In antiquity and throughout the Middle Ages, Christians had generally adhered to the Augustinian view that fate was simply God's natural order (Gregersen, 2003, p. 362), and similar determinism was found in the other world religions of that time (Alles & Ellwood, 2007). If the outcome of any course of action was already predestined or predetermined, it mattered little what choices we made as mere human beings. 'Chance had no place in God's world' (Gregersen, 2003, p. 362) and nobody had any concept of risk, except in some marginal contexts (Giddens, 2002, p. 21). Then a new idea emerged. The idea that advantages could only be gained if something was at stake and that these stakes were not only a matter of costs, directly tradable against the advantages, but a matter of regrettable decisions if a loss occurred that we had hoped to avoid (Luhmann, 1993, p. 11). Although the notion of divine will was not yet openly questioned, as that would have had dire consequences for the individuals doing so (Figure 2.13), we were starting to realise our own agency (Sztompka, 1993, pp. 191–193).

The etymological origin of the word 'risk' is ambiguous, with suggestions ranging from the Arabic word 'risq' (meaning anything that has been given to you by God),

Figure 2.13 Galileo Galilei—punished for questioning the order of things.

the Latin word 'risicum' (referring to the challenges to sailors), the Greek word 'rhiza' (referring to hazards of sailing) and so on (Althaus, 2005, p. 570). Regardless, the word risk is suggested to have come into English through Spanish or Portuguese (Giddens, 2002, p. 21), which were the primary maritime nations of fifteenth and sixteenth-century Europe (Wolf, 1997). The Babylonians had invented basic risk transfer schemes thousands of years earlier, which later spread to the Phoenicians, Indians, Greeks, Romans and so on (Trenerry, 2009, pp. 45—49). However, it was among the maritime merchants of that time that early forms of modern insurance emerged (Luhmann, 1993, p. 9). As the oceans were dangerous and the loss of both ships and cargo was common, these traders organised to share the cost of each loss within a larger group. From being relatively uncommon but catastrophic for the lone unfortunate merchant, the cost of losses became more frequent but bearable for each merchant in the collective.

The invention of such insurance schemes was paralleled by the emergence of basic notions of likelihood and probability among some true Renaissance men of fifteenth and sixteenth-century Italy (Figure 2.14), who seems to have mainly been focussing

Figure 2.14 Gambling was a primary driver for the invention of risk.

on gambling (Bernstein, 1996, pp. 39—56). The following century saw a sharp increase in our understanding of these concepts, with the first scientific publication in the Netherlands, the development of the first theory of probability in France and the birth of modern statistics in England (Bernstein, 1996, pp. 57—84). However, their application for maritime insurance still initially resembled gambling at the coffee houses that mushroomed in London (Martin, 1876, pp. 117—118). One of these coffee houses was Lloyds, which turned out to become the world's most well-known insurance and reinsurance market today.

It was also in the mid-seventeenth century that the first notion of risk as a combination of the likelihood or probability of an event and its consequences emerged. It came out of the Port-Royal monastery in France, with the landmark statement that 'the fear of an evil ought to be proportionate, not only to its magnitude, but also to its probability' (Arnauld & Nicole, 1861, p. 368). By the mid-eighteenth century, much of the mathematical foundation for our modern approaches to risk was already in place (Bernstein, 1996). Risk calculations were applied in a growing range of societal sectors (Zinn, 2008, p. 9), and the notion of fate had increasingly been replaced by the idea of our ability to master uncertainty with the use of probability (Althaus, 2005, p. 570). The idea of risk was, in other words, a cornerstone of modernity itself (Beck, 1992; Luhmann, 1993). Or, as Bernstein explains in his opus on the history of risk:

> *The revolutionary idea that defines the boundary between modern times and the past is the mastery of risk: the notion that the future is more than a whim of the gods and that men and women are not passive before nature. Until human beings discovered a way across that boundary, the future was a mirror of the past or the murky domain of oracles and soothsayers who held a monopoly over knowledge of anticipated events*
>
> ***(Bernstein, 1996, p. 1).***

Our understanding of risk grew, and our toolbox to address forward-looking development challenges became progressively more sophisticated. We applied our relatively novel appreciation of our own agency, changing our world based on decisions made by either formally or informally weighing the likelihood and consequence of what could happen in different alternatives. Advances in most spheres of human life ensued at an unprecedented and increasing speed, and the notion of risk seems to have been one of the requisites for this great transformation (Bernstein, 1996).

Over time, the focus on the meaning of risk slowly shifted from the older entrepreneurial concept of risk, emphasising the potential advantages of decisions and ventures, towards our contemporary use of risk as something negative (Althaus, 2005, p. 570). This change has been increasingly noticeable since the beginning of the nineteenth century. Perhaps as a result of our growing realisation of the naivety in our optimistic view on science and progress of that time, when we experienced their limits and pitfalls regarding environmental degradation, social upheaval and so on. Although the contemporary financial risk management of banks and insurance companies still adheres to the older view that risk can be either positive or negative, most other sciences differentiate between risk as something negative and opportunity as something positive (Renn, 2008, pp. 1−2). This latter view also represents an almost universal lay notion of risk, which is an essential indicator of its usefulness. I agree, in other words, with Hearn (2012, p. 209) when championing 'garden-variety' concepts, as opposed to the exotic:

> *... the ultimate test of a social science concept is not its beauty or arresting comparisons, but rather its utility − does it help us identify and make sense of a distinct range of phenomena, in all their variability? And one of the surest, though by no means failsafe, indicators of usefulness, is the way a concept' beds down' into ordinary language, becoming a part of common sense. While common sense definitely needs to be examined and challenged at points, we should not lightly abandon normal everyday conceptions, [...], because these will encode the practical experience of myriad speakers, whose collective wisdom deserves respect, though not reverence* **(Hearn, 2012, p. 209).**

This book is about managing this 'garden-variety' of risk to sustainable development. We know now that there is neither such risk if there are no hazards nor if there is no vulnerability (Wisner et al., 2004). We also know that we have choices for our actions and that these choices determine both likelihoods and consequences of potential, undesirable courses of events (Renn, 2008). Our contemporary world is more complex than ever before in human history. Although our knowledge and analytic capacity vastly exceed any previous society on Earth, we are also increasingly aware of the limits to our capacity for foresight. As we are notoriously poor in predicting our future (Taleb, 2007), the foundation for sustainable development must be based on how well we can manage risk and develop our resilience. In other words, we must be able to manage uncertainty, complexity and ambiguity (Renn et al., 2011).

Figure 2.15 Shanghai—a growing megacity in a precarious location. 28.5 million inhabitants in 2022 and counting.

Conclusion

We have conquered our planet not only in the sense of populating all of its habitable corners but also in the sense of our increasing influence on its vital systems (Kates et al., 1990). This influence has grown so significant in modern times that a new geological period has been suggested—the Anthropocene—in which humankind is reckoned as a major geological force (Crutzen & Stoermer, 2000). Regardless of the academic debate around this suggestion, we must realise that we have become capable of both great deeds and destruction. We have pushed onto the edge of a cliff. However, we must also remember that we have faced numerous challenges over millennia, and we have prevailed so far. Sometimes at high costs, sometimes with entire civilisations collapsing, we have adapted and moved on. Thus, the main issue may not be so much the survival of our species but what costs we will pay this time. This issue can only be determined by how and when we will start substantially addressing humankind's current core challenges.

We have gone from having a global population of around seven million at the time of the foundation of the first cities (Kremer, 1993, p. 683) to having 55 cities in the world with more than seven million inhabitants each, and the largest city—Tokyo—having more than five times that population (United Nations, 2019) (Figure 2.15). This situation was utterly impossible in agrarian society, let alone in hunter-gatherer society. The great transformations that took us between these categories of societies (or paradigms or epistemes, if you will) and into industrial society fundamentally altered our relationships with the environment and among ourselves. Considering the grave state we gradually realise we are in, we are in dire need of a new great transformation. Perhaps we are in the middle of one already (Chapter 16). It is difficult to tell. However, we can expect by studying past social change that such transformation is conditional on a complex mix of economic, political and ideological factors. We must attempt to align such factors by mindful and intentional action, although the end result is impossible to fully foresee and comprehend.

Our future is uncertain, our world is dynamic and complex, and our understanding of both is ambiguous. We are only human beings. We cannot predict the future. However, we have the ability to anticipate what could happen and estimate how likely that is to happen and what consequences would arise. And more importantly, we can use the outcome of such exercise to guide our decisions for a more sustainable future. We invented risk during the last great transformation, but we urgently need to apply this discovery, not only for a specific installation or societal sector but more holistically, to guide our overall decisions for sustainability. We have greater capacity today for understanding and addressing our challenges than ever before in human history. The question is: How will we use it?

References

Alles, G. D., & Ellwood, R. S. (2007). Free will and determinism. In *Encyclopedia of world religions*. Facts on File.

Althaus, C. E. (2005). A disciplinary perspective on the epistemological status of risk. *Risk Analysis, 25*, 567–588.

Arnauld, A., & Nicole, P. (1861). *Port-Royal Logic. James Gordon.*

Bairoch, P. (1982). International industrialization levels from 1750 to 1980. *Journal of European Economic History, 11*, 269–333.

Bar-Yosef, O. (1998). On the nature of transitions: The middle to Upper Palaeolithic and the Neolithic revolution. *Cambridge Archaeological Journal, 8*, 141–163.

Bar-Yosef, O. (2002). The Upper Paleolithic revolution. *Annual Review of Anthropology, 31*, 363–393. https://doi.org/10.2307/4132885

Bar-Yosef, O. (2017). Multiple origins of agriculture in Eurasia and Africa. In M. Tibayrenc, & F. J. Ayala (Eds.), *On human nature: Biology, psychology, ethics, politics, and religion* (pp. 297–331). Elsevier. https://doi.org/10.1016/b978-0-12-420190-3.00019-3

Beck, U. (1992). *Risk society: Towards a new modernity*. Sage Publications.

Bernstein, P. L. (1996). *Against the gods: The remarkable story of risk.* John Wiley & Sons.

Borgerhoff Mulder, M., Bowles, S., Hertz, T., Bell, A., Beise, J., Clark, G., Fazzio, I., Gurven, M., Hill, K., Hooper, P. L., Irons, W., Kaplan, H., Leonetti, D., Low, B., Marlowe, F., McElreath, R., Naidu, S., Nolin, D., Piraino, P., ... Wiessner, P. (2009). Intergenerational wealth transmission and the dynamics of inequality in small-scale societies. *Science, 326*, 682–688. https://doi.org/10.1126/science.1178336

Bowles, S., & Choi, J. K. (2013). Coevolution of farming and private property during the early Holocene. *Proceedings of the National Academy of Sciences of the U S A, 110*(22), 8830–8835. https://doi.org/10.1073/pnas.1212149110

Bowles, S., & Choi, J.-K. (2019). The Neolithic agricultural revolution and the origins of private property. *Journal of Political Economy, 127*(5), 2186–2228.

Buckley, F. H., & Rasmusen, E. B. (2000). The uneasy case for the flat tax. *Constitutional Political Economy, 11*, 295–318.

Childe, V. G. (1950). The urban revolution. *Town Planning Review, 21*, 3.

Childe, V. G., & Shennan, S. (2009). *The prehistory of European Society*. Spokesman Books.

Cohen, M. N., & Crane-Kramer, G. M. M. (Eds.). (2007). *Ancient health: Skeletal indicators of agricultural and economic intensification*. University Press of Florida.

Cohen, M. R., & Nagel, E. (1934). *An introduction to logic and scientific method*. Harcourt, Brace and Company.

Crutzen, P. J., & Stoermer, E. F. (2000). The Anthropocene. *Global Change Newsletter, 41*, 17–18.
Cunliffe, B. W. (2005). *Iron age communities in Britain: An account of England, Scotland and Wales from the seventh century BC until the Roman conquest*. Routledge.
Desvaux, M. (2009). Footprints to the future. *Medicine, Conflict and Survival, 25*, 221–245.
Dunham, A. L. (1955). *The industrial revolution in France, 1815-1848*. Exposition Press.
Eaton, S. B. (2006). The ancestral human diet: What was it and should it be a paradigm for contemporary nutrition. *Proceedings of the Nutrition Society, 65*, 1–6. https://doi.org/10.1079/PNS2005471
Eberhard, D. M., Simons, G. F., & Fennig, C. D. (Eds.). (2022). *Ethnologue: Languages of the world* (25 ed.). SIL International http://www.ethnologue.com.
Eisenstein, E. L. (1980). *The printing press as an agent of change: Communications and cultural transformations in early-modern Europe (1)*. Cambridge University Press.
Emberling, G. (2003). Urban social transformations and the problem of the first city: New research from Mesopotamia. In M. L. Smith (Ed.), *The social construction of ancient cities* (pp. 254–268). Smithsonian.
Eriksson, A., Betti, L., Friend, A. D., Lycett, S. J., Singarayer, J. S., von Cramon-Taubadel, N., Valdes, P. J., Balloux, F., & Manica, A. (2012). Late Pleistocene climate change and the global expansion of anatomically modern humans. *Proceedings of the National Academy of Sciences, 109*, 16089–16094. https://doi.org/10.1073/pnas.1209494109
Fischer-Kowalski, M., & Haberl, H. (1997). Tons, joules, and money: Modes of production and their sustainability problems. *Society & Natural Resources, 10*, 61–85.
FitzGerald, C. P. (1981). Reflections on the character of the industrial revolution in China. *The Australian Journal of Chinese Affairs*, 65–71.
Gellner, E. (1983). *Nations and nationalism*. Blackwell Publishers.
Gellner, E. (1989). *Plough, sword and book: The structure of human history*. University of Chicago Press.
Giddens, A. (2002). *Runaway world: How globalization is reshaping our lives*. Profile Books.
Goring-Morris, A. N., & Belfer-Cohen, A. (2011). Neolithization processes in the levant: The outer envelope. *Current Anthropology, 52*, S195–S208. https://doi.org/10.1086/658860
Gosch, S. S., & Stearns, P. N. (2008). *Premodern travel in world history*. Routledge.
Green, C. M. (1939). *Holyoke, Massachusetts: A case history of the industrial revolution in America*. Yale University Press.
Gregersen, N. H. (2003). Risk and religion: Toward a theology of risk taking. *Zygon: Journal of Religion & Science, 38*, 355–376. https://doi.org/10.1111/1467-9744.00504
Gupta, A. K. (2004). Origin of agriculture and domestication of plants and animals linked to early Holocene climate amelioration. *Current Science, 87*, 54–59.
Haas, J. (2001). Warfare and the evolution of culture. In G. M. Feinman, & T. D. Price (Eds.), *Archaeology at the millennium* (pp. 329–350). Springer. https://doi.org/10.1007/978-0-387-72611-3_9
Haberl, H., Fischer-Kowalski, M., Krausmann, F., Martinez-Alier, J., & Winiwarter, V. (2011). A socio-metabolic transition towards sustainability? Challenges for another great transformation. *Sustainable Development, 19*, 1–14. https://doi.org/10.1002/sd.410
Hall, J. A. (1986). *Powers and liberties: The causes and consequences of the rise of the west*. University of California Press.
Hassan, R. (2003). Network time and the new knowledge epoch. *Time & Society, 12*, 225–241. https://doi.org/10.1177/0961463X030122004
Haywood, J. (2011). *The new atlas of world history: Global events at a glance*. Princeton University Press.
Hearn, J. (2012). *Theorizing power*. Palgrave Macmillan.
Hearn, J. (2021). Reframing the history of the competition concept: Neoliberalism, meritocracy, modernity. *The Journal of Historical Sociology, 34*, 375–392. https://doi.org/10.1111/johs.12324
Higham, C. (1996). *The bronze age of Southeast Asia*. Cambridge University Press.
Hodder, I. (2014). Çatalhöyük: The leopard changes its spots. A summary of recent work. *Anatolian Studies, 64*, 1–22. https://doi.org/10.1017/s0066154614000027
Hooke, R. L. (2000). On the history of humans as geomorphic agents. *Geology, 28*, 843–846.
Kates, R. W., Turner, B. L., II, & Clarke, W. C. (1990). The great transformation. In B. L. Turner, II, W. C. Clarke, R. W. Kates, J. F. Richards, J. T. Mathews, & W. B. Meyer (Eds.), *The great transformation* (pp. 1–17). Beacon Press.

Kelly, K. D. (1979). The independent mode of production. *Review of Radical Political Economics, 11*, 38–48.
Kremer, M. (1993). Population growth and technological change: One million B.C. to 1990. *Quarterly Journal of Economics, 108*, 681–716. https://doi.org/10.2307/2118405
Kuijt, I. (2000). People and space in early agricultural villages: Exploring daily lives, community size, and architecture in the late pre-pottery Neolithic. *Journal of Anthropological Archaeology, 19*, 75–102. https://doi.org/10.1006/jaar.1999.0352
Li, M., Wiedmann, T., Fang, K., & Hadjikakou, M. (2021). The role of planetary boundaries in assessing absolute environmental sustainability across scales. *Environment International, 152*, 106475. https://doi.org/10.1016/j.envint.2021.106475
Luhmann, N. (1993). *Risk: A sociological theory*. Walter de Gruyter.
Mann, M. (1986). *The sources of social power: Volume 1, A history of power from the beginning to AD 1760*. Cambridge University Press.
Manzanilla, L. (Ed.). (1997). *Emergence and change in early urban societies*. Springer. https://doi.org/10.1007/978-1-4899-1848-2
Martin, F. (1876). *The history of Lloyd's and of marine insurance in Great Britain*. Macmillan & Co.
Mellars, P. (2006). Why did modern human populations disperse from Africa ca. 60,000 years ago? A new model. *Proceedings of the National Academy of Sciences, 103*, 9381–9386. https://doi.org/10.1073/pnas.0510792103
Miller, D. E., & Van Der Merwe, N. J. (1994). Early metal working in sub-Saharan Africa: A review of recent research. *The Journal of African History, 35*, 1–36.
North, D. C., & Thomas, R. P. (1971). The rise and fall of the manorial system: A theoretical model. *The Journal of Economic History, 31*, 777–803.
Polanyi, K. (2001). *The great transformation*. Beacon Press.
Pryor, F. L. (2004). From foraging to farming: The so-called Neolithic Revolution. *Research in Economic History, 22*, 1–39.
Reba, M., Reitsma, F., & Seto, K. C. (2016). Spatializing 6,000 years of global urbanization from 3700 BC to AD 2000. *Scientific Data, 3*, 160034. https://doi.org/10.1038/sdata.2016.34
Renn, O. (2008). *Risk governance: Coping with uncertainty in a complex world*. Earthscan.
Renn, O., Klinke, A., & van Asselt, M. B. A. (2011). Coping with complexity, uncertainty and ambiguity in risk governance: A synthesis. *AMBIO: A Journal of the Human Environment, 40*, 231–246. https://doi.org/10.1007/s13280-010-0134-0
Richardson, S. (2012). Early Mesopotamia: The presumptive state. *Past & Present, 215*, 3–49.
Rockström, J., Steffen, W., Noone, K., Persson, Å., Chapin, F. S., III, Lambin, E. F., Lenton, T. M., Scheffer, M., Folke, C., Schellnhuber, H. J., Nykvist, B., de Wit, C. A., Hughes, T. P., van der Leeuw, S., Rodhe, H., Sörlin, S., Snyder, P. K., Costanza, R., Svedin, U., … Foley, J. A. (2009). A safe operating space for humanity. *Nature, 461*, 472–475. https://doi.org/10.1038/461472a
Rushforth, S. (1992). The legitimation of beliefs in a hunter-gatherer society: Bearlake Athapaskan knowledge and authority. *American Ethnologist, 19*, 483–500. https://doi.org/10.1525/ae.1992.19.3.02a00040
Shennan, S. (2018). *The first farmers of Europe: An evolutionary perspective*. Cambridge University Press. https://doi.org/10.1017/9781108386029
Smith, T. C. (1955). *Political change and industrial development in Japan: Government enterprise*. Stanford University Press.
Smith, G. R., & Archer, R. (2020). Climate, population, food security: Adapting and evolving in times of global change. *The International Journal of Sustainable Development and World Ecology, 27*(5), 419–423. https://doi.org/10.1080/13504509.2020.1712558
Steffen, W., Richardson, K., Rockström, J., Cornell, S. E., Fetzer, I., Bennett, E. M., Biggs, R., Carpenter, S. R., de Vries, W., de Wit, C. A., Folke, C., Gerten, D., Heinke, J., Mace, G. M., Persson, L. M., Ramanathan, V., Reyers, B., & Sörlin, S. (2015). Planetary boundaries: Guiding human development on a changing planet. *Science, 347*(6223), 1259855. https://doi.org/10.1126/science.1259855
Sztompka, P. (1993). *The sociology of social change*. Blackwell.
Tainter, J. A. (2000). Problem solving: Complexity, history, sustainability. *Population and Environment, 22*, 3–41.

Taleb, N. N. (2007). *The black swan: The impact of the highly improbable*. Random House.
Taylor, P. J. (2012). Extraordinary cities: Early 'City-ness' and the origins of agriculture and states. *International Journal of Urban and Regional Research, 36*, 415–447. https://doi.org/10.1111/j.1468-2427.2011.01101.x
Timmermann, A., & Friedrich, T. (2016). Late Pleistocene climate drivers of early human migration. *Nature, 538*(7623), 92–95. https://doi.org/10.1038/nature19365
Trenerry, C. F. (2009). *The origin and early history of insurance: Including the contract of bottomry*. Lawbook Exchange.
United Nations. (2019). *World urbanization prospects: The 2018 revision. United Nations.*
Veblen, T. (2003). *Imperial Germany and the industrial revolution*. Batoche Books.
Weir, W. (2005). *50 weapons that changed warfare*. Career Press.
Weisdorf, J. L. (2005). From foraging to farming: Explaining the Neolithic revolution. *Journal of Economic Surveys, 19*, 561–586. https://doi.org/10.1111/j.0950-0804.2005.00259.x
Weiss, E., & Zohary, D. (2011). The Neolithic Southwest Asian founder crops. *Current Anthropology, 52*(S4), S237–S254. https://doi.org/10.1086/658367
Wisner, B., Blaikie, P. M., Cannon, T., & Davis, I. R. (2004). *At risk: Natural hazards, people's vulnerability and disasters*. Routledge.
Withers, C. W. J. (2007). *Placing the Enlightenment: Thinking geographically about the age of reason*. University of Chicago Press.
Wolf, E. (1997). *Europe and the people without history*. University of California Press.
Woolf, G. (2020). *The life and death of ancient cities: A natural history*. Oxford University Press.
Xu, S., Kangwanpong, D., Seielstad, M., Srikummool, M., Kampuansai, J., & Jin, L. (2010). Genetic evidence supports linguistic affinity of Mlabri - a hunter-gatherer group in Thailand. *BMC Genetics, 11*, 18. https://doi.org/10.1186/1471-2156-11-18
Zeder, M. A. (2011). The origins of agriculture in the Near East. *Current Anthropology, 52*(S4), S221–S235. https://doi.org/10.1086/659307
Zhao, Z. (2011). New Archaeobotanic data for the study of the origins of agriculture in China. *Current Anthropology, 52*(S4), S295–S306. https://doi.org/10.1086/659308
Zinn, J. O. (2008). Introduction. In J. O. Zinn (Ed.), *Social theories of risk and uncertainty: An introduction* (pp. 1–17). Blackwell Publishing.

CHAPTER 3

Our growing awareness of sustainability challenges

Introduction

There is no doubt that the development of industrial society has left our world in a grave state. Although we have faced substantial sustainability challenges in the past, our current situation is qualitatively and quantitatively different. It is qualitatively different in the much greater complexity of our contemporary challenges and quantitatively different in their much grander scale, in both spatial and temporal terms, than ever before. The direct cause-and-effect relationships that generated the first sustainability challenge of having too many mouths to feed with the available food that could be hunted or gathered in the local area have been replaced by complex nonlinear networks of numerous cause-and-effect relationships generating sustainability challenges that are much more difficult to both comprehend and address. Especially since these relationships often cut across sectorial, geographical and administrative boundaries that have become so ingrained in how we look upon and understand our world that they restrict our ability to grasp the whole picture (Becker, 2021). We have gone from our past local sustainability challenges to sustainability challenges experienced both globally and locally.

Many of these sustainability challenges have not gone unnoticed, even if we have yet failed to mobilise adequate action to address most of them. Ozone depletion and particular chemical pollutants are notable exceptions, which we will return to in the next chapter (Chapter 4). This chapter overviews our growing awareness of our contemporary sustainability challenges, as discussed on two strings of world conferences over the last five decades and in the global policy dialogue leading up to the Sustainable Development Goals (SDGs) for 2015—30. It is, in other words, focusing on the evolution of ideas concerning safety and sustainability in the international community, which I suggest could be seen as coming together into an implicit vision of a new and fundamentally different kind of society compared to Gellner's (1989) 'Industria' (Chapter 2)—a vision of 'Sustainia', if you like (Chapter 16).

Our challenges as discussed at world conferences

The state of the world has not gone unnoticed by the international community. There has been a string of conferences and summits about our sustainability challenges and what

Sustainability Science, Second Edition
ISBN 978-0-323-95640-6, https://doi.org/10.1016/B978-0-323-95640-6.00013-0

we should do about them. There are two groups of world conferences particularly central to the focus of this book. First, we have five world conferences focusing explicitly on sustainable development, dating back five decades. Secondly, we have three more recent world conferences focusing on managing risk, aiming to facilitate sustainable development. There is also a long list of climate change conferences held more or less annually since 1995 within the United Nations Framework Convention on Climate Change (UNFCCC) and several other policy dialogues, which I have no room to describe here. However, I return to some of their more essential outcomes later in the chapter concerning Agenda 2030 and the SDGs. The following sections briefly introduce these two strings of world conferences, focusing on the development of ideas concerning our sustainability challenges and what to do about them.

Stockholm, Rio, Johannesburg and back to Rio and Stockholm again

The world conferences on sustainable development started in 1972, although that term was not used back then. They have been held more or less every ten years, except for 1982. However, the United Nations General Assembly approved the World Charter for Nature that year, declaring humankind to be a part of nature since life depends on the uninterrupted functioning of natural systems. There has been a whole range of other meetings or summits focusing on aspects of sustainable development. Still, I limit myself to introducing the five main world conferences on our sustainability challenges—sometimes called Earth Summits—in Stockholm, Rio de Janeiro, Johannesburg, Rio de Janeiro and most recently, back in Stockholm again in 2022.

Stockholm 1972

After a few decades of focusing mainly on the fragile peace after World War II, the United Nations answered the suggestion from Sweden to hold a first world conference on development and the environment. It was called the United Nations Conference on the Human Environment and was held in Stockholm in 1972 (Figure 3.1). Although the height of the Cold War resulted in the Soviet Union and most of its allies not attending, the conference still brought together representatives of 113 states (86% of the 132 member states of the UN then), both developed and developing, and several UN agencies, other intergovernmental organisations and non-governmental organisations (UNCHD, 1972). The conference placed the environment on the global agenda and produced a declaration with 26 principles and an action plan with 109 recommendations (UNCHD, 1972).

When looking at the principles of the Declaration of the United Nations Conference on the Human Environment, also referred to as the Stockholm Declaration, it is clear that the conference defined the human environment broader than more traditional notions of the environment. The first principle focuses on the fundamental

Figure 3.1 The opening of the first Earth Summit in 1972 in Stockholm. *(Photo by UN/Yutaka Nagata.)*

right to freedom, equality and adequate living conditions in an environment that permits a life of dignity and well-being. It explicitly states that apartheid, racial segregation, discrimination and all forms of oppression and domination must be condemned and eliminated. The last principle calls for the abolition of weapons of mass destruction. The declaration calls for reducing and preventing pollution to ensure that serious or irreversible damage is not inflicted upon ecosystems or human health. It balances the need to conserve both renewable and nonrenewable natural resources with the need for socio-economic development, all with clear rhetoric of justice and equality. Demographic policies are presented as essential to limit population growth and excessive population pressures on the environment, but only as long as they do not infringe on fundamental human rights. Human settlements and urbanisation must be planned to avoid adverse environmental effects and obtain maximum social, economic and environmental benefits for all. The declaration explicitly states that we, as human beings, have a solemn responsibility to protect and improve the environment for present and future generations, which denotes a clear link to sustainability. However, that term was not used at that time.

The declaration states the importance of integrating conservation into development planning and clarifies that resources are needed. For instance, it brings up the need for aid and adequate levels and stability of prices on raw materials, as that was (and still is) essential for the economies of so many developing countries. Moreover, the declaration states that the environmental policies of all countries should improve and not hamper the present or future development of developing countries or the attainment of better living conditions for all. Environmental challenges due to underdevelopment and what it refers to as natural disasters are portrayed as best addressed through accelerated development supported by substantial financial aid and technical assistance to developing countries—at the same time as referring to the UN Charter and international law when stating that countries have the sovereign right to exploit their natural resources, as well as the responsibility to ensure that areas beyond their borders are not adversely affected by it. The declaration makes it also up to each country to decide on its own environmental standards, though it mentions the possibility that specific criteria could be agreed upon internationally (Box 3.1). It specifies that although each country should have its own national institutions for natural resource management, they should cooperate on equal terms to effectively control, prevent, reduce and eliminate adverse environmental effects and develop international law regarding pollution and other environmental impacts.

The Stockholm Declaration highlights the utmost importance of research, innovation and education. It specifies that science and technology must be applied to the identification, avoidance and control of environmental risks and the solution of environmental problems for the common good of humankind. Research and development in environmental problems must be promoted everywhere, especially in developing countries, and the open access to and transfer of knowledge and solutions must be facilitated and supported. Environmental education for both young and adults is, thus, presented as essential for the informed opinion and responsible conduct of individuals, communities and private companies in protecting and improving the full human dimension of the environment, giving particular consideration to the underprivileged.

BOX 3.1 The creation of the environmental conscience of the UN system

One of the 1972 Stockholm conference results was the recommendation to create a small secretariat for environmental action and coordination within the United Nations system. This outcome led, already later that year, to the establishment of the United Nations Environment Programme (UNEP), which since then has grown into a large international organisation with its headquarters in Nairobi, Kenya, six regional offices and numerous country offices. UNEP's mission is 'to provide leadership and encourage partnership in caring for the environment by inspiring, informing, and enabling nations and peoples to improve their quality of life without compromising that of future generations'. Its work encompasses assessing environmental conditions and trends, developing international and national environmental instruments, strengthening institutions for the wise management of the environment, facilitating the transfer of knowledge and technology for sustainable development, and encouraging new partnerships and mindsets within civil society and the private sector. UNEP's current priority areas are climate change, disasters and conflicts; ecosystem management; environmental governance; harmful substances; and resource efficiency.

Rio de Janeiro 1992

Twenty years after Stockholm, the turn had come to Brazil. Compared to its predecessor, the United Nations Conference on Environment and Development in Rio de Janeiro in 1992 was a sizeable event (Figure 3.2). The conference, often called the Earth Summit, had representatives from 172 states (96% of the 179 member states of the UN then), including 108 heads of state or government and from thousands of non-governmental organisations. It resulted in the Rio Declaration on Environment and Development, which I focus on in this brief section, and in Agenda 21, a nonbinding action plan for sustainable development. The United Nations Framework Convention on Climate Change (UNFCCC) and the Convention on Biological Diversity (CBD) were also opened for signatures at the conference, which are vital legally binding agreements concerning two central aspects of the sustainability of humankind.

The conference had been preceded by a decade of intense work laying the foundation for the concept of sustainable development as we know it. First, the International Union for the Conservation of Nature and Natural Resources (IUCN) published their World Conservation Strategy (WCS), declaring conservation to be intrinsically linked to development and the alleviation of poverty of hundreds of millions of people (IUCN, 1980). In other words, unless the health and productivity of the planet are protected, all humankind is at risk. Then, a few years later, the World Commission on Environment and Development (WCED) was created and constituted as an independent body by the General Assembly of the United Nations. After years of work, they published in 1987 the highly influential report called 'Our Common Future', which is considered the founding source for the concept of sustainable development as inherently linking social, environmental and economic aspects (WCED, 1987). The table was, in other words, set for a fruitful conference (Figure 3.3).

Figure 3.2 Group photo of world leaders at the Earth Summit in Rio de Janeiro, 1992. *(Photo by UN Photo/Michos Tzovaras.)*

Figure 3.3 Sustainable development entails social, environmental and economic aspects. *(Source: Dusit/Shutterstock.com.)*

The resulting Rio Declaration on Environment and Development has 27 principles (UNCED, 1992), spanning broad areas like the environment, development and peace. It places us, humans, at the centre of concern for sustainable development and states that we are entitled to healthy and productive lives in harmony with nature. The declaration reaffirms the UN Charter and principles of international law when stating that each country has the sovereign right to exploit its own resources and the responsibility to ensure that activities within its borders do not cause damage to areas beyond.

The Rio Declaration affirms the right to development, which must be fulfilled to meet present and future generations' developmental and environmental needs. For sustainable development to be achieved, environmental protection must constitute an integral part of the development, and the two cannot be considered in isolation. It states that all countries should cooperate in the essential task of eradicating poverty, as an essential requirement for sustainable development, to better meet the needs of the majority of the global population. The Rio Declaration prioritises the particular situation and needs of developing countries, particularly the least developed countries. However, international actions should also aim to address the interests and needs of all countries.

The Rio Declaration asserts that countries should cooperate to conserve, protect and restore the health and integrity of the Earth's ecosystems, but makes it clear that countries differ in their contributions to global environmental degradation and have thus different responsibilities. Moreover, developed countries should acknowledge that the technologies and financial resources they command also demand them to shoulder a more significant part of the responsibilities to address our global sustainability challenges. It declares that countries should cooperate in developing capacity for engendering sustainable development by exchanging scientific and technical knowledge and improving the development, adaptation, diffusion and transfer of innovative solutions to our sustainability challenges.

The declaration asserts that sustainable development requires reducing and eliminating unsustainable production and consumption patterns. Although the promotion of appropriate demographic policies is still intended mainly for the state, it clearly states that environmental challenges are best addressed with the participation of all concerned citizens. The declaration gives examples of what such participation could entail on different levels. The vital role of women in environmental management and development is highlighted, and their full participation is declared essential to achieve sustainable development. It is then interesting to count the number of women in the official photos of the world leaders at the conference (Figure 3.2). Similarly, the participation of youth and indigenous people is explicitly emphasised, the former for being our future leaders and the latter for their knowledge and traditional practices for sustainability.

The Rio Declaration states that each country should have effective national environmental legislation, with environmental standards, objectives and priorities that reflect the environmental and developmental context in which they apply. The legislation should cover liability and compensation for pollution and other environmental impacts. The declaration also asserts that national authorities should promote the internalisation of environmental costs into auditing regimes, thus making the polluter pay the cost of the pollution. Yet, it reduces the weight of this requirement by adding that it must be in relation to the public interest and without distorting international trade and investment. The Rio Declaration states that environmental impact assessment should be undertaken for potential activities that may have significant adverse impacts on the environment, that the precautionary principle should be widely applied, and that lack of complete scientific certainty should not be used as a reason for not implementing cost-effective measures to prevent environmental degradation.

The Rio Declaration states that countries should work together to promote a supportive and open international economic system, leading to sustainable development through economic growth and better management of environmental degradation in all countries. Unilateral actions to deal with environmental challenges outside the importing country should be avoided, and addressing transboundary or global environmental problems should, as far as possible, be based on international consensus. Moreover, countries should collaborate to deter or prevent the relocation of any activities and substances which are harmful to the environment and human health to other countries. The declaration states that if activities in one country may have significant adverse effects on the environment in another, they should consult each other early and in good faith. It is also applicable to natural disasters or other emergencies that are likely to produce sudden harmful transboundary effects on the environment, and the international community is obliged to make every effort to assist the countries affected. The declaration asserts that countries and people should cooperate in a spirit of partnership to fulfil the declaration's principles and further develop international law in sustainable development.

The Rio Declaration explicitly states that development, environmental protection and peace are intrinsically linked, and armed conflict is inherently destructive to sustainable development. Countries are, thus, obliged to resolve all their environmental disputes peacefully, protect the environment and natural resources of people under oppression, domination and occupation, and respect international law protecting the environment during armed conflict.

Johannesburg 2002

In 2002, it was South Africa's turn to host the world conference. It was ten years after the United Nations Conference on Environment and Development in Rio de Janeiro, and the event is often referred to as Rio+10 or Earth Summit 2002. Its formal name is, however, the World Summit on Sustainable Development, held in Johannesburg under the African leadership of both the host country and the UN (Figure 3.4). The conference grew even further with 191 states (100% of the member states of the UN then) and many intergovernmental organisations and non-governmental organisations (United Nations, 2002c). This time, however, under President George W. Bush's leadership, the United States boycotted the conference by not sending a participating delegation, only a brief visit by the Secretary of State.

Although the resulting declaration, called the Johannesburg Declaration on Sustainable Development or just the Johannesburg Declaration (United Nations, 2002a), clearly builds on the declarations from the two previous world conferences, it is this time more generally written and gives even less clear direction. However, the declaration was joined by an implementation plan, providing additional guidelines (United Nations, 2002b).

Figure 3.4 Former South African President Nelson Mandela and the Secretary-General of the UN Kofi Annan in Johannesburg, 2002. *(Photo by UN Photo/Evan Schneider.)*

Also, the UN Secretary-General Kofi Annan proposed the WEHAB initiative, highlighting key sustainability challenges concerning water and sanitation, energy, health and environment, agriculture and biodiversity and ecosystem management.

The Johannesburg Declaration has 37 paragraphs divided under the headings 'From our origins to our future', 'From Stockholm to Rio de Janeiro to Johannesburg', 'The challenges we face', 'Our commitment to sustainable development', 'Multilateralism is the future' and 'Making it happen!' (United Nations, 2002a). It spends ten paragraphs setting the scene by reaffirming commitments and collective responsibilities and referring to past conferences and documents.

The following five paragraphs summarise our sustainability challenges. The declaration states that we must recognise that our security and sustainability depend on eradicating poverty and closing the ever-increasing gap between rich and poor. It also asserts that we must change our consumption and production patterns and protect our natural resources as the environment continues to suffer. The declaration spells out biodiversity loss, depleted fish stocks, pollution, desertification, increasingly frequent and devastating natural disasters and climate change as evident sustainability challenges. It also specifies globalisation to be an added dimension to these challenges since what the Johannesburg Declaration brings up as positive effects in terms of rapid integration of markets, mobility of capital and significant increases in investment flows are unevenly distributed around the world and are met with adverse effects, mainly to the disadvantage of developing countries (United Nations, 2002a).

The declaration then continues by presenting our commitment to sustainable development in 15 paragraphs, stating the strength in our diversity, the need for solidarity and cooperation over boundaries, the importance of gender equality, the vital role of indigenous people, the importance of regional cooperation, the unique circumstances of small island states, the importance of developed countries to reach the internationally agreed levels of official development assistance, the importance of increased income-generating opportunities, and the crucial roles of the private sector to contribute to sustainable development and to enforce corporate accountability (United Nations, 2002a). The Johannesburg Declaration asserts that global society has the means to address our sustainability challenges but recognises that sustainable development requires a long-term perspective, broad-based participation and improved governance at all levels.

The Johannesburg Declaration asserts our resolve to promptly increase access to basic requirements, such as water and sanitation, food, shelter, health care and energy, and to protect biodiversity. It declares our will to cooperate to increase access to financial resources and open markets and ensure capacity-building, technology transfer, education and training to permanently banish underdevelopment. The declaration focuses on conditions that pose severe threats to sustainable development, such as chronic hunger and malnutrition, armed conflict and occupation, natural disasters, terrorism, organised crime and illicit drugs, illegal arms trade, trafficking, corruption, intolerance and xenophobia, and endemic diseases.

The Johannesburg Declaration then focuses on the importance of multilateral solutions to our sustainability challenges in three paragraphs. It asserts the need for more effective, democratic and accountable international institutions, reaffirming the commitment to the principles of the UN Charter and international law while arguing for stronger multilateralism. The declaration states our commitment to monitor the progress towards achieving development goals before focusing the final four paragraphs on how to make it happen. It asserts the need for an inclusive process in which we commit ourselves to act together, united by a shared determination to save our planet, promote human development and achieve universal prosperity and peace. The declaration then ends by asserting a commitment to the Plan of Implementation of the World Summit on Sustainable Development and by pledging determination to realise our collective hope for sustainable development.

Rio de Janeiro 2012

To revitalise the momentum created by the first Rio conference, the next world conference on sustainable development went back to Brazil, referred to as Rio+20. However, its official name is United Nations Conference on Sustainable Development, which happened in Rio de Janeiro in 2012 (UNCSD, 2012). It was a gigantic event, then the largest in the history of the United Nations, with 192 states (99.5% of the 193 member states in the UN then), 105 Heads of state, government or their deputies, 487 government ministers, as well as numerous intergovernmental organisations, non-governmental organisations and private companies. Including all the hundreds of side events, almost 50,000 people participated in one way or another in the conference, not including the more or less equal number of protesters in downtown Rio de Janeiro (Guardian, 2012) (Figure 3.5).

The two main themes of the conference were a green economy, which stirred most of the protests, and the institutional framework for sustainable development. The Rio+20 conference was built on past conferences and established ideas concerning sustainable development. Its preparations centred on seven priority areas: jobs, energy, cities, food security and agriculture, water, oceans and disasters. While the three preceding conferences resulted in relatively condensed declarations complemented by more detailed plans, the Rio+20 conference resulted in only one substantial document called 'The Future We Want' (UNCSD, 2012). This document, of 53 pages and 283 paragraphs,

Figure 3.5 One of the protest slogans on the streets around Rio+20.

is divided into six main sections called 'I. Our common vision', 'II. Renewing political commitment', 'III. Green economy in the context of sustainable development and poverty eradication', 'IV. Institutional framework for sustainable development', 'V. Framework for action and follow-up' and 'VI. Means of implementation'.

The first main section consists of 13 paragraphs reaffirming the commitment to an economically, socially, and environmentally sustainable future for our planet and present and future generations, to accelerate the achievement of agreed development goals, to the UN Charter, Human Rights Law and other international law, and to strengthen international cooperation to address the persistent challenges related to sustainable development. The document recognises that people are at the centre of sustainable development and that striving for a just, equitable and inclusive world entails combating poverty, unsustainable production and consumption patterns and depletion of natural resources. It also recognises that this can only be achieved through broad participation and good governance on all levels. In short, the first main section is intended to lay the foundation to address the two themes of (1) a green economy in the context of sustainable development and poverty eradication and (2) an institutional framework for sustainable development.

The second main section focuses 42 paragraphs on renewing political commitment. It is divided into three subsections called 'A. Reaffirming the Rio Principles and past action plans', 'B. Advancing integration, implementation and coherence', and 'C. Engaging major groups and other stakeholders'. The document recalls the Stockholm Declaration and reaffirms the declarations and action plans of the first Rio conference and the Johannesburg conference, as well as a long list of other international documents and conventions, such as the United Nations Millennium Declaration, the Doha Declaration on Financing for Development, the UNFCCC and the Convention on Biological Diversity. However, the document recognises uneven progress since the first Rio Declaration in 1992 for many reasons and emphasises the need to take concrete measures to accelerate progress in implementing previous commitments. The document emphasises a more conducive enabling environment, stronger international cooperation and broader participation.

Although the document acknowledges examples of progress towards sustainable development on different levels, it voices concern over the global levels of poverty and unemployment, as well as over climate change. It reaffirms the importance of supporting developing countries and urges all to refrain from using unilateral economic, financial or trade measures outside the UN Charter and international law that may hamper their development. The document asserts that special support must be given to people living in areas affected by occupation, complex humanitarian emergencies and terrorism. It acknowledges the territorial integrity and political independence of states and the importance of conforming to international law. It recognises that each country faces specific sustainability challenges and emphasises in particular African countries, least developed countries, landlocked developing countries, small island developing countries and middle-income countries. The Rio+20 document highlights that sustainable development must benefit and involve all. Such broad participation requires the involvement of not only all states but also private companies, universities and civil society, as well as including women and men, youth and children, and indigenous people. In short, the document calls for holistic and integrated approaches to sustainable development that will guide humanity to live in harmony with nature and lead to efforts to restore the health and integrity of Earth's ecosystems.

The third main section includes 19 paragraphs describing the ideas of a green economy as a critical tool for sustainable development. A green economy is an economy that 'results in improved human wellbeing and social equity, while significantly reducing environmental risks and ecological scarcities' (UNEP, 2011, p. 2). That is to say, an economy that is low carbon, resource efficient and socially inclusive. The paragraphs generally repeat the statements of previous sections, but in relation to a green economy, with some more specific additions.

The fourth main section focuses 29 paragraphs on developing the institutional framework for sustainable development, with five subsections called 'A. Strengthening the three dimensions of sustainable development', 'B. Strengthening intergovernmental arrangements for sustainable development', 'C. Environmental pillar in the context of sustainable development', 'D. International financial institutions and United Nations operational activities' and 'E. Regional, national, subnational and local levels'. After asserting the need for strengthening the overall institutional framework for sustainable development, the Rio+20 document reaffirms the role of the United Nations General Assembly, commits to strengthening the Economic and Social Council, declares to establish a High-level political forum, commits to strengthening UNEP substantially, and reaffirms the role of international financial institutions. It calls upon international organisations to further mainstream sustainable development in their respective mandates, programmes, strategies and decision-making processes. The document reaffirms the importance of broadening and strengthening the participation of developing countries in international economic decision-making. It acknowledges and encourages regional, national and local initiatives for sustainable development.

The fifth main section of the Rio+20 document is the most substantive by far and focuses on presenting a framework for action and follow-up. It includes 148 paragraphs divided into two subsections called 'A. Thematic areas and cross-sectoral issues', with 141 paragraphs, and 'B. Sustainable development goals', with seven paragraphs. The document includes 24 thematic areas and cross-sectoral themes, many with considerable links in between. These include poverty eradication and the WEHAB areas from 10 years earlier but rephrased as water and sanitation, energy, health and population, food security, nutrition and sustainable agriculture, and biodiversity. They also include sustainable cities and human settlements, which, together with the former, often are referred to as the WEHAB+ and a range of areas and themes related to human industry, that is, sustainable consumption and production, sustainable transport, chemicals and waste, mining, and sustainable tourism. Climate change is included, as well as desertification, land degradation and drought. The areas and themes also focus on particular geographical areas—that is, oceans and seas, forests and mountains—and specifically challenging contexts—that is, small island developing countries, least developed countries, landlocked developing countries and Africa—and highlight the importance of regional efforts again. Education, gender equality and women's empowerment are included as vital pathways for sustainable development, as well as disaster risk reduction, which for the first time is explicitly included as a central aspect of sustainable development (Figure 3.6). The second subsection focuses on SDGs, with paragraphs reaffirming the commitment to the Millennium Development Goals and committing to setting a limited number of action-oriented and universally applicable SDGs for our future.

The sixth and last main section of the Rio+20 document focuses on means of implementation, including 32 paragraphs divided into five subsections called 'A. Finance', 'B. Technology', 'C. Capacity-building', 'D. Trade', and 'E. Registry of commitments'. It reaffirms previous commitments to fund development and calls on all countries to prioritise sustainable development by allocating resources following national priorities and needs. The document recognises the need for strong financial support to developing countries in their efforts to promote sustainable development. It establishes an intergovernmental process under the auspices of the General Assembly to assess financing needs, consider the effectiveness, consistency and synergies of existing instruments and frameworks and evaluate additional initiatives. New financial sources and novel partnerships are encouraged—especially South-South, triangular, and public-private partnerships—although global financial and economic challenges are recognised as hindering factors. The document also stresses the importance of fighting corruption and illicit financial flows on all levels.

The document emphasises the importance of developing, adapting, disseminating and transferring environmentally sound technologies, especially in and to developing

Figure 3.6 Managing disaster risk is emphasised as a central requisite for sustainable development. *(Source: Paul Saad | Dreamstime.com.)*

countries. In short, to make vital information, knowledge and solutions available to all. The importance of capacity-building is emphasised as a crucial pathway to sustainable development, and trade is reaffirmed as an engine for development and sustained economic growth.

Stockholm 2022

Following the idea of revitalising interest in sustainability by commemorating earlier events, the most recent world conference returned to Stockholm in 2022, where it all started in 1972, and is referred to as Stockholm+50. After steady growth over time in the size of the previous world conferences, this time, it did not manage to attract the same attention and interest. Around ten heads of state or government participated, which is only a tenth of what Rio+20 mustered, and a bit more than 90 government ministers (UNEP, 2022b). All in all, 155 states were represented (UNEP, 2022a), which only comprised 80% of the 193 member states of the UN at that time and a lower proportion than ever before. Several dignitaries were still speaking at the main event, 5000 people participated in person (UNEP, 2022a), and hundreds of side events organised by organisations from the state, market and civil society spread across the city. Moreover, the world conference itself was preceded by a series of five regional consultations during the preceding 2 months (Asia and the Pacific, Latin America and the Caribbean, Africa, Europe and North America, and West Asia), which allowed thousands of people to voice their concerns and engage in the dialogue. These preparatory processes, as well as the main conference and several side events, had an explicit focus on youth. Arguably more so than the previous world conferences.

The focus of the Stockholm+50 conference was to highlight the urgency of our sustainability challenges to get leaders to accelerate the implementation of the Agenda 2030 and the SDGs. The conference had three thematic leadership dialogues, in addition to a series of more conventional plenary sessions dealing with procedural matters and allowing a general debate primarily comprising state representatives describing past, present and future commitments to sustainable development. These leadership dialogues focused on (1) Reflecting on the urgent need for actions to achieve a healthy planet and prosperity of all, the timely theme of (2) Achieving a sustainable and inclusive recovery from the Covid-19 pandemic, and (3) Accelerating the implementation of the environmental dimension of Sustainable Development in the context of the Decade of Action, which the UN launched in 2020 to accelerate efforts towards the SDGs.

The outcome document of the Stockholm+50 conference is again different from its predecessors'. It is called *Stockholm+50: a healthy planet for the prosperity of all — our responsibility, our opportunity* and comprises four pages of text, including footnotes. It is, as such, similar in length to the formal declarations of the first Stockholm conference and the Johannesburg conference and considerably shorter than the first Rio Declaration, although structured differently and not accompanied by a more extensive document clarifying what to do and how (i.e. the Action Plan for the Human Development, Agenda 21, and the Plan of Implementation of the World Summit on Sustainable Development). It is not that the Stockholm+50 conference is without documentation—quite the contrary. In addition to the reports from the regional consultations mentioned above, a youth-focused report, two UN-focused reports, and a much-needed report on reducing the environmental impact of the world conference itself, a group of leading experts from the Stockholm Environment Institute (SEI) and the Council on Energy, Environment and Water (CEEW) published an independent scientific report ahead of the conference that synthesised scientific evidence and analysed the intertwined human and environmental crisis facing our world (SEI & CEEW, 2022). It is just that the outcome document is very vague and not having the same agreed appended document to lean on. The explanation for this apparent shortcoming is that there was already a framework in place: The Agenda 2030 and the SDGs, which I will return to later in the chapter.

The outcome document is referred to as a concept note. It comprises 14 paragraphs divided into three main sections called 'I. The mandate — Stockholm+50 international meeting', 'II. The thematic focus — a healthy planet for the prosperity of all', and 'III. The expectation — our responsibility, our opportunity to ensure continued prosperity for all'.

The first section has four paragraphs. The first and fourth refer to the United Nations General Assembly resolutions that decided to convene the conference and determined its modalities. The other two paragraphs state the expectation of clear and concrete recommendations for action at all levels, including strengthened cooperation. They list all types of actors invited as observers to the conference, ranging from international organisations to indigenous peoples' organisations and from private companies to academic institutions. The section is, thus, largely procedural.

The second section also has four paragraphs, but somewhat more substantive than the previous section. The first paragraph refers back to the first conference in Stockholm, laments the increasing challenges for humanity since then, and points out the imbalances in sustainable development's economic, social and environmental pillars. The second paragraph of this section focuses on the COVID-19 pandemic as both a challenge for development and an example of how planetary and other threats lead to a systemic crisis of human development and insecurity. The third section asserts that science clearly demonstrates the urgent need for transformation of socio-economic systems (including measures of progress and well-being, true costs of economic products, targeted pro-poor subsidies, sustainable consumption that addresses both under- and over-consumption, circular production practices, investment in education, gender equality and women's rights). It also asserts that countries must be enabled to increase fiscal space, invest in sustainable infrastructure, rebuild key sectors and value chains, grow green and decent jobs, and create sustainable and equitable development. The last paragraph of the second section challenges humankind to leave the unsustainable path we have been on since the first

Stockholm conference. It challenges us to build on the previous outcome documents and to accelerate the implementation of the commitments already made.

The final section of the outcome document has six paragraphs, mainly stating the various opportunities provided by the world conference but also explicitly stating that the conference is intended to complement already established spaces and agreements for addressing our sustainability challenges. The first and second paragraphs propose Stockholm+50 to provide opportunities to (1) rebuild relationships of trust for strengthened cooperation and solidarity, (2) accelerate the system-wide actions needed to recover and build forward from the pandemic, (3) connect and build bridges across agendas, (4) rethink conceptions and measures of progress and well-being, and (5) explore emerging areas in support of a healthy planet. The third paragraph reiterates that Stockholm+50 offers an opportunity to (6) accelerate action to implement Agenda 2030 and the SDGs (SDGs), (7) look further ahead to a 50-year timeframe, (8) think and act beyond the silos of individual challenges and (9) mark a milestone in how we conceive and deliver on human well-being, capabilities and freedoms. The fourth and fifth paragraphs focus on the leadership dialogues, which are presented as a start of more lasting processes to align already established global frameworks and processes for sustainability and engage youth. The final paragraph points towards the unprecedented possibilities of technology, which, if used responsibly, can help to ensure prosperity for all. The following 50 years will be crucial for accomplishing that.

The evolution of ideas concerning sustainable development

From reading the outcome documents, it is clear that the five world conferences focusing on sustainable development build on each other. The foundation for sustainable development was put in place already at the first Stockholm Declaration, with its references to balancing conservation with social and economic development to protect and improve the environment for present and future generations. However, the term sustainable development was not mainstream until just before the first Rio conference, which was primarily built on the *Our Common Future* report from a few years earlier (WCED, 1987). Although sustainable development, resting on the three pillars of economic development, social development and environmental protection, took hold and continues to be the central concept since then, it is interesting to note the decline in references to the environment since the first Rio Declaration (Figure 3.7).

It is also clear that the outcome documents are all products of their time, reflecting the dreams and fears of the people involved in drafting them. The Stockholm Declaration was drafted in 1972, just a few years after political representation for non-Whites was abolished entirely in Apartheid South Africa, a string of civil rights acts had been passed in the United States, and we were in the final stage of the colonial period and under the constant Cold War threat of global annihilation. It is thus not surprising that the declaration entails statements against apartheid, racial segregation, oppression, weapons of mass

Figure 3.7 Word clouds of the Declaration of the United Nations Conference on the Human Environment from 1972, the Rio Declaration on Environment and Development from 1992, the Johannesburg Declaration of Sustainable Development from 2002, the Future We Want from 2012, and Stockholm+50: a healthy planet for the prosperity of all — our responsibility, our opportunity from 2022 (constructed in monkeylearn.com/word-cloud). Visualising the 50 most commonly used words and word combinations, except common words, and sizing them after their relative frequency.

destruction and so on. Similarly, the Rio Declaration from 1992 emphasises the importance of free trade, which is not surprising either. The Cold War was just over, and neoliberalism had come out victorious, appearing to be the only way forward. This view had become more nuanced in time for the drafting of the Johannesburg Declaration in 2002, which declares free trade and globalisation not only as beneficial for sustainable development but also as potentially detrimental.

Even if the outcome documents are products of their time, it is interesting to note how little has changed in the challenges described, especially from the last conferences. However, several areas have a clear progression over time regarding focus and content. For instance, climate change is mentioned in the outcome document of the first Stockholm conference (UNCHD, 1972, p. 20), stated as a challenge demanding intense study in the outcome documents of the first Rio conference, a main focus of and recurrent theme throughout the outcome documents from the Johannesburg and Rio+20 conferences, and indeed mentioned among the primary challenges in the much shorter outcome document of the Stockholm+50 conference. Similarly, the importance of disaster risk for sustainable development has also increased incrementally over time, from being mentioned in 1972 to being included as a thematic area and cross-sectoral theme in 2012 and the Agenda 2030 and SDGs that the most recent conference in Stockholm was intended to complement. More on that later in the chapter.

Other key aspects of sustainable development have emerged over time, such as gender equality, which was not included in any way in 1972 but has been a critical focus and recurrent theme in the three following conferences since 1992. Specific geographical focus has also emerged over time, with Africa, least developed countries and small island developing countries getting particular attention since 1992, and landlocked developing countries since 2002. Middle-income countries are mentioned in the outcome document of the Rio+20 conference, which is understandable considering the growing environmental impact of the rapidly developing economies of our world. However, middle-income counties failed to make it to the final list of thematic areas and cross-sectoral issues.

The inclusion of increasingly diverse actors has also developed over time, both individually and institutionally. In addition to including women since 1992, other groups are also explicitly mentioned as vital for sustainable development. Although youth involvement is mentioned already in the outcome document from the first Stockholm conference, children and youth have been a central focus since the first Rio conference, as well as indigenous people. The more institutional actors have also expanded over time, from mainly focusing on states and intergovernmental organisations in Stockholm to increasing inclusion of non-governmental organisations and focus on public-private partnerships since the first Rio conference and to increasing direct involvement of private companies since Johannesburg. This latter shift is also clear in the descriptions of funding sources, with an increasing focus on trade and untraditional partnerships as complements to aid and development cooperation over time.

The current view of the international community on our sustainability challenges can be summarised by the Agenda 2030 and the 17 SDGs (Figure 3.8), which we will return to later. However, it is easy to see that these goals have been added and adapted over time, which is visible in the string of world conferences, even if the details and emphasis have been relocated from the outcome documents of the first four conferences to the documents of the Agenda 2030, which the Stockholm+50 was intended to draw increased attention to. The other main new way to address our sustainability challenges, in addition to updating the current institutional framework for sustainable development, was to introduce the green economy. It is, again, interesting to note that the international community sticks to the beaten track at best and seems to be collectively unable to think new thoughts at worst. All agree that the situation is dire, and we have not adequately addressed it within the current economic and institutional system. However, the outcome is still not in any substantial way questioning that economic and institutional system.

Yokohama, Kobe and Sendai

Three world conferences have focused on disasters and risk so far, although the next is due in 2025. These conferences have explicitly focused on the role of risk and disasters in relation to sustainable development and are essential to consider when considering how ideas concerning our sustainability challenges have evolved over the last decades. All three conferences were held in Japan—in Yokohama, Kobe and Sendai. The following sections briefly introduce these world conferences and elaborate on how the main ideas concerning disaster risk in relation to sustainable development have evolved over time.

Figure 3.8 The 17 sustainable development goals (SDGs).

Yokohama 1994

With growing recognition of the global plight of disasters, the 1990s were declared by the United Nations General Assembly as the International Decade for Natural Disaster Reduction (IDNDR). The primary objective of this initiative was to decrease the loss of life and property and the resulting social and economic disruption caused by what was then referred to as natural disasters. To review the midterm achievements and chart an action programme for the future, the IDNDR organised a world conference hosted by the Government of Japan. The event was called the World Conference on Natural Disaster Reduction and was held in Yokohama in 1994. There were delegations from 147 states present, as well as from 41 international organisations and 47 non-governmental organisations, with around 3500 registered delegates (El-Sabh, 1994, p. 334). It was the first time policymakers, practitioners and academics focusing on disaster risk—representing the state, market or civil society—got together from all over the world. The resulting document includes the Yokohama Message, consisting of a two-page and ten-paragraph statement, and the Yokohama Strategy and Plan of Action, outlining principles, plan of action and follow-up for a safer world (IDNDR, 1994) (Figure 3.9).

The Yokohama Message affirms that human and economic losses in natural disasters have increased and that the poor are usually the most vulnerable. Disaster prevention, mitigation, preparedness and response are stated as contributing to and gaining from sustainable development, and the three former are better than the latter at reducing disaster losses. The message affirms that regional and international cooperation and assistance are vital for meeting this purpose, though the necessary information, knowledge and technology often are available at low cost. It states that community participation should be encouraged to gain insight into the factors that favour or hinder prevention and

Figure 3.9 Yokohama, 100 years after the 1923 Great Kanto Earthquake and Tsunami killed 120,000–140,000 people.

mitigation, promote the preservation of the environment for future generations' development, and find effective and efficient means to reduce the impact of disasters. The Yokohama Message then spends its final four paragraphs outlining the Yokohama Strategy and Plan of Action for the rest of the decade and beyond.

The other part of the outcome of the conference, the Yokohama Strategy and Plan of Action, starts with an introduction recognising increasing disaster losses, recalling the General Assembly resolutions launching the IDNDR and adopting an integrated approach to disaster management, recognising that sustainable development cannot be achieved without reducing disaster losses, reaffirming the first Rio Declaration and emphasising the need to pay special attention to the least developed-, landlocked- and small island developing countries (IDNDR, 1994). The rest of the document is divided into three sections called 'I. Principles', 'II. Plan of Action' and 'III. Follow-up Action'.

Ten principles summarise the state-of-the-art understanding of risk and disaster management at that time and are presented in full in Box 3.2. The first section then continues

BOX 3.2 The 10 principles of the Yokohama strategy

1. Risk assessment is a required step for the adoption of adequate and successful disaster reduction policies and measures.
2. Disaster prevention and preparedness are of primary importance in reducing the need for disaster relief.
3. Disaster prevention and preparedness should be considered as integral aspects of development policy and planning at national, regional, bilateral, multilateral and international levels.
4. The development and strengthening of capacities to prevent, reduce and mitigate disasters is a top priority area to be addressed during the decade *to provide a strong basis for follow-up activities to the decade.*
5. Early warnings of impending disasters and their effective dissemination using telecommunications, including broadcast services, are key factors to successful *disaster prevention and preparedness.*
6. Preventive measures are most effective when they involve participation at all levels, from the local community through the national government to the *regional and international levels.*
7. Vulnerability can be reduced by the application of proper design and patterns of development focused on target groups, by appropriate education and *training of the whole community.*
8. The international community accepts the need to share the necessary technology to prevent, reduce and mitigate disaster; this should be made freely available *and in a timely manner as an integral part of technical cooperation.*
9. Environmental protection as a component of sustainable development consistent with poverty alleviation is imperative in the prevention and mitigation of *natural disasters.*

(Continued)

BOX 3.2 The 10 principles of the Yokohama strategy—cont'd

10. *Each country bears the primary responsibility for protecting its people, infrastructure*, and other national assets from the impact of natural disasters. The international community should demonstrate the strong political determination required to mobilise adequate and make efficient use of existing resources, including financial, scientific and technological means, in the field of natural disaster reduction, bearing in mind the needs of developing countries, particularly *the least developed countries.*

by presenting the basis for the strategy in seven paragraphs, mainly expanding the arguments presented in the Yokohama Message. For instance, while the natural phenomena causing disasters are, in most cases, beyond human control, vulnerability is a result of human activity and can be reduced. Large-scale urban concentrations are particularly fragile because of their complexity and the accumulation of population and infrastructures in limited areas. It is also interesting to note that the document includes explicit references to patterns of consumption, production and development that have the potential to increase vulnerability to natural disasters and that sustainable development can contribute to reducing this vulnerability.

The first section of the Yokohama Strategy and Plan of Action then continues by presenting an assessment of the status of disaster reduction midway into the IDNDR and a strategy for the year 2000 and beyond. The former summarises the assessment into nine bullet points, including areas such as the limited spread of awareness of the potential benefits of disaster reduction and disaster reduction not systematically being part of development policies. It is also important to note that the document explicitly states the strong need to strengthen the resilience of local communities as a central conclusion of the assessment. On the other hand, the strategy consists of 18 rather general bullet points, including, for instance, developing a global culture of prevention and effective national legislation and administrative action.

The second section presents a plan of action in the form of three lists of activities, including 18 bullet points at the community and national levels, 12 at the regional and subregional levels and 18 at the international level. For example, to upgrade the resistance of critical infrastructure on the national level, strengthen early warning on the regional level, and integrate disaster reduction into development assistance on the international level. Finally, the third section focuses on follow-up actions and presents nine bullet points on how to do this. For instance, by requesting the General Assembly to consider adopting a resolution endorsing the Yokohama Strategy or recommending the inclusion of a subitem entitled 'Implementation of the outcome of the World Conference on Natural Disaster Reduction' in the provisional agenda of the General Assembly under the item entitled 'Environment and Sustainable Development'.

Kobe 2005

Then in early 2005, just a month after the devastating Indian Ocean Tsunami, it was time for a second world conference in Japan. This time it was held in Kobe under the name of the World Conference on Disaster Reduction and proved to be highly effective in creating a global momentum for what was now called disaster risk reduction. The time and location of the event were deliberately chosen to generate support for the topic, being the location of one of the most expensive natural disasters in the world almost on the day a decade earlier (Figure 3.10). However, having the 2004 Indian Ocean Tsunami close in mind is generally believed to have made all the difference. The conference was a massive event, with around 4000 participants from 168 states, 78 observer organisations, 161 non-governmental organisations and 154 media organisations (ISDR, 2005c, p. 96). The interest was immense, and the public segment of the conference attracted around 40,000 participants and visitors (ISDR, 2005c, p. 96).

The conference's outcome includes the Hyogo Declaration and the Hyogo Framework for Action 2005–15. There was outstanding commitment behind these documents, with 168 states standing behind them from the start and the United Nations General Assembly endorsing them in a resolution the same year. The declaration contains one page of introduction, with condolences to all people affected by the Indian Ocean Tsunami and recognition that reducing disaster risk is a requisite for sustainable development, and one and a half pages with eight main statements (ISDR, 2005a). The Hyogo Declaration explicitly states that disasters have a tremendously detrimental impact on efforts at all levels to eradicate poverty and are a significant challenge to sustainable development. Disaster risk reduction, sustainable development and poverty eradication are, thus, intrinsically linked.

The Hyogo Declaration asserts the importance of involving all actors, from both public and private spheres, as well as from civil society and the scientific community. It

Figure 3.10 Kobe Earthquake Memorial Park. *(Photo by 663 highland, shared on the Creative Commons.)*

recognises that a culture of disaster prevention and resilience must be fostered at all levels and that we, as human beings, are far from powerless in reducing the risk of disasters, giving examples of risk reduction measures. The declaration affirms the primary responsibility of states to protect people and property and provides examples of priority areas. For example, national legal and policy frameworks, community-based disaster risk reduction, and support to capacity development for disaster risk reduction in disaster-prone developing countries, particularly in the least developed countries and small island developing countries.

The declaration announces the adoption of the Hyogo Framework for Action and emphasises the importance of it being translated into concrete actions at all levels. It also recognises the need for follow-up, developing progress indicators, and developing information-sharing mechanisms. The Hyogo Declaration then ends with a call for action from all actors, and a thank you statement to the conference host.

The official name of the Hyogo Framework for Action is the 'Hyogo Framework for Action 2005—15: Building the Resilience of Nations and Communities to Disasters' (ISDR, 2005b). It is a highly influential document with 19 pages and 34 paragraphs, divided into four main sections called 'I. Preamble', 'II. World Conference on Disaster Reduction: Objectives, expected outcome and strategic goals', 'III. Priorities for action 2005—15' and 'IV. Implementation and follow-up'.

The first main section focuses on the challenges posed by disasters and the lessons learnt from implementing the Yokohama Strategy. The framework states that disaster risk is a global concern and increasing due to various factors, such as urbanisation, socio-economic conditions, environmental degradation and climate change. It asserts that disaster risk arises when hazards interact with physical, social, economic and environmental vulnerabilities. Efforts to reduce disaster risks must be systematically integrated into policies, plans and programmes for sustainable development and poverty reduction and supported through bilateral, regional and international cooperation. The framework recalls the Yokohama Strategy but identifies significant challenges for the coming years in ensuring more systematic action to address disaster risks in the context of sustainable development. It stresses the importance of proactive disaster risk reduction and allocating more resources for implementation. Specific gaps and challenges are identified in five main areas: (1) governance: organisational, legal and policy frameworks; (2) risk identification, assessment, monitoring and early warning; (3) knowledge management and education; (4) reducing underlying risk factors; and (5) preparedness for effective response and recovery.

The second main section of the framework presents the objectives of the world conference, as well as the expected outcome and three strategic goals for the implementation of the framework. The aim of the framework is the 'substantial reduction of disaster losses, in lives and in the social, economic and environmental assets of communities and countries' (ISDR, 2005b, p. 3). It sets out to do that by more effective integration of disaster risk reduction into sustainable development, stronger capacities for disaster risk reduction on all levels, and systematic incorporation of risk reduction approaches into the design and implementation of preparedness, response and recovery programmes.

The third main section of the Hyogo Framework for Action presents priorities for action to meet the strategic goals and generate its expected outcome. It starts by presenting general considerations to take into account, such as increased commitment to disaster risk reduction, the primary responsibility of states, and the importance of applying a multi-hazard approach and integrating a gender perspective in disaster risk reduction. It is also important to note that the framework describes disaster risk reduction as an issue of sustainable development and focuses on developing resilience. The Hyogo Framework for Action then presents five priority areas, each with a list of key activities. The five priority areas mirror the lessons learnt from the Yokohama Strategy and are presented in Box 3.3.

The key activities of the first priority area for action focus on national legal and institutional frameworks, resources and community participation. In contrast, the second focuses on risk assessments, early warning, capacity development and regional and emerging risks. The key activities of the third priority area focus on information management, education and training, research and public awareness, and the fourth on environmental and natural resource management, socio-economic development, land use planning and other technical measures. Finally, the key activities of the fifth priority area focus on developing preparedness for effective disaster response and recovery on all levels.

The fourth and final main section focuses on the implementation and follow-up of the framework. It is divided into six subsections called 'A. General considerations', 'B. States', 'C. Regional organisations and institutions', 'D. International organisations', 'E. The International Strategy for Disaster Reduction', and 'F. Resource mobilisation'. Each subsection contains several paragraphs guiding the implementation and follow-up, for instance, stimulating multidisciplinary and multisector cooperation, prioritising least developed countries, designating national coordination mechanisms for disaster risk reduction and so on.

BOX 3.3 The five priority areas for action of the Hyogo Framework for Action

1. Ensure that disaster risk reduction is a national and a local priority with a strong institutional basis for implementation.
2. Identify, assess and monitor disaster risks and enhance early warning.
3. Use knowledge, innovation and education to build a culture of safety and resilience at all levels.
4. Reduce the underlying risk factors.
5. Strengthen disaster preparedness for effective response at all levels.

Sendai 2015

The third conference was called the UN World Conference on Disaster Risk Reduction and was held in Sendai in 2015, commemorating the devastating 2011 Tōhoku Earthquake and Tsunami (Figure 3.11). While there are many similarities with the previous two conferences, the process leading up to this conference and its outcome documents are more explicitly aligned with the global policy dialogue on sustainable development. This alignment was not only facilitated by the increasing focus on disaster risk in the outcome documents of the Rio+20 conference 3 years earlier but, more importantly, by the Sendai process coinciding with the global policy processes behind the post-2015 development agenda. The outcome documents, which I soon come back to, are thus tailored to work hand in hand with the other post-2015 agreements, including Agenda 2030, the Paris Agreement on Climate Change, the Addis Ababa Action Agenda on Financing for Development, the New Urban Agenda, and ultimately the SDGs for 2015–30.

The conference in Sendai was even bigger than the one in Kobe, with over 6500 participants from 187 states and numerous international organisations and non-governmental organisations (United Nations, 2015a). Moreover, the public forum attached to the conference had a staggering 143,000 visitors throughout the conference, making it several times larger than the previous UN gatherings in Japan.

The outcome documents of the conference are the one-page Sendai Declaration (UNISDR, 2015) and the Sendai Framework for Disaster Risk Reduction 2015–30

Figure 3.11 Sendai Airport after the 2011 Tōhoku Earthquake and Tsunami. *(Source: Ministry of Land, Infrastructure, Transport and Tourism (Japan), licensed under the Creative Commons Attribution 3.0 IGO.)*

(United Nations, 2015b), both immediately adopted by the states present at the conference and endorsed by the UN General Assembly directly after. The Sendai Framework was, as such, the first major international agreement of the post-2015 development agenda, but anticipated the outcomes of Agenda 2030 and the SDGs for a closer fit than ever before. The Sendai Declaration consists of four paragraphs, recognising the increasing impact of disasters and their complexity and declaring renewed determination to reduce disaster losses, acknowledging the vital role of the Hyogo Framework for Action for establishing global commitment, explicitly acknowledging the importance of collective efforts to make the world safer for the benefit of the present and future generations, and thanking Japan for hosting the conference and for its commitment to disaster risk reduction in the global development agenda. It merely sets the stage for the main outcome document—the Sendai Framework.

The Sendai Framework for Disaster Risk Reduction 2015–30 (United Nations, 2015b) comprises 19 pages and 50 paragraphs structured in six main sections called 'I. Preamble', 'II. Expected outcome and goal', 'III. Guiding principles', 'IV. Priorities for Action', 'V. Role of stakeholders' and 'VI. International cooperation and global partnership'. Although this structure resembles its predecessor's, the Sendai Framework is written in a more goal-oriented language.

The first main section—the Preamble—starts by describing the conference as a unique opportunity for countries to adopt a concise, focused, forward-looking and action-oriented post-2015 framework for disaster risk reduction that builds on the strengths and accomplishments of the Hyogo Framework for Action and the regional and national strategies and platforms. It then focuses on the challenges posed by disasters during the last ten years and on the lessons learnt, gaps identified and future challenges from implementing the Hyogo Framework for Action. In short, much has been done, but much work remains.

The second main section states the expected outcome and goal of implementing the framework and presents seven global targets. The Sendai Framework's expected outcome is similar to its predecessor's. It reads: 'The substantial reduction of disaster risk and losses in lives, livelihoods and health and in the economic, physical, social, cultural and environmental assets of persons, businesses, communities and countries' (United Nations, 2015b, p. 12). To attain that expected outcome, the Sendai Framework asserts that the following goal must be pursued:

> *Prevent new and reduce existing disaster risk through the implementation of integrated and inclusive economic, structural, legal, social, health, cultural, educational, environmental, technological, political and institutional measures that prevent and reduce hazard exposure and vulnerability to disaster, increase preparedness for response and recovery, and thus strengthen resilience*
>
> ***(United Nations, 2015b, p. 12).***

It then explicitly states that achieving that goal requires capacity development, particularly in the least developed countries, small island developing states, landlocked developing countries and African countries, and middle-income countries facing specific challenges. It includes mobilising international support for implementation following each state's national priorities. The Sendai Framework then takes it further than its predecessors by setting seven global targets to assess progress against (United Nations, 2015b, p. 12):

(a) Substantially reduce global disaster mortality by 2030, aiming to lower the average per 100,000 global mortality rate in the decade 2020–30 compared to the period 2005–15;

(b) Substantially reduce the number of affected people globally by 2030, aiming to lower the average global figure per 100,000 in the decade 2020–30 compared to the period 2005–15;

(c) Reduce direct disaster economic loss in relation to the global gross domestic product (GDP) by 2030;

(d) Substantially reduce disaster damage to critical infrastructure and disruption of basic services, among them health and educational facilities, including through developing their resilience by 2030;

(e) Substantially increase the number of countries with national and local disaster risk reduction strategies by 2020;

(f) Substantially enhance international cooperation with developing countries through adequate and sustainable support to complement their national actions for implementation of the present Framework by 2030;

(g) Substantially increase the availability of and access to multi-hazard early warning systems and disaster risk information and assessments to people by 2030.

While the international community failed to agree on absolute targets and settled for the relative targets listed above, the Sendai Framework states that appropriate indicators will complement the targets.

The third main section of the Sendai Framework draws on its predecessors when laying down several guiding principles for disaster risk reduction, such as preventing and reducing disaster risk primarily being the responsibility of each state, involving all sectors of government as well as market and civil society actors, and focusing on all hazards. It explicitly acknowledges that even if the drivers of disaster risk span all spatial scales—from local to global—disaster risk, as such, has specific local characteristics that must be understood for any risk-reducing measures to be effective.

The fourth main section of the Sendai Framework states its priorities for action. It builds on the experience of implementing the Hyogo Framework for Action and reshuffles much of the same priorities under four new headings:

Priority 1: Understanding disaster risk.

Priority 2: Strengthening disaster risk governance to manage disaster risk.

Priority 3: Investing in disaster risk reduction for resilience.

Priority 4: Enhancing disaster preparedness for effective response and to 'Build Back Better' in recovery, rehabilitation and reconstruction.

The Sendai Framework then presents several key activities to implement under each priority, divided into activities on national and local levels and on global and regional levels. These priorities for action and their lists of activities are comprehensive and constitute almost half the framework in terms of pages (United Nations, 2015b, pp. 14–22).

The fifth main section of the Sendai Framework is devoted to acknowledging the roles and importance of different actors. While it reiterates the state's primary responsibility, reducing disaster risk is here explicitly portrayed as a shared responsibility between a range of actors—including civil society, academia, business and media—each with different roles to fulfil for effective disaster risk reduction. The framework also highlights the essential contributions of particular groups that usually are listed as particularly vulnerable but whose input is vital for the successful implementation of the framework, such as women, children and youth, persons with disabilities, elderly people, indigenous people and migrants (United Nations, 2015b, p. 23).

The last main section of the Sendai Framework focuses on the international cooperation and global partnership necessary for its implementation. It is divided into four subsections, focusing on general considerations, means of implementation, support from international organisations, and follow-up actions. It highlights the need for enhanced provision of means and technology transfer to assist developing countries, especially the least developed countries, small island developing states, landlocked developing countries and African countries, and middle-income countries facing specific challenges. It highlights the importance of North-South cooperation, complemented by South-South and triangular cooperation. It reaffirms the need for international support, improved access and use of available science and technology, and integration of disaster risk reduction into development assistance across sectors. The framework lists many international organisations as support providers and effectively reforms UNISDR into the United Nations Office for Disaster Risk Reduction (UNDRR) (United Nations, 2015b, pp. 25–26).

The evolution of ideas concerning disaster risk

Although the global political commitment to the outcome of the second and third world conferences on disaster risk reduction was much greater and more explicit than for the first, the actual differences in ideas between the Yokohama, Hyogo and Sendai conferences are even less pronounced than between the five world conferences on sustainable development presented earlier. The development of the Sendai Framework was based on the lessons learnt from the Hyogo Framework for Action, which in turn was based on a review of the implementation of the Yokohama documents, addressing the identified challenges mainly by reemphasising the contents and making aspects more explicit and arguably more focused on implementation. The focus on natural disaster reduction was conceptually abandoned already in Kobe for a focus on disaster risk reduction, with a number of related terminological changes. Otherwise, much of the content remains the same (Figure 3.12). However, some developments are worth mentioning.

First, the outcome documents from the Kobe conference leave the past focus on natural disasters in favour of a broader focus on all types of disaster risk and addresses the conceptual error of seeing disasters as natural when they are indeed constructed in societal processes (Chapter 7). This shift is maintained in the Sendai Framework. The outcome documents of the last two conferences are also more explicit in their focus on risk and proactive measures to reduce risk. However, the outcome documents of the Yokohama conference also stress the importance of prevention but in other wordings. In addition, the Hyogo Framework for Action and the Sendai Framework explicitly emphasise the importance of a multi-hazard approach to disaster risk, as it is not until including all relevant types of events that we can fully assess and appreciate these challenges. They also stress the importance of a multisector approach, emphasising the need to see society as a whole and not address disaster risk in seemingly isolated parts. Related to this, the Hyogo Framework for Action and the Sendai Framework also stress more explicitly the need for including a more comprehensive range of actors in disaster risk reduction—public and private, as well as from civil society and academia. The Sendai Framework also stresses the

Figure 3.12 Word clouds of the Yokohama Message and the Yokohama Strategy and Plan of Action from 1994 and of the Hyogo Declaration and the Hyogo Framework for Action from 2005. *(Constructed in Wordle (www.wordle.net.))*

need to involve groups usually considered particularly vulnerable. Moreover, although the outcome documents from the Yokohama conference mention the need to strengthen the resilience of local communities, the concept of resilience is much more integrated into the outcome documents of the Hyogo conference before becoming one of the most central concepts in the Sendai Framework. Finally, the Sendai Framework specifies global targets to support the assessment of the progress in achieving its expected outcome and goal. It is arguably structured in a more actionable way to facilitate implementation. It is also more explicitly aligned with the global policy dialogue for sustainable development.

To summarise, the current view of the international community is that disaster risk is a central challenge for sustainable development and that disaster risk essentially is a result of failed development. Managing risk is, in other words, a requisite for sustainable development and must be integrated with economic development, social development and environmental protection.

Our agenda 2030 and sustainable development goals

Following the two strings of world conferences presented above over time, it is interesting to note the converging trajectories of ideas. Although the outcome documents of both the first conference in Stockholm and the Yokohama conference already include clear references linking sustainable development and disaster risk, these links have become increasingly explicit over time, resulting in an increasing awareness worldwide that managing risk and resilience is intrinsically linked to sustainable development. Both in the sense of the former being requisites for the latter and in the latter setting the scene for the former. Although less explicitly discussed during the Stockholm+50 conference, the framework that the conference was intended to accelerate action for was explicitly designed to integrate these issues.

This framework is called Agenda 2030, or the 2030 Agenda, and constitutes the current global framework for sustainable development adopted by the member states of the United Nations in 2015—the same year as the Sendai Framework for Action and 3 years of intense work after the Rio+20 documents. This current framework for sustainable development is fundamentally different from its predecessor—the Millennium Development Goals—which focused exclusively on the challenges of developing countries and the assistance of affluent countries while letting the latter off the hook to continue their unsustainable ways. Agenda 2030, on the other hand, is entirely inclusive and demands everyone to contribute. For the first time, the world has common goals to strive towards, regardless of where you happen to be on the planet.

Agenda 2030 comprises 17 SDGs, each focusing on particular areas (Figure 3.8). However, the SDGs are interconnected in such a way that accomplishing one requires meeting others, which in turn demands yet others to be met. For instance, achieving good health and well-being (SDG3) requires zero hunger (SDG2) and clean water and sanitation (SDG6), which in turn demand gender equality (SDG5) and healthy ecosystems (SDGs 14 and 15) and so on. The number of such chains of interconnections between the goals is immense, and the preparatory science report for Stockholm+50 calls for systematically analysing them to facilitate finding synergies, more effective sequencing of activities and positive tipping points (SEI & CEEW, 2022, p. 22). Each goal has several targets and indicators—some more specific, measurable and achievable than others. In total, the 17 SDGs have 169 targets and 232 unique indicators attached to them. However, of all the agreed targets for sustainable development since the first Stockholm conference in 1972, only about 10% have been achieved or even seen significant progress (SEI & CEEW, 2022, p. 11). In other words, we have faltered not in our aspirations but in our actions. Unfortunately, this lack of action is ongoing, with study after study reporting insufficient progress or pace towards achieving the SDGs (e.g., Andriamahefazafy et al., 2022; Mensi & Udenigwe, 2021; Rajapakse, 2022). It seems like something else is needed than another world conference.

Conclusion

We live in a complex world, and the impact of human activity on our environment has accelerated exponentially over the last centuries. We have generated incredible economic growth, and we have never had such high life expectancy and adult literacy rates before in history nor so low infant mortality. However, hundreds of millions are still left behind, and the environmental mortgage we have accumulated to back this unprecedented development has caught up with us. We are increasingly aware of our challenges but seem so far to be collectively incapacitated to make any substantial turns from our path towards escalating trouble. The international community lists what needs to be addressed and done but cannot seem to come up with any viable suggestions for how to do it. At least not any suggestions that really mobilise all. Taken together, the world conferences and the Agenda 2030 describe a future society fundamentally different from our current society, which we come back to in Chapter 16, yet we seem collectively unwilling to change much at all.

We seem stuck in a giant global prisoner's dilemma (Chapter 13). Everybody involved wants to see global changes, but only as long as they can keep on reaping the benefits of continuing on the beaten track. However, such an egocentric mindset is understandable when all must be mobilised for change for the real benefits to materialise. If some people benefit from freeloading, it may just as well be me. Understanding that dilemma makes it relatively straightforward to understand why it is so difficult for the international community to think new thoughts and question our contemporary world order. Why talk about shifting an economic and institutional system that has created and maintained power and wealth in the hands of the people around the table? This is a real predicament for the future of humankind.

References

Andriamahefazafy, M., Touron-Gardic, G., March, A., Hosch, G., Palomares, M. L. D., & Failler, P. (2022). Sustainable development goal 14: To what degree have we achieved the 2020 targets for our oceans. *Ocean & Coastal Management, 227*, 106273. https://doi.org/10.1016/j.ocecoaman.2022.106273

Becker, P. (2021). Fragmentation, commodification and responsibilisation in the governing of flood risk mitigation in Sweden. *Environment and Planning C: Politics and Space, 39*(2), 393–413. https://doi.org/10.1177/2399654420940727

El-Sabh, M. I. (1994). World conference on natural disaster reduction. *Natural Hazards, 9*, 333–352. https://doi.org/10.1007/BF00690745

Gellner, E. (1989). *Plough, sword and book: The structure of human history*. University of Chicago Press.

Guardian, T. (2012). Rio+20 summit opens amid protests. *The Guardian*. Retrieved 2022-03-14 from https://www.theguardian.com/environment/video/2012/jun/21/rio20-summit-opens-protests-video.

IDNDR. (1994). *Yokohama strategy and plan of action for a safer world: Guidelines for natural disaster prevention, preparedness and mitigation.*

ISDR. (2005a). *Hyogo declaration: Extract from the final report of the world conference on disaster reduction (A/CONF.206/6)*. International Strategy for Disaster Reduction.

ISDR. (2005b). *Hyogo framework for action 2005–2015: Building the resilience of nations and communities to disasters.*

ISDR. (2005c). *World conference on disaster reduction: Proceedings of the conference.*

IUCN. (1980). *World conservation strategy: Living resource conservation for sustainable development.*

Mensi, A., & Udenigwe, C. C. (2021). Emerging and practical food innovations for achieving the Sustainable Development Goals (SDG) target 2.2. *Trends in Food Science & Technology, 111*, 783–789. https://doi.org/10.1016/j.tifs.2021.01.079

Rajapakse, J. (Ed.). (2022). *Safe water and sanitation for a healthier world*. Springer. https://doi.org/10.1007/978-3-030-94020-1

SEI, CEEW. (2022). *Stockholm+50: Unlocking a better future*. Stockholm Environment Institute. https://doi.org/10.51414/sei2022.011

UNCED. (1992). *The Rio declaration on environment and development.*

UNCHD. (1972). *Report of the United nations conference on the human environment.*

UNCSD. (2012). *The future we want.*

UNEP. (2011). *Towards a green economy: Pathways to sustainable development and poverty eradication – a synthesis for policy makers.*

UNEP. (2022a). *Greening of Stockholm +50*. UNEP.

UNEP. (2022b). *Stockholm+50, media advisory*, 27 May 2022 https://www.stockholm50.global/news-and-stories/press-releases/stockholm50-international-meeting.

UNISDR. (2015). *Sendai declaration*. UNISDR.

United Nations. (2002a). *Johannesburg declaration on sustainable development.*

United Nations. (2002b). *Plan of implementation of the world summit on sustainable development. United Nations.*

United Nations. (2002c). *Report of the world summit on sustainable development.*

United Nations. (2015a). *Sendai: UN conference adopts new, people-centred disaster risk reduction strategy*. https://news.un.org/en/story/2015/03/493782-sendai-un-conference-adopts-new-people-centred-disaster-risk-reduction-strategy.

United Nations. (2015b). *Sendai framework for disaster risk reduction 2015–2030. United Nations.*

WCED. (1987). *Our common future*. Oxford University Press.

CHAPTER 4

Our boundaries for sustainability

Introduction

Our current sustainability challenges have not gone unnoticed. As we saw in the previous chapter, they have attracted increasing attention among decision-makers for over half a century. However, the attention has yet to transform into substantial concrete action to address the challenges at hand and ensure our solutions do not generate new ones that would undermine the ability of future generations to meet their needs—just as fossil fuels have done (Chapter 2). The scientific community has been trying to describe our sustainability challenges and suggest ways to address them, at least since the Enlightenment (Grober, 2007). For instance, Thomas Malthus describes the challenge of population growth already at the end of the 18th century, when suggesting 'that the power of population is indefinitely greater than the power in the earth to produce subsistence for man' (Malthus, 1798, p. 4), or the first scientific proof of the greenhouse effect by Svante Arrhenius a century later (Arrhenius, 1896). However, decision-makers have, so far, either not understood the gravity of the problems or at least prioritised other and often short-term political or economic gains over the sustainability of humankind. To motivate decision-makers to prioritise the vital over the valuable, the scientific community has realised that it is not enough just to describe sustainability challenges and explain their causes and effects in relative terms. It may be more motivating to also communicate absolute boundaries over which there would be cataclysmic consequences for human life and societies. So, while the international community has become increasingly concerned about the current and anticipated state of the world, the scientific community has not only endeavoured to inform the global dialogue with knowledge about the challenges per se but also with estimations of such boundaries.

This chapter attempts to briefly introduce a framework for understanding boundaries for the sustainable development of humankind. First, in relation to a set of planetary boundaries for our sustainability suggested by influential Earth system scientists—the environmental ceiling for our development, if you like. Then it presents a social scientific critique that complements the idea of planetary boundaries with the notion of social boundaries necessary for sustainability concerning the needs of human beings that cannot be ignored in our efforts to stay within the planetary boundaries.

Sustainability Science, Second Edition
ISBN 978-0-323-95640-6, https://doi.org/10.1016/B978-0-323-95640-6.00002-6

Planetary boundaries for sustainability

Our contemporary sustainability challenges are best described by a group of influential researchers, calling themselves Earth systems scientists and proposing a set of nine planetary boundaries within which there is a safe operating space for humanity (Rockström et al., 2009a). These are climate change, ozone depletion, novel entities (chemical pollution), atmospheric aerosol loading, ocean acidification, altered nitrogen and phosphorous cycles, biodiversity loss, land use change, and water (freshwater use). Unfortunately, we have already exceeded several of them (Li et al., 2021; Persson et al., 2022) (Figure 4.1).

Climate change

The most publicly known planetary boundary is climate change, which Arrhenius showed to be connected to human activity more than 120 years ago (Arrhenius, 1896) and Keeling showed to be escalating and confirmed the link to human activity around 60 years ago (e.g., Pales & Keeling, 1965). It is important to note that the basic understanding of climate change has not only been restricted to science but has been in public

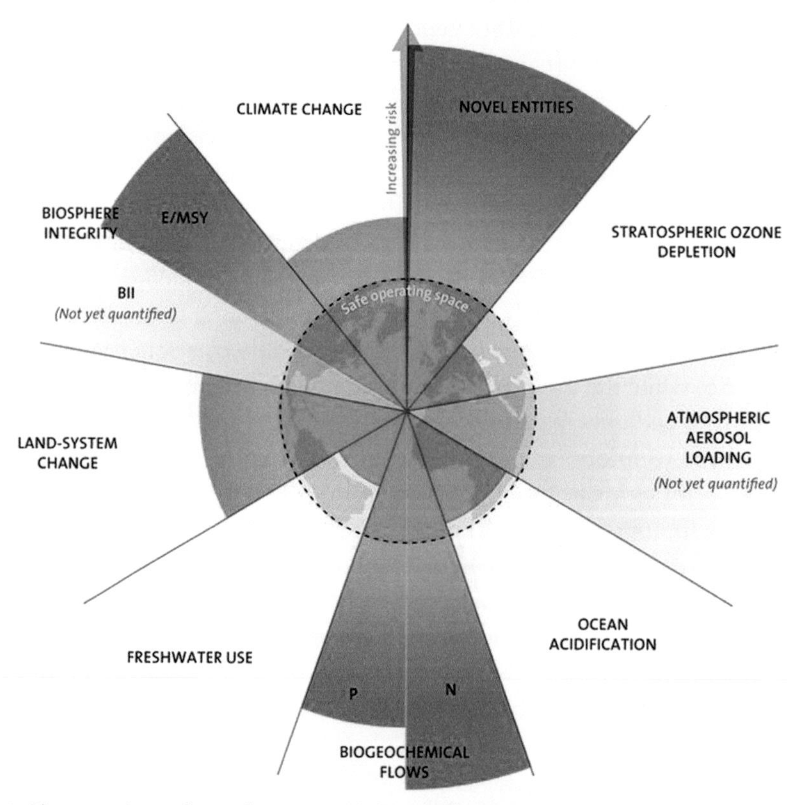

Figure 4.1 Planetary boundaries for sustainability, 2022 update (Note: chemical pollution is here called novel entities). *(Source: Stockholm Resilience Centre/Azote.)*

discourse for over a century (Figure 4.2). It is also staggering to find that the oil and gas giant ExxonMobil not only knew already in the late 1970s that their industry was causing climate change but had advanced models that accurately anticipated global warming that has been consistent with subsequent observations and were at least as capable as later academic and government models (Supran et al., 2023). Yet, they consistently contradicted their own findings, skilfully manipulating the public for decades, together with others in the petroleum industry, to ignore the problem. Hence, we have so far not managed to address this challenge, and we have already crossed the suggested planetary boundary, indicating that it is now very urgent to address climate change with vigour if we are to survive on this planet. Climate change is one of the main processes of change that are redrawing our risk landscape and a central focus of Chapter 6.

Ozone depletion

Another well-known planetary boundary is ozone depletion in the atmosphere, caused by the emission of halogen gases and other ozone-depleting substances (ODS) through human activity. While being a health risk when emitted and exposing human beings on the surface of our planet, ozone is also naturally produced 15–30 km up in the atmosphere (i.e. the stratosphere), where it is vital for protecting the world below from harmful solar radiation (Figure 4.3). A weakened ozone layer means an increased risk of skin cancer and plant damage—without an ozone layer, life on Earth would be impossible. This sustainability challenge has been known since the landmark work of Crutzen, Molina and Rowland in the 1970s (e.g., Crutzen, 1974; Rowland & Molina, 1975), for which they won the Nobel Prize in 1995. It is also the global sustainability challenge humankind has so far proved capable of managing. The reaction of the international community, when faced with the problem, resulted in the most widely ratified UN treaties ever—effectively banning its main causes. This action seems to have been effective in halting and even reversing ozone depletion, although there are still patches of the ozone layer that have not fully recovered. Protecting the ozone layer also helps protect Earth's ability to sequester carbon, which is vital to mitigate climate change.

Novel entities

The third planetary boundary was originally called chemical pollution but was later renamed novel entities to highlight that it concerns substances entirely created and introduced into our environment by ourselves. For instance, synthetic organic pollutants, radioactive materials, endocrine disrupters, genetically modified organisms, nanomaterials and microplastics. It is a well-known planetary boundary, though notoriously difficult to quantify (Rockström et al., 2009b). We know they are prevalent and increasing, but their effects remain largely unknown. For instance, microplastics are found in our food, drinks, water and air, at the same time as many potential health issues—such as internal

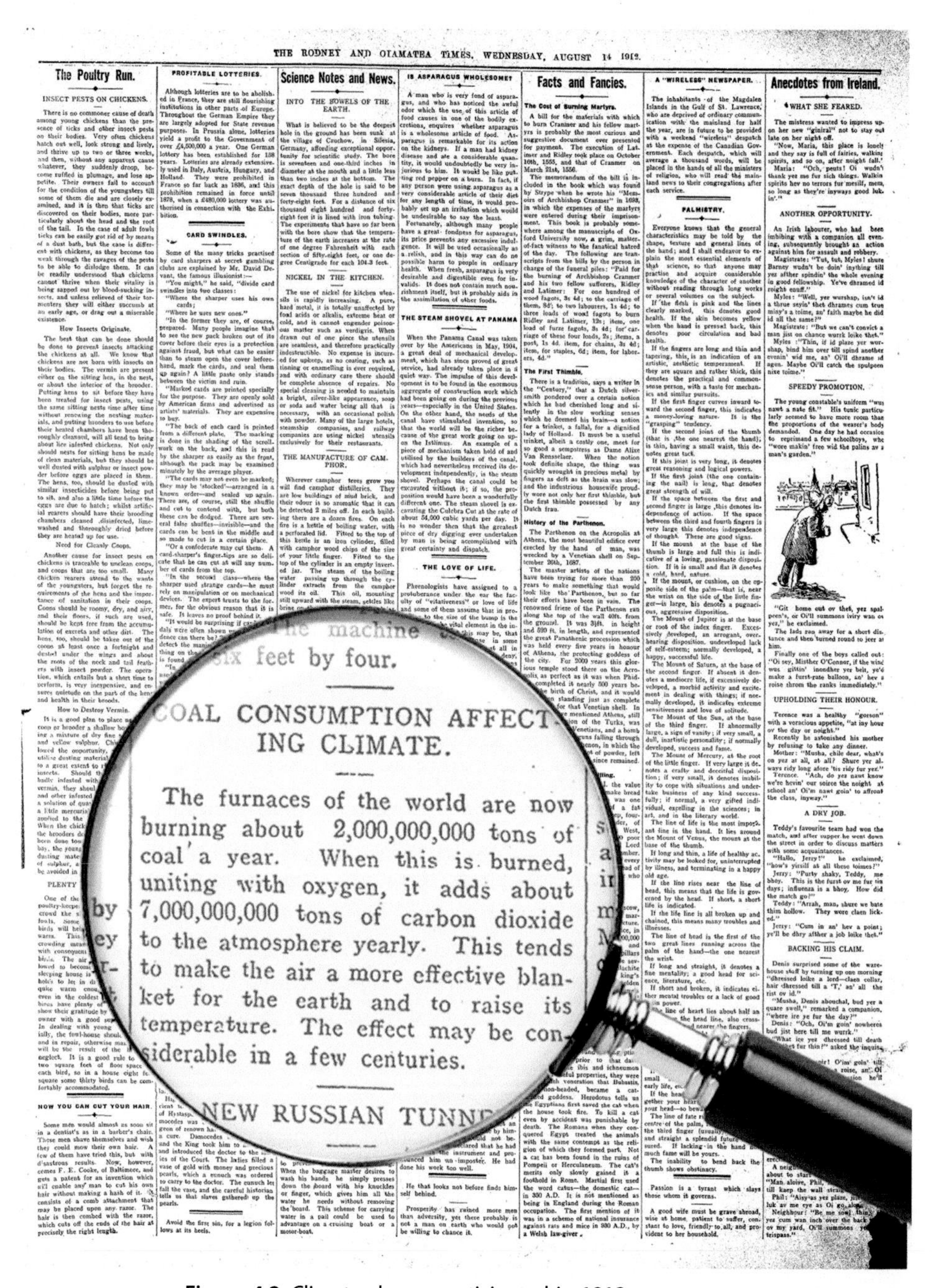

THE RODNEY AND OTAMATEA TIMES, WEDNESDAY, AUGUST 14 1912.

The Poultry Run. | PROFITABLE LOTTERIES. | Science Notes and News. | IS ASPARAGUS WHOLESOME? | Facts and Fancies. | A "WIRELESS" NEWSPAPER. | Anecdotes from Ireland.

COAL CONSUMPTION AFFECTING CLIMATE.

The furnaces of the world are now burning about 2,000,000,000 tons of coal a year. When this is burned, uniting with oxygen, it adds about 7,000,000,000 tons of carbon dioxide to the atmosphere yearly. This tends to make the air a more effective blanket for the earth and to raise its temperature. The effect may be considerable in a few centuries.

NEW RUSSIAN TUNN

Figure 4.2 Climate change anticipated in 1912 newspaper.

Figure 4.3 The ozone layer is Earth's main protection against damaging ultraviolet radiation.

exposure, how they are processed by a living organism, interaction with the immune system, if they can affect the placenta, foetus and brain—are largely unexplored (Vethaak & Legler, 2021, p. 674). Here, the scientific community is most worried about new entities with irreversible and additive effects on biological organisms, such as reduced fertility and genetic damages (e.g., Marques-Pinto & Carvalho, 2013). It is therefore particularly worrying to conclude that humanity is now operating outside this planetary boundary and that the increasing rate of proliferation, production and pollution of novel entities far exceed our ability even to assess and monitor their effects (Persson et al., 2022).

Atmospheric aerosol loading

Another kind of pollution is aerosols in the atmosphere, which is a less-known planetary boundary that is equally difficult to quantify (Rockström et al., 2009b). Atmospheric aerosol loading is what the scientific community calls the overall concentration of particles in the air and atmosphere (Rockström et al., 2009b). Ample scientific studies indicate adverse health effects and connections to regional climate change (e.g., Butt et al., 2016; Hu et al., 2020; Tiwari & Saxena, 2021). Aerosols affect atmospheric circulation patterns, cloud formation, and the amount of solar radiation absorbed or reflected in the atmosphere. They have even been suggested to alter monsoon systems in tropical regions (e.g., Tsai et al., 2016). While there are substantial differences in atmospheric aerosol loading between the world's regions, it is generally worst in their most densely populated areas (Figure 4.4). However, human activity is not only producing aerosols directly—through pollution from factories, traffic and so on—but is also exacerbating the generation of particle loading indirectly through human-induced climate change and poorly managed land and water resources (Tegen et al., 2004), which result in dryer soil

Figure 4.4 Aerosols creating a beautiful but harmful golden haze in the late afternoon in Bangkok.

conditions that in turn facilitate airborne dust. The scientific community has so far not been able to set a specific threshold at which global-scale effects would occur, but that should not delay urgent action as local and regional effects are already severe. It is estimated that outdoor air pollution contributes to 4.5 million premature deaths annually (Global Burden of Disease Collaborative Network, 2021), out of which only the combustion of fossil fuels contributes to one million deaths per year (McDuffie et al., 2021).

Ocean acidification

Ocean acidification is an even less well-known planetary boundary in the mainstream, driven mainly by the same primary process as climate change—'The Other CO_2 Problem' (Doney et al., 2009), if you will. When the level of carbon dioxide increases in the atmosphere, a substantial part of it dissolves in our oceans, lakes and rivers and transforms into carbonic acid that increases the acidity of the water. This, in turn, inhibits the calcifying processes so crucial for the creation of shells and skeletons for myriads of marine organisms, such as corals and planktons, which are the start of most marine and coastal food chains. In short, ocean acidification over the planetary boundary would destroy entire ecosystems we depend on for survival. While the oceans have so far acted as a tremendous sink for much of our emissions of carbon dioxide (Iida et al., 2021), thus limiting the warming as such, the reductions in pH, increases in inorganic carbon, and alterations in acid-base chemistry are on the way to reach levels with uncharted, yet potentially devastating impacts on marine life and ecosystems, as well as on the human communities depending on them (Doney et al., 2020).

Altered nitrogen and phosphorous cycles

Human activity has altered not only the carbon cycle but also the critical nitrogen and phosphorous cycles, which are vital for both the production and decomposition of organic material. These alterations in biogeochemical flows are caused mainly by the crop and livestock production of modern agriculture, starting to create significant and rapidly growing nutrient surpluses in the first half of the last century (Bouwman et al., 2013, p. 20882). While nitrogen and phosphorous substances are vital nutrients for plants, such as trees, crops and algae, too much of them has dire consequences on ecosystems. Too much nitrogen causes acidification of land and water, and too much of either nitrogen or phosphorous in our oceans, lakes, or rivers causes excessive plant growth that diminishes oxygen below the limit for animal life (eutrophication) (Figure 4.5). Also, ecosystems on land are subject to the impact of too many nutrients as many species that are specialised to live in less fertile soil get overgrown by other plants as the fertility of the soil changes.

The disturbance of nitrogen and phosphorous cycles are now major sources of biodiversity loss and threatens vital ecosystems worldwide, as our capacity for extracting and spreading these substances, mainly as fertilisers, surpasses all natural processes on Earth (Rockström et al., 2009b). The planetary boundaries for both nitrogen and phosphorous cycles are already exceeded (Li et al., 2021). We are also on our way to using up all global phosphorous reserves. Considering the limited availability of phosphorous material, this is expected to happen within this century and would also have dire consequences for life on our planet (Cordell et al., 2009)—especially since imbalances between carbon, nitrogen, and phosphorous may, in themselves, have detrimental effects on ecosystems and the future carbon sequestration capacity of Earth (Peñuelas et al., 2013). These complex sustainability challenges require urgent and radical transformations of our agricultural systems (Bouwman et al., 2013).

Figure 4.5 Algal bloom in the Bay of Finland. *(Photo by Ronja Addams-Moring, shared on the Creative Commons.)*

Biodiversity loss

Connected to several of the planetary boundaries above is biodiversity loss. Without reflection, losing some plant or animal species may not seem so alarming. Species have gone extinct throughout history. Of the four billion species that are estimated to have ever evolved on our planet, 99% are no longer here (Barnosky et al., 2011, p. 51). However, extinction has normally been more or less balanced with speciation—the creation of new species through evolution—except in a handful of periods of mass extinction. The previous one was the extinction of the dinosaurs some 66 million years ago, which is believed to have been caused by the rapid cooling of the planet as a result of the cataclysmic impact of a large meteorite (Barnosky et al., 2011; Condamine et al., 2021). Most probably, the meteorite leaving the massive Chicxulub impact crater at the tip of the Yucatán Peninsula in Mexico (Figure 4.6) (Condamine et al., 2021), releasing energy equivalent to ten billion Hiroshima bombs (Bermudez et al., 2022) that scorched the Earth and generated a global tsunami with an initial height of 4.5 km (2.8 miles) (Range et al., 2022) and massive earthquakes that lasted for weeks or even months (Bermudez et al., 2022). The current rate of extinction is, however, unprecedented. At least for the entire period we have fossil records (Hooper et al., 2012, p. 105). For instance, the current extinction rate of freshwater snails and slugs is 1000 times higher than even conservative estimates of our last mass extinction event (Neubauer et al., 2021). It is obvious that we are now in a sixth mass extinction period (Barnosky et al., 2011). We are losing species much faster than new can evolve, which will have dire consequences for the world, as we know it, for millions of years to come (e.g., Neubauer

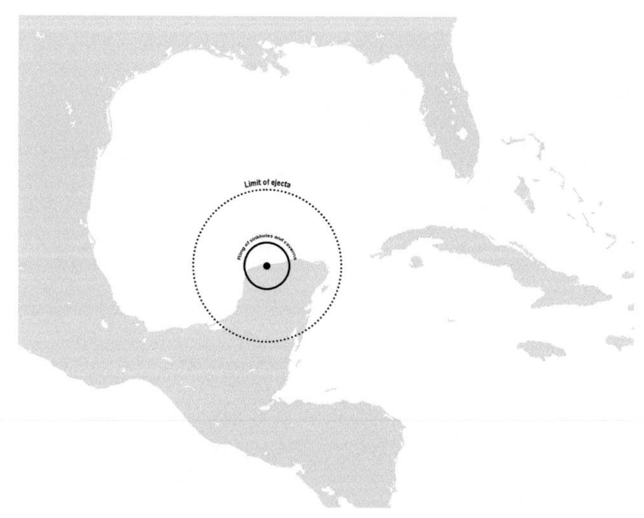

Figure 4.6 The location of the Chicxulub impact crater of the meteorite that probably wiped out the dinosaurs.

et al., 2021). We increasingly realise that biodiversity loss is a major driver of ecosystem change, along with climate change, pollution and other factors (Hooper et al., 2012). Our current biodiversity loss potentially undermines a whole range of ecosystems services that we depend upon for our lives, well-being and livelihoods (Cardinale et al., 2012). This planetary boundary is unfortunately already vastly exceeded (Li et al., 2021), and urgent action is needed.

Land use change

Land use is vital for meeting a range of human needs—such as food, fuel, fibre and shelter—but changing land use affects Earth's biogeochemistry, biogeophysics, biodiversity, and climate (Hurtt et al., 2020, p. 5449). Land use change is, thus, a planetary boundary in the sense of local and regional transformations of forests, wetlands and other natural ecosystems into agricultural lands and urban areas, with intricate links to other planetary boundaries, such as biodiversity loss, freshwater use, climate change and altered nitrogen and phosphorous cycles. While it took a millennium to double the proportion of agricultural land to 7%–8% just before the start of the Industrial Revolution, humankind has since then appropriated another 30% of the total area of the world for our crops and livestock (Hurtt et al., 2020, p. 5427). Land use change is driven primarily by forestry and agricultural expansion and intensification, but the expansion of urban areas also has significant and irreversible impacts (Gao & O'Neill, 2020). Studies estimate urban areas to cover more than 800,000 km^2 already (Huang et al., 2021) and to increase to between 1.1 million and 3.6 million km^2 over the 21st century (Gao & O'Neill, 2020). That corresponds to urban areas of the size between one and three South Africa, after expanding up to eight times the area of Germany from the present.

Crossing this planetary boundary threatens to undermine human well-being and sustainability, as we may reach a tipping point above which the impact on biodiversity and fundamental regulatory cycles cause rapid environmental change (Rockström et al., 2009b). For instance, studies indicate that increasing logging, tilling, and infrastructure construction in the Amazon rainforest eventually reaches a point that permanently tips the entire rainforest into a semiarid savannah (Cumming et al., 2012; Lovejoy & Nobre, 2019). Such change would have terrifying consequences on global regulatory processes concerning climate, biogeochemistry and so on. It is, therefore, particularly alarming when studies suggest that global land use changes are even four times greater than previously estimated (Winkler et al., 2021).

Water

Finally, the last planetary boundary in this framework was originally called freshwater use but has been revised into a more general planetary boundary for water (Gleeson et al., 2020). This planetary boundary is attracting increasing attention in the mainstream as it affects biodiversity, ecosystems, food security, health, climate and so on. While nearly

Figure 4.7 Our taste for avocados in Europe results in rivers running dry in Chile.

70% of the world is covered by water, only 0.5% is available freshwater. Human activity is now the dominant driving force for altering the global freshwater cycle (Rockström et al., 2009b), and around 25% of all rivers in the world run dry before reaching their end as a result of our water use (Figure 4.7) (Molden et al., 2007). In addition to surface water and groundwater available to us (blue water), freshwater also exists as soil moisture (green water) and as frozen and atmospheric water (Gleeson et al., 2020).

The deterioration of global water resources threatens us in three different ways. First of all, deforestation and soil degradation cause loss of soil moisture and are threatening biomass production on land and thus also the capture of carbon in the atmosphere (Rockström et al., 2009b). This green water is also vital for climate regulation, which makes disturbances in water cycles cause local and regional changes in climate (Gleeson et al., 2020). Finally, human activity is also changing runoff water volumes and patterns, threatening the water supply and ecosystems that many depend on downstream (Rockström et al., 2009b). Water scarcity is also expected to become an increasingly important factor for conflict in more arid areas (cf. Selby & Hoffmann, 2012; Unfried et al., 2022).

Local effects of global processes

Although the planetary boundaries are conceptualised on the global level, several of them are applicable on more local scales in the sense of accentuating local and regional sustainability challenges. Climate change in the Sahel region of West Africa has, for instance, already resulted in dramatic declines in average rainfall and temperature in an already dry and hot region (Sambou et al., 2021; Sissoko et al., 2011). Ozone depletion is suggested to exacerbate regional climate change in Australia, New Zealand, and Patagonia (Bornman et al., 2019; Thompson et al., 2011). Chemical and aerosol pollution are common close to cities and industries, especially in rapidly developing countries, while the world's coral reefs are worst affected by ocean acidification. The Baltic Sea receives annually about one million tons of nitrogen and 50,000 tons of phosphorous from human activity surrounding it (Zadeh, 2018), leaving in summers an area about the size of Latvia or Sri Lanka without sufficient oxygen in the water to sustain animal life (Conley, 2012). Biodiversity loss has been highest in the tropical subregions of the Americas (Almond et al., 2020). Europe has the highest proportion of land used for settlements, production systems, and infrastructure (EEA, 2013). Large parts of the Middle East are experiencing an increasing risk of extreme water scarcity (WRI, 2022).

A social foundation for sustainability

The planetary boundaries set an environmental ceiling for our development. However, policies aimed at ensuring we stay within them may, if poorly designed, hinder the development of the poor and disadvantaged. With this insight, Raworth (2012) suggests complementing the planetary boundaries with social boundaries to ensure no one's development is held back. Such social boundaries act as a foundation for building a sustainable society concerning sufficient food security, adequate water and sanitation, education, and so on (Figure 4.8). Raworth (2012) suggests basing these social boundaries on human rights and lists 11 factors to measure.

Our world cannot be considered sustainable with up to 811 million people facing hunger in 2020 (FAO et al., 2021, p. 125)—161 million more than the year before, mainly due to the effects of the Covid-19 pandemic swiftly destroying years of development (cf. FAO et al., 2012, p. 8). Even if the global nutritional status of children has improved, it is not acceptable that 22% of all children under five years old are so malnourished that they are stunted in their growth (FAO et al., 2021, p. 31). Especially not when it is increasingly evident that malnutrition is not only causing illness and death but is also impeding cognitive development, causing irreversible learning disabilities for children, and reducing work capacity and productivity (FAO et al., 2021). Similarly, the world cannot be considered sustainable, with 25% of the global population still not having access to safely managed drinking water, including almost 5% consuming water from unprotected dug wells or springs and 1.6% directly from a river, stream, lake, pond, dam or irrigation canal (WHO & UNICEF, 2021, pp. 28–29). That is 122 million people drinking untreated surface water, or the same population as 12 Sweden. It is even worse concerning sanitation, with 46% still not having access to safely managed sanitation services, including 8% using open pit latrines, hanging latrines or bucket latrines, and more than 6% practising open defecation (WHO & UNICEF, 2021, pp. 48–49). That is 494 million people still disposing of their faeces openly in fields, forests, open bodies of water, beaches or other places, with all that entails for health and security. Only China and India have larger populations than that, corresponding to ten Colombia or Spain or 100 Ireland or the Central African Republic practising open defecation.

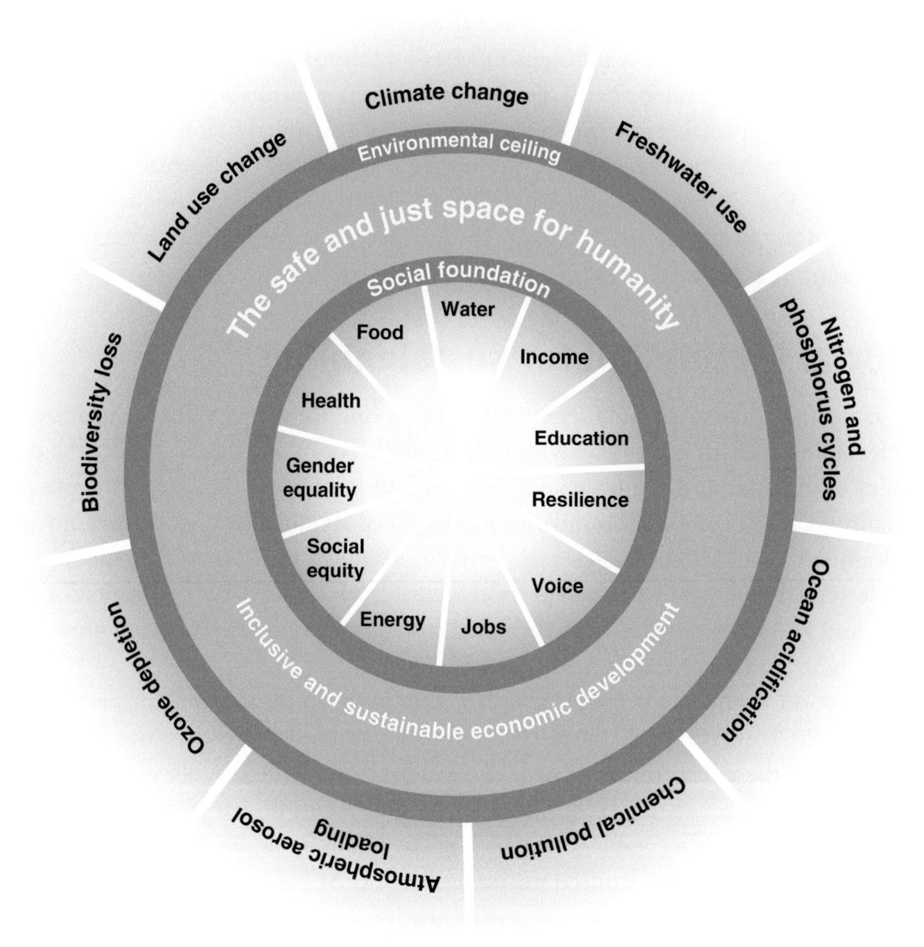

Figure 4.8 A safe and just space for humanity. *(Source: Oxfam. The 11 dimensions of the social foundation are illustrative and are based on governments' priorities for Rio+20. The nine dimensions of the environmental ceiling are based on the planetary boundaries set out by Rockström et al (2009b)).*

The notion of social boundaries is a crucial addition to the planetary boundaries framework, though the list of social boundaries can be made longer than Raworth initially suggested. For instance, rights to life and security are not explicitly included, nor are shelter and housing and other human rights included in international law. While her factors have been operationalised in relation to the Sustainable Development Goals for 2015–30 (SDGs) (O'Neill et al., 2018), I think it is a missed opportunity not to include most, if not all, SDGs in such a holistic framework for sustainable development (Chapter 3). The 17 SDGs are clearly linked to human rights, with over 90% of their goals and targets corresponding to widely established human rights obligations. Moreover, the SDGs also include more or less explicit links to the nine planetary boundaries. In addition to the obvious, such as water (SDG 6), climate change (SDG 13), biodiversity loss (SDGs 14 and 15) and land use change (SDG 15), the other planetary boundaries are included in the targets of the SDGs. For instance, ocean acidification is the explicit focus of SDG 14.3 (minimise and address the impacts of ocean acidification). At the same time, altered nitrogen and phosphorous cycles are included in SDG 14.1 (prevent and significantly reduce marine pollution of all kinds, particularly from land-based activities, including marine debris and nutrient pollution). Novel entities are addressed in SDG 3.9 (substantially reduce the number of deaths and illnesses from hazardous chemicals and air, water and soil pollution and contamination) and SDG 12.4 (achieve the environmentally sound management of chemicals and all wastes throughout their life cycle, under agreed international frameworks, and significantly reduce their release to air, water and soil to minimise their adverse impacts on human health and the environment). The same two targets plus SDG 11.6 (reduce the adverse per capita environmental impact of cities, including by paying particular attention to air quality, municipal and other waste management) apply to the air pollution part of atmospheric aerosol loading. Finally, although ozone depletion is not mentioned explicitly in the goals or targets, SDG 12.4 covers that too.

Regardless of what is ultimately included, the doughnut-shaped framework helps show that we must stay within many planetary and social boundaries for any development to be sustainable (Figure 4.8). Raworth (2012) also explicitly argues that staying within these boundaries is particularly challenging because the social foundation and environmental ceiling are interdependent. For example, increasing food security by introducing or increasing the use of fertilisers may, without appropriate knowledge and care, lead to local disruptions in the nitrogen and phosphorous cycles that harm aquatic ecosystems on which the poor depend. Innovation is thus at the heart of sustainable development and sharing and using new knowledge and solutions. We have overcome many sustainability challenges in the past, but never before have they been so tightly linked to each other, making efforts to deal with one aggravating others.

Conclusion

We are increasingly grasping the dire current state of the world. We are also becoming more and more capable of anticipating a range of future states depending on what courses of action we put into practice from now on. Earth system scientists provide convincing input to decision-makers concerning critical planetary boundaries, monitor how we fare in relation to them and become increasingly good at communicating the results to policymakers and the public. Yet, there are few signs of any significant change of direction—especially not concerning the consumption and voting patterns of the citizens of liberal democracies, who contribute the most to our sustainability challenges through their own consumption and by not using their voting power to put politicians in place who are ready to prioritise sustainability over short-term political and economic gains. We have the power to change direction but seem collectively incapacitated to do so.

In short, it does not look good. While we have been making steady progress concerning several social boundaries, recent pandemics and conflicts have proven how fragile hard-earned development can be for the poorest and most disadvantaged. Moreover, the lion's share of the increasing stress humankind places on our planet is not related to also allowing them to live within fundamental social boundaries but instead to the seemingly ever-increasing consumption of the people who already have it all. I believe the economist Jeffrey Sachs is correct in an interview concerning our sustainability challenges: 'If you're not scared, you're not well informed' (Sachs, 2011). We have painted ourselves into a corner, and getting out means many uncomfortable steps and a lot of repainting to make it right. The first thing is, however, to realise that we have our backs against the wall and that we need to stop painting the same way we have been doing for far too long. It is clear that although nobody on the planet seems to be against sustainable development, few grasp the scale of transformation it entails.

References

Almond, R. E. A., Grooten, M., & Petersen, T. (Eds.). (2020). *Living planet report 2020: Bending the curve of biodiversity loss*. WWF.

Arrhenius, S. P. S. (1896). On the influence of carbonic acid in the air upon the temperature of the ground. *The London, Edinburgh and Dublin Philosophical Magazine and Journal of Science, 41*, 237–276.

Barnosky, A. D., Matzke, N., Tomiya, S., Wogan, G. O. U., Swartz, B., Quental, T. B., Marshall, C., McGuire, J. L., Lindsey, E. L., Maguire, K. C., Mersey, B., & Ferrer, E. A. (2011). Has the Earth's sixth mass extinction already arrived. *Nature, 471*, 51–57. https://doi.org/10.1038/nature09678

Bermudez, H., Vega, F. J., Martini, M., DePalma, R., Ross, C., Bolivar, L., Vega-Sandoval, F. A., Gulick, S. P. S., Stockli, D. F., De Palma, M., & Cui, Y. (2022). The Chicxulub mega-earthquake: Evidence from Colombia, Mexico, and the United States. In *Geological Society of America* (Vol 54). https://doi.org/10.1130/abs/2022am-377578

Bornman, J. F., Barnes, P. W., Robson, T. M., Robinson, S. A., Jansen, M. A. K., Ballaré, C. L., & Flint, S. D. (2019). Linkages between stratospheric ozone, UV radiation and climate change and their implications for terrestrial ecosystems. *Photochemical & Photobiolical Sciences, 18*(3), 681–716. https://doi.org/10.1039/c8pp90061b

Bouwman, L., Goldewijk, K. K., Van Der Hoek, K. W., Beusen, A. H. W., Van Vuuren, D. P., Willems, J., Rufino, M. C., & Stehfest, E. (2013). Exploring global changes in nitrogen and phosphorus cycles in agriculture induced by livestock production over the 1900-2050 period. *Proceedings of the National Academy of Sciences of the U S A, 110*(52), 20882–20887. https://doi.org/10.1073/pnas.1012878108

Butt, E. W., Rap, A., Schmidt, A., Scott, C. E., Pringle, K. J., Reddington, C. L., Richards, N. A. D., Woodhouse, M. T., Ramirez-Villegas, J., Yang, H., Vakkari, V., Stone, E. A., Rupakheti, M., Praveen, P. S., van Zyl, P. G., Beukes, J. P., Josipovic, M., ... Spracklen, D. V. (2016). The impact of residential combustion emissions on atmospheric aerosol, human health, and climate. *Atmospheric Chemistry and Physics, 16*(2), 873–905. https://doi.org/10.5194/acp-16-873-2016

Cardinale, B. J., Duffy, J. E., Gonzalez, A., Hooper, D. U., Perrings, C., Venail, P., Narwani, A., Mace, G. M., Tilman, D., Wardle, D. A., Kinzig, A. P., Daily, G. C., Loreau, M., Grace, J. B., Larigauderie, A., Srivastava, D. S., & Naeem, S. (2012). Biodiversity loss and its impact on humanity. *Nature, 486*, 59. https://doi.org/10.1038/nature11148

Condamine, F. L., Guinot, G., Benton, M. J., & Currie, P. J. (2021). Dinosaur biodiversity declined well before the asteroid impact, influenced by ecological and environmental pressures. *Nature Communications, 12*(1), 3833. https://doi.org/10.1038/s41467-021-23754-0

Conley, D. J. (2012). Ecology: Save the Baltic Sea. *Nature, 486*, 463–464. https://doi.org/10.1038/486463a

Cordell, D., Drangert, J.-O., & White, S. (2009). The story of phosphorus: Global food security and food for thought. *Global Environmental Change, 19*, 292–305. https://doi.org/10.1016/j.gloenvcha.2008.10.009

Crutzen, P. J. (1974). Estimates of possible future ozone reductions from continued use of fluoro-chloromethanes (CF2Cl2, CFCl3). *Geophysical Research Letters, 1*(5), 205–208. https://doi.org/10.1029/gl001i005p00205

Cumming, G. S., Southworth, J., Rondon, X. J., & Marsik, M. (2012). Spatial complexity in fragmenting Amazonian rainforests: Do feedbacks from edge effects push forests towards an ecological threshold. *Ecological Complexity, 11*, 67–74. https://doi.org/10.1016/j.ecocom.2012.03.002

Doney, S. C., Busch, D. S., Cooley, S. R., & Kroeker, K. J. (2020). The impacts of ocean acidification on marine ecosystems and reliant human communities. *Annual Review of Environment and Resources, 45*, 83–112.

Doney, S. C., Fabry, V. J., Feely, R. A., & Kleypas, J. A. (2009). Ocean acidification: The other CO2 problem. *Annual Review of Marine Science, 1*, 169–192. https://doi.org/10.1146/annurev.marine.010908.163834

EEA. (2013). *Environmental topics. Land use.*

FAO, WFP, & IFAD. (2012). The State of Food Insecurity in the World 2012: Economic growth is necessary but not sufficient to accelerate reduction of hunger and malnutrition.

FAO, IFAD, UNICEF, WFP, & WHO. (2021). The state of food security and nutrition in the world 2021. FAO. https://doi.org/10.4060/cb4474en.

Gao, J., & O'Neill, B. C. (2020). Mapping global urban land for the 21st century with data-driven simulations and shared socioeconomic pathways. *Nature Communications, 11*(1), 2302. https://doi.org/10.1038/s41467-020-15788-7

Gleeson, T., Wang-Erlandsson, L., Zipper, S. C., Porkka, M., Jaramillo, F., Gerten, D., Fetzer, I., Cornell, S. E., Piemontese, L., Gordon, L. J., Rockström, J., Oki, T., Sivapalan, M., Wada, Y., Brauman, K. A., Flörke, M., Bierkens, M. F. P., Lehner, B., Keys, P., … Famiglietti, J. S. (2020). The water planetary boundary: Interrogation and revision. *One Earth, 2*(3), 223–234. https://doi.org/10.1016/j.oneear.2020.02.009

Global Burden of Disease Collaborative Network. (2021). *Global burden of disease study 2019 (GBD 2019).* Institute for Health Metrics and Evaluation.

Grober, U. (2007). *Deep roots: A conceptual history of 'sustainable development' (Nachhaltigkeit). Discussion papers// Beim Präsidenten, Emeriti Projekte* (p. 2007-002). Wissenschaftszentrum Berlin für Sozialforschung.

Hooper, D. U., Adair, E. C., Cardinale, B. J., Byrnes, J. E. K., Hungate, B. A., Matulich, K. L., Gonzalez, A., Duffy, J. E., Gamfeldt, L., & O'Connor, M. I. (2012). A global synthesis reveals biodiversity loss as a major driver of ecosystem change. *Nature, 486*, 105–108. https://doi.org/10.1038/nature11118

Huang, X., Huang, J., Wen, D., & Li, J. (2021). An updated MODIS global urban extent product (MGUP) from 2001 to 2018 based on an automated mapping approach. *International Journal of Applied Earth Observation and Geoinformation, 95*, 102255. https://doi.org/10.1016/j.jag.2020.102255

Hurtt, G. C., Chini, L., Sahajpal, R., Frolking, S., Bodirsky, B. L., Calvin, K., Doelman, J. C., Fisk, J., Fujimori, S., Klein Goldewijk, K., Hasegawa, T., Havlik, P., Heinimann, A., Humpenöder, F., Jungclaus, J., Kaplan, J. O., Kennedy, J., Krisztin, T., Lawrence, D., … Zhang, X. (2020). Harmonization of global land use change and management for the period 850–2100 (LUH2) for CMIP6. *Geoscientific Model Development, 13*(11), 5425–5464. https://doi.org/10.5194/gmd-13-5425-2020

Hu, W., Wang, Z., Huang, S., Ren, L., Yue, S., Li, P., Xie, Q., Zhao, W., Wei, L., Ren, H., Wu, L., Deng, J., & Fu, P. (2020). Biological aerosol particles in polluted regions. *Current Pollution Reports, 6*(2), 65–89. https://doi.org/10.1007/s40726-020-00138-4

Iida, Y., Takatani, Y., Kojima, A., & Ishii, M. (2021). Global trends of oceanocean CO2 sink and ocean acidification: An observation-based reconstruction of surface ocean inorganic carbon variables. *Journal of Oceanography, 77*(2), 323–358. https://doi.org/10.1007/s10872-020-00571-5

Li, M., Wiedmann, T., Fang, K., & Hadjikakou, M. (2021). The role of planetary boundaries in assessing absolute environmental sustainability across scales. *Environment International, 152*, 106475. https://doi.org/10.1016/j.envint.2021.106475

Lovejoy, T. E., & Nobre, C. (2019). Amazon tipping point: Last chance for action. *Science Advances, 5*(12), eaba2949. https://doi.org/10.1126/sciadv.aba2949

Malthus, T. (1998). *An essay on the principle of population.* J. Johnson.

Marques-Pinto, A., & Carvalho, D. (2013). Human infertility: Are endocrine disruptors to blame. *Endocrine Connections, 2*(3), R15–R29. https://doi.org/10.1530/EC-13-0036

McDuffie, E., Martin, R., Yin, H., & Brauer, M. (2021). *Global burden of Disease from major air pollution sources (GBD MAPS):* A global approach *(Research Report 210).* Health Effects Institute.

Molden, D., Frenken, K., Barker, R., de Fraiture, C., Mati, B., Svendsen, M., Sadoff, C., & Finlayson, C. M. (2007). Trends in water and agricultural development. In D. Molden (Ed.), *Water for food, water for life: A comprehensive assessment of water management in agriculture* (pp. 57–89). Earthscan.

Neubauer, T. A., Hauffe, T., Silvestro, D., Schauer, J., Kadolsky, D., Wesselingh, F. P., Harzhauser, M., & Wilke, T. (2021). Current extinction rate in European freshwater gastropods greatly exceeds that of the late Cretaceous mass extinction. *Communications Earth & Environment, 2*(1), 1–7. https://doi.org/10.1038/s43247-021-00167-x

O'Neill, D. W., Fanning, A. L., Lamb, W. F., & Steinberger, J. K. (2018). A good life for all within planetary boundaries. *Nature Sustainability, 1*(2), 88–95. https://doi.org/10.1038/s41893-018-0021-4

Pales, J. C., & Keeling, C. D. (1965). The concentration of atmospheric carbon dioxide in Hawaii. *Journal of Geophysical Research, 70*(24), 6053–6076. https://doi.org/10.1029/jz070i024p06053

Peñuelas, J., Poulter, B., Sardans, J., Ciais, P., van der Velde, M., Bopp, L., Boucher, O., Godderis, Y., Hinsinger, P., Llusia, J., Nardin, E., Vicca, S., Obersteiner, M., & Janssens, I. A. (2013). Human-induced nitrogen-phosphorus imbalances alter natural and managed ecosystems across the globe. *Nature Communications, 4*, 2934. https://doi.org/10.1038/ncomms3934

Persson, L., Carney Almroth, B. M., Collins, C. D., Cornell, S., de Wit, C. A., Diamond, M. L., Fantke, P., Hassellöv, M., MacLeod, M., Ryberg, M. W., Søgaard Jørgensen, P., Villarrubia-Gómez, P., Wang, Z., & Hauschild, M. Z. (2022). Outside the safe operating space of the planetary boundary for novel entities. *Environmental Science & Technology, 56*(3), 1510–1521. https://doi.org/10.1021/acs.est.1c04158

Range, M. M., Arbic, B. K., Johnson, B. C., Moore, T. C., Titov, V., Adcroft, A. J., Ansong, J. K., Hollis, C. J., Ritsema, J., Scotese, C. R., & Wang, H. (2022). The Chicxulub impact produced a powerful global tsunami. *AGU Advances, 3*(5). https://doi.org/10.1029/2021av000627

Raworth, K. (2012). *A safe and just space for humanity: Can we live within the Doughnut.*

Rockström, J., Steffen, W., Noone, K., Persson, Å., Chapin, F. S., III, Lambin, E. F., Lenton, T. M., Scheffer, M., Folke, C., Schellnhuber, H. J., Nykvist, B., de Wit, C. A., Hughes, T. P., van der Leeuw, S., Rodhe, H., Sörlin, S., Snyder, P. K., Costanza, R., Svedin, U., … Foley, J. A. (2009a). A safe operating space for humanity. *Nature, 461*, 472–475. https://doi.org/10.1038/461472a

Rockström, J., Steffen, W., Noone, K., Persson, Å., Chapin, F. S., III, Lambin, E. F., Lenton, T. M., Scheffer, M., Folke, C., Schellnhuber, H. J., Nykvist, B., de Wit, C. A., Hughes, T. P., van der Leeuw, S., Rodhe, H., Sörlin, S., Snyder, P. K., Costanza, R., Svedin, U., … Foley, J. A. (2009b). Planetary boundaries: Exploring the safe operating space for humanity. *Ecology and Society, 14*, 32.

Rowland, F. S., & Molina, M. J. (1975). Chlorofluoromethanes in the environment. *Reviews of Geophysics, 13*(1), 1. https://doi.org/10.1029/rg013i001p00001

Sachs, J. D. (2011). *Interview for ODI during the 4th OECD world forum in Delhi.*

Sambou, M. G., Pohl, B., Janicot, S., Famien, A. M., Roucou, P., Badiane, D., & Gaye, A. T. (2021). Heat waves in spring from Senegal to Sahel: Evolution under climate change. *International Journal of Climatology, 41*(14), 6238–6253. https://doi.org/10.1002/joc.7176

Selby, J., & Hoffmann, C. (2012). Water scarcity, conflict, and migration: A comparative analysis and reappraisal. *Environment and Planning C: Government and Policy, 30*(6), 997–1014. https://doi.org/10.1068/c11335j

Sissoko, K., Keulen, H., Verhagen, J., Tekken, V., & Battaglini, A. (2011). Agriculture, livelihoods and climate change in the west african Sahel. *Regional Environmental Change, 11*, 119–125. https://doi.org/10.1007/s10113-010-0164-y?LI=true#page-1

Supran, G., Rahmstorf, S., & Oreskes, N. (2023). Assessing ExxonMobil's global warming projections. *Science, 379*(6628), eabk0063. https://doi.org/10.1126/science.abk0063

Tegen, I., Werner, M., Harrison, S. P., & Kohfeld, K. E. (2004). Relative importance of climate and land use in determining present and future global soil dust emission. *Geophysical Research Letters, 31*(5), 1–4. https://doi.org/10.1029/2003gl019216

Thompson, D. W. J., Solomon, S., Kushner, P. J., England, M. H., Grise, K. M., & Karoly, D. J. (2011). Signatures of the Antarctic ozone hole in Southern Hemisphere surface climate change. *Nature Geoscience, 4*, 741. https://doi.org/10.1038/ngeo1296

Tiwari, S., & Saxena, P. (Eds.). (2021). *Air pollution and its complications*. Springer International Publishing. https://doi.org/10.1007/978-3-030-70509-1

Tsai, I., Wang, W., Hsu, H., & Lee, W. (2016). Aerosol effects on summer monsoon over Asia during 1980s and 1990s. *Journal of Geophysical Research: Atmospheres, 121*(19), 761–776. https://doi.org/10.1002/2016jd025388

Unfried, K., Kis-Katos, K., & Poser, T. (2022). Water scarcity and social conflict. *Journal of Environmental Economics and Management, 113*, 102633. https://doi.org/10.1016/j.jeem.2022.102633

Vethaak, A. D., & Legler, J. (2021). Microplastics and human health. *Science, 371*(6530), 672–674. https://doi.org/10.1126/science.abe5041

WHO, & UNICEF. (2021). Progress on household drinking water, sanitation and hygiene 2000–2020: Five years into the SDGs. WHO and UNICEF.

Winkler, K., Fuchs, R., Rounsevell, M., & Herold, M. (2021). Global land use changes are four times greater than previously estimated. *Nature Communications, 12*(1), 2501. https://doi.org/10.1038/s41467-021-22702-2

WRI. (2022). *Aqueduct water risk atlas*. World Resources Institute. https://www.wri.org/applications/aqueduct/water-risk-atlas.

Zadeh, S. M. (2018). Nutrients. In J. Mateo-Sagasta, S. M. Zadeh, & H. Turral (Eds.), *More people, more food, worse water?: A global review of water pollution from agriculture* (pp. 53–75). FAO and IWMI.

CHAPTER 5

Our disturbances, disruptions and disasters

Introduction

Connected with the long-term challenges to our society's sustainability, there are various events with more immediate impact. Although being mere symptoms of the unsustainable situation that we are in, these events disturb, disrupt and destroy in various ways the possibilities for development for individuals, communities and societies worldwide. Sustainable development demands, therefore, the capacity to manage the risk of a wide range of such events (Haimes, 1992, pp. 101–106). More dramatic, and often sudden, initiating events may give rise to highly destructive courses of events, often referred to as emergencies (e.g., Huang et al., 2021), disasters (e.g., Bosher et al., 2021) or catastrophes (e.g., Lu, 2018). Most scholars use these terms synonymously to signify quantitative differences in scale, while others assign qualitatively different meanings to them (Quarantelli, 2000). Regardless of the label, these are well understood as posing significant threats to sustainable development (Becker, 2009, p. 12; Schipper & Pelling, 2006, p. 20). Reducing their risk is included as a thematic area and cross-sectoral issue in the outcome document of the last two Earth Summits in Rio and Stockholm (Chapter 3).

However, our predisposition for the spectacular should not make us forget the many smaller events, which may seem relatively insignificant on their own but whose cumulative impact on society in many ways vastly surpasses the few and dramatic. For instance, in 2010, the earthquake in Haiti raised the total global death toll in disasters to around 337,000 people (CRED, 2021), while armed conflict directly killed about 55,000 people (Global Burden of Disease Collaborative Network, 2011). This death toll is obviously terrible enough, but consider, then, that more than 900,000 people died from malaria, 1.28 million in road traffic accidents, 1.37 million from HIV/AIDS and 1.93 million from diarrhoeal diseases that same year (Global Burden of Disease Collaborative Network, 2011). Still, these horrific numbers do not even come close to the top three global causes of death that year, namely, respiratory diseases (3.51 million), cancer (8.33 million) and cardiovascular diseases (15.85 million) (Global Burden of Disease Collaborative Network, 2011). All such smaller events may erode sustainable development like water drops on stone (Figure 5.1).

The different kinds of events can be categorised in many ways based on their magnitude, frequency, speed of onset and so on. However, for this book, I choose to present them organised after the processes that initiate them. This choice generates the following

Sustainability Science, Second Edition
ISBN 978-0-323-95640-6, https://doi.org/10.1016/B978-0-323-95640-6.00014-2

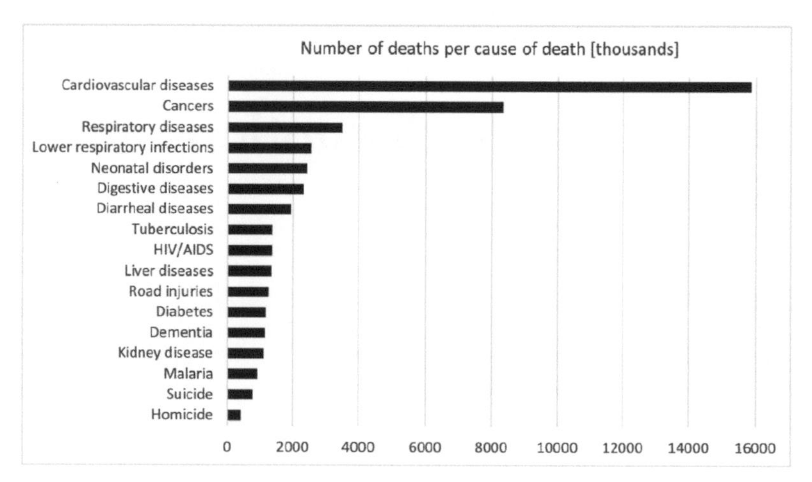

Figure 5.1 Top 17 causes of death in 2010.

five overall types: hydrometeorological events, geological events, biological events, accidental events and antagonistic events. However, none of these types of events would be of any significance if they would not involve something that we as humans value, nor if that is not adversely affected by them. For instance, a landslide happening in the heart of the Brooks Range in northern Alaska, far away from any human activity or property, would not be noticed regardless of size. The measles virus has not been considered a problem in Sweden for decades as a result of effective public health programmes (Follin et al., 2008)—even if increasing unwillingness to vaccinate children may change that since at least 95% vaccination coverage is required for herd immunity—while hundreds of thousands still die of measles in other parts of the world (Mulholland et al., 2020, p. 1783). In other words, for an event to have any significance for sustainable development, it requires at least one thing humans value, which is vulnerable to the impact of the event, and a hazard to set the event in motion.

Hydrometeorological events

Hydrometeorological events have to do with water and weather in various forms. They include heavy rainfall, snowfall and hail; floods; droughts; mudflows and avalanches; storms, cyclones and tornados; heat waves; and cold spells. Their initiating hazards are, in scientific terms, generally phenomena of atmospheric, hydrological or oceanographic origin (Twigg, 2004, p. 15), but are often aggravated and even initiated by human activity—not only by creating vulnerable conditions but also by significant human influence on the hazard processes themselves.

There are, in principle, four key dimensions of hydrometeorological hazards—precipitation, wind speed, temperature and location—determining how we perceive and what we call each event. For instance, heavy rainfall, high wind speed and low temperature are typically referred to as a blizzard, while no precipitation and high wind speed over drylands may be referred to as a sandstorm. However, one event may encompass several kinds of incidents, making it difficult to draw any conclusive boundaries between what is what. An example is Hurricane Sandy on the US East Coast in 2012, which included a cyclone, severe rainfall and snowfall, hailstorms, floods, mudflows and storm surges. As stated earlier, it is essential to remember that without such hazards occurring in locations with something human beings value, which is vulnerable to their impact, there is no problem. In other words, if Hurricane Sandy had never threatened to reach any shores or disturb transatlantic transportation, it would have been little more than an entry in the statistics of meteorological institutes.

Heavy precipitation

Heavy rainfall is not only a causal factor of floods but of several other incidents that disturb, disrupt and destroy what human beings value. Heavy rainfall frequently damages crops and has major impacts on society and the economy (Goswami et al., 2006, p. 1442). It causes erosion of exposed fertile topsoil, further undermining future agricultural production (Pimentel et al., 1995). Heavy rainfall also slows down our transportation systems (Al Hassan & Barker, 1999; Keay & Simmonds, 2005), and even damages them as road and rail embankments are eroded (Cerdà, 2007). Rainfall is, in other words, also an aggravating factor of events that are categorised as geological, biological and accidental. For instance, rainfall-induced landslides (Alexander, 1993, p. 244; Cardinali et al., 2006), disease-carrying mosquitoes multiplying in rainwater pools (Pascual et al., 2008) and increased road traffic accidents (Keay & Simmonds, 2006). That said, rainfall is essential for our lives on this planet and is a vital resource for us to with care. Most places on Earth get precipitation in some form or another. Only when it falls too intensely, for too long or too often, or too little and too infrequent, in relation to each specific context, does it lead to negative consequences (Figure 5.2). Therefore, it is particularly worrying that rainfall patterns are expected to change substantially in a changing climate (IPCC, 2021), which I will return to later.

Although common in fewer places of our world than rain, heavy snowfall and hail also affect individuals, communities and societies. Snowfall by impeding or blocking transportation systems and disrupting communication lines (Perry & Symons, 1980), as well as increasing the risk for accidents (Andrey et al., 2003), and hail by destroying crops (Vinet, 2001), damaging property (Hohl et al., 2002), and even injuring or killing people (Oliver, 2005). Again, snowfall is not only a menace but also necessary for providing water to a large proportion of the global population who lives in glacier- or snow-fed river basins. For instance, the rivers fed by the ice and snow of the Himalayas and the Hindu Kush mountains provide water to almost a quarter of the global population (Molden et al., 2022, p. 223). Snowfall and hail are only disruptive or destructive when falling

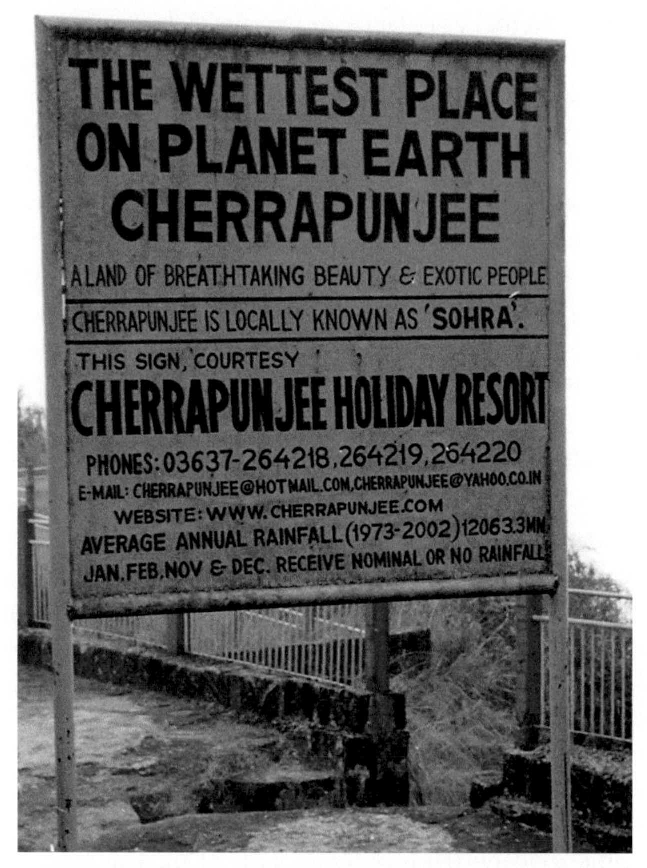

Figure 5.2 Cherrapunji in India gets an annual average of 12,000 mm and holds the world records of 26,461 mm of rain in one year and 9300 mm in one month. *(Photo by R. Mehra, shared on the Creative Commons.)*

too intensely or for too long in relation to a particular context. For instance, 10 cm of snow falling on a January afternoon in northern Finland would not likely adversely impact the functioning of society, while similar snowfall severely disrupted air, rail and road transportation all over the eastern half of England on 18 January 2013. I know because I was stuck in it.

Floods

Heavy rainfall and meltwater are obvious contributing factors to floods, but the processes behind these devastating events are much more complex. Floods are the most common type of events recorded as disasters worldwide (CRED, 2021). They happen on all continents and span from disturbing a few households to affecting millions of people and damaging millions of hectares of agricultural land (e.g., O'Grady et al., 2011). However, floods may also be vital for wetlands, biodiversity and certain farming practices

Figure 5.3 Flood recession sorghum, a crucial resort in many places in Africa with pronounced climate variability. *(Shared by Rik Schuiling on the Creative Commons.)*

(Figure 5.3), which makes flood risk management particularly challenging. Floods may be slow and gradual or rapid and sudden. They may be over as quickly as they came or last for months. There are at least five main types of floods (Box 5.1), but it is important to note that one particular flood event may be a combination of several types.

The first type of flood results from insufficient drainage in relation to rainfall, causing water to flow to and accumulate in local topographical lows before eventually receding through the available natural or man-made drainage (Falconer et al., 2009, p. 199). This type of flood is referred to as pluvial floods (Falconer et al., 2009) or sometimes as flash floods (Kusky, 2005), although this latter term is also being used to describe sudden and dramatic floods (Evans, 2005). Pluvial floods are particularly problematic in urban areas. The hard roof and ground surfaces make water runoff extremely quick, and drainage systems may not be designed to cope with extreme rainfall (Golding, 2009, p. 7; Sörensen et al., 2016). Although this has caused such floods sometimes to be referred to as urban floods, they are still pluvial floods in this overall categorisation (Figure 5.4).

The second main type of flood results from too much rain and meltwater finding its way to, and flowing through, rivers and other watercourses, exceeding their capacity for

BOX 5.1 Five types of flood

Pluvial flood—caused by insufficient drainage from local topographical lows.
Fluvial flood—caused by too much water in a watercourse.
Coastal flood—caused by storm surge, tsunami, land subsidence or sea level rise.
Groundwater flood—caused by groundwater rising and reaching the surface.
Breaching flood—caused by water breaching natural or man-made retention barriers.

Figure 5.4 Flooded Kenyan village. *(Photo by Magnus Hagelsteen.)*

conveying it within their normal edges (Jha et al., 2012). This type of flood is referred to as fluvial floods or river floods (Jha et al., 2012) and is generally slower and easier to forecast than pluvial floods (Han, 2011). The third main type of flood is coastal floods, resulting from storm surges, tsunamis, land subsidence or sea level rise (Han, 2011). It is also possible to forecast. However, both fluvial and coastal floods are capable of great destruction, as we have tended to settle and build along rivers and shores for millennia. More than 50% of the global population lives closer than 3 km from a freshwater body (Kummu et al., 2011), and about 10% in coastal areas less than 10 m above sea level (United Nations, 2017). The apparent reason for this is that water and other resources in and around rivers, lakes and oceans are central to the lives and livelihoods of so many people, with 37% of the global population living in coastal communities (United Nations, 2017). However, it is not only that we place ourselves in harm's way for the benefits of living close to water. We have also exacerbated the floods themselves by disconnecting and draining floodplains vital for retaining and diffusing floodwater (Schneider, 2010). If not managed better, this urge for land is expected to continuously aggravate the situation as the combination of demographic and climatic changes place increasing numbers of people and properties at risk (Askman et al., 2018; Becker, 2018; Hall et al., 2005).

The fourth main type of flood results from the groundwater table below rising and reaching the surface, usually after long periods of heavy sustained rainfall (Jha et al., 2012, p. 61). Groundwater floods are thus often entirely masked by other types of floods (Evans, 2005, p. 11), and more of an aggravating factor as the groundwater blocks surface water infiltration and fills up stormwater drainage from underground (Jha et al., 2012, p. 61). Although more uncommon, groundwater floods can also be caused by substantial reductions in groundwater usage (Jha et al., 2012). Groundwater processes are slow, with considerable time lags, often delaying the impact of groundwater floods and protracting their duration to weeks or months (Jha et al., 2012) (Box 5.2).

BOX 5.2 The endless floods of Pikine

Most people see floods as something unpredictable and random; as unfortunate anomalies from everyday life bringing disruption and destruction to individuals, households and communities. However, millions worldwide live in locations where floods are more or less completely predictable and systematic. One of these places is Pikine, a rapidly growing suburb of Dakar in Senegal, where a combination of changing weather patterns, increasing population pressures, and failed urban development has created one of the worst examples of unsustainable development concerning floods in the world.

Pikine was initially established due to a deliberate policy of the colonial authorities to manage the rapid population growth of post–World War II Dakar (Ba et al., 2009). People protested against moving tens of thousands of people to what was then far from the city without ready plans for transportation, education or medical services. The choice of name is a telling sign of what the local population thought of the place. 'Pikini' means 'Nothing' in Wolof, the most common language of Senegal, and nobody could imagine that Pikine once would be a suburb of Dakar (Ba et al., 2009).

After growing slowly for two decades, droughts in the 1970s and crises following the externally required Structural Adjustment Programmes in the 1980s and 1990s caused a literal explosion of people moving into Pikine (Ba et al., 2009). The official population of Pikine was 140,000 in 1971 (Ba et al., 2009), but outgrew the population of Dakar itself in 2013 and is currently over 1.5 million (ANSD, 2013). However, the real population is undoubtedly much larger and still growing. As Pikine expanded, the authorities could not keep up, leaving a large part without enforced spatial plans and building codes and functioning electricity, water and sanitation, drainage and so on. This failed development, in turn, caused hundreds of thousands of people to settle in local low points and on dry lakebeds. When weather patterns changed around a decade ago, increasing amounts of rainfall during the rainy season cause the groundwater level to rise again, saturating the ground and keeping floodwater up to waist or even chest height for extended periods. Some places have been flooded for up to half a year, every year since 2005, causing children to have lived more or less half their lives in an environment with water, humidity and mould, as well as with pollution, faeces and garbage floating around inside and outside their homes. The floods have been estimated to directly affect between a third and half of the population of Pikine (ANSD, 2010, p. 103; Hungerford et al., 2019) and urbanisation and climate change are continuing to aggravate the situation.

Finally, the fifth main type of floods results from the failure or collapse of man-made structures or natural formations retaining or conveying water, such as artificial water management systems (Jha et al., 2012, p. 62) or landslide-produced dams (Cui et al., 2009). These floods may span across scales but potentially produce very rapid and destructive events. For instance, the Johnstown flood of 1889 with more than 2000 deaths (Dieck, 2008), the Huaraz landslide dam failure in 1941 with 4–6000 casualties (Schuster et al., 2002, p. 12), or the Banqiao dam failure in 1975 with more than 200,000 deaths (Lind et al., 2004, p. 89). These floods often result from geological or technological events, which will be discussed later in this chapter. Others may be caused by a glacial blockage or ice jams, in which temporary obstructions retain masses of water that flood upstream areas and cause very rapid and destructive downstream flash floods when the obstructions are eventually breached (O'Connor & Costa, 2004).

Droughts

Another destructive hydrometeorological event related to rainfall is drought, although the processes behind it are more complex than just the amount of rain (Nalbantis & Tsakiris, 2009, pp. 881–882). There are four main types of droughts (Box 5.3): meteorological, hydrological, agricultural and socioeconomic (Keyantash & Dracup, 2002, p. 1168; Wilhite & Glantz, 1985). Droughts can affect a large proportion of our world (Figure 5.5), but they have a very different impact depending on context (e.g., Alexander, 1993, pp. 150–153). Smaller localised droughts may be disruptive for the unfortunate, affected farmer but may not significantly affect the overall food security of society. However, more significant droughts can potentially affect entire regions with food insecurity and even famine as a result. Droughts are creeping events, allowing their consequences to be forecasted months in advance. They may be regular, and they may last for years. Drought may be a normal and recurring feature of our climate in many parts of the world. However, the food crisis that often follows can be viewed as a 'silent tsunami' of inconceivable destruction and suffering that is not getting sufficient attention in mass media or the international community. This lack of attention is particularly frustrating as we are producing more food on a global scale than needed to feed everybody if it was more equally distributed. Food insecurity and famines are, in other words, rarely caused by a lack of food but by a lack of access to food (Sen, 1982).

BOX 5.3 Four types of drought

Meteorological drought—caused by insufficient rainfall in relation to a long-term average.
Hydrological drought—caused by insufficient water stored on and in the ground.
Agricultural drought—caused by insufficient soil moisture and/or soil nutrients.
Socioeconomic drought—caused by societal food demands exceeding the available supply.

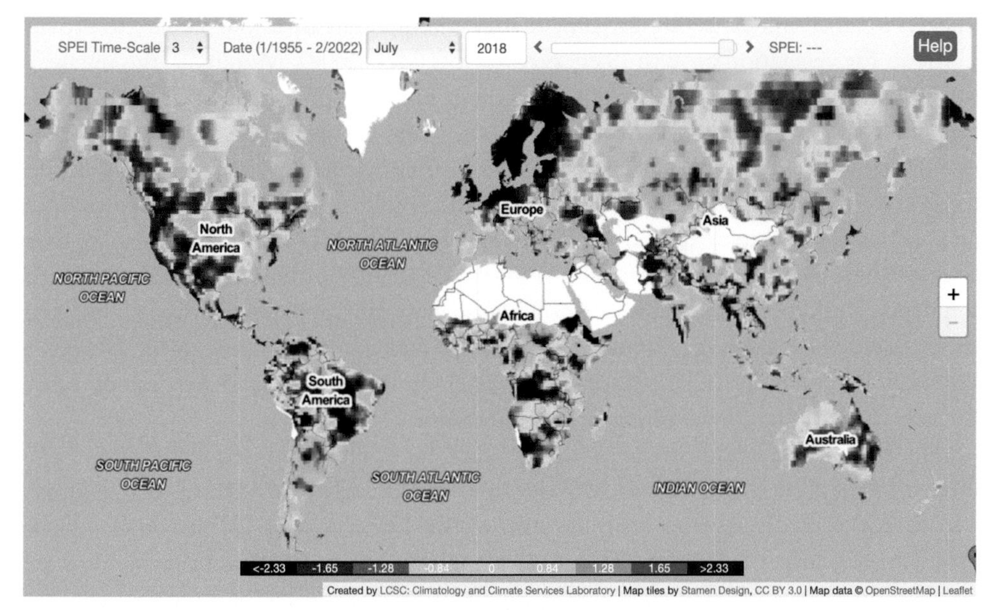

Figure 5.5 The spatial distribution of drought in July 2018 from the Global Drought Monitor, with intense drought all over northwestern Europe (3-month SPEI <0 represents drought).

Meteorological drought refers, in general, to a deficit of rainfall in relation to a long-term average and is what the lay conception of drought most commonly refers to (Slegers & Stroosnijder, 2008, pp. 376–377). However, only looking at rainfall in relation to an annual or seasonal average may hide extended dry periods between short periods of heavy rain. Dry periods may cause partial or total crop failure, even if the average does not indicate meteorological drought, depending on when they occur (Slegers & Stroosnijder, 2008). It is, in other words, possible for an area to experience severe floods and droughts in the same year.

Hydrological drought, on the other hand, generally refers to a deficit in water stored on and in the ground (Keyantash & Dracup, 2002, p. 1168). In more scientific words, to a significant decrease in water availability in the land phase of the hydrological cycle (Nalbantis & Tsakiris, 2009, p. 882). This type of drought may thus be triggered not only by insufficient rainfall but also by increased surface water and groundwater use.

Agricultural drought refers to a deficit of soil moisture and/or soil nutrients limiting vegetation growth (Slegers & Stroosnijder, 2008, p. 377). It is, in other words, possible to get agricultural drought, regardless of sufficient rainfall, due to insufficient water infiltration capacity, water-holding capacity or nutrient availability (Slegers & Stroosnijder, 2008). Various land degradation processes negatively impact the physical soil properties that determine these factors. For example, reduced vegetation cover leads to reduced amounts of organic matter in the soil, which diminishes its water-holding capacity

(Slegers & Stroosnijder, 2008). Organic matter and minerals are also necessary to provide nutrients for vegetation growth (Slegers & Stroosnijder, 2008). Moreover, the uptake of one nutrient depends on the availability of other nutrients in the soil, which means that vegetation growth is limited by the nutrient most scarcely available. When more nutrients are taken up than returned to the soil, its nutrient availability will gradually deplete (Slegers & Stroosnijder, 2008). Reduced vegetation cover also exacerbates soil erosion, as rain and wind can batter exposed soil (Pimentel et al., 1995). Unsustainable farming practices also heavily aggravate soil erosion (Pimentel et al., 1995). Soil erosion decreases the soil depth, further reducing its water-holding capacity and increasing surface sealing processes that reduce the water infiltration capacity and increase water runoff (Slegers & Stroosnijder, 2008, p. 377). Ultimately, these processes may lead to desertification, further complicating any attempts at soil restoration (Figure 5.6).

Finally, socioeconomic drought is sometimes considered a consequence of the other three types of droughts. Unless societal demands consistently exceed the available supply, a socioeconomic drought occurs only together with one or more of the other types of droughts (Keyantash & Dracup, 2002, p. 1168). However, as populations grow and water use patterns change, the demand may change and create a socioeconomic drought even with unchanged rainfall, water levels and soil moisture/nutrients. Prices of food increase on the market, tipping families into poverty and leaving the poor malnourished.

Human activity is a major factor in all four types of droughts. We are affecting rainfall patterns, overusing surface water and groundwater, reducing vegetation cover and applying unsustainable farming practices, and increasing our demands for agricultural produce.

Figure 5.6 Drought and desertification in Somaliland. *(Photo by Oxfam East Africa, shared on the Creative Commons.)*

Mudflows and avalanches

Rainfall and snowfall on mountain slopes can, under unfortunate circumstances, accumulate large masses of saturated mud or snow that can result in destructive mudflows or avalanches (Alexander, 1993). Although having roots in precipitation, mudflows also have geological roots and are presented with other landslides later in this chapter. Avalanches are rapid downhill flows of snow that are generally restricted to mountain areas, although suitable local topographic conditions may also be found in lowland areas (Luckman, 2010, p. 27). They rarely start in forest terrain, but when picking up their speed, they may overrun even the densest forest, leaving avalanche tracks straight through it (Luckman, 2010). The location and spatial extent of avalanches depend on meteorological and climatic conditions, as well as on slope angle (typically 25−55 degrees) and other terrain characteristics determining where snow can accumulate and flow (Luckman, 2010). The size of avalanches varies from a few cubic metres to hundreds of thousands and even millions of cubic metres (Luckman, 1977, p. 32). Avalanches can consist of different types of snow, from powder snow to wet slush, and are triggered when the forces on the snowpack exceed its cohesive strength (Alexander, 1993, p. 185). It could be spontaneous, as the weight of the snowpack becomes too heavy or the snow change structure due to climatic factors, or triggered by external forces, like a skier or animal, explosives, and so on. Although it is possible to monitor the risk of avalanches, the speed of onset is very sudden, and the duration is limited to the seconds or minutes it takes for the snow to reach its downhill destination. Avalanches have the potential for great destruction, and there are examples of snow avalanches killing hundreds of people and destroying buildings and infrastructure (Höller, 2009) (Box 5.4).

Storms, cyclones and tornados

In addition to the challenges with precipitation and water, there are also challenges associated with high wind speed. The wind results from differences in atmospheric pressure, causing air to move from locations with higher pressure to locations with lower pressure to equalise the differences. A rotating planet with a warm equator and cold poles, combined with a range of lesser drivers, creates a complex system that can generate winds of various speeds and forms. Although everyday winds are vital for much in our world, there are winds with immense destructive force (Figure 5.7), such as storms, cyclones and tornados.

The word storm is commonly used in everyday language for any strong winds, but the meteorological definition of a storm requires a wind speed over 24.5 m/s (89 km/h or 55 mph). Storms can happen more or less anywhere on the planet. They range from being local and short-lived to affecting entire countries and lasting for more than a week. Storms build up and are possible to both forecast and track in the same way we forecast and track everyday weather. They are nonetheless capable of destruction and happen

BOX 5.4 The white menace

The Alps are stunningly beautiful, with their pointed peaks, sheer rock faces and lush valleys. However, as snowfall and gravitation combine, steep slopes become treacherous, silently accumulating the wrath of ancient gods in the form of avalanches. Although two-thirds of all avalanche casualties in Austria since 1950 have been happening in the backcountry, mainly in relation to skiing, the remaining third have been spontaneously released avalanches that affect communities more directly (Höller, 2009). Something that the village of Galtür had to experience the hard way on 23 February 1999.

The year 1999 started with a series of snowstorms dumping their loads on the Austrian Alps of western Tyrol. However, what would have been an excellent start to the skiing season turned ugly, as variations in temperature created a think layered snowpack that set the scene for unprecedented avalanches that would baffle scientists and mountain rescuers alike. Then in the morning of February 23, the unstable snowpack on the mountain above Galtür suddenly snapped and came roaring down at around 300 km/h or 190 mph (Dinwiddie et al., 2011, p. 308). 170,000 tons of snow and ice crashed into the village, killing 31 people, ploughing through areas considered safe from avalanches and causing avalanche hazard zones to be reconsidered (Dinwiddie et al., 2011).

Figure 5.7 Cyclone Winston slams Fiji in 2016. *(NASA image by Jeff Schmaltz, LANCE/EOSDIS Rapid Response.)*

rather frequently when the season is right. Wind cause soil erosion, and strong winds over water create waves and storm surges that flood coastal areas, hamper maritime activities and erode coastlines. Strong winds break or uproot trees, destroy crops, damage buildings and infrastructure and injure or kill people as debris flies around and trees fall. Strong winds can also carry snow, sand or other particles, limiting visibility and increasing their destructive forces in the form of blizzards and sandstorms. Storms are also associated with heavy rainfall, snowfall or hail, depending on conditions, as well as with lightning.

Lightning has tremendous destructive power and strikes around 40–50 times per second globally (Christian et al., 2003, p. 1). Lightning is unevenly distributed around the world, with approximately 70% occurring in the tropics (NOAA, 2013), making the common European reference to the likelihood of being struck by lightning concerning improbable things less applicable in other places. For example, one area around the small village of Kifuka in the Democratic Republic of Congo receives around 158 lightning bolts per square kilometre and year (NOAA, 2013), which is about 1500 times more than the Scottish Highlands (Christian et al., 2003, p. 5).

Cyclones, or hurricanes or typhoons as they are also called depending on where on the planet they form (Box 5.5), are rotating storms with particular destructive force (Alexander, 1993, p. 154; Longshore, 2008, pp. 397–398). They develop over warm ocean water, initiated as a cluster of thunderstorms, often referred to as a tropical disturbance. If a tropical disturbance draws sufficient energy from the water, it turns into a tropical depression, which may, in turn, continue to grow into a tropical storm and finally into a cyclone. These steps can take as little as half a day, up to a couple of days, if happening at all. A mature cyclone typically lasts two to six days (Longshore, 2008) but could potentially last two or three weeks. A cyclone's path is difficult to forecast precisely for extended periods. Yet, cyclones travel westward, typically bending north and then northeast in the Northern Hemisphere and south and then southeast in the Southern Hemisphere (Figure 5.8). This pattern is connected to Earth's rotation, which explains why all cyclones rotate counter-clockwise in the Northern Hemisphere and clockwise in the Southern Hemisphere (Shultz et al., 2005, p. 21). Cyclones are, as such, giant conveyors of energy from the warmer, energy-rich areas at least 5 degrees latitude away from the equator, where they form, towards cooler areas or land.

Mature cyclones generally move with a speed of 10–40 km/h (6–25 mph) (Longshore, 2008, p. 423) and have a radius of between less than 222 km, which is referred to as very small, and more than 888 km, which is referred to as very large (Liu & Chan, 2002, p. 2135). Cyclones produce winds stronger than 33 m/s (119 km/h or 74 mph),

BOX 5.5 Cyclones, hurricanes or typhoons

Cyclone is the scientific name of all rotating storms, regardless of where they are on the planet. It is also, generally, the local name for such storms over the Pacific Ocean south of the equator and over the Indian Ocean. However, the same kind of rotating storms is called hurricanes over the Atlantic and Northeast Pacific Ocean (east of the dateline) and typhoons over the Northwest Pacific Ocean (west of the dateline). In other words, they are the same thing, called different names.

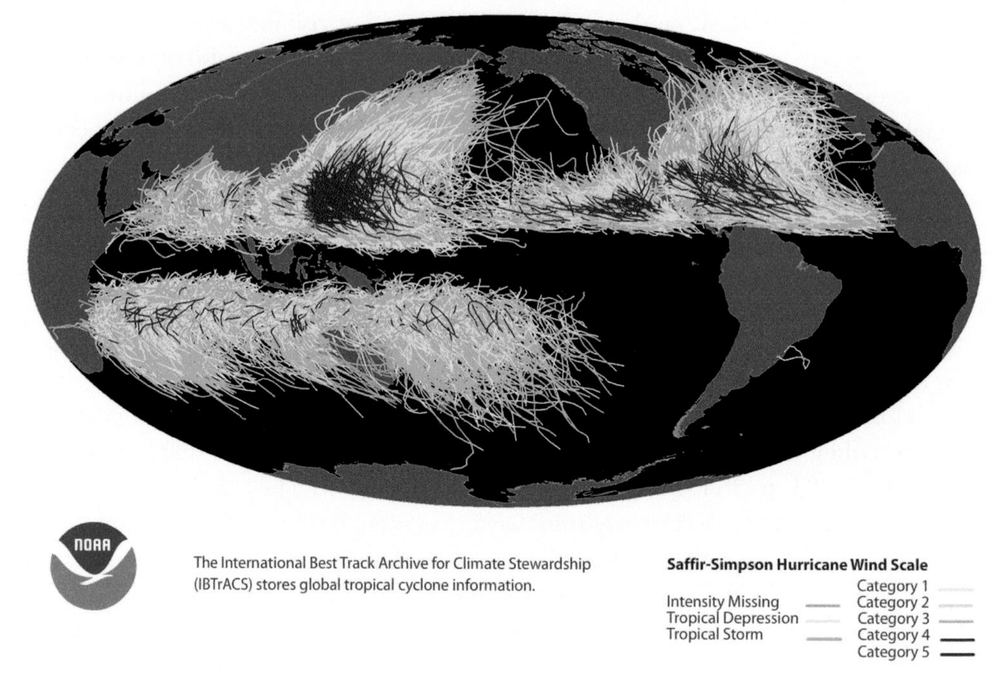

Figure 5.8 Cyclones tracks since the 1850s, as registered by IBTrACS. *(National Centres for Environmental Information, NOAA.)*

with Cyclone Winston being one of the strongest cyclones ever to make landfall (Figure 5.7), with wind speeds up to 80 m/s (285 km/h or 177 mph) when battering Fiji in 2016. When we finally managed to get out of Suva, the capital of Fiji, we found that the destruction was virtually total up to 40 km (25 miles) away from its path. Although the destructive potential of a cyclone is mainly connected to wind speed, the affected area's size and vulnerability to the impact of strong winds are also crucial for the total destruction.

Although a complex combination of contextual factors determines vulnerability to the impact of cyclones, there is an important tool developed for Western societies to communicate the destructive capacity of cyclones. It is called the Saffir—Simpson wind scale and indicates the destruction of different intensity cyclones in societies with similar infrastructure and built environments as the United States (Table 5.1). Cyclones also carry vast amounts of humid air that can transform into 237—576 mm (7—17 in.) of precipitation when cooled off (Longshore, 2008, p. 423), which together with storm surge further increases their destructive capacity in both intensity and spatial extent.

Although cyclones have the most intense destructive capacity in absolute terms, it is another type of rotating storm with the greatest damage potential locally. A tornado is a tightly organised column of rotating winds centred on an area with exceptionally low air pressure (Longshore, 2008, p. 394). This air pattern creates a funnel of extreme winds and

Table 5.1 The Saffir–Simpson Wind Scale, indicating wind speed and damage potential in the North American context (NOAA, 2013).

Category	Sustained winds	Types of damage due to hurricane winds
1	74–95 mph 64–82 kt 119–153 km/h	Very dangerous winds will produce some damage: well-constructed frame homes could damage roofs, shingles, vinyl siding and gutters. Large branches of trees will snap, and shallowly rooted trees may be toppled. Extensive damage to power lines and poles likely will result in power outages that could last a few to several days.
2	96–110 mph 83–95 kt 154–177 km/h	Extremely dangerous winds will cause extensive damage: well-constructed frames homes could sustain major roof and siding damage. Many shallowly rooted trees will be snapped or uprooted and block numerous roads. Near-total power loss is expected with outages that could last several days to a week.
3 (Major)	111–129 mph 96–112 kt 178–208 km/h	Devastating damage will occur: well-built framed homes may incur major damage or removal of roof decking, and gable ends. Many trees will be snapped or uprooted, blocking numerous roads. Electricity and water will be unavailable for several days to weeks after the storm passes.
4 (Major)	130–156 mph 113–136 kt 209–251 km/h	Catastrophic damage will occur: well-built framed homes can sustain severe damage with the loss of most of the roof structure and/or some exterior walls. Most trees will be snapped or uprooted and power poles downed. Fallen trees and power poles will isolate residential areas. Power outages will last weeks to possibly months. Most of the area will be uninhabitable for weeks or months.
5 (Major)	157 mph or higher 137 ht or higher 252 km/h or higher	Catastrophic damage will occur: a high percentage of framed homes will be destroyed, with total roof failure and wall collapse. Fallen trees and power poles will isolate residential areas. Power outages will last for weeks to possibly months. Most of the area will be uninhabitable for weeks or months.

pressure drops, destroying all it can in its path, sucking it up into the air until the centrifugal and gravitational forces overcome the tornado's power, and it is thrown out and down again. Tornados of different intensities occur globally, generally between 20 and 60 degrees latitudes, but are by far the most frequent and intense in the North American Midwest (Berz et al., 2001, p. 455). Significant tornados also happen in Argentina, Australia, Bangladesh, Canada, China, Central Europe, South Africa, Uruguay and so on but generally not with the same destructive capacity as in the United States (Figure 5.9). Tornados sometimes spawn in groups with tremendous destructive power. Cyclones may also generate tornados and groups of tornados. While most cyclones can generate no more than ten tornados, the 1967 Hurricane Beulah spawned 115 tornados across Texas (Alexander, 1993, p. 172).

Most tornados have a diameter of between 20 and 50 m, a path on the ground of between 100 and 3000 m, and wind speeds between 25 and 50 m/s (89−180 km/h or 55−112 mph) (Longshore, 2008, p. 394). However, there are examples of tornados that have been more than 1000 m wide, with paths up to 300 km and estimated wind speeds of over 139 m/s (500 km/h or 311 mph) (Berz et al., 2001, p. 455). Tornados are, in other words, localised in both spatial and temporal terms but can be extremely intense (Figure 5.10).

Heat waves and cold spells

Even without precipitation or strong winds, extreme temperature is also challenging for individuals, communities and society. Heat waves are often forgotten when considering

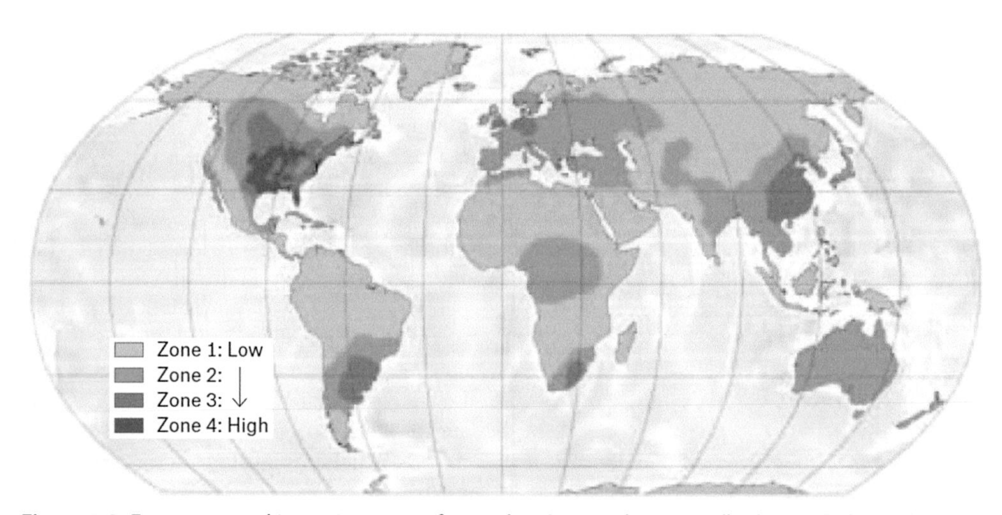

Figure 5.9 Frequency and intensity zones of tornados. *(Copyright 2011, Münchener Rückversicherungs-Gesellschaft, Königinstrasse 107, 90902 München, Germany.)*

Figure 5.10 A category F5 tornado as it approaches Elie, Canada, in 2007. *(Photo by Justin1569 shared on the Creative Commons.)*

hydrometeorological events that challenge sustainable development. However, heat waves do have immense destructive capacity (Figure 5.11), which is strongly exemplified by the almost 15,000 excess deaths in France during the European heat wave of August 2003 (Poumadère et al., 2005, p. 1483)—representing a 60% increase of mortality in France during that period (Haines et al., 2006, p. 2103). There are, unfortunately, many more examples of heat waves. What they all have in common is that heat waves are intrinsically social, affecting disproportionally the elderly, disabled, socially isolated and so on (Klinenberg, 2002). Who is affected by heat waves is undoubtedly a significant factor in why they are so often ignored or forgotten, regardless of the actual death toll (Box 5.6). In fact, while heat waves claim several times more deaths than earthquakes, tornados and floods combined in the USA (Klinenberg, 2002, p. 17), the attention devoted to them is still just a fraction compared to the others. This lack of attention is particularly frustrating since many heat-related deaths are preventable (Haines et al., 2006, p. 2103). A common argument concerning heat waves is that the people who perish in them would probably have died soon anyway. Still, when studying mortality patterns in and after significant heat waves, it is clear that such arguments only hold to a relatively limited degree (Kaiser et al., 2007; Le Tertre et al., 2006).

In addition to the expected global increase in temperature due to climate change, human activity is also exacerbating the risk of heat waves more locally as a result of what is referred to as urban heat islands. Urban heat islands are simply urban areas with increased temperatures compared to their rural surroundings. They are caused by a complex combination of differences between urban and rural areas in terms of higher heat absorption and retention properties of building materials, multiple surfaces for absorption and reflection of sunlight of tall buildings, lower evapotranspiration properties of hard surfaces, wind-blocking, excess heat from energy consuming processes, pollution and so on

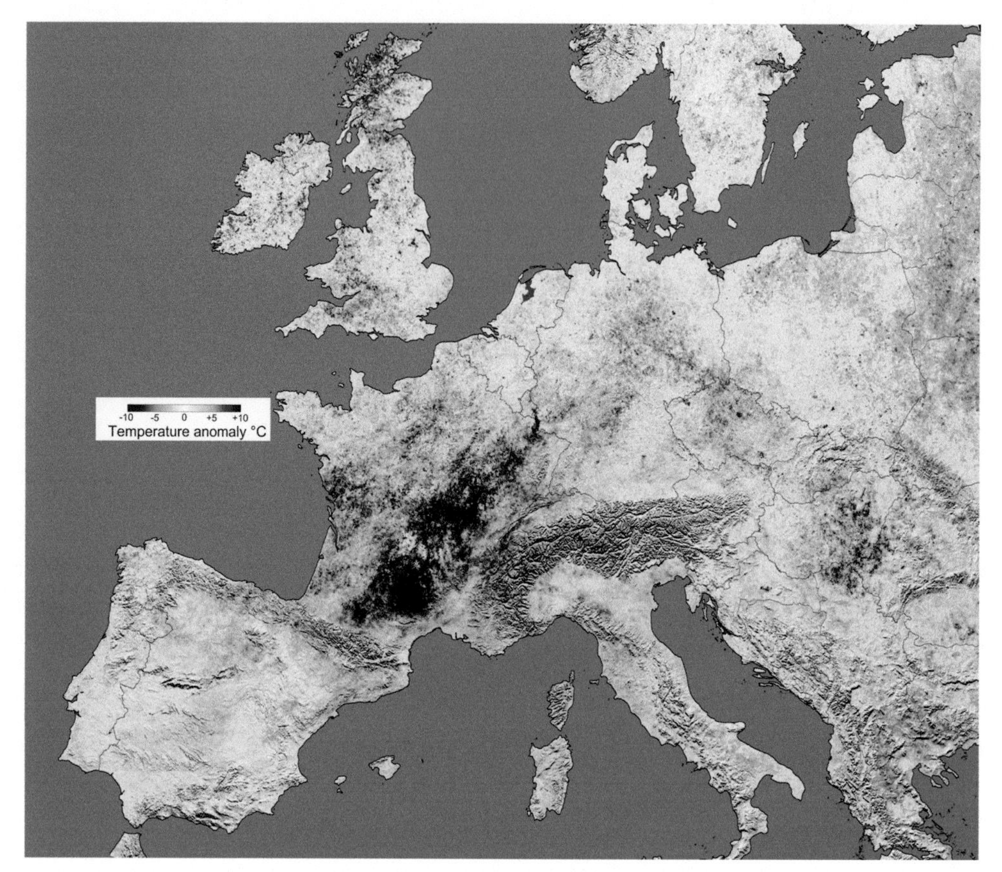

Figure 5.11 2003 heat wave temperature variations relative to 2001 temperatures in Europe.

(Oke, 1982). Unfortunately, urban areas are also the most vulnerable to heat waves. They have more people socially isolated than rural areas and are typically more dependent on critical infrastructure that may also be affected by heat waves, such as electricity and road and rail transport (McEvoy et al., 2012). On the other hand, rural areas are more at risk of wildfires that may be aggravated and even initiated by heat waves.

The other extreme in temperature is similar to heat waves but with greater potential for affecting infrastructure—here called a cold spell. Cold spells are periods of unusually low temperatures in relation to a particular context. For instance, temperatures around freezing are not much of a problem in inland Russia but are claiming lives in South Africa (AFP, 2007). Again, it is the poor and isolated who suffer and die—silently and invisibly. Cold spells cause water pipes to freeze and rupture, making metal brittle. They can, under particular circumstances, cause the formation of a thick layer of ice onto exposed surfaces, sometimes referred to as an ice storm, which often results in power cuts as power cables break under the weight of the ice (Figure 5.12).

BOX 5.6 The silent discriminate killer

In mid-July 1995, Chicago experienced a week of record-high temperatures that would write itself into history as a major disaster and revealer of inequalities in urban America. Klinenberg states in his landmark study of this terrible event that '[h]eat waves are slow, silent, and invisible killers whose direct impact on health is difficult to determine' (Klinenberg, 2002, p. 26). However, the event caused over 600 excess deaths and 3300 excess emergency health care visits, only in Chicago (Dematte et al., 1998), and the people suffering and dying were not in any way from all walks of life. Most of the causalities were poor and elderly living in the city centre, without air conditioning or the means to have it on, refraining from opening their windows or being outside during dark hours due to fear of crime (Klinenberg, 2002). Older men were worse affected than older women due to the generally greater social networks of the latter group (Klinenberg, 2002). African-Americans were worse affected than Caucasians, while Hispanics were less affected than both other groups, further indicating that both poverty and social inclusion mattered, as the Hispanics at that time lived in areas with higher social cohesion (Klinenberg, 2002). Statistically, only 26% of the deaths would have happened in the near future regardless of the heat wave (Kaiser et al., 2007), clearly indicating that many lives would have been possible to save with more attention focused on the problem.

Figure 5.12 Ice cover after one of the great ice storm in Versoix, Switzerland. *(Photo by Pintopc, shared on the Creative Commons.)*

Wildfires

Wildfires are the last type of event I have categorised as a hydrometeorological event, although factors such as vegetation type and topography are also important. A wildfire is an uncontrolled fire in the vegetation of an area (Figure 5.13). They are also commonly referred to as forest fire, bushfire, grassfire, veldfire and so on, depending on the type of vegetation burning and where on the planet you are. Wildfires can occur anywhere on

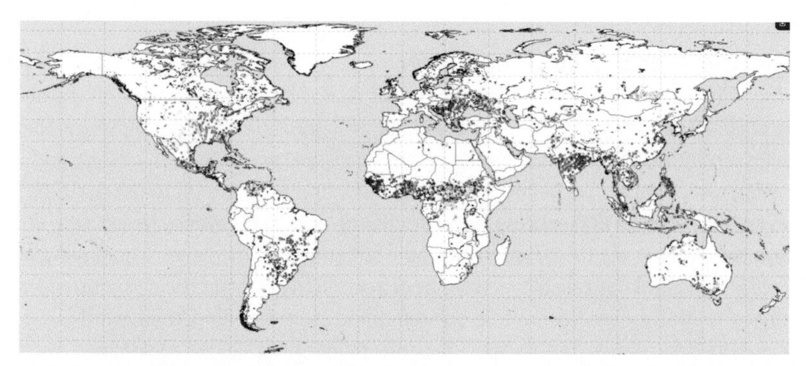

Figure 5.13 Ongoing wildfires on 29 March 2022, as observed by EU's Copernicus global fire monitoring (the lighter yellow, the higher radiative power).

Figure 5.14 Wildfire threatening Estreito da Calheta on Madeira, Portugal.

Earth where plants grow but are particularly frequent and intense in areas with substantial biomass and extended dry periods.

Wildfires come in all sizes and shapes, from just a few trees burning to the Black Saturday bushfires in Australia in 2009, which burnt 350,000 ha (Cameron et al., 2009, p. 11) and killed 173 people (McLennan & Handmer, 2012, p. 4). They can be initiated by lightning, sparks from trains, people, self-igniting plants and so on and can be slow and predictable or rapid and unpredictable. In unfortunate circumstances, a wildfire can move faster than 20 km/h (12 mph) and jump wide rivers (Cheney & Sullivan, 2008), making it very difficult to manage (Figure 5.14).

Geological events

Geological events include earthquakes, tsunamis, landslides and erosion, volcano eruptions and sinkholes. Their initiating hazards are, in scientific terms, generally phenomena of geophysical, geomorphological and hydrogeological origin (Twigg, 2004, p. 15).

Although we have little or no influence over the movements of tectonic plates or the location of mantle plumes, human activity is not only involved in these types of events by creating vulnerable conditions but also by directly influencing the hazard processes behind them.

Earthquakes and tsunamis

Earthquakes are tremors of the Earth's crust caused by sudden energy releases in its interior (Dinwiddie et al., 2011, p. 202). Most often as a result of strains accumulated in rock, exceeding its elastic limit and causing it to rupture (Sinvhal, 2010, p. 24). They happen very often, much more often than we tend to believe. Approximately every 20 s, there is an earthquake that can be felt by people somewhere on the planet (USGS, 2013). Luckily, only a few of these earthquakes are powerful enough to cause damage to us on the surface. These earthquakes are of a magnitude of five and over on the so-called Richter scale and generally occur along fault lines between the tectonic plates that make up the crust of our planet (Figure 5.15).

Earthquakes happen suddenly and may last from a few seconds to several minutes in any affected location (Dinwiddie et al., 2011, p. 202). The energy released diffuses in all directions away from the rupture itself, often referred to as the hypocentre, travelling as seismic waves of variable speed mainly due to the Earth's varying density, rigidity and elasticity (Sinvhal, 2010, p. 24). The first place on the surface to be affected is generally the point vertically above the rupture, referred to as the epicentre. The seismic waves continue to travel outward, weakening as the energy dissipates in an increasing volume of Earth's crust. The seismic waves can be represented on a map as more or less concentric circular shapes around the epicentre. However, it is essential to note that their damage

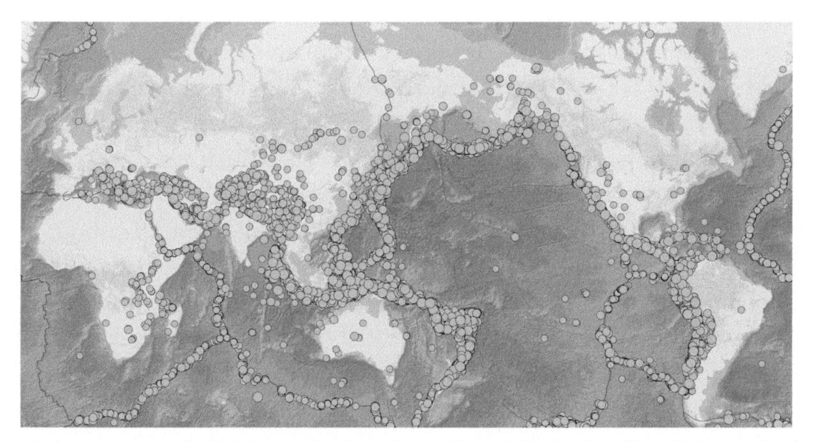

Figure 5.15 The location of the 18,733 earthquakes with a magnitude of 5 and more on the Richter scale between 2010 and 2020 (the greater magnitude, the larger circle). *(USGS.)*

may follow a very different logic due to spatial variation in local geology (Borcherdt & Glassmoyer, 1992) and vulnerability regarding building types and quality, community preparedness and so on. The spatial extent of earthquake damage may, thus, be more complex to map—not so much for the little earthquake that causes one painting to fall from the wall, but for more significant earthquakes that may cause damage as far away as 500 km (Sinvhal, 2010, p. 69) and even further if a tsunami is generated (Alexander, 1993).

While the 2004 Indian Ocean tsunami abruptly placed the word *tsunami* on everybody's mind, it comes from Japanese and means 'harbour wave' (Alexander, 1993, p. 80). It is also Japan that has been affected by most tsunamis in the world, with 364 recorded events before 30 March 2022 (NOAA, 2022), followed by the USA (272) and Indonesia (258). In general, the Pacific Ocean is where most tsunamis occur. Tsunamis are fast-travelling ocean waves with a very long wavelength and low amplitude in deep water. They are caused by the sudden displacement of large volumes of water. Tsunamis are often referred to as 'seismic sea waves'. Yet, they may not only be caused by earthquakes but by violent undersea volcanic eruptions or landslides (Alexander, 1993, p. 80), as well as by landslides above sea level (Fritz et al., 2009) or glacier calvings (Nettles & Ekström, 2010, p. 475). However, all of these are often effects of or related to earthquakes, maintaining the strong link between tsunamis and earthquakes.

When a large tsunami wave of less than 1 m height over the deep sea but with a 200 km wavelength and a speed of up to 700 km/h (435 mph) approaches shallower water, it slows down and starts to compress. As all the water in the wave has to go somewhere, the height of the wave grows tremendously, resulting in a massive wave or a series of enormous waves arriving over hours. It can be up to 30 m high along the coasts closest to the earthquake and up to 10 m as far as 2000 km away. The figures of this example are from the 2004 Indian Ocean tsunami (Wang & Liu, 2010), which killed hundreds of thousands of people—mainly in Indonesia, Sri Lanka, India and Thailand, but also as far away as on the African coast. Another particularly devastating tsunami occurred together with the 2011 Tōhoku earthquake outside the coast of Japan, which did not only destroy countless settlements along the coast but caused a nuclear disaster by damaging the Fukushima Daiichi Nuclear Power Plant. Such tsunamis can not be felt by fishermen out at sea but leave them to find their towns and villages on the coast destroyed when coming back. Hence, the Japanese name tsunami: Harbour wave.

Although parts of the processes behind the formation of tsunamis are still not entirely known, it is generally believed that an earthquake-induced tsunami requires an earthquake of considerable magnitude (Sinvhal, 2010, p. 147). Earthquake magnitude has traditionally been measured on the Richter scale, even if other newer scales compete among scientists worldwide. The Richter scale is a logarithmic scale representing the shaking amplitude and energy released in an earthquake (Box 5.7). This logarithmic scale means an earthquake measuring 6.0 on the Richter scale has ten times higher shaking

BOX 5.7 The Richter scale

<2.0: Not generally felt by people; happens continually.

2.0–2.9: Felt slightly by some people; happens more than a million times per year.

3.0–3.9: Shakes indoor objects slightly, but causes damage very rarely; happens over 100,000 times per year.

4.0–4.9: Shakes indoor objects, but causes none to minimal damage; happens around 10,000–15,000 times per year.

5.0–5.9: Causes damage of varying severity to poorly constructed buildings, and may cause a few casualties; happens 1000–1500 times per year.

6.0–6.9: Causes moderate-to-severe damage to poorly constructed buildings, modest damage to well-constructed buildings, and slight damage to earthquake-resistant structures. Death toll ranges from none to 25,000; happens 100–150 times per year.

7.0–7.9: Causes damage to most buildings, some to partially or completely collapse. Even well-constructed buildings are likely to be damaged. Death toll ranges from none to 250,000; happens 10–20 times per year.

8.0–8.9: Major damage to most buildings and many structures likely to be destroyed. Even earthquake-resistant buildings are likely to receive moderate to heavy damage. Death toll ranges from 1000 and up; happens one time per year.

>9.0: More or less total destruction of buildings and infrastructure. Permanent changes in topography. Death toll usually from 50,000 and up; happens one time per 10–50 years.

amplitude and 31.6 times higher energy release than an earthquake measuring 5.0. Luckily, this scale also roughly corresponds to the frequency of earthquakes of different magnitudes, but in reverse. It means that an earthquake that measures 6.0 on the Richter scale is approximately ten times less likely than an earthquake that measures 5.0 (USGS, 2013). Although the damage potential of an earthquake is linked to its magnitude, it is essential to note that any estimated average earthquake effects presented for any magnitude scale are profoundly contextual. For instance, the 2010 earthquake in Haiti measured 7.1 on the Richter scale, killed up to 300,000 people and left the country in ruins (Figure 5.16), while the M8.8 earthquake in Chile, only six weeks later, released 350 times more energy but resulted in 525 deaths. Again, vulnerability determines the event more than the hazard.

Finally, earthquakes epitomise the natural hazard, and the lay notion of earthquakes is essentially about processes entirely out of our control. Even among leading experts, who are proponents of the idea of vulnerability as a determining factor of risk and disaster, the hazard processes behind earthquakes are still seen as out of our hands. However, while the movement of tectonic plates, over which we have little or no influence, is clearly the most significant cause of earthquakes, human activity can play a role in initiating or even creating earthquakes. For instance, the several 100 million tons of water behind

Figure 5.16 Destruction in Haiti 2010. *(Photo by Magnus Hagelsteen.)*

the Zipingpu Dam in China is believed by some leading scientists to have been a cause of the 2008 Sichuan earthquake, and there are dozens of more widely recognised smaller reservoir-triggered earthquakes around the world (Kerr & Stone, 2009). Also, mining and injecting or extracting liquids or gas are known to induce and trigger earthquakes (McGarr et al., 2002, pp. 650–651). There seem, in other words, to be few definite boundaries to our human agency.

Landslides and erosion

Other types of events that I categorise as geological, although also often having hydrometeorological roots, are landslides and erosion. The former comprises all types of gravitational movements of mass that include a substantial component of rock or soil, that is, rockfall, earthflow and mudflow (Dinwiddie et al., 2011, p. 237). As gravitation is a global force, so is the geographical distribution of landslides. However, landslide hazard is not equally distributed worldwide (Figure 5.17), as several factors interact to determine their location, spatial extent, magnitude and likelihood. For instance, landslide probability has been suggested to be determined by factors such as slope angle, elevation, earth properties, vegetation cover and drainage (Dai et al., 2002, p. 67).

Other processes are causing increased landslide risk, in addition to the main factor of a rapidly growing number of people, buildings and infrastructure being located in landslide-prone areas. First, there are processes occurring on the surface, such as steepening or heightening the slope profile, removing lateral or vertical support, and increasing the load (Alexander, 1993, p. 243). These may be caused by natural processes—such as faulting, tectonic uplift and erosion—or by human activity—such as artificial slopes, excavation for construction, landfill, dams and buildings. Secondly, there are processes occurring beneath the surface, such as the disintegration of rock

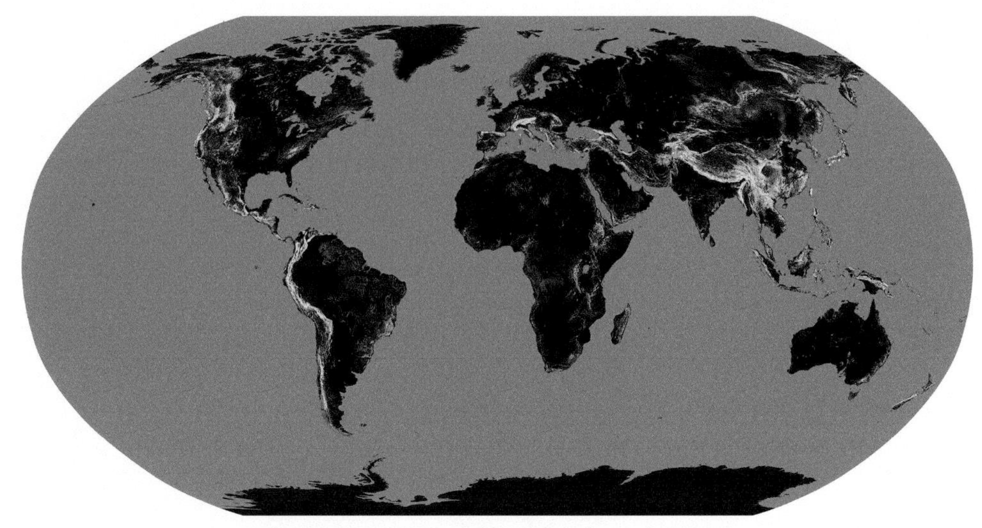

Figure 5.17 Estimated landslide hazard. *(NASA.)*

and soil cohesion, deforestation and water infiltration (Alexander, 1993). Again, these may be caused by natural processes—such as weathering, wildfires and splash erosion—or by human activity—such as tilling, vegetation clearing and inadequate drainage systems. There are also several triggers of landslides, which also are related to natural processes or human activity. More immediate causes include vibrations from heavy traffic or trains, thawing, heavy rainfall and earthquakes (Figure 5.18). Landslides can also occur without any apparent trigger but as the sudden result of a combination of long-term changes (Alexander, 1993). Human activity plays, in other words, a very central role in causing landslides as we alter the topography and hydrology of our environment, increase the load by new buildings and infrastructure, reduce vegetation cover, till and plough, and operate heavy and vibrating machinery.

Landslides span from small and insignificant to massive, involving hundreds of million cubic metres of material (Fauqué & Tchilinguirian, 2002; Schuster et al., 2002). Landslides have, thus, the capacity for great destruction. There are examples of landslides destroying entire villages or even towns, such as the 2010 Gansu landslide in China, killing around 1500 people (Dinwiddie et al., 2011, p. 237) or the 1962 Nevados Huascaran landslide in Peru, killing between 4000 and 5000 people in nine small towns (Schuster et al., 2002, p. 15). Landslides can, as presented earlier, even cause a tsunami if hitting water under the right circumstances. Landslides can also block the flow of rivers (Box 5.8), creating dams that may flood upstream areas and destroy everything downstream as the dam is eventually breached and a violent wall of water comes crashing down (Hung et al., 2002; Schuster et al., 2002, pp. 11—12).

Figure 5.18 Earthquake-induced landslide in El Salvador in 2001.

One of the main causal processes behind landslides deserves attention in its own right. Erosion is not as spectacular as most other events presented in this chapter. Yet, it has an immense impact on the lives of millions of people and the functioning of society (Boardman, 2021). It can strip the fertile topsoil—which was tens, hundreds and even thousands of years in the making—from agricultural land in just a few years and even months in extreme cases, making it a significant challenge for sustainable development worldwide (Toy et al., 2002, p. 1). Aside from being a causal process behind many landslides and behind reduced soil quality and agricultural droughts, as presented earlier, erosion also leads to the eutrophication of lakes, sedimentation of rivers and other watercourses, reduction of water reservoir capacity, and muddy floods of communities and infrastructure (Boardman et al., 2009). Coastal erosion also eats into the coastline, potentially threatening communities, infrastructure and so on (e.g., Larsen et al., 2008, p. 444). While erosion is a natural process of removing, transporting and depositing soil and rock from one location to another, it is profoundly influenced by human activity (Toy et al., 2002, p. 19).

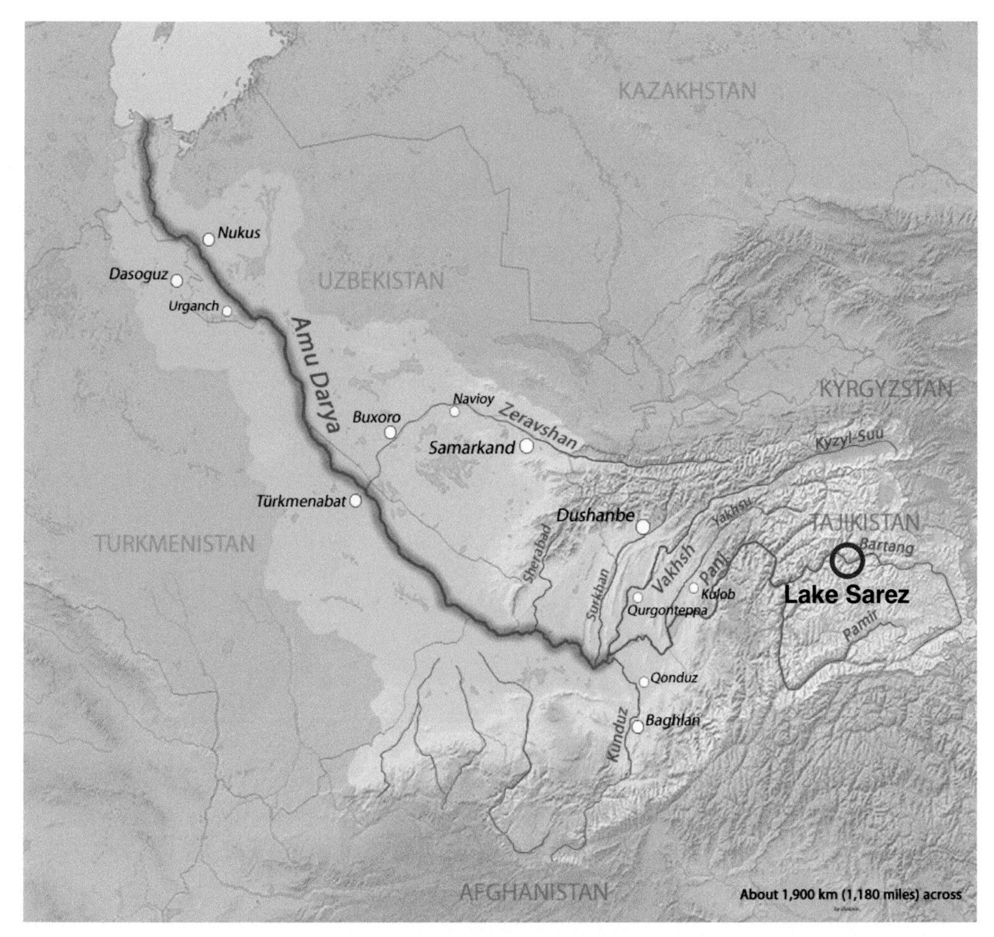

Figure 5.19 Location of Lake Sarez in the Amu Darya River basin. *(Based on Shannon1, Creative Commons.)*

Water and wind are the two leading causes of erosion, representing about 56% and 28% of the total degraded land (Blanco & Lal, 2008, pp. 2—3). Other less important causes are gravitational erosion and exfoliation. The latter is when the sun is rapidly heating the rock, which cools down again at night, causing it to crack. Water erosion is very simplistically caused by rainfall, river flow, waves (wave action, hydraulic action and abrasion), corrosion, glacier movement, thawing and so on. In contrast, wind erosion is caused by wind picking up loose particles (deflation) that batter the ground as they fly by (abrasion), causing additional particles to become loose and fly away. The water or wind then transports the material and deposits it elsewhere. No material ever disappears. It is just moved to another location—for good and for bad.

BOX 5.8 Living beneath a wall of water

In the winter of 1911, an earthquake triggered a massive landslide in the Pamir Mountains of southeastern Tajikistan. The landslide completely blocked the Bartang River, resulting in a dam of rock and gravel called Usoi Dam, containing 17 million cubic metres of water under the name of Lake Sarez (Risley et al., 2006a). As the dam filled up and the water started to flow down the river again, people resumed their lives and livelihoods that depend so much on the river in an otherwise dry landscape. Upstream the river starts in the northeastern tip of Afghanistan and makes a bend through eastern Tajikistan before flowing into Lake Sarez. Downstream Bartang joins river Panj, at the border with Afghanistan, which flows along the border as one of the main tributaries to the great river Amu Darya that runs through Turkmenistan and Uzbekistan before reaching the Aral Sea (Figure 5.19). Millions of people live along these rivers downstream, and a breach of the Usoi dam would have cataclysmic consequences (Risley et al., 2006b). Although investments have been made for an early warning system, it is doubtful if that will have much effect for the 18–25 m high wall of water that would be expected to come crashing down the first 180 km below Lake Sarez or the around 10 m high flash flood that would affect areas several hundred km further downstream (Risley et al., 2006b). Even if some people got to safety, all they have and know would be wiped from the face of the Earth. Even more moderate floods caused by landslide-induced overflows of the dam would affect tens of thousands of people just in the upper parts of the rivers (Risley et al., 2006a).

These natural processes have been shaping the topography of our world since the first rains and wind hundreds of millions of years ago. However, erosion rates have multiplied many times during the development of human society. First, early hunter-gatherers started to burn down forests to force out their prey. Then, the Neolithic Revolution brought more slash-and-burn practices, irrigation and tilling before the Industrial Revolution exponentially accelerated our capacity for using natural resources and moving soil and rock. About 35% of soil erosion is attributed to overgrazing, 30% to deforestation and 28% to excessive cultivation (Blanco & Lal, 2008, pp. 8–10), where the remaining 7% is attributed to industrial activity, infrastructure construction, urbanisation and so on. Also, coastal erosion is aggravated by human activity (Figure 5.20), not only by climate change-induced sea level rise but also by more local activities. For instance, the extraction of sand and gravel, reducing sedimentary inputs through dams, building jetties and breakwaters that interfere with longshore current sediment transport, and destruction of dunes to build houses, marinas, hotels and resorts (Sanjaume & Pardo-Pascual, 2005).

Although coastal erosion is a significant challenge for sustainability in many locations of the world, it may also have benefits in other circumstances. Finding any positive aspects of coastal erosion of small, densely populated islands continuously losing ground is difficult. However, in other cases, eroded material will be deposited elsewhere, and if that location happens to be a flood-prone coastline, it reduces the risk of coastal floods.

Figure 5.20 Coastal erosion threatening Happisburgh, the United Kingdom. *(Photo by Andrew Dunn, shared on the Creative Commons.)*

In our world with finite resources and depending on what we perceive as valuable and important to protect, it may be better to allow erosion somewhere to protect other more valuable locations (Dawson et al., 2009); working with nature instead of trying to control nature, which in the case of coastal erosion, is a losing game (Becker & Payo, 2013).

Volcano eruptions

Volcanoes are openings in the Earth's crust where the magma from beneath can erupt onto the surface (Dinwiddie et al., 2011, p. 84), pushed outwards under high pressure and emitted in the form of lava, volcanic ash and gases (Alexander, 1993, p. 90). They come in different types and shapes. Volcanoes are generally found in the interfaces between tectonic plates and in the interior areas of tectonic plates where the crust is stretched thin or above a so-called mantle plume (Figure 5.21). Over 1500 volcanoes have erupted in the last 10,000 years, and more than 500 in recorded history (Peterson & Tilling, 2000, p. 968). The vast majority of the currently active volcanoes and the more recent eruptions are located around the tectonic plate underneath the Pacific Ocean—thus referred to as the Pacific Ring of Fire.

Volcano eruptions span from being small and calm to large and explosive. The direct impact of hot lava, volcanic ash and gases is immensely destructive but most often geographically limited by the topography around the volcano. There are many examples of entire villages or towns being consumed, such as the classic case of Pompeii or the 1928 Etna eruption that destroyed the town of Mascali (Duncan et al., 1996). There are also examples of extremely violent eruptions and literal explosions of entire mountains. For example, the 1980 eruption of Mount St Helens in the United States (Figure 5.22), which blast released more energy than 400 Hiroshima bombs (7 Mt (Kieffer, 1981, p. 570) compared to 0.016 Mt (Wilson, 2007, p. 168)). This cataclysmic explosion is only surpassed by the 1883 eruption of Krakatoa in Indonesia, which blast created a

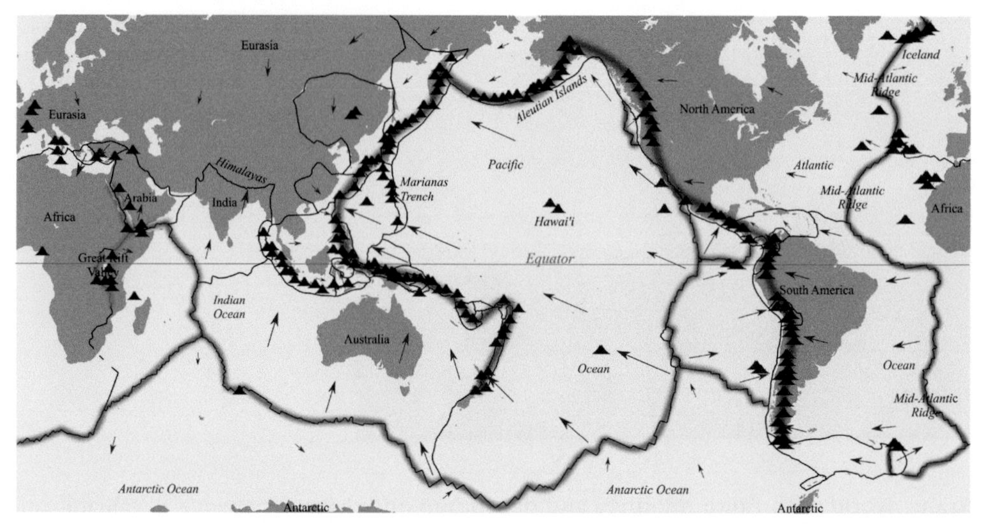

Figure 5.21 Tectonic plates and volcanoes around the world. *(By Astroskiandhike, shared on the Creative Commons.)*

Figure 5.22 Mount St Helens before and after the explosive eruption in 1980. *(Photos by Harry Glicken, USGS.)*

shockwave that reverberated seven times around the world and was recorded by barographs up to 5 days after the explosion (Abercromby et al., 1888). Volcanoes may also cause massive flash floods when occurring underneath glaciers (Smellie, 2000). They emit clouds of ash and gases that may disturb modern aviation and even create climate anomalies that impact agricultural production on entire continents (Zeilinga de Boer & Sanders, 2002). Massive volcanic eruptions are also suggested to have caused the mass extinction event around 252 million years ago (Zhang et al., 2021), which may have wiped out up to 96% of all species (Barnosky et al., 2011, p. 51). Hence, sometimes referred to as 'the Great Dying' (Penn et al., 2018).

Sinkholes

The last type of events I categorise as geological is also often forgotten, although they can be dramatic. Sinkholes are gradual or sudden depressions or holes in the ground caused by subsurface hollows (Figure 5.23), created by carbonate rock dissolving or spreading loose material further down in underground cracks and cavities. A sinkhole is, in other words, formed when the roof of an expanding cavity becomes too weak to support the forces on it (Chan, 1994, p. 6). As with most other events presented in this chapter, sinkholes vary in size and shape. From a small dent in a road to China's 660 m deep and 600 m wide Xiaozhai Tiankeng sinkhole (Waltham et al., 2005, p. 64). People only perish in sudden sinkholes of considerable depths, but gradual and shallow sinkholes also disrupt development, as buildings and infrastructure get damaged.

Although the processes behind the formation of sinkholes are natural, they are heavily influenced by human activity (Alexander, 1993, pp. 283–284). First, buildings and infrastructure increase the load placed upon the roofs of sinkholes, increasing the likelihood of collapse. Second, water seepage from either leaking pipes or reservoirs on the surface may cause or expand sinkholes, as additional water may initiate or exacerbate the natural processes (Buttrick & van Schalkwyk, 1998, p. 173). Ironically, extracting groundwater has even more dramatic effects on sinkhole formation. It is because lowering the groundwater table and hydrostatic pressure may increase the movement of loose material and water, thus increasing subsurface erosion, reducing the support under arching roofs, and increasing depressions and cracks that can damage water pipes and facilitate further surface water infiltration (Buttrick & van Schalkwyk, 1998) (Box 5.9).

Biological events

Biological events include diseases and epidemics, pests and invasive species. Their initiating hazards are, in scientific terms, generally phenomena of organic origin or conveyed

Figure 5.23 A sinkhole in the middle of a residential area.

BOX 5.9 The geological-political legacy of apartheid

South Africa is a dynamic developing country in a challenging transition. The economic purchasing power per person has doubled since the final years of Apartheid (World Bank, 2022), while the Human Development Index (HDI) for South Africa has increased marginally since then (UNDP, 2022). South Africa is, thus, not only one of the few countries in the world that have not experienced human development in this period. It also indicates an overall deterioration in the noneconomic indicators of HDI. Next to the mounting impact of HIV/AIDS on life expectancy, this deterioration is mainly caused by policies making a minor part of the population increasingly wealthy while the majority becomes increasingly poor. This inequality has obvious roots in Apartheid, but it is important to note that it has been growing ever since. Another lingering legacy of Apartheid is the still almost completely ethnically segregated townships generally located near the traditional formal towns. In addition to all challenges with poverty, poor infrastructure, inadequate social services, HIV/AIDS, violent crime and so on, a staggering part of these townships in Mpumalanga, Gauteng, North West, Kwa-Zulu Natal, Eastern Cape and Western Cape provinces are located on carbonate rock. That is partly because of the geology of South Africa, but considering how rare it is to find any traditional towns in these areas, it is difficult not to see the past political situation in South Africa as a major cause for this unsustainable situation concerning sinkholes. The risk of sinkholes increases with growing population pressures that increase the load on these fragile lands, the number of leaking water pipes, the vibrations from traffic, the groundwater usage and so on. This situation is political dynamite for a country so divided.

by biological vectors, including pathogenic microorganisms, toxins and bioactive substances (Twigg, 2004, p. 15). This type of event is particularly diverse and complicated, which makes it almost impossible to present in the available space in this book. Without further due, I do my best.

Disease and epidemics

Diseases are abnormal conditions that affect our bodies or minds, limiting our capacity to live full lives and may lead to death. Although they are often forgotten when considering events that disturb, disrupt or destroy sustainable development, diseases are by far the biggest killers. They comprise almost all leading causes of death worldwide (Figure 5.24). Diseases are often categorised as communicable or noncommunicable, depending on if they can be transmitted between people or not. We tend to worry the most about communicable diseases, such as epidemics. Yet, noncommunicable diseases are killing far more people globally, including number 1, 2, 3, 6, 7, 9 and 10 of the ten leading causes of death (Figure 5.24). It is, however, essential to note that there are substantial differences in mortality between affluent and poor countries, with communicable diseases being number 1, 2, 5, 6, 8 and 9 of the leading causes of death in low-income countries (Figure 5.24). It is particularly horrific to find neonatal conditions killing newborn children on top of the causes of death in low-income countries.

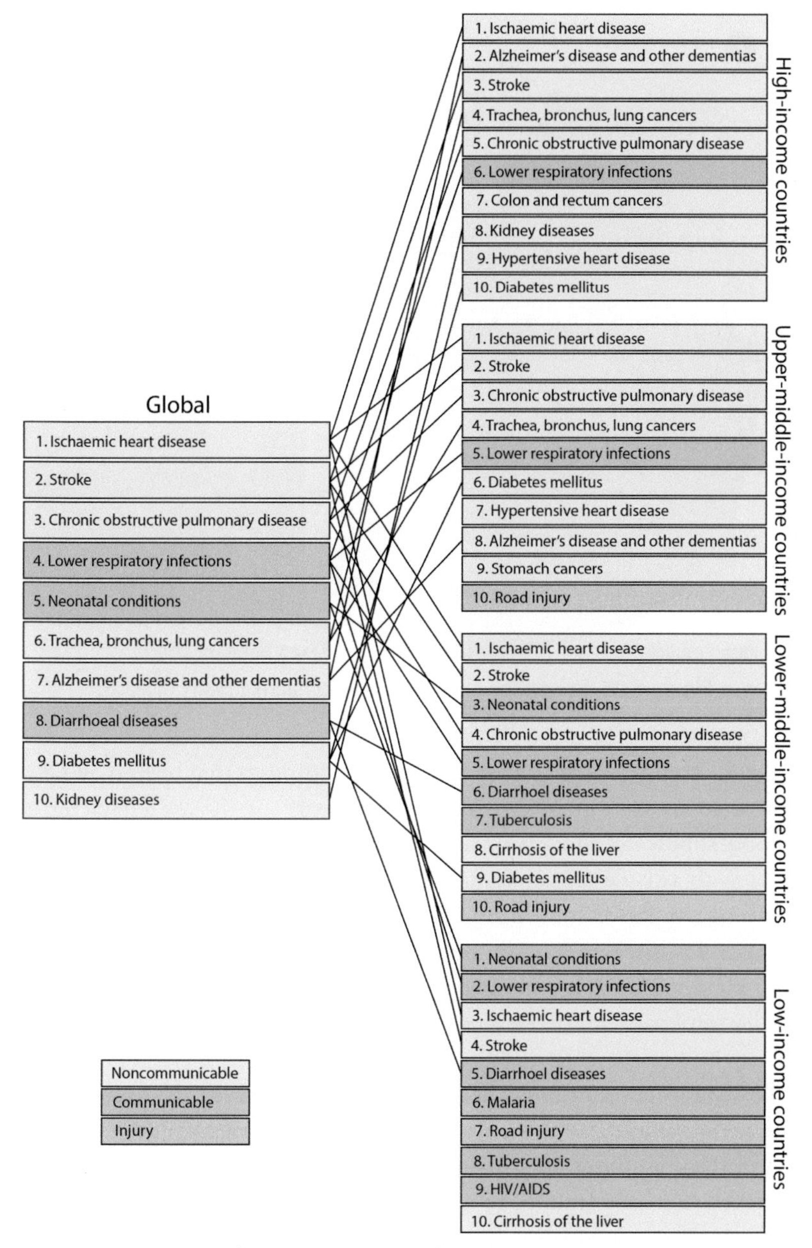

Figure 5.24 The leading causes of death in 2019 (WHO, 2022).

It is impossible to do all noncommunicable diseases justice in this kind of book, so I have to limit myself to briefly clarifying the top killers in the table. First of all, ischaemic heart disease is characterised by the reduced blood supply to the heart muscle (Figure 5.25), usually due to the clogging of blood vessels (Shephard, 1981, p. 15). In

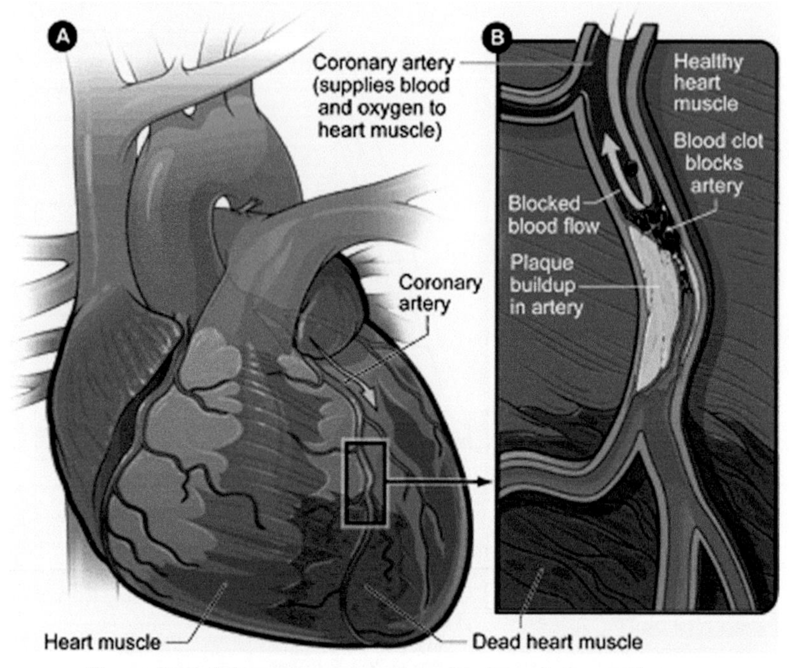

Figure 5.25 Schematic explanation of ischaemic heart disease.

lay terms, this is often referred to as a heart attack, although that term may be used more narrowly in medicine. Similarly, stroke is a category of disease characterised by the reduced blood supply to the brain. These top killers are highly related to your diet and lifestyle, with risk factors including fatty food, obesity and smoking. Smoking is also a significant risk factor for chronic obstructive pulmonary disease—the third worst noncommunicable disease—continuously narrowing airways over time (Pride & Milic-Emili, 2003, p. 159). It includes emphysema and chronic bronchitis and is caused by inhaling harmful particles or gas from smoking, air pollution, asbestos, stone dust and so on.

The following noncommunicable disease on the global list of causes of death is trachea, bronchus or lung cancers. It is a group of diseases characterised by uncontrolled cell growth in tissue in different parts of the airways. If left untreated, the cancer grows and can spread, eventually leading to death. Similarly to chronic obstructive pulmonary disease, risk factors include inhaling harmful particles or gas in combination with a genetic predisposition. Diabetes mellitus, or just diabetes as it is referred to outside of medicine, is a group of diseases affecting the metabolism, leading to excessive blood sugar levels, either because the body is not producing enough insulin or because the body is not reacting to the insulin it produces. Intense research is trying to answer what triggers Type 1 diabetes, which mainly children get and seems unrelated to lifestyle and primarily caused by genetic predisposition. Type 2 diabetes, on the other hand, is primarily driven by lifestyle factors and genetic predisposition.

The last category of noncommunicable disease among the global top 10 killers is kidney diseases, which have also been described as the most neglected chronic diseases (e.g., Luyckx et al., 2018). Kidney diseases comprise a broad range of conditions in which the kidneys are damaged and cannot filter our blood as well as they should, which may lead to anaemia, increased risk of infections, stroke and heart disease. Common risk factors for kidney disease include high blood pressure, diabetes and obesity.

Considering the influence of obesity on the majority of these top noncommunicable killers, it is worrying to reflect on the rapid increase of obesity on a global scale. Since 2000, there are more obese than underweight adults in the world (Delpeuch et al., 2009), and the number keeps rising steadily. By 2030, it is anticipated that more than 1 billion adults will be obese (WOF, 2022, p. 19) and that obesity affects more than 250 million children and adolescents (WOF, 2022, p. 57).

The area of communicable diseases is also too vast and complex to present in any sufficient way in this book. Communicable diseases result from the infection, presence and growth of different pathogens in the human body. They are, thus, distinct from noncommunicable diseases in that the often long and gradual build-up is exchanged with a more rapid disease pattern, that environmental, lifestyle and genetic risk factors are less important (except for sexually transmitted diseases), and that some of them have the potential to cause epidemics. Communicable diseases spread in different ways depending on the pathogen. Some spread through physical contact, either direct human contact or contact with infected objects or through inhalation of contagious droplets or aerosol particles in the air through coughing, sneezing, breathing and so on. Other communicable diseases spread through consuming infected food or drinks, exchanging body fluids, or through a vector, such as a mosquito or a fly. When looking at the tables with the top ten causes of death in low-income counties or the world in total, most ways of transmission are represented (Figure 5.24).

The number one killer among communicable diseases is lower respiratory infections or pneumonia, as it is more commonly called—even if there are other subdivisions of lower respiratory infections in strict medical terms. Several infections can lead to pneumonia. They generally spread through saliva, or other mucus from the respiratory system, exchanged through direct human contact, coughing or sneezing, and so on. Pneumonia can also be caused by influenza and is thus a potential effect of epidemic influenza that we are so worried about. Like COVID-19 or the 1918—19 Spanish Flu (Figure 5.26), which is estimated to have killed more than 50 million people globally (Byrne, 2008, p. 306). Although our recent pandemics have not caused nearly the same death toll relative to the population, they have significantly impacted our globalised world. For instance, COVID-19 killed around 6.2 million people in the first 2 years of the pandemic but pushed hundreds of millions into extreme poverty and hunger.

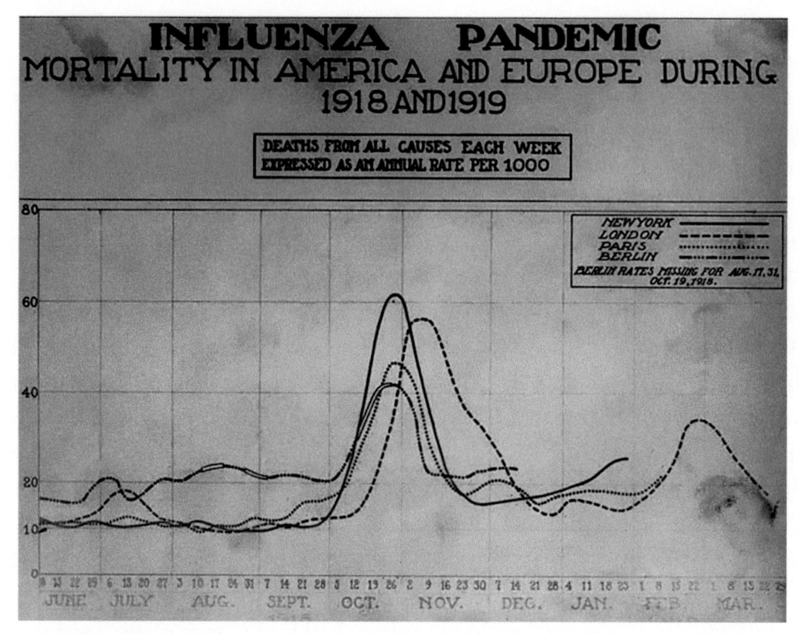

Figure 5.26 Mortality in selected cities during the Spanish Flu pandemic.

The next category of communicable diseases on the list, and the leading cause of death in low-income countries, is neonatal conditions (Figure 5.24). Although the neonatal period is only the first 28 days after birth, it accounts for 44% of all deaths in children under five years old (Liu et al., 2015). It is important to note that while listed as communicable, neonatal conditions also include noncommunicable conditions, such as premature birth and birth asphyxia. The communicable conditions include, for instance, sepsis, neonatal pneumonia, tetanus and neonatal diarrhoea (Liu et al., 2015, p. 432). In other words, neonatal conditions include categories of communicable diseases that are also leading causes of death for the population in general.

One of these categories comprises diarrhoeal diseases—the third biggest killer among communicable diseases globally, but mainly killing people in the less affluent parts of the world. It is a group of diseases causing recurrent loose or liquid bowel movements that, in turn, lead to dehydration and salt imbalances. It is important to note that noncommunicable diseases can cause diarrhoea too, such as Crohn's disease or some cancers. Still, they are dwarfed by their communicable counterparts in terms of impact. Communicable diarrhoeal diseases include, for instance, cholera, salmonella, amoebic infection and rotaviral infection. They are generally transmitted via contaminated food or drinks or through insufficient hygiene. Although most often simply treatable, and the death toll has decreased significantly in recent decades (GBD, 2018), diarrhoea is still the second most common cause of death in children younger than five years old globally (WHO,

2022). It is incredibly frustrating and a clear symptom that our current world order has failed to generate sufficient sustainable development on a global scale, even if great leaps in public health have been taken in various places worldwide. Although possible to prevent with moderate investments, several diarrhoeal diseases can also generate epidemics and must be continuously monitored and prepared for.

Other important communicable diseases that do not make it to the global list but are leading causes of death in less developed parts of the world are tuberculosis, malaria and HIV/AIDS. Tuberculosis, or TB as it is commonly called, is a disease in which various strains of mycobacteria attack the lungs or, in less common cases, other organs, leading to death if untreated for more than half of the infected people developing the disease. Tuberculosis generally spreads via droplets of saliva in the air when an infected person coughs or sneezes, which can be a person without any other symptoms, since only about 10% of the infected ever develop tuberculosis (Schaible, 2009, p. 329). This high proportion of asymptomatic carriers makes tuberculosis very difficult to prevent with other means than mass vaccinations.

Malaria is a vector-borne disease transmitted from one infected person to a new person via a biting mosquito. The disease is a microorganism that harbours in the infected person's liver. Many different kinds of mosquitoes can carry malaria. There are five different types of malaria organisms with varying capacities for causing death and different susceptibilities to different prophylaxes. The available prophylaxes that reduce the risk of contracting the disease are expensive and limited to the affluent, so the most common preventive measures include chemically treated mosquito bed nets, insect repellents and behavioural change campaigns to limit exposure to the nocturnal mosquitoes (Figure 5.27).

Finally, HIV stands for human immunodeficiency virus and is the cause of AIDS, which stands for acquired immunodeficiency syndrome. HIV/AIDS is a disease in our human immune system, breaking it down and making the ill person much more susceptive and affected by infections, including infections that do not usually affect people with functioning immune systems. HIV is transmitted primarily through unsafe sexual practices, contaminated needles or blood transfusions, and from mother to child during pregnancy, birth, and, to some extent, breastfeeding. Lifestyle is, thus, a major factor in contracting the disease. HIV/AIDS is, however, not only killing over 860 million people per year globally (Global Burden of Disease Collaborative Network, 2021) but undermining the livelihoods of many more families since the ill get increasingly weak and orphaning millions of children. HIV/AIDS is spread unevenly over the world (Figure 5.28). It was also the disease with the most considerable growth in death toll between 1990 and 2010, with a staggering 390% increase (Lozano et al., 2013, p. 2105). However, the trend has luckily been broken since then, with a 47% decline in AIDS-related deaths between 2010 and 2020 (UNAIDS, 2021, p. 15).

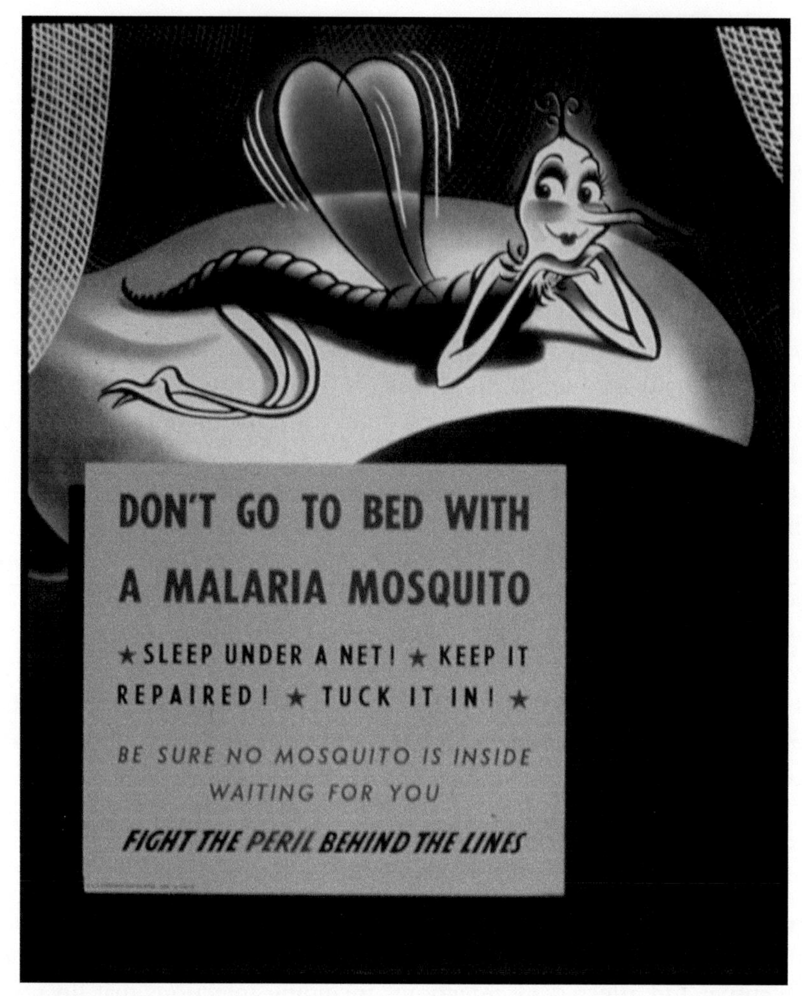

Figure 5.27 World War II malaria prevention campaign in the United States.

Pest and invasive species

The second main group of biological events is not directly targeting us as human beings, but our agricultural produce, animals or ecosystems—either as different kinds of pests, directly destroying our harvests or animals, or as invasive species, disturbing and altering the ecosystems we depend on. Pests can be a plant disease, insect, bird or other destructive agent. Depending on the kind of pest, they can affect just a few unfortunate farmers or cause food insecurity in entire regions. Pests also have massive financial implications, both in direct losses and in managing and preventing these events. Examples of events in which pests have had severe impacts on societies are the 2004 locust invasions in the Sahel region of West Africa (Figure 5.29), the 2001 outbreak of foot and mouth disease in the United

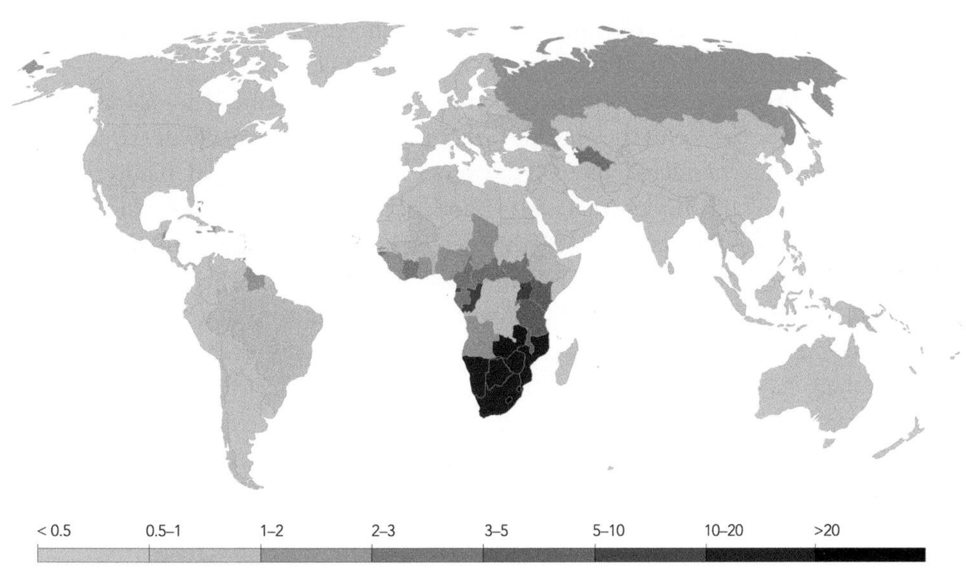

Figure 5.28 Estimated HIV/AIDS prevalence [%] in 2020 (UNAIDS, 2021, p. 13).

Figure 5.29 Swarm of locusts near Satrokala, Madagascar, in 2014. *(Photo by Iwoelbern, shared on the Creative Commons.)*

Kingdom, and the first half of twentieth-century Panama disease that made the then most commonly grown commercial banana more or less extinct.

The term invasive species is here used to group non-native plant or animal species that adversely affect the ecosystems they invade ecologically, environmentally or economically. Invasive species are considered one of the most important causes of biodiversity loss (Courchamp et al., 2003; Linders et al., 2019), and are, therefore, essential for sustainability. Although invasive species may spread by themselves, human activity is fundamental in understanding the much greater speed of contemporary ecological change than throughout history. On top of all the examples of us intentionally introducing new species to an area for one purpose or another—such as the cane toad to Australia or the Canada goose to Europe—there is a whole range of ways in which we unwittingly facilitate for invasive species. For example, commercial forestry or wildfires may open space for newcomers. Travellers bring spores and seeds home. Foreign animals escape from fur farms or game reserves. Our ships bring marine species from afar in the ballast water.

Accidental events

Accidental events include domestic accidents, industrial, infrastructure and transport accidents, environmental impact and pollution, and financial crises. They are, in scientific terms, generally phenomena of unintentional human origin and are often accidental but automatic consequences of the development of society. They are growing in potential magnitude as our innovations get increasingly sophisticated. There is a growing awareness of this challenge, and leading universities are starting to focus on potentially extinction-level risks concerning new technology, such as artificial intelligence, nanotechnology and biotechnology. An important example is the Centre for the Study of Existential Risk at the University of Cambridge, which I believe is a much-needed and groundbreaking initiative to look ahead into our uncertain future.

Accidental events are sometimes referred to as just anthropogenic events in the sense of being caused by human activity. However, I differentiate between unintended and intended events and call the former accidental events and the latter antagonistic events presented later. I want to remind you that human activity is central to initiating or aggravating all other types of events presented in this chapter.

Domestic accidents

Domestic accidents include all those unfortunate events that happen to individuals or groups in the private sphere, such as falls (Box 5.10), drowning, domestic fires and accidental poisoning. These events are often forgotten when considering events that impact sustainability, and they may seem trivial at first sight. However, more than 1.1 million people die globally per year due to the four kinds of accidents just mentioned (Global

BOX 5.10 The plight of an ageing population

Sweden is not only the world champion in collecting statistics, but it is also one of the five countries in the world, with more than 20% of the population being 65 years or older. The ageing population is active and contributes to society in more ways than measurable in financial terms. However, getting older changes the accident exposure patterns away from what society generally focuses on and towards the private and invisible. More than four times as many people in Sweden die, and seven times more people need health care per year due to falling than in a road traffic accident (MSB, 2011). The total cost of injuries in relation to fall accidents is estimated to be a staggering 0.7% of the Gross Domestic Product (GDP), and the responsible governmental authority is determined to do something about it (MSB, 2011).

Burden of Disease Collaborative Network, 2021), and the total costs for societies are staggering. While fire safety is a formal concern in planning and building—although poorly implemented in many parts of the world—most other types of domestic accidents are not. For a society to manage its resources sustainably, it is essential to also include the mundane in managing risk and resilience.

Industrial, infrastructure and transport accidents

Industrial, infrastructure and transport accidents are sudden events that impact individuals, communities and society, causing injury or even death and disrupting or destroying property and the environment. Industrial accidents range from fires and explosions to chemical releases and nuclear meltdowns. They can be limited and stable or massive and dynamic, potentially happening in any industrial area and affecting anything from the immediate surroundings to entire countries or even regions. Examples of industrial accidents are the 1984 Bhopal gas leak that affected around 300,000 people and left 2500 dead (CRED, 2013), and the 2010 Deepwater Horizon explosion, fire and oil spill (Figure 5.30), which killed 11 people and caused damages amounting to an astonishing USD 91,000 million in claims (MacAlister, 2013). To put that figure into perspective, it is roughly the same amount of money as the entire Gross Domestic Product (GDP) of Morocco that year (World Bank, 2022)—a country with more than 30 million people and a GDP larger than 70% of the countries in the world.

Infrastructure accidents are similar to industrial accidents but occur in infrastructure installations and systems. They are also often sudden events that disrupt the functioning of society, and some have great destructive capacity. Infrastructure accidents include bridge collapse, dam failure, levee breaches, power outages and blackouts, and pipe ruptures and can be categorised into two main groups. The first group are infrastructure accidents that primarily cause the failure of the flow or function that the infrastructure is intended to generate. All other consequences are indirectly caused as those flows and

Figure 5.30 Deepwater Horizon disaster in 2010.

functions fail. For instance, the 2003 Northeast blackout in Canada and the USA cut electricity for approximately 50 million people and cost between USD 4 and 10 billion only in the US (Anderson et al., 2007, pp. 183–184). The second group are infrastructure accidents that cause widespread destruction across sectors as a direct consequence of the accident itself, such as the 1975 Banqiao Dam failure mentioned earlier. We are, in other words, not only expecting our infrastructure to provide what is needed for the everyday functioning of society but also to outlast potential events, even if we know deep down that is an impossible expectation to fulfil.

Transport accidents include accidents on road and rail, at sea, and in the air. They may include hazardous material or a large number of causalities. However, the kind of transport accident with the greatest death toll is everyday road traffic accidents, killing a few each time in a myriad of tragic incidents. Around 1.3 million people are estimated to die on our roads every year, and road traffic injuries are the leading cause of death for children and young adults aged 5–29 (Global Burden of Disease Collaborative Network, 2021). After the number of deaths due to road traffic accidents had been increasing for decades, with a 46% increase in the death toll from 1990 to 2010 (Lozano et al., 2013, p. 2109), the trend seems to have been broken, and we have seen a more stable death toll since then (Global Burden of Disease Collaborative Network, 2021). In addition to the shocking number of lives lost, road traffic accidents cause unimaginable costs to societies in terms of suffering, lost productivity, health care and so on.

Although all other types of transport accidents together only kill around 5% of the people killed in road traffic accidents (Lozano et al., 2013), they are capable of fewer but more disastrous events that shake society to its core. For instance, the 1981 Bihar train disaster in India that killed between 500 and 800 people (Spignesi, 2005, pp. 181–183), the 1987 Doña Paz ferry disaster in the Philippines that claimed more than 4000 lives (Eyers, 2013, p. 183), or the 1996 Charkhi Dadri air disaster over India that killed 351 people (Ripley & Fitch, 2004, p. 100).

Environmental impact and pollution

Accidental events are not only sudden and dramatic but can be gradual and invisible. Environmental impact and pollution include long-term events caused by human activity that are detrimental to and undesirable for sustainable development (Johnson et al., 1997)—most often thought of as the disrupting and destruction of ecosystems, extinction of wildlife, contamination of air and water, soil erosion and so on, which have negative indirect consequences on people and societies, as we are all so dependent on these resources. Here, the planetary boundaries presented in Chapter 4 come in as estimations of how much environmental impact on a global scale our Earth can take before collapsing. The environmental impact of human activity also increases the risk of several other types of events presented in this chapter, such as drought through soil erosion, floods through sedimentation of eroded material in watercourses, landslides through deforestation, storm surge and tsunami through cutting down mangroves, and so on. It is also increasingly clear that the environmental impact of human activity in various ways, both directly or indirectly, causes or fuels conflicts as resources become scarce and people move around (Warner et al., 2010)—even if political and economic factors may outweigh the impact of actual local environmental and demographical conditions (Raleigh & Urdal, 2007).

In addition to all these more indirect impacts of environmental degradation, it is increasingly revealed that it also directly affects human health on a scale not appreciated earlier. Pollution is increasingly recognised as a significant factor in infant mortality (Proietti et al., 2013). It is on a global scale as big of a health problem as malaria or tuberculosis (Mills-Knapp et al., 2012). Microscopic soot particles in the urban air cause low birth weight (Dadvand et al., 2012) and contribute to several noncommunicable diseases presented earlier. Heavy metals from poor battery recycling, lead smelting, mining and dumpsites cause neurological, gastrointestinal and cardiovascular problems, cancer, miscarriages, stillbirths, congenital disabilities and so on (Mills-Knapp et al., 2012, pp. 13–30). Asbestos used in building materials and car parts, as well as on mining sites, becomes a major cause of diseases in the respiratory system if not taken proper care of when handling it (Mills-Knapp et al., 2012, pp. 17–25). The list can be made very long, but the main message is that pollution is not only a problem of environmental degradation but also for our health and that there are immense differences in exposure between the rich and the poor (Figure 5.31).

Figure 5.31 The poor are disproportionately exposed to pollution. *(Photo by Magnus Hagelsteen.)*

Financial crises

Financial crises are the last category of accidental events that may undermine development. Financial crises are a group of different events in which some assets, for various reasons, suddenly lose a considerable part of their economic value, causing the market to revaluate and change its market behaviour and resulting in stock market crashes, financial bubbles, currency crises and banking panics. It is important to note that financial crises impact economic value—how valuable we perceive something on the market—and not necessarily how valuable the assets are for other purposes. For instance, just because the financial value of a house plummets 50%, it will still keep the cold winds out the same way as before. However, as modern capitalism is the prevailing paradigm of our time, financial value is crucial for the functioning of society, and radical fluctuations in it are likely to disturb, disrupt and even destroy many people's lives as the effects cascade throughout society.

On a macro-level, financial crises cause increased unemployment, reduced tax revenues, budget cuts, reduced quality or disbanded services, increased foreign debts and so on. When translating that into micro-level impact, it is clear that the social consequences can be immense. Scores of people lose their jobs, and property prices drop. While they cannot find other employment or afford to pay their mortgage, they have to move but cannot sell their house as nobody wants to buy it for the price they need. Many sociologists—ever since Durkheim's (2005) seminal work on suicide in the late nineteenth century—have emphasised that economic crises can create distress, not only as people may experience real financial problems but also as they may perceive being deprived of what they expected from their future. This distress can lead to increased antisocial behaviour, crime, physical and mental health problems, drug use and suicide

(Ragnarsdóttir et al., 2012, p. 2). It has also been suggested that financial crises fuel racism (Balibar, 1991), which is a rather persuasive idea when looking around Europe and North America today.

As one of the great ironies of human existence, financial crises happen in systems entirely made by ourselves. Yet, we have not managed to come up with ways to prevent them effectively. We seem to know more about preventing and mitigating most other types of events presented in this chapter, which include factors over which we have less control. Perhaps it is just that. Financial crises may happen just because we think we are in control, leaving us bogged down in debates between almost diametrically opposed suggestions for how to manage them. Perhaps our inability to stabilise our volatile financial system is because strong forces are not particularly interested in that. Extreme fluctuations are detrimental for the many but beneficial for the few who exploit the fluctuations to generate wealth incredibly fast. But, with high risk comes great danger.

Antagonistic events

Antagonistic events include violent conflict, terrorism and crime. They are, in other words, generally phenomena of intentional human origin.

Violent conflict

Violent conflict is a group of events where at least two parties differ in opinion over something, and at least one engages in activities seeking to persuade the opposing parties through force. There are many ways of categorising such organised conflicts. The Department of Peace and Conflict Research at Uppsala University suggests a highly influential one, dividing them into state-based armed conflicts, one-sided violence, and non-state conflicts (Pettersson et al., 2021).

State-based armed conflicts are contested incompatibilities that concern government and/or territory, where the use of armed force between two parties—of which at least one must be a state—results in at least 25 battle-related deaths in a year (Pettersson et al., 2021, see online appendix for definitions). Wars are armed conflicts with at least 1000 battle-related deaths in a year (Pettersson et al., 2021). The number of such state-based conflicts fluctuates annually, but there has been a negative trend with increasing numbers of such conflicts since 2010 (Pettersson et al., 2021, p. 811). In 2010, there were 31 state-based conflicts, compared to a record-high of 56, including eight wars, in 2020 (Figure 5.32) (UCDP, 2022). It amounts to a 77% increase. State-based armed conflicts can be categorised further into interstate armed conflict—with two or more states involved—intrastate armed conflict—with the government of a state and internal opposition groups involved—and internationalised intrastate armed conflict—which is an intrastate armed conflict with troop intervention from other states.

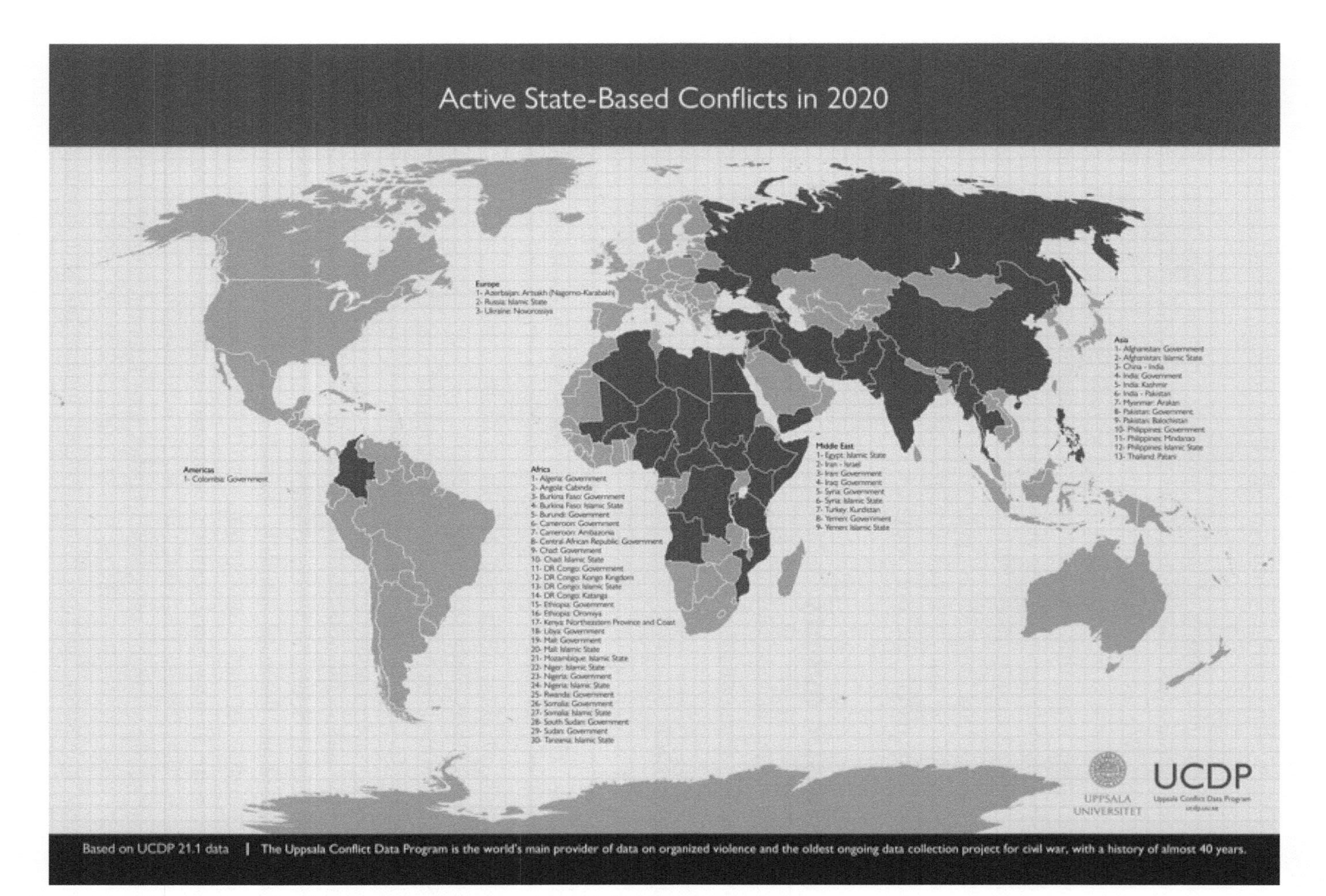

Figure 5.32 The location of active state-based conflicts in 2020 (UCDP, 2022).

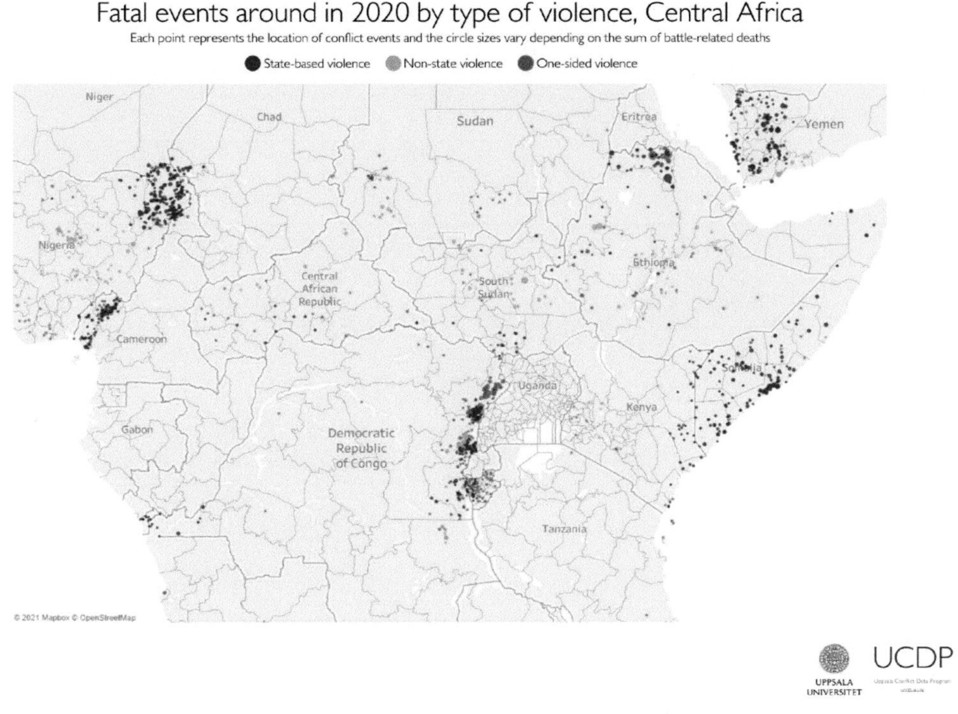

Figure 5.33 Concentrations of one-sided violence in Africa (UCDP, 2022).

There was no ongoing interstate armed conflict, 22 intrastate, and nine internationalised intrastate armed conflicts in 2010, while there were three interstate, 28 intrastate, and 25 internationalised intrastate armed conflicts in 2020 (Pettersson et al., 2021). While such numbers of intrastate armed conflicts can be found for most of the last 50 years, the number of internationalised intrastate armed conflicts is unprecedented (Pettersson et al., 2021, p. 812).

The next category of conflict is one-sided violence. It involves the use of armed force by the government of a state or a formally organised group against civilians, which results in at least 25 deaths in a year (excluding extrajudicial killings in custody). There were 22 cases of such one-sided violence in 2010, compared to 41 in 2020 (UCDP, 2022)—an 86% increase. Although one-sided violence occurs on all inhabited continents, it is currently most frequent in the central latitudes of Africa (Figure 5.33). Infamous actors engaging in one-sided violence in 2020 are, for instance, Islamic State, Al-Shabaab, the gangs of Port-au-Prince in Haiti, and specific governments, such as the Governments of Ethiopia, Eritrea and Burkina Faso (UCDP, 2022).

Finally, a non-state conflict is defined as 'the use of armed force between two organised armed groups, neither of which is the government of a state, which results in at least 25 battle-related deaths in a year' (Pettersson et al., 2021, see online appendix). It is a

Figure 5.34 The scourge of non-state violence in Mexico (UCDP, 2022).

category of violent conflict that has literally exploded in recent years, from 29 non-state conflicts in 2010 to 72 such conflicts in 2020 (UCDP, 2022). That amounts to a 148% increase. Furthermore, the fatalities in this kind of conflict have increased even more in the same period, first substantially driven by non-state conflicts in the Middle East before being surpassed by the more recent surge in this kind of violence in the Americas (Pettersson et al., 2021, pp. 813–814). This surge has primarily been driven by violent Mexican cartels (Figure 5.34), but non-state violence is also rising in Brazil (Pettersson et al., 2021, p. 813).

People are killed regardless of the type of conflict, both directly—as a consequence of violent acts—and indirectly—as conflicts destroy people's means of subsistence and displace them from their homes. For instance, the conflicts in the Democratic Republic of the Congo are estimated to have directly killed tens of thousands between 1997 and 2010 (UCDP, 2022). Still, the total death toll is in the millions— even if the top estimates of 5.4 million may be an overestimation (Human Security Report Project, 2011). Conflicts have, in other words, social, psychological, economic and environmental impacts that are difficult to appreciate from afar.

Terrorism

Closely linked to conflict, terrorism is a particular way to force an opponent to change opinion or behaviour by inducing fear through violent force, often deliberately attacking non-state targets. It is a highly contested term with no commonly agreed definition. It sometimes includes acts of states, such as one-sided violence presented earlier and deliberately targeting civilians in armed conflict. However, it is generally seen as committed by a party in a conflict that does not have the means to fight the opposing side using conventional means of combat. This narrower sense of terrorism is, thus, focusing on covert activities instead of open confrontation, making it difficult for the opposing side to effectively manage this threat by continuing to apply more traditional warfare.

Terrorism can be domestic or international. An attack may succeed or fail, causing widespread death and destruction or not. The National Consortium for the Study of Terrorism and Responses to Terrorism (START) defines a terrorist attack 'as the threatened or actual use of illegal force and violence by a non-state actor to attain a political, economic, religious, or social goal through fear, coercion, or intimidation' (START, 2022). Their Global Terrorism Database includes more than 200,000 registered terrorist incidents between 1970 and 2019 (and counting). To be included in this database, an incident must be intentional, violent and perpetrated by subnational actors (START, 2022). The database does, in other words, not include acts of state terrorism. In addition, at least two of the following three criteria must be present for an incident to be included (START, 2022):

1. The act must be aimed at attaining a political, religious, social or economic goal (mere pursuit of profit is not sufficient but must involve the pursuit of more profound systemic economic change).
2. There must be evidence of an intention to coerce, intimidate or convey some other message to a larger audience than the immediate victims.
3. The act must be outside the context of legitimate warfare, that is, outside the parameters permitted by international humanitarian law (Figure 5.35).

For example, around 8500 terrorist incidents occurred in 2019, killing over 20,000 people (START, 2022). About 36% of the incidents occurred in South Asia, 25% in the Middle East and North Africa, 23% in Sub-Saharan Africa, 9% in South America, and 8% in Southeast Asia (START, 2022). It leaves 2% for Western Europe and 1% for North America, two regions with great fear of terrorism among policymakers and the public. The remaining 1% of the terrorist attacks in 2019 were shared by Eastern Europe, East Asia, Central America and Caribbean, Australasia and Oceania, and Central Asia (START, 2022). It may be worth noting that 37 times as many people committed suicide than were killed by terrorism in 2019 (Global Burden of Disease Collaborative Network, 2021; START, 2022), and to compare the death toll of terrorism to most of the other types of events presented in this chapter may belittle the problem. However, it is

Figure 5.35 Nobody will ever forget 11 September 2001. *(Photo by Robert, shared on the Creative Commons.)*

important to note that terrorism not only kills and destroys the direct targets but has vast indirect impacts on the economy, behaviour and so on. That is the whole idea of terrorism.

Crime

Crime is here used to describe various events in which laws are broken, which undermine the functioning of society and may cause harm or even death to people. It is impossible to do such multifaceted issue justice in just a short introduction. Still, it is essential not to forget that crime is a central challenge for sustainable development. First of all, interpersonal violence is estimated to kill more than 400,000 people annually (Global Burden of Disease Collaborative Network, 2021) and hurt many more. The prevalence of rape and gender-based violence is staggering (Spowart, 2020), and the impact of crime or fear of crime on human well-being is immense (Lorenc et al., 2012). In addition to these human costs, crime is also costly in economic terms, including victim costs (e.g., direct losses for the victims, health care, lost earnings), criminal justice system costs (e.g., police, courts, prisons), crime career costs (i.e., lost productive workforce), and intangible costs (e.g., pain and suffering, fear, stress) (McCollister et al., 2010).

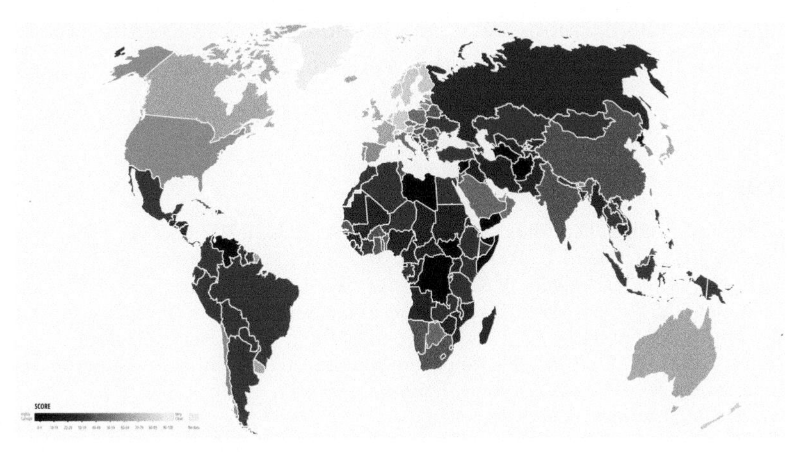

Figure 5.36 Corruption Perceptions Index 2021. *(Source: Copyright Transparency International. All Rights Reserved. For more information, visit http://www.transparency.org.)*

However, it is not only violent crime that hampers sustainable development. Economic crime costs societies enormous amounts of money, and corruption undermines their functioning. Not only is money diverted away from essential services, but trust and cohesion are also undermined. The nongovernmental organisation Transparency International monitors corruption globally and publishes its Corruption Perceptions Index annually, in which countries are scored on how corrupt their public sectors are perceived to be (Figure 5.36). The result is a powerful message to policymakers, investors, aid donors and the public. However, it is important to remember that behind each number is a frustrating reality of many individuals trying to get by, which is not directly captured in the index.

Conclusion

The symptoms of the unsustainable state we are in are many and growing. Various destructive events continuously undermine sustainable development. Some are sudden and spectacular, while others are gradual and mundane. Giddens may be right when stating that '[w]e live in a world where hazards created by ourselves are as, or more, threatening than those that come from the outside' (Giddens, 2002, p. 34)—especially if considering the actual impact of human activity on many of the hazards traditionally seen as being entirely determined by natural processes. When also considering human activity as almost exclusively determining how vulnerable individuals, communities or societies are to the impact of various hazards, it is clear that we have not only the capacity to reduce the risk of all these types of symptomatic events but the moral obligation to assist others with less capacity to do so. Sustainable development is, in other words, dependent on us making conscious decisions and taking intentional actions to reduce the risk of anything that could divert us away from the world that we want—for now, and for the future.

References

Abercromby, R., Archibald, E. D., Bonney, T. G., Geikie, A., Judd, J. W., Lockyer, J. N., Strachey, S. R., Wharton, W. J. L., Evans, F. J., Russell, F. A. R., Scott, R. H., Stokes, G. G., & Strachey, L.-G. T. G. E. N. (1888). *The eruption of Krakatoa: And subsequent phenomena.* Trübner & Co.

AFP. (2007). *At least 13 dead in South African cold snap.* Retrieved March 30, 2022 from https://www.terradaily.com/reports/At_Least_13_Dead_In_South_African_Cold_Snap_999.html.

Al Hassan, Y., & Barker, D. J. (1999). The impact of unseasonable or extreme weather on traffic activity within Lothian region, Scotland. *Journal of Transport Geography, 7*, 209–213.

Alexander, D. E. (1993). *Natural disasters.* Routledge.

Anderson, C. W., Santos, J. R., & Haimes, Y. Y. (2007). A risk-based input–output methodology for measuring the effects of the August 2003 Northeast blackout. *Economic Systems Research, 19*, 183–204. https://doi.org/10.1080/09535310701330233

Andrey, J., Mills, B., Leahy, M., & Suggett, J. (2003). Weather as a chronic hazard for road transportation in Canadian cities. *Natural Hazards, 28*, 319–343.

ANSD. (2010). *Situation Economique et Sociale du Senegal en 2009.*

ANSD. (2013). *Projection de la population de la region de Dakar 2013-2025.*

Askman, J., Nilsson, O., & Becker, P. (2018). Why people live in flood-prone areas in Akuressa, Sri Lanka. *International Journal of Disaster Risk Science, 9*, 143–156. https://doi.org/10.1007/s13753-018-0167-8

Ba, M., Diagne, B., Dieng, A., Diouf, L., Diop, O., Faye, F., Guèye, T., Guèye, O., Guibbert, J.-J., Kandji, K., Lemare-Boly, S., Mbengue, M., Ndiaye, P. B., Ndione, E. S., Ndoye, F., Niang, M., Sall, M., & Touré, C. (2009). *Pikine aujourd'hui et demain: Diagnostic participatif de la ville de Pikine (Dakar, Sénégal).* Enda Graf Sahel.

Balibar, E. (1991). Racism and crisis. In E. Balibar, & I. M. Wallerstein (Eds.), *Race, nation, class: Ambiguous identities* (pp. 217–227). Verso.

Barnosky, A. D., Matzke, N., Tomiya, S., Wogan, G. O. U., Swartz, B., Quental, T. B., Marshall, C., McGuire, J. L., Lindsey, E. L., Maguire, K. C., Mersey, B., & Ferrer, E. A. (2011). Has the Earth's sixth mass extinction already arrived. *Nature, 471*, 51–57. https://doi.org/10.1038/nature09678

Becker, P. (2009). Grasping the hydra: The need for a holistic and systematic approach to disaster risk reduction. *Jàmbá: Journal of Disaster Risk Studies, 2*, 12–24.

Becker, P. (2018). Dependence, trust, and influence of external actors on municipal urban flood risk mitigation: The case of Lomma Municipality, Sweden. *International Journal of Disaster Risk Reduction, 31*(1004), 1004–1012. https://doi.org/10.1016/j.ijdrr.2018.09.005

Becker, P., & Payo, A. (2013). Changing the paradigm: A requisite for safe and sustainable coastal societies. *Journal of Geography & Natural Disasters, S1.*

Berz, G., Kron, W., Loster, T., Rauch, E., Schimetschek, J., Schmieder, J., Siebert, A., Smolka, A., & Wirtz, A. (2001). World map of natural hazards—a global view of the distribution and intensity of significant exposures. *Natural Hazards, 23*, 443–465. https://doi.org/10.1023/A:1011193724026?LI=-true#page-1

Blanco, H., & Lal, R. (2008). *Principles of soil conservation and management.* Springer.

Boardman, J. (2021). How much is soil erosion costing us. *Geography, 106*(1), 32–38. https://doi.org/10.1080/00167487.2020.1862584

Boardman, J., Shepheard, M. L., Walker, E., & Foster, I. D. L. (2009). Soil erosion and risk-assessment for on- and off-farm impacts: A test case using the Midhurst area, west Sussex, UK. *Journal of Environmental Management, 90*, 2578–2588. https://doi.org/10.1016/j.jenvman.2009.01.018

Borcherdt, R. D., & Glassmoyer, G. (1992). On the characteristics of local geology and their influence on ground motions generated by the Loma Prieta earthquake in the San Francisco Bay region, California. *Bulletin of the Seismological Society of America, 82*, 603–641.

Bosher, L., Chmutina, K., & van Niekerk, D. (2021). Stop going around in circles: Towards a reconceptualisation of disaster risk management phases. *Disaster Prevention and Management: An International Journal, 30*(4/5), 525–537. https://doi.org/10.1108/dpm-03-2021-0071

Buttrick, D., & van Schalkwyk, A. (1998). Hazard and risk assessment for sinkhole formation on dolomite land in South Africa. *Environmental Geology, 36*, 170–178.

Byrne, J. P. (2008). *Encyclopedia of pestilence, pandemics, and plagues*. Greenwood Press.

Cameron, P. A., Mitra, B., Fitzgerald, M., Scheinkestel, C. D., Stripp, A., Batey, C., Niggemeyer, L., Truesdale, M., Holman, P., Mehra, R., Wasiak, J., & Cleland, H. (2009). Black Saturday: The immediate impact of the February 2009 bushfires in Victoria, Australia. *Medical Journal of Australia, 191*, 11–16.

Cardinali, M., Galli, M., Guzzetti, F., Ardizzone, F., Reichenbach, P., & Bartoccini, P. (2006). Rainfall induced landslides in December 2004 in South-western Umbria, central Italy: Types, extent, damage and risk assessment. *Natural Hazards and Earth System Sciences, 6*, 237–260.

Cerdà, A. (2007). Soil water erosion on road embankments in eastern Spain. *Science of the Total Environment, 378*, 151–155. https://doi.org/10.1016/j.scitotenv.2007.01.041

Chan, Y. C. (1994). *Factors affecting sinkhole formation*. Civil Engineering Department.

Cheney, P., & Sullivan, A. (2008). *Grassfires: Fuel, weather and fire behaviour*. CSIRO Publishing.

Christian, H. J., Blakeslee, R. J., Boccippio, D. J., Boeck, W. L., Buechler, D. E., Driscoll, K. T., Goodman, S. J., Hall, J. M., Koshak, W. J., Mach, D. M., & Stewart, M. F. (2003). Global frequency and distribution of lightning as observed from space by the Optical Transient Detector. *Journal of Geophysical Research, 108*, 1–15. https://doi.org/10.1029/2002JD002347

Courchamp, F., Chapuis, J.-L., & Pascal, M. (2003). Mammal invaders on islands: Impact, control and control impact. *Biological Reviews, 78*, 347–383. https://doi.org/10.1017/S1464793102006061

CRED. (2013). *EM-DAT: The international disaster database*. Centre for Research on the Epidemiology of Disasters.

CRED. (2021). *EM-DAT: The international disaster database*. Centre for Research on the Epidemiology of Disasters.

Cui, P., Zhu, Y.-y., Han, Y.-s., Chen, X.-q., & Zhuang, J.-q. (2009). The 12 May Wenchuan earthquake-induced landslide lakes: Distribution and preliminary risk evaluation. *Landslides, 6*, 209–223. https://doi.org/10.1007/s10346-009-0160-9?LI=true#page-1

Dadvand, P., Parker, J., Bell, M. L., Bonzini, M., Brauer, M., Darrow, L., Gehring, U., Glinianaia, S. V., Gouveia, N., Ha, E.-H., Leem, J. H., van den Hooven, E. H., Jalaludin, B., Jesdale, B. M., Lepeule, J., Morello-Frosch, R., Morgan, G. G., Pesatori, A. C., Pierik, F. H., ... Woodruff, T. J. (2012). Maternal exposure to particulate air pollution and term birth weight: A multi-country evaluation of effect and heterogeneity. *Environmental Health Perspectives*. https://doi.org/10.1289/ehp.1205575

Dai, F. C., Lee, C. F., & Ngai, Y. Y. (2002). Landslide risk assessment and management: An overview. *Engineering Geology, 64*, 65–87. https://doi.org/10.1016/S0013-7952(01)00093-X

Dawson, R. J., Dickson, M. E., Nicholls, R. J., Hall, J. W., Walkden, M. J. A., Stansby, P. K., Mokrech, M., Richards, J., Zhou, J., Milligan, J., Jordan, A., Pearson, S., Rees, J., Bates, P. D., Koukoulas, S., & Watkinson, A. R. (2009). Integrated analysis of risks of coastal flooding and cliff erosion under scenarios of long term change. *Climatic Change, 95*, 249–288. https://doi.org/10.1007/s10584-008-9532-8

Delpeuch, F., Maire, B., Monnier, E., & Holdsworth, M. (2009). *Globesity: A planet out of control*. Earthscan, 6813.

Dematte, J. E., O'Mara, K., Buescher, J., Whitney, C. G., Forsythe, S., McNamee, T., Adiga, R. B., & Ndukwu, I. M. (1998). Near-fatal heat stroke during the 1995 heat wave in Chicago. *Annals of Internal Medicine, 129*, 173–181.

Dieck, H. (2008). *The Johnstown flood*. Penn State University Press.

Dinwiddie, R., Lamb, S., & Reynolds, R. (2011). *Violent earth*. Dorling Kindersley.

Duncan, A. M., Dibben, C., Chester, D. K., & Guest, J. E. (1996). The 1928 eruption of Mount Etna volcano, Sicily, and the destruction of the town of Mascali. *Disasters, 20*, 1–20. https://doi.org/10.1111/j.1467-7717.1996.tb00511.x

Durkheim, E. (2005). *Suicide: A study in sociology*. Routledge.

Evans, G. W. (2005). *Flood risk assessment*. Estates Gazette.

Eyers, J. (2013). *Final voyage: The world's worst maritime disasters*. Bloomsbury.

Falconer, R. H., Cobby, D., Smyth, P., Astle, G., Dent, J., & Golding, B. W. (2009). Pluvial flooding: New approaches in flood warning, mapping and risk management. *Journal of Flood Risk Management, 2*, 198–208. https://doi.org/10.1111/j.1753-318X.2009.01034.x

Fauqué, L., & Tchilinguirian, P. (2002). Villavil rockslides, Catamarca province, Argentina. In S. G. Evans, & J. V. DeGraff (Eds.), *Catastrophic landslides: Effects, occurrence, and mechanisms* (pp. 303–324). Geological Society of America.

Follin, P., Dotevall, L., Jertborn, M., Khalid, Y., Liljeqvist, J.-Å., Muntz, S., Qvarfordt, I., Söderström, A., Wiman, Å., Åhrén, C., Österberg, P., & Johansen, K. (2008). Effective control measures limited measles outbreak after extensive nosocomial exposures in January-February 2008 in Gothenburg, Sweden. *Euro Surveillance, 13*(7-9), 1–5.

Fritz, H. M., Mohammed, F., & Yoo, J. (2009). Lituya bay landslide impact generated mega-tsunami 50th anniversary. *Pure and Applied Geophysics, 166*, 153–175. https://doi.org/10.1007/s00024-008-0435-4?LI=true#page-1

GBD. (2018). Estimates of the global, regional, and national morbidity, mortality, and aetiologies of diarrhoea in 195 countries: A systematic analysis for the global burden of disease study 2016. *The Lancet Infectious Diseases, 18*(11), 1211–1228. https://doi.org/10.1016/S1473-3099(18)30362-1

Giddens, A. (2002). *Runaway world: How globalization is reshaping our lives*. Profile Books.

Global Burden of Disease Collaborative Network. (2011). *Global burden of disease study 2010 (GBD 2010)*. Institute for Health Metrics and Evaluation.

Global Burden of Disease Collaborative Network. (2021). *Global burden of disease study 2019 (GBD 2019)*. Institute for Health Metrics and Evaluation.

Golding, B. W. (2009). Long lead time flood warnings: Reality or fantasy. *Meteorological Applications, 16*, 3–12. https://doi.org/10.1002/met.123

Goswami, B. N., Venugopal, V., Sengupta, D., Madhusoodanan, M. S., & Xavier, P. K. (2006). Increasing trend of extreme rain events over India in a warming environment. *Science, 314*, 1442–1445.

Haimes, Y. Y. (1992). Sustainable development: A holistic approach to natural resource management. *IEEE Transactions on Systems, Man and Cybernetics, 22*, 413–417.

Haines, A., Kovats, R. S., Campbell-Lendrum, D., & Corvalan, C. (2006). Climate change and human health: Impacts, vulnerability, and mitigation. *The Lancet, 367*, 2101–2109. https://doi.org/10.1016/S0140-6736(06)68933-2

Hall, J. W., Sayers, P. B., & Dawson, R. J. (2005). National-scale assessment of current and future flood risk in England and Wales. *Natural Hazards, 36*, 147–164. https://doi.org/10.1007/s11069-004-4546-7?LI=true#page-1

Han, D. (2011). *Flood risk assessment and management*. Bentham Science Publishers.

Hohl, R., Schiesser, H.-H., & Knepper, I. (2002). The use of weather radars to estimate hail damage to automobiles: An exploratory study in Switzerland. *Atmospheric Research, 61*, 215–238.

Höller, P. (2009). Avalanche cycles in Austria: An analysis of the major events in the last 50 years. *Natural Hazards, 48*, 399–424. https://doi.org/10.1007/s11069-008-9271-1?LI=true#page-1

Huang, D., Wang, S., & Liu, Z. (2021). A systematic review of prediction methods for emergency management. *International Journal of Disaster Risk Reduction, 62*, 102412. https://doi.org/10.1016/j.ijdrr.2021.102412

Human Security Report Project. (2011). *Human security Report 2009/2010: The causes of Peace and the shrinking costs of war*.

Hungerford, H., Smiley, S., Blair, T., Beutler, S., Bowers, N., & Cadet, E. (2019). Coping with floods in Pikine, Senegal: An exploration of household impacts and prevention efforts. *Urban Science, 3*(2), 54. https://doi.org/10.3390/urbansci3020054

Hung, J.-J., Lee, C.-T., & Lin, M.-L. (2002). Tsao-ling rockslides, Taiwan. In S. G. Evans, & J. V. DeGraff (Eds.), *Catastrophic landslides: Effects, occurrence, and mechanisms* (pp. 91–115). Geological Society of America.

IPCC. (2021). *Climate change 2021: The physical science basis*. Cambridge University Press.

Jha, A. K., Bloch, R., & Lamond, J. (2012). *Cities and flooding: A guide to integrated urban flood risk management for the 21st century*. World Bank.

Johnson, D. L., Ambrose, S. H., Bassett, T. J., Bowen, M. L., Crummey, D. E., Isaacson, J. S., Johnson, D. N., Lamb, P., Saul, M., & Winter-Nelson, A. E. (1997). Meanings of environmental terms. *Journal of Environmental Quality, 26*, 581–589. https://doi.org/10.2134/jeq1997.00472425002600030002x

Kaiser, R., Le Tertre, A., Schwartz, J., Gotway, C. A., Daley, W. R., & Rubin, C. H. (2007). The effect of the 1995 heat wave in Chicago on all-cause and cause-specific mortality. *American Journal of Public Health, 97*(Suppl. 1), S158–S162. https://doi.org/10.2105/AJPH.2006.100081

Keay, K., & Simmonds, I. (2005). The association of rainfall and other weather variables with road traffic volume in Melbourne, Australia. *Accident Analysis & Prevention, 37*, 109–124.

Keay, K., & Simmonds, I. (2006). Road accidents and rainfall in a large Australian city. *Accident Analysis & Prevention, 38*, 445–454.

Kerr, R. A., & Stone, R. (2009). Seismology. A human trigger for the great quake of Sichuan. *Science, 323*, 322.

Keyantash, J., & Dracup, J. A. (2002). The quantification of drought: An evaluation of drought indices. *Bulletin of the American Meteorological Society, 83*, 1167–1180.

Kieffer, S. W. (1981). Blast dynamics on Mt. St. Helens on May 18, 1980. *Nature, 291*, 568–570.

Klinenberg, E. (2002). *Heat wave: A social autopsy of disaster in Chicago*. University of Chicago Press.

Kummu, M., de Moel, H., Ward, P. J., & Varis, O. (2011). How close do we live to water? A global analysis of population distance to freshwater bodies. *PLoS One, 6*, e20578. https://doi.org/10.1371/journal.pone.0020578

Kusky, T. M. (2005). Flood. In *Encyclopedia of earth science* (pp. 151–153). Facts on File.

Larsen, P. H., Goldsmith, S., Smith, O., Wilson, M. L., Strzepek, K., Chinowsky, P., & Saylor, B. (2008). Estimating future costs for Alaska public infrastructure at risk from climate change. *Global Environmental Change, 18*, 442–457. https://doi.org/10.1016/j.gloenvcha.2008.03.005

Le Tertre, A., Lefranc, A., Eilstein, D., Declercq, C., Medina, S., Blanchard, M., Chardon, B., Fabre, P., Filleul, L., Jusot, J.-F., Pascal, L., Prouvost, H., Cassadou, S., & Ledrans, M. (2006). Impact of the 2003 heatwave on all-cause mortality in 9 French cities. *Epidemiology, 17*, 75–79.

Linders, T. E. W., Schaffner, U., Eschen, R., Abebe, A., Choge, S. K., Nigatu, L., Mbaabu, P. R., Shiferaw, H., & Allan, E. (2019). Direct and indirect effects of invasive species: Biodiversity loss is a major mechanism by which an invasive tree affects ecosystem functioning. *Journal of Ecology, 107*(6), 2660–2672. https://doi.org/10.1111/1365-2745.13268

Lind, N., Hartford, D., & Assaf, H. (2004). Hydrodynamic models of human stability in a flood. *Journal of the American Water Resources Association, 40*, 89–96. https://doi.org/10.1111/j.1752-1688.2004.tb01012.x

Liu, K. S., & Chan, J. C. L. (2002). Synoptic flow patterns associated with small and large tropical cyclones over the western North Pacific. *Monthly Weather Review, 130*, 2134–2142.

Liu, L., Oza, S., Hogan, D., Perin, J., Rudan, I., Lawn, J. E., Cousens, S., Mathers, C., & Black, R. E. (2015). Global, regional, and national causes of child mortality in 2000-13, with projections to inform post-2015 priorities: An updated systematic analysis. *Lancet, 385*(9966), 430–440. https://doi.org/10.1016/S0140-6736(14)61698-6

Longshore, D. (2008). *Encyclopedia of hurricanes, typhoons, and cyclones*. Facts on File.

Lorenc, T., Clayton, S., Neary, D., Whitehead, M., Petticrew, M., Thomson, H., Cummins, S., Sowden, A., & Renton, A. (2012). Crime, fear of crime, environment, and mental health and wellbeing: Mapping review of theories and causal pathways. *Health & Place, 18*, 757–765. https://doi.org/10.1016/j.healthplace.2012.04.001

Lozano, R., Naghavi, M., Foreman, K., Lim, S., Shibuya, K., Aboyans, V., Abraham, J., Adair, T., Aggarwal, R., Ahn, S. Y., Alvarado, M., Anderson, H. R., Anderson, L. M., Andrews, K. G., Atkinson, C., Baddour, L. M., Barker-Collo, S., Bartels, D. H., Bell, M. L., … Hoen, B. (2013). Global and regional mortality from 235 causes of death for 20 age groups in 1990 and 2010: A systematic analysis for the global burden of disease study 2010. *The Lancet, 380*, 2095–2128. https://doi.org/10.1016/S0140-6736(12)61728-0

Lu, X. (2018). Online communication behavior at the onset of a catastrophe: An exploratory study of the 2008 Wenchuan earthquake in China. *Natural Hazards, 91*(2), 785–802. https://doi.org/10.1007/s11069-017-3155-1

Luckman, B. H. (1977). The geomorphic activity of snow avalanches. *Geografiska Annaler—Series A: Physical Geography, 59*, 31—48. https://doi.org/10.2307/520580

Luckman, B. H. (2010). Dendrogeomorphology and snow avalanche research. In M. Stoffel, M. Bollschweiler, D. R. Butler, & B. H. Luckman (Eds.), *Tree rings and natural hazards, a state-of-the-art* (pp. 27—34). Springer.

Luyckx, V. A., Tonelli, M., & Stanifer, J. W. (2018). The global burden of kidney disease and the sustainable development goals. *Bulletin of the World Health Organization, 96*(6), 414—422D. https://doi.org/10.2471/BLT.17.206441

MacAlister, T. (2013). *The Guardian—BP hit by new $34bn Deepwater Horizon claim*. Retrieved December 19, 2020.

McCollister, K. E., French, M. T., & Fang, H. (2010). The cost of crime to society: New crime-specific estimates for policy and program evaluation. *Drug and Alcohol Dependence, 108*, 98—109. https://doi.org/10.1016/j.drugalcdep.2009.12.002

McEvoy, D., Ahmed, I., & Mullett, J. (2012). The impact of the 2009 heat wave on Melbourne's critical infrastructure. *Local Environment, 17*, 783—796. https://doi.org/10.1080/13549839.2012.678320

McGarr, A., Simpson, D., & Seeber, L. (2002). Case histories of induced or triggered seismicity. In W. H. K. Lee, P. Jennings, C. Kisslinger, & H. Kanamori (Eds.), *International handbook of earthquake and engineering seismology*. Academic Press.

McLennan, B. J., & Handmer, J. W. (2012). Reframing responsibility-sharing for bushfire risk management in Australia after Black Saturday. *Environmental Hazards, 11*, 1—15. https://doi.org/10.1080/17477891.2011.608835

Mills-Knapp, S., Traore, K., Ericson, B., Keith, J., Hanrahan, D., & Caravanos, J. (2012). *The world's worst pollution problems: Assessing health risks at hazardous waste sites*.

Molden, D. J., Shrestha, A. B., Immerzeel, W. W., Maharjan, A., Rasul, G., Wester, P., Wagle, N., Pradhananga, S., & Nepal, S. (2022). The great glacier and snow-dependent rivers of Asia and climate change: Heading for troubled waters. In A. K. Biswas, & C. Tortajada (Eds.), *Water resources development and management: Water security under climate change* (pp. 223—250). Springer. https://doi.org/10.1007/978-981-16-5493-0_12

MSB. (2011). *Nyhetsarkiv. Antalet fallolyckor fortsätter att öka*.

Mulholland, K., Kretsinger, K., Wondwossen, L., & Crowcroft, N. (2020). Action needed now to prevent further increases in measles and measles deaths in the coming years. *Lancet, 396*(10265), 1782—1784. https://doi.org/10.1016/S0140-6736(20)32394-1

Nalbantis, I., & Tsakiris, G. (2009). Assessment of hydrological drought revisited. *Water Resources Management, 23*, 881—897. https://doi.org/10.1007/s11269-008-9305-1?LI=true#page-1

Nettles, M., & Ekström, G. (2010). Glacial earthquakes in Greenland and Antarctica. *Annual Review of Earth and Planetary Sciences, 38*, 467—491. https://doi.org/10.1146/annurev-earth-040809-152414

NOAA. (2013). *Science on a sphere*. Annual Lightning Flash Rate Map.

NOAA. (2022). *Global historical Tsunami database*. https://www.ngdc.noaa.gov/hazel/view/hazards/tsunami/event-data.

O'Connor, J. E., & Costa, J. E. (2004). *The world's largest floods, past and present: Their causes and magnitudes*. USGS.

O'Grady, D., Leblanc, M., & Gillieson, D. (2011). Use of ENVISAT ASAR Global Monitoring Mode to complement optical data in the mapping of rapid broad-scale flooding in Pakistan. *Hydrology and Earth System Sciences, 15*, 3475—3494.

Oke, T. R. (1982). The energetic basis of the urban heat island. *Quarterly Journal of the Royal Meteorological Society, 108*, 1—24.

Oliver, J. E. (2005). *Encyclopedia of world climatology*. Springer.

Pascual, M., Cazelles, B., Bouma, M. J., Chaves, L. F., & Koelle, K. (2008). Shifting patterns: Malaria dynamics and rainfall variability in an African highland. *Proceedings of the Royal Society B: Biological Sciences, 275*, 123—132.

Penn, J. L., Deutsch, C., Payne, J. L., & Sperling, E. A. (2018). Temperature-dependent hypoxia explains biogeography and severity of end-Permian marine mass extinction. *Science, 362*(6419), eaat1327. https://doi.org/10.1126/science.aat1327

Perry, A. H., & Symons, L. (1980). The economic and social disruption arising from the snowfall hazard in Scotland—the example of January 1978. *Scottish Geographical Magazine, 96*, 20–25. https://doi.org/10.1080/00369228008736446

Peterson, D. W., & Tilling, R. I. (2000). Lava flow hazards. In H. Sigurdsson, B. Houghton, S. R. McNutt, H. Rymer, & J. Stix (Eds.), *Encyclopedia of volcanoes* (pp. 957–972). Academic Press.

Pettersson, T., Davies, S., Deniz, A., Engström, G., Hawach, N., Högbladh, S., & Öberg, M. S. M. (2021). Organized violence 1989–2020, with a special emphasis on Syria. *Journal of Peace Research, 58*(4), 809–825. https://doi.org/10.1177/00223433211026126

Pimentel, D., Harvey, C., Resosudarmo, P., Sinclair, K., Kurz, D., McNair, M., Crist, S., Shpritz, L., Fitton, L., & Saffouri, R. (1995). Environmental and economic costs of soil erosion and conservation benefits. *Science, 267*, 1117–1123.

Poumadère, M., Mays, C., Le Mer, S., & Blong, R. (2005). The 2003 heat wave in France: Dangerous climate change here and now. *Risk Analysis, 25*, 1483–1494. https://doi.org/10.1111/j.1539-6924.2005.00694.x

Pride, N. B., & Milic-Emili, J. (2003). Lung mechanics. In P. M. A. Calverley, W. MacNee, N. B. Pride, & S. I. Rennard (Eds.), *Chronic obstructive pulmonary disease* (pp. 151–174). CRC Press.

Proietti, E., Röösli, M., Frey, U., & Latzin, P. (2013). Air pollution during pregnancy and neonatal outcome: A review. *Journal of Aerosol Medicine and Pulmonary Drug Delivery, 26*, 9–23. https://doi.org/10.1089/jamp.2011.0932

Quarantelli, E. L. (2000). *Emergencies, disaster and catastrophes are different phenomena* (pp. 1–5). Disaster Research Center. Preliminar.

Ragnarsdóttir, B. H., Bernburg, J. G., & Ólafsdóttir, S. (2012). The global financial crisis and individual distress: The role of subjective comparisons after the collapse of the Icelandic economy. *Sociology*. https://doi.org/10.1177/0038038512453790

Raleigh, C., & Urdal, H. (2007). Climate change, environmental degradation and armed conflict. *Political Geography, 26*, 674–694. https://doi.org/10.1016/j.polgeo.2007.06.005

Ripley, R. F., & Fitch, J. L. (2004). The efficacy of standard aviation English. In M. A. Turney (Ed.), *Tapping diverse talent in aviation: Culture, gender, and diversity*. Ashgate.

Risley, J. C., Walder, J. S., & Denlinger, R. P. (2006a). Usoi dam wave overtopping and flood routing in the Bartang and Panj Rivers, Tajikistan. *Natural Hazards, 38*, 375–390.

Risley, J. C., Walder, J. S., & Denlinger, R. P. (2006b). *Usoi dam wave overtopping and flood routing in the Bartang and Panj rivers, Tajikistan*. USGS.

Sanjaume, E., & Pardo-Pascual, J. E. (2005). Erosion by human impact on the Valencian coastline (E of Spain). *Journal of Coastal Research*, 76–82.

Schaible, U. E. (2009). *Mycobacterium tuberculosis* and his comrades. In U. E. Schaible, & A. Haas (Eds.), *Intracellular niches of microbes a pathogens guide through the host cell* (pp. 327–354). Wiley-VCH.

Schipper, E. L. F., & Pelling, M. (2006). Disaster risk, climate change and international development: Scope for, and challenges to, integration. *Disasters, 30*, 19–38. https://doi.org/10.1111/j.1467-9523.2006.00304.x

Schneider, E. (2010). Floodplain restoration of large European rivers, with examples from the Rhine and the Danube. In M. Eiseltová (Ed.), *Restoration of lakes, streams, floodplains, and bogs in Europe: Principles and case studies* (pp. 185–223). Springer. http://link.springer.com/10.1007/978-90-481-9265-6_11.

Schuster, R. L., Salcedo, D. A., & Valenzuela, L. (2002). Overview of catastrophic landslides of South America. In S. G. Evans, & J. V. DeGraff (Eds.), *Catastrophic landslides: Effects, occurrence, and mechanisms* (pp. 1–34). Geological Society of America.

Sen, A. K. (1982). *Poverty and famines: An essay on entitlement and deprivation*. Oxford University Press.

Shephard, R. J. (1981). *Ischaemic heart disease and exercise*. Croom Helm.

Shultz, J. M., Russell, J., & Espinel, Z. (2005). Epidemiology of tropical cyclones: The dynamics of disaster, disease, and development. *Epidemiologic Reviews, 27*, 21–35.

Sinvhal, A. (2010). *Understanding earthquake disasters*. Tata McGraw Hill Education.

Slegers, M. F. W., & Stroosnijder, L. (2008). Beyond the desertification narrative: A framework for agricultural drought in semi-arid east Africa. *AMBIO: A Journal of the Human Environment, 37*, 372–380.

Smellie, J. L. (2000). Subglacial eruption. In H. Sigurdsson, B. Houghton, S. R. McNutt, H. Rymer, & J. Stix (Eds.), *Encyclopedia of volcanoes* (pp. 403–420). Academic Press.

Sörensen, J., Persson, A., Sternudd, C., Aspegren, H., Nilsson, J., Nordström, J., Jönsson, K., Mottaghi, M., Becker, P., Pilesjö, P., Larsson, R., Berndtsson, R., & Mobini, S. (2016). Re-thinking urban flood management—time for a regime shift. *Water, 8*, 332–346. https://doi.org/10.3390/w8080332

Spignesi, S. J. (2005). *Catastrophe!: the 100 greatest disasters of all time*. New York: Citadel.

Spowart, S. (2020). Global sexual violence. In A. J. Masys, R. Izurieta, & M. Reina Ortiz (Eds.), *Global health security* (pp. 163–186). Springer. https://doi.org/10.1007/978-3-030-23491-1_8

START. (2022). *Global Terrorism Database*. https://www.start.umd.edu/gtd/.

Toy, T. J., Foster, G. R., & Renard, K. G. (2002). *Soil erosion: Processes, predicition, measurement, and control*. John Wiley & Sons.

Twigg, J. (2004). *Disaster risk reduction: Mitigation and preparedness in development and emergency programming*. Overseas Development Institute.

UCDP. (2022). *Uppsala conflict data program*. https://ucdp.uu.se.

UNAIDS. (2021). *Global AIDS update 2021: Confronting inequalities*. UNAIDS.

UNDP. (2022). *Human development data center*. https://hdr.undp.org/en/data.

United Nations. (2017). Factsheet: People and oceans. In *The Ocean conference*. New York: United Nations, 5–9 June 2017.

USGS. (2013). *Earthquake hazards program. Earthquake facts and statistics*.

Vinet, F. (2001). Climatology of hail in France. *Atmospheric Research, 56*, 309–323.

Waltham, T., Bell, F. G., & Culshaw, M. G. (2005). *Sinkholes and subsidence karst and cavernous rocks in engineering and construction*. Springer.

Wang, X., & Liu, P. L. F. (2010). An analysis of 2004 Sumatra earthquake fault plane mechanisms and Indian Ocean tsunami. *Journal of Hydraulic Research*, 147–154. https://doi.org/10.1080/00221686.2006.9521671. March 2006.

Warner, K., Hamza, M., Oliver-Smith, A., Renaud, F., & Julca, A. (2010). Climate change, environmental degradation and migration. *Natural Hazards, 55*, 689–715. https://doi.org/10.1007/s11069-009-9419-7?LI=true#page-1

WHO. (2022). *Global health estimates*. https://www.who.int/data/global-health-estimates.

Wilhite, D. A., & Glantz, M. H. (1985). Understanding: The drought phenomenon: The role of definitions. *Water International, 10*(3), 111–120. https://doi.org/10.1080/02508068508686328

Wilson, W. (2007). The winning weapon? Rethinking nuclear weapons in light of Hiroshima. *International Security, 31*, 162–179. https://doi.org/10.1162/isec.2007.31.4.162

WOF. (2022). *World obesity Atlas 2022*. World Obesity Federation.

World Bank. (2022). *World development indicators database*. https://data.worldbank.org.

Zeilinga de Boer, J., & Sanders, D. T. (2002). *Volcanoes in human history: The far-reaching effects of major eruptions*. Princeton University Press.

Zhang, H., Zhang, F., Chen, J. B., Erwin, D. H., Syverson, D. D., Ni, P., Rampino, M., Chi, Z., Cai, Y. F., Xiang, L., Li, W. Q., Liu, S. A., Wang, R. C., Wang, X. D., Feng, Z., Li, H. M., Zhang, T., Cai, H. M., Zheng, W., & Shen, S. Z. (2021). Felsic volcanism as a factor driving the end-Permian mass extinction. *Science Advances*, 7(47). https://doi.org/10.1126/sciadv.abh1390. eabh1390.

CHAPTER 6

Our dynamic risk landscape

Introduction

The different types of destructive events that may disturb, disrupt or destroy people's lives, livelihoods and even lifeworlds (Chapter 5) are not evenly distributed in space and unchanging over time. So far, we have deliberated over how our past defines our present (Chapter 2) and considered our increasing awareness of how contemporary actions, or rather inaction, determine what futures are possible for our children and coming generations (Chapters 3 and 4). While hydrometeorological events—such as floods and heat waves—have increasingly challenged human societies over time, the exacerbation of risk has, up to this point, not been substantially driven by climate change. There is no doubt that climate change is a major determining factor of our future, but it is not the only process of change to consider. Similarly, geological events—such as earthquakes and volcano eruptions—have always occurred but killed very few and affected the functioning of society far less in hunter-gatherer and early agrarian societies. It was not until later in the history of humankind that people started to live in heavy masonry, multi-storey buildings in densely populated areas, conducive to catastrophic consequences. For instance, the Antioch earthquakes in 115 AD and 526 AD are estimated to have killed around a quarter of a million people each time, and the infamous eruption of Mount Vesuvius in 79 AD buried Pompeii and its entire population of about 20,000 people. The examples are many and continue into our present and, unfortunately, also into the future. The same processes of change that aggregated people in towns and cities also exacerbated the risk of epidemics, which, through a more recent change process, are more likely to turn into pandemics.

In short, behind each type of destructive event we currently encounter around the world and can anticipate for the future, there are processes of change that continuously reshape risk. These change processes are not only adding to the complexity of the world but also its dynamic character. In other words, they further complicate any attempts to reduce risk to sustainable development and demand approaches to risk and sustainability that can take into account such complexity and dynamic change.

There is a wide range of processes that are continuously changing our complex world. Some of them are largely independent of human activity, while most are at least partly determined by what we do on Earth. I focus the following sections on presenting several of the most influential change processes regarding the impact on risk and resilience. These are environmental degradation, climate change, demographic and socioeconomic

Sustainability Science, Second Edition
ISBN 978-0-323-95640-6, https://doi.org/10.1016/B978-0-323-95640-6.00006-3

processes, globalisation, increasing complexity and changing antagonistic threats. In other words, I have deliberately left out plate tectonics and other geological and hydrological processes that, although having a significant impact on our planet over time, are comparatively stable. The processes of change that I do include are full of uncertainty and need to be incorporated in any serious attempt to manage risk for sustainable development.

Environmental degradation

As presented earlier in this book, environmental degradation is a significant sustainability challenge and threat to our well-being and survival. It is an overall process of change that, in various ways, alters or disturbs the environment in what we perceive to be deleterious or undesirable ways (Johnson et al., 1997, p. 584). Although environmental degradation may occur naturally, human activity is both initiating and increasingly aggravating environmental degradation, taking it to unprecedented levels. Our air is increasingly polluted, and deforestation threatens long-term oxygen turnover and disrupts vital biogeochemical cycles. Our water is polluted or used up, and our soil is degraded and eroded. This degradation may not only directly impact health through breathing unhealthy air or drinking unhealthy or too little water but also generate indirect impacts as environmental degradation increases the risk of a range of events presented earlier (Chapter 5).

For instance, environmental degradation is increasing the risk of floods through a range of mechanisms, such as reducing the natural coastal flood protection of mangroves (Temmerman et al., 2012) or diminishing the water-carrying capacity of riverbeds through sedimentation of eroded soil (Boardman et al., 2009). Drought risk is exacerbated by reductions in vegetation cover and accelerated soil erosion. The risk of different kinds of storms increases as deforestation diminishes the dampening effects of forests on the wind. Reducing vegetation cover increases the risk of avalanches and is also a key factor for landslides and erosion. Pollution is a major causal factor of diseases and poisonings that cause the death and suffering of millions of people, and ozone depletion is increasing the risk of skin cancer. Environmental degradation is, in other words, continuously changing our risk landscape and has been identified as one of the ten threats to humanity listed by the High-level Panel on Threats, Challenges and Change of the United Nations (HPTCC, 2004).

Climate change

Archaeological findings indicate that climate change has been a determinant in human development as far back as the ice ages (Tauger, 2021) and beyond. Studies identify, for instance, climate change as a primary contributing factor to our early migrations out of Africa (Eriksson et al., 2012). At the height of the last ice age, 20,000 years ago, sea levels were 120 m lower than today (Rahmstorf, 2012), which most certainly facilitated our early migrations further afield in the world (Shaver, 2011, p. 130). While our climate has been changing back and forth ever since the formation of our atmosphere, we are seeing and foreseeing more rapid changes in our climate now than ever before (Figure 6.1).

The Intergovernmental Panel on Climate Change (IPCC) defines climate change as changes in both averages and variability of weather patterns that persist for an extended period, typically decades or longer (IPCC, 2021). Changes in climate over time may be caused by natural internal processes of our Earth, such as El Niño or the North Atlantic Oscillation, or by external forces, such as variations in solar radiation or human-induced changes in the composition of the atmosphere or land use. Although our climate is determined by a very complex system of factors and any predictions are fraught with uncertainty, it is now 'unequivocal that human influence has warmed the atmosphere, ocean and land' (IPCC, 2021, p. 6) and that such human-induced climate change is a fundamental challenge for our future.

Human-induced climate change is primarily caused by increasing concentrations of carbon dioxide and other greenhouse gases—such as methane and nitrous oxide—in

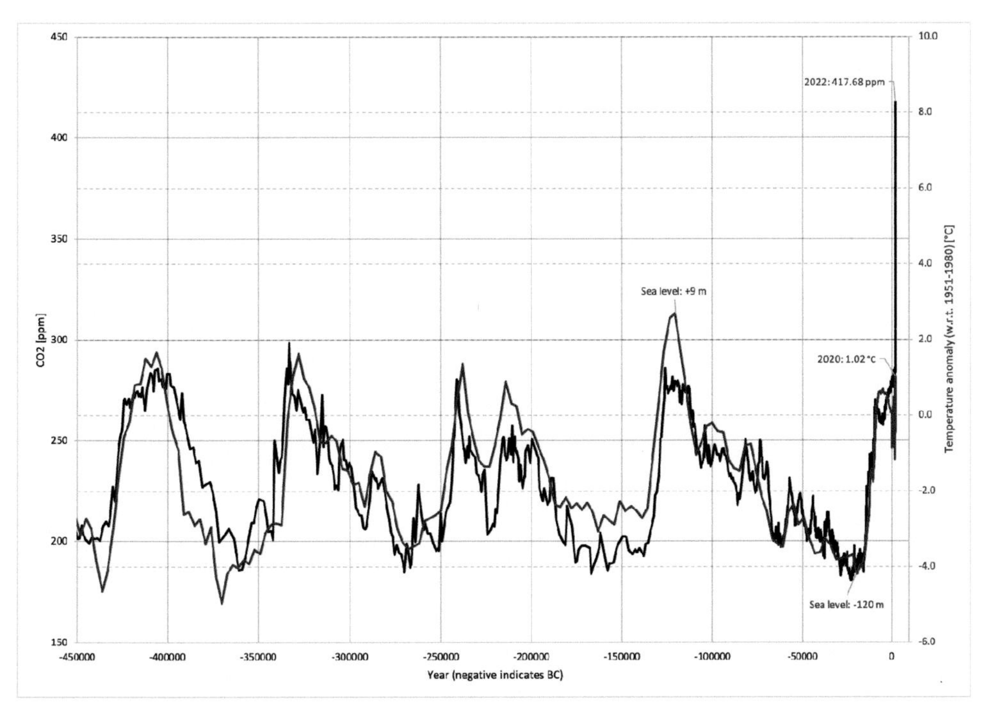

Figure 6.1 Changes in atmospheric CO_2 concentrations and temperature, with reference to 1951—80 baseline, since 450,000 BCE. *(CO_2 concentrations from National Oceanic and Atmospheric Administration (NOAA) and global average temperature anomalies from NASA/GISS/GISTEMP v4 (year 1880—2021) and data combined from multiple sources by the 2° Institute (year 450,000 BC-1879).)*

the atmosphere. Fluctuations in the concentration of carbon dioxide have happened before, and it is striking how closely linked it has been to fluctuations in temperature (Figure 6.1), making it immediately apparent that some relationship exists between temperature and carbon dioxide in the atmosphere. It is, therefore, terrifying to think about what temperatures we may be heading for if merely looking at the current concentration of carbon dioxide in Figure 6.1 and waiting for the lag time in our climate system to pass. Although the climate projections of the scientific community are much more elaborate and somewhat more conservative, we are simply heading for an unprecedented situation for as long as humankind has existed.

Regardless of what climate deniers claim, both convinced and bought, there is ample evidence showing that carbon dioxide is the principal control knob governing the temperature on Earth (IPCC, 2021). After fluctuating between 260 and 280 ppm throughout our entire agrarian history, since the beginning of the Neolithic Revolution (Chapter 2), the trend broke with the onset of the industrial revolution, and the increase in the concentration of carbon dioxide in the atmosphere has escalated ever since (Figure 6.2). The current concentration of carbon dioxide when writing this chapter is

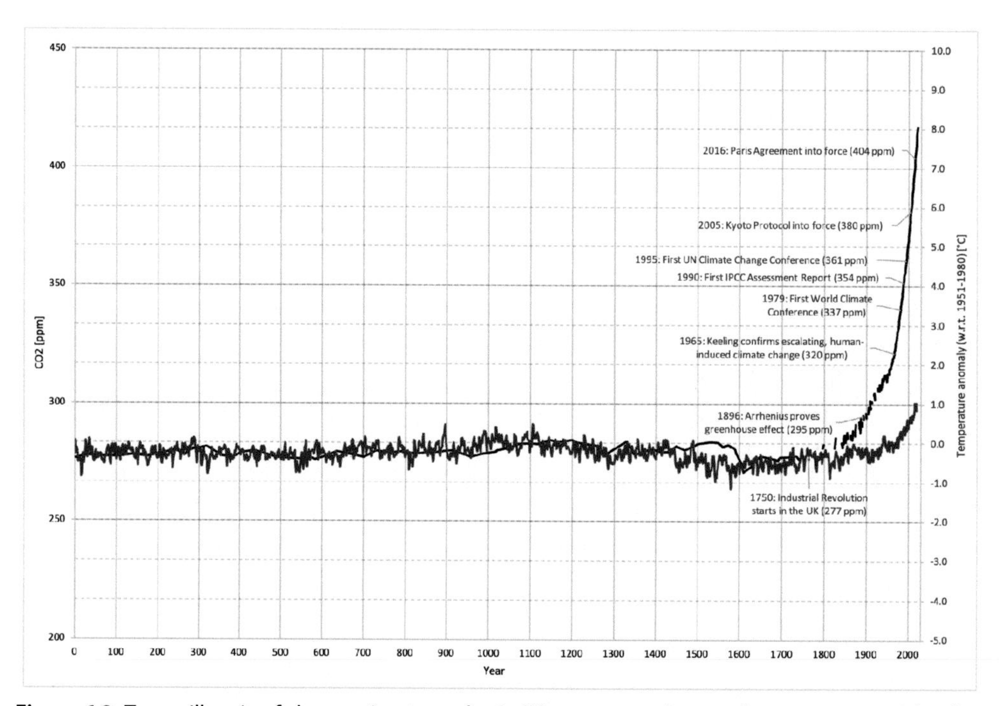

Figure 6.2 Two millennia of changes in atmospheric CO_2 concentrations and temperature, with reference to 1951–80 baseline. *(CO₂ concentrations from National Oceanic and Atmospheric Administration (NOAA) and global average temperature anomalies from NASA/GISS/GISTEMP v4 (year 1880–2021) and data combined from multiple sources by the 2° Institute (year 0–1879).)*

around 425 ppm. It is rising much faster than in any period in the geological records (Lacis et al., 2010, p. 359). Notwithstanding our increasing knowledge and attention to the problem over the last 50 years, we have let the concentration of carbon dioxide become higher than it has been for millions of years.

Our climate system is very complex, and many interconnected factors influence it, in addition to the emissions of greenhouse gases. Many of these factors are also affected by human activity, such as ozone depletion, deforestation and freshwater use. Studies also suggest that excess heat in urban areas not only creates urban heat islands but may warm up more distant rural areas (Zhang et al., 2013). Thus, climate change is a process of change affecting us globally and regionally, with substantial geographical variations. For instance, the IPCC climate projections suggest the Arctic be affected by the most significant increases in mean temperatures (IPCC, 2021), which is likely to cause a substantial reduction of the ice sheet and rising global sea levels. For instance, if the Greenland ice sheet melts more or less completely, it would cause the global sea level to rise around 7 m (IPCC, 2007, p. 752). This melting is obviously not expected to happen overnight, but it is still a scary thought, as many hundreds of millions of people live in such low-lying areas (McGranahan et al., 2007; Small & Nicholls, 2003, p. 593), including a substantial part of the population of Shanghai, New York, Mumbai, London, Buenos Aires and Copenhagen (Figure 6.3). However, if the ice on Antarctica melts, we expect a 60 m sea level rise (Conway, 2010), which would make Bangladesh and the Netherlands disappear, and Westminster of London into excellent cod fishing waters. These cataclysmic scenarios may be somewhat distant in time. Still, several countries would have to be abandoned with only a sea level rise over 1 m, such as the Maldives, Marshall Islands, Tuvalu, Kiribati and Tokelau (Rahmstorf, 2012), which is expected in a not-so-distant future with the current speed of climate change. Especially as the impacts of sea level rise are not only inundated land but also coastal erosion (Lewis, 1990; Rahmstorf, 2012), as changing sea level influences natural coastal sedimentary balances.

Climate change is not only expected in terms of rising mean temperatures and sea levels, although these impacts are the most certain (IPCC, 2022). It is also expected to affect various hydrometeorological events presented earlier (Chapter 5). Climate change is expected to generate more frequent, intense and protracted heat waves (IPCC, 2022). Heat stress has been estimated to potentially reduce the labour capacity during peak months to 80% in 2050 and 40% in 2200 under established climate scenarios in some areas (Dunne et al., 2013). Changes in rainfall patterns are also anticipated, resulting in more frequent, intense and protracted floods (IPCC, 2022; Kundzewicz et al., 2010). This change is also expected in many places with decreased annual rainfall due to more intense rainfall but with more extended periods between (IPCC, 2022). Climate change is expected to increase the frequency and duration of droughts and multiply the global land area subjected to extreme droughts from around 1% now to 30% by the end of this century (Burke et al., 2006). Some studies indicate less frequent cyclones

Figure 6.3 Bangkok, one of the major cities most vulnerable to sea-level rise in the world.

in total (Knutson et al., 2010). However, science is consistently showing that climate change results in more frequent intense cyclones (Knutson et al., 2010, 2020; Nordhaus, 2006; Webster et al., 2005), and that the most intense cyclones are getting both stronger (Elsner et al., 2008) and more frequent (Knutson et al., 2010, 2020). Increasing storms are, in turn, leading to increasing storm surge (Knutson et al., 2020; Hallegatte et al., 2011; von Storch & Woth, 2008)—in combination with sea level rise (Knutson et al., 2020)—as well as increased precipitation rates within 100 km of the storm centre (Knutson et al., 2010).

Climate change affects not only hydrometeorological events but also geological and biological events. Changing rainfall and wind patterns influence erosion (Borrelli et al., 2020) and groundwater levels, thus also affecting landslides (Jakob & Lambert, 2009) and sinkhole formation (Gombert et al., 2010). Climate change also affects disease patterns, moving malaria, dengue and other vector-borne diseases into areas where people have little or no immunity and increasing the burden of diarrhoeal and cardiovascular diseases (Costello et al., 2009). Animal and plant diseases are also expected to increase and undermine people's livelihoods and nutrition (Costello et al., 2009). However, climate change is not only a process that affects us through increasing frequency, intensity and duration of various types of destructive events but it also affects everyday life for all of us.

There have been numerous conferences and several international treaties. There is no lack of scientific input. While only a few sporadic scientific journal articles concerning climate change can be found[1] per year in the early- and mid-20th century, the scientific interest started to increase in the 1970s (119 articles), multiplied in the 1980s (790 articles) before taking off in the 1990s (11,490 articles), 2000s (56,593 articles) and 2010s (203,016 articles) to reach present-day levels of over 40,000 articles per year and still growing. Yet, nothing significant has so far been done to address climate change—regardless of virtually all countries in the world being signatories of the Paris Agreement, which states that global warming should be kept to well below 2 °C, relative to pre-industrial times, and if possible limited to 1.5 °C. Even at 1.5 °C of global warming, there is an increasing risk of extreme events unprecedented in the observational record (IPCC, 2021, p. 19). However, 1.5 °C is the threshold of what the planet can safely absorb, and every fraction of a degree more vastly increases the risk of irreversible changes to the climate (Hoegh-Guldberg et al., 2019). We have already reached between 0.84 and 1.10 °C (Marotzke et al., 2022, p. 147), so there is not much left before we endeavour into genuine peril. Unfortunately, research indicates that we will reach 1.5 °C warming in the next decade regardless of what we do (Marotzke et al., 2022). While this is alarming, it is essential to remember that every tenth of a degree over that threshold counts for what life on Earth will be like for our children and future generations. It is indeed high time for action. You have the power as consumers and citizens. Change your travel and consumption patterns, and vote for politicians prioritising sustainability over short-term political gains.

Demographic and socioeconomic processes

In addition to climate change, we also have processes of change continuously rewriting our population map, with profound effects on risk and sustainability. First of all, we have population growth. The slow-growing or more or less stable global population of our historical past is long gone. While it took us 200,000 years to reach our first billion in global population and the growth from 1 billion to 2 billion took 110 years (Kremer, 1993, p. 683), our last full billion took 11 years, from 2012 to 2023 (United Nations, 2019a). We are, in other words, more than 8 billion people on our planet now and counting. Luckily, the growth rate is decreasing, and the world population is expected to stabilise between 10 and 11 billion in the second half of this century (United Nations, 2019a). Let us hope so, or that Thomas Malthus was wrong when describing his pessimistic relationship between population and the Earth's productive capacity (Malthus, 1798, p. 4). Irrespective

[1] In Scopus, search term 'climate change', 27/04/2022.

of which, providing for even the most basic human rights of that many billion people demands us to become much more efficient than we currently are.

There is no doubt that a growing population puts increasing strains on our already overly exploited planet. Especially since we have taken on the moral obligation to ensure that everybody gets their fundamental human rights satisfied, as proclaimed in international law. In other words, that we meet the social foundation for sustainable development as described in Chapter 4 (Raworth, 2012). However, population growth is heavily linked to socioeconomic development, with a sharp drop in the number of children per woman as human development indicators improve (Figure 6.4).

In very simplified terms, population growth happens over the generations when parents get more children, who go on to have children than themselves. In a world with constant life expectancy and no deaths before reproducing, that would mean that each couple of parents—fully acknowledging that there are many other ways of organising private life—would need two children for the population to be stable. However, life expectancy is unfortunately different and changing depending on where you are on the planet. Not everybody survives to have children, can have children or wants to have children. It means that each woman needs to have about 2.1 children, on average in the more developed countries, for the populations not to drop in the absence of migration (Espenshade et al., 2003). This rate is referred to as the replacement fertility rate, and although many of the most developed countries are far below it, such as Singapore (1.27), South Korea (1.35) and Germany (1.48), immigration and increasing life expectancy are

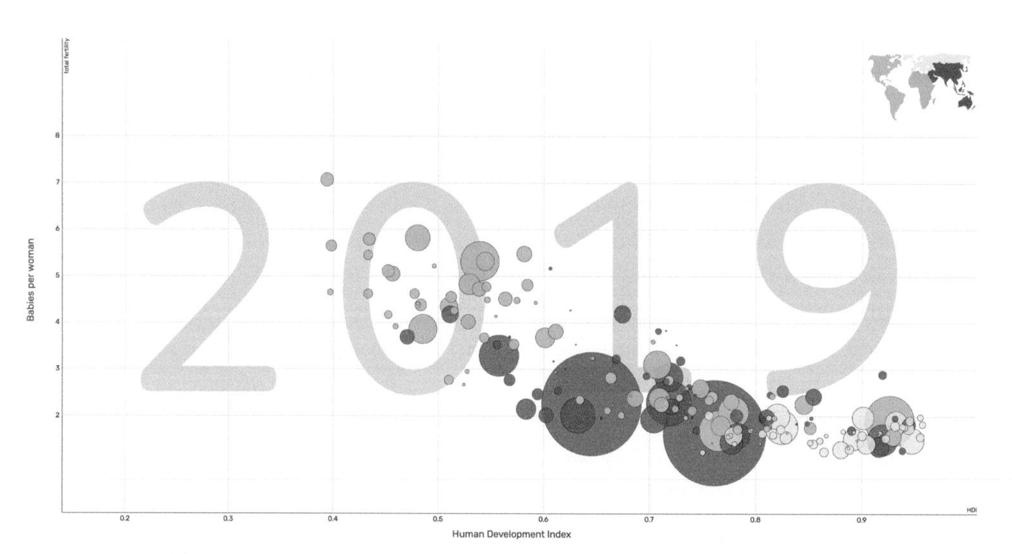

Figure 6.4 Total fertility and Human Development Index in 2019, where each dot is a country, colour signifies region, and size indicates population size. *(Adapted based on free material from www.gapminder.org.)*

still stabilising or increasing their populations. It is far worse for the countries having both low fertility rates and out-migration, such as many countries in Eastern Europe, for example, Poland (1.29), Hungary (1.41) and Bosnia and Herzegovina (1.41), and also in Southern Europe, in the wake of financial downturn, for example, Portugal (1.24), Greece (1.30) and Cyprus (1.33) (United Nations, 2019a).

In 2022, half the countries on our planet had fertility rates below 2.1 children per woman and the other half above 2.1 children per woman (United Nations, 2019a). However, there were still 32 countries with fertility rates higher than four children per woman, out of which all but Iraq (4.04) and Timor-Leste (4.86) were in Africa. It is important to note, though, that the replacement fertility rate varies more for developing countries—spanning from the same as for developed countries to 3.4 children per woman in extreme cases (Espenshade et al., 2003). Such extremes are mainly connected to low and sometimes declining life expectancy, with many premature deaths in diseases and armed conflicts, such as certain periods in Afghanistan, Burundi and Sierra Leone (Espenshade et al., 2003, p. 577). Nevertheless, even with higher replacement fertility rates, it is in many of these countries that we also find the highest fertility rates, such as Niger with 6.86 children per woman, Somalia with 5.71 and the Democratic Republic of the Congo with 5.52 children per woman in 2022 (United Nations, 2019a).

Most of the already large populations in the world, with more than 100 million people, were approaching or had already passed their replacement fertility rates in 2022, such as China (1.66), India (2.2), the United States (1.89), Indonesia (2.22), Brazil (1.65), Bangladesh (1.95), Russia (1.79), Mexico (2.02) and Japan (1.53) (United Nations, 2019a). Left among such giants that are still growing due to their fertility rates are Pakistan (3.12), Ethiopia (3.56), Philippines (2.75) and Egypt (2.98), with Nigeria (5.11) and the Democratic Republic of the Congo (5.52) still growing at staggering paces. With 217 million inhabitants, Nigeria was already larger than Brazil in 2022 and is projected to outgrow Indonesia in 2039 and the United States in 2047 (United Nations, 2019a). Considering that it is almost exclusively in Africa that we will find high fertility rates in the future (Figure 6.5), it is also there that the largest part of our future population growth will happen.

The global population has more than tripled since 1950 (United Nations, 2019a), but the total economic purchasing power increased around 14 times during that same period (World Bank, 2022). The lion's share of this economic growth has been concentrated in urban areas (Bairoch, 1991), which, together with other factors, are driving people to move into cities and towns virtually everywhere in the world (Satterthwaite et al., 2009, pp. 15—16; Wang et al., 2022). Since 2007, most of the global population has been living in urban areas for the first time in human history, and the entire population growth for the rest of the century is forecasted to occur there (United Nations, 2019b).

Aside from the historical reason of military protection, cities and towns exist because of other advantages of concentrating human activity in smaller areas (O'Flaherty, 2005,

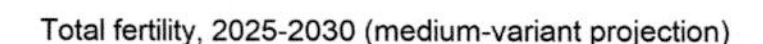

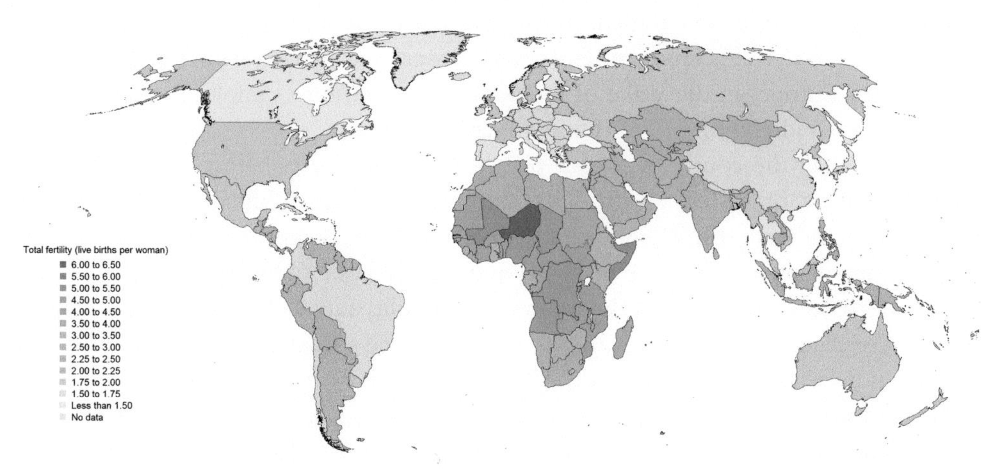

Figure 6.5 Map of projected fertility rates 2025–30. *(© 2019 UN, DESA, Population Division. Licensed under Creative Commons licence CC BY 3.0 IGO.)*

pp. 12–32), resulting in increased efficiency as it facilitates the finer division of labour and the flows of people, goods and ideas (Glaeser, 1998). However, concentrating human activity is also associated with sustainability challenges, further exacerbated by other ongoing change processes, such as climate change and globalisation.

The concentration of people and capital increases the vulnerability to and potential losses in disrupting events (Freeman, 2003, pp. 34–35; Quarantelli, 2003, p. 212). However, such concentration also can cause calamity in itself, as the complexity of society increases (Luhmann, 1998) and intensifies both the likelihood and consequence of critical failures and destructive events. This issue is dealt with in a separate section on increasing complexity below. Concentrating human activity also means concentrating the consumption of resources and increasing the ecological footprint (Rees & Wackernagel, 2008), which results in cities and towns appropriating the natural resources of vast and dispersed ecosystems (Folke et al., 1997). This connection and other rural-urban linkages make a distinct rural-urban divide unfeasible (see Tacoli, 2006). Even if old central cities often lend their names to metropolitan areas, they contain a shrinking proportion of the wealth and population in these wider regions (Fainstein & Campbell, 2002).

Although people have been moving into cities and towns since the first Mesopotamian cities thousands of years ago (Chapter 2), the emergence of urbanisation as a key process of change is connected to the more substantial demographic change starting with the industrial revolution. When looking at the world's ten largest cities in 1500, long before the start of the industrial revolution, only one was located in what we now call the Western world, while in 1900, only one was not (Table 6.1). However, as the industrial revolution spread, urbanisation reached more and more countries, and in 2020 none of the ten largest cities

Table 6.1 The top 10 cities of the world in terms of population over time. 1500—950 (Chandler, 1987) and 2020 (United Nations, 2019b).

Rank	1500	1800	1900	1950	2020
1	Beijing	Beijing	London	New York	Tokyo
2	Vijayanagar	London	New York	London	Delhi
3	Cairo	Guangzhou	Paris	Tokyo	Shanghai
4	Hangzhou	Tokyo	Berlin	Paris	São Paulo
5	Tabriz	Constantinople	Chicago	Shanghai	Mexico City
6	Constantinople	Paris	Vienna	Moscow	Dhaka
7	Gaur	Naples	Tokyo	Buenos Aires	Cairo
8	Paris	Hangzhou	St Petersburg	Chicago	Beijing
9	Guangzhou	Osaka	Manchester	Ruhr	Mumbai
10	Nanjing	Kyoto	Philadelphia	Kolkata	Osaka

was located in the Western world (Table 6.1). In 1930, New York was the only city with more than 10 million inhabitants (Ziv & Cox, 2007), often called a megacity, while Tokyo joined in 1950 and Mexico City in 1975 (United Nations, 2019b). Today we have 32 megacities worldwide, and several more are approaching 10 million inhabitants fast. It is also important to note the substantially higher urban growth rates in developing countries, resulting in only one in five megacities located in what is now considered the most developed countries (very high human development).

Urbanisation is not only drawing people into megacities but cities and towns of various sizes. What is clear is that the current level of urbanisation varies substantially depending on where you are in the world (Figure 6.6), from 14% of the population living in urban areas in Burundi to 100% in Singapore in 2022 (United Nations, 2019b). However, what is also clear is that urbanisation rates also vary much, from a 2.5% average

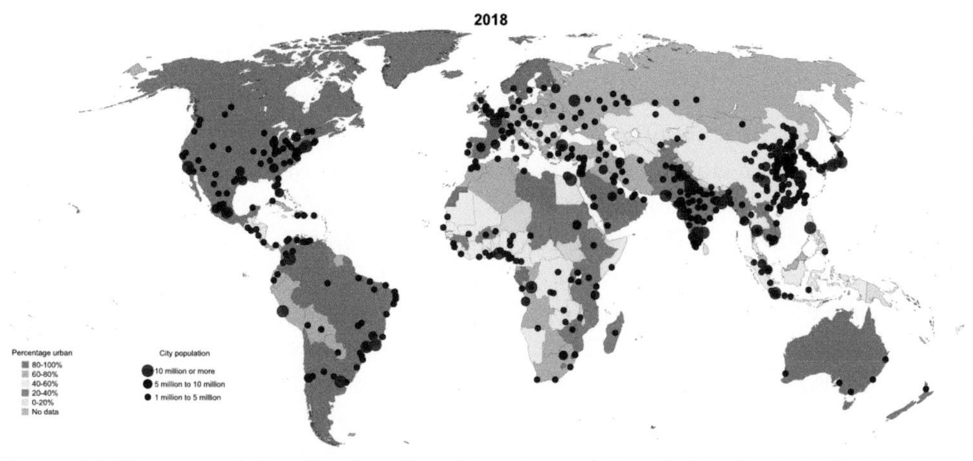

Figure 6.6 Urban population (% of total) and location and size of cities in 2018. *(© 2018 United Nations, DESA, Population Division. Licensed under Creative Commons licence CC BY 3.0 IGO.)*

annual rate of change in the percentage of the urban population in Burundi to −0.61% in Western Samoa between 2020 and 2025 (United Nations, 2019b). Very high average annual rates of change of the urban population of 4% or more can be found in 17 African countries and Syria, while a negative change is only occurring in seven Eastern European countries, Lebanon, Japan and five small island developing states or territories for the same period (United Nations, 2019b). 'Urbanisation uproots some of the people all the time, and all of the people some of the time, but it does not uproot all people all the time' (Gellner, 1989, p. 92).

Again, Africa is experiencing the most remarkable relative changes, although some of the least developed Asian countries are also in for the same ride. Generally, it is the same group of countries that experiences intense urbanisation that also experiences rapid population growth. This link might be common sense, but it may also be worrying. It may indicate that the traditional drivers of urbanisation—such as employment and economic opportunity—are relative and subjective notions in our globalised world. Relative in the sense of still being considered an opportunity—even if urban poverty has been highly underestimated (Satterthwaite, 1997) and rapidly increasing (Crush et al., 2012)—and subjective in the sense of the urban attraction being primarily built on the image of the city and not on actual circumstances.

In addition to population growth and urbanisation, there is a continuous escalation of the gap between rich and poor, both between and within countries. Gaps between rich and poor countries probably started to emerge already from the onset of early state formation. Yet, life expectancy was initially evenly distributed across the world and growing at a glacial pace from just over 20 years at the outset of the Neolithic Revolution to around 30 years in 1400 (Preston, 1995, p. 30; Simon, 1995, p. 8). Remember that Europe had been a relatively backward part of the world for a long time then (Wolf, 1997). However, the subsequent colonisation and the industrial revolution marked a sharp growth of socioeconomic gaps—first between the centres and the peripheries of empires and later between industrialised and agrarian countries.

In 1800, when the industrial revolution had just started to spread outside of the United Kingdom, most countries were still relatively similar in terms of life expectancy, with only marginal regional differences (Figure 6.7). Western economies had generally been starting to develop earlier, and it is important to note that most countries with the highest income per person were colonial powers. However, in 1809, Sweden had a lower life expectancy than what is now called the Democratic Republic of the Congo and Zimbabwe and lower income per person than South Africa and Uruguay. A century later, Sweden had the highest life expectancy in the world, with 58.5 years, while little had changed in the Democratic Republic of the Congo and Zimbabwe. South Africa and Uruguay had also seen economic development, but no life expectancy increase and Uruguay still had a higher income per person than Sweden. Socioeconomic development was, in other words, generally isolated to the Western world, accompanied by

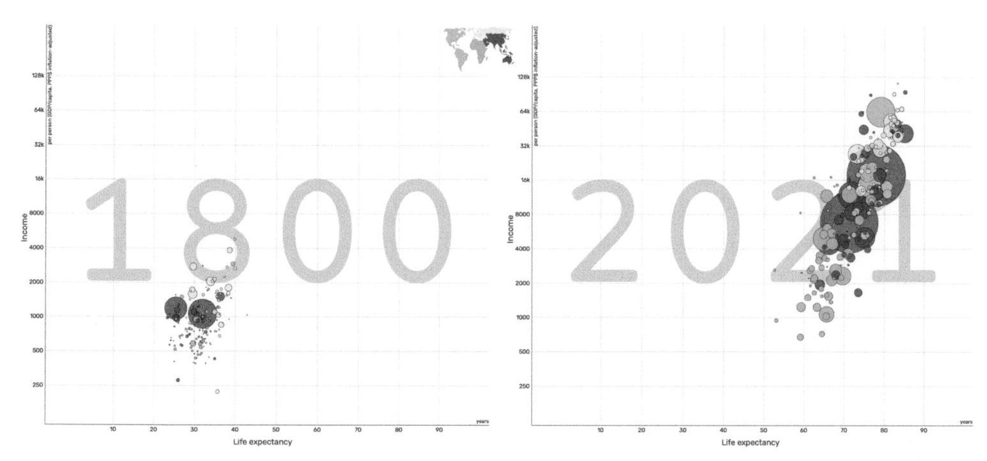

Figure 6.7 Income per person and life expectancy in 1800 and 2021, where each dot is a country, colour signifies region and size indicates population size. *(Adapted based on free material from www.gapminder.org.)*

some economic growth among its neighbours, that is, Latin America, North Africa and the Middle East.

Then in the 1920s, 1930s and 1940s, life expectancy started to increase more generally around the world, with the notable temporary exception of the countries ravaged by World War II. Economic development was still mainly isolated to the same countries as earlier, although there were exceptions, and Africa and Asia were largely left behind. The general increase in life expectancy continued, and then in the 1950s and 1960s, most countries in Latin America, North Africa and the Middle East experienced rapid development that has, in many ways, more or less closed the socioeconomic gap to Europe. East Asia and the Pacific started on a similar trajectory in the 1970s and 1980s, and South Asia in the 1990s, and most of the countries are rapidly closing the gap to all but the most developed countries (Figure 6.7). However, even if the last decades have seen positive trends for several African countries, Africa is still largely left behind, together with a handful of war-torn countries, such as Afghanistan, Yemen and Haiti, and isolated North Korea.

In short, development has not at all been fair since the start of the industrial revolution. Chad is still on a similar average income level as Sweden in 1810, Afghanistan as Sweden in 1850, Haiti as Sweden in 1880, and Armenia as Sweden in 1950. In 1800, the average income per person in the richest country was 21 times that of the poorest country, and both were European (the Netherlands and Romania). In 2021, the average income per person in the richest country was 168 times higher than that of the poorest country, with apparent regional differences.

However, these averages hide the actual extent of economic inequality, as the wealth within a country is rarely distributed equally among its population. For instance, the richest 1% of the people of the United States owned around 32% of all wealth at the end of 2021, while the bottom half of the population owned 2.6% (FRED, 2022), and 11.4% of the US population live officially in poverty (Shrider et al., 2021, p. 14). This share represents 37.2 million people, more than the total population in all but 38 countries worldwide. However, 58 countries have even more unequal income distribution than the United States. These are mainly African (28 countries) and Latin American (25 countries), but also include Turkmenistan, the Philippines, Iran, North Korea and Turkey (Gapminder, 2019).

The picture becomes even more shocking when looking at global inequality in individual wealth. It was estimated in early 2022 that the richest ten men had six times more wealth than the poorest 3.1 billion people (Ahmed et al., 2022). 3.1 billion people! To put that in perspective, let us visualise that we let these two groups of people stand in two lines next to each other, with equal space between each person. If the line with the ten men reaches across the endline of a basketball court, the other line reaches 116 times around the world at the equator.

Although the world has seen examples of unprecedented developmental leaps in recent decades, the gap between rich and poor continues to grow, both between and within countries. Gaps that Nobel Prize laureate Stiglitz (2012) and others point out may have detrimental consequences for our future. For instance, Stewart (2000, 2002) argues convincingly that inequalities between culturally defined groups, referred to as horizontal inequalities, provide a fertile setting for armed conflict. Scarcity has also been identified as a significant factor for the success of mythmaking and scapegoating that ultimately may stimulate violence (Armstrong, 2001, pp. 184–186; Gellner, 1997, Van Evera, 1994, pp. 26–33). Moreover, income inequality within countries has also been associated with various social issues, such as the proportion of people with mental illness, social trust, child well-being, homicide rates, imprisonment rates, educational attainment, social mobility and so on (Wilkinson, 2005; Wilkinson & Pickett, 2009).

Globalisation

Although some people have been moving to distant places for millennia—exploring, trading, migrating or conquering—the vast majority stayed put without much of a notion of the wider world. Today, the situation is entirely different. People generally still live locally, with only a small but growing part of the global population having the means to travel far. But everywhere you go, people are increasingly and in various ways directly or indirectly connected to each other and the world. The change process behind this transformation is called globalisation, and few corners of our world have not been affected by it (Figure 6.8).

Globalisation is characterised by four types of transformation (Held et al., 1999, p. 17). First, globalisation involves an extension of political, economic and social activities across

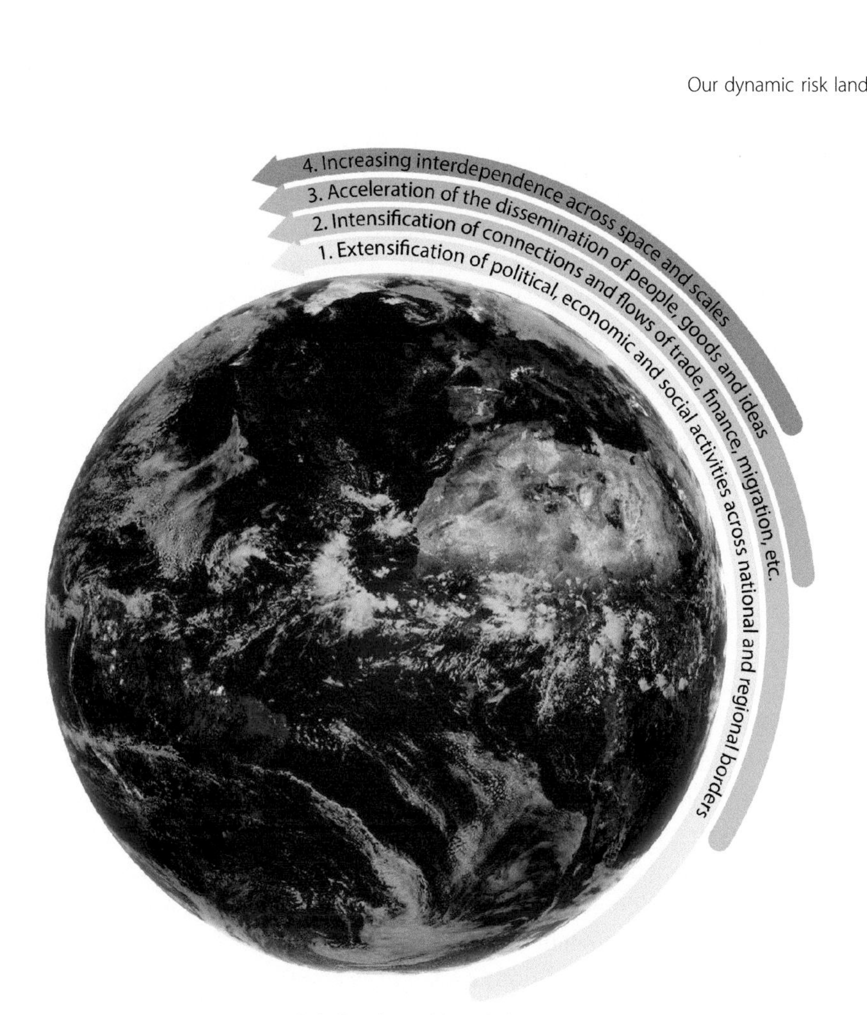

Figure 6.8 Our globalised world and the processes behind it.

national and regional borders. Second, globalisation is marked by the intensification of connections and flows of trade, finance, migration and so on. Third, globalisation is also characterised by the increasing speed of global interactions and processes, such as systems of transport and communication, which accelerate the dissemination of people, goods and ideas. Finally, the increasing extensity, intensity and velocity of global interactions increase our interdependence in such a way that 'the effects of distant events can be highly significant elsewhere and specific local developments can come to have considerable global consequences' (Held & McGrew, 2001, p. 324). In short, globalisation is a process of change that continuously extends, intensifies and accelerates our global interconnectedness in such a way that we are increasingly interdependent on each other.

When precisely globalisation started has been heavily debated for decades. Regardless of how far back its roots can be traced, globalisation emerged as an important change process with the industrial revolution and its innovations in transport and communication. Steamships and railroads reduced the time, cost and risk of travel and transportation, and telegraph lines started to crisscross even the widest oceans, spatially and temporally shrinking the world (Held et al., 1999; Ritzer & Dean, 2015). Today, we talk about modern society as a network society. Still, even if the modern state may be under threat in places—either by globalisation (Hettne et al., 1998, pp. 397–399) or because it is not yet firmly established everywhere—it is still the universal idea of sovereignty as well as the major power container of our contemporary world (Billig, 1995, pp. 175–176; Mann, 2013). Critics argue that globalisation is an exaggeration and several transformations traditionally believed to be connected to globalisation may actually strengthen some states (Mann, 2013). Although highly influential, globalisation should, in other words, not be overstated, such as the wholesale idea of the decreasing importance of territory or the idea that it is only a process towards homogeneity. Territory is still of utmost importance in the world (Agnew, 2009; Kohl, 2019), and globalisation appears to be quite the opposite of homogenisation, with new identities emerging through growing interaction among the people affected by it, in what Eriksen (1994) refers to as creolisation and Tomlinson (1999) calls hybridisation. Regardless of the name we assign to such cultural processes, ample empirical data support the perspective that globalisation is more about cultural creativity than uniformity if scratching a bit on the surface.

Even if the modern state may still be a force to reckon with, globalisation is increasing the importance of cities (Figure 6.9). These cities, in the metropolitan sense, are not only places where the state, market and civil society coexist but also hubs in networks of flows of people, capital, goods and services (Castells, 2010). Regardless if it is the hubs that run the networks (Sassen, 1995) or the networks that generate the hubs (Castells, 2010), cities depend on effective and efficient flows between as well as within them, constantly, regardless of disturbance or disruption, now and in the future. If a city, by any circumstance, is disconnected, it initially causes problems for others in the network, but quite

Figure 6.9 Tokyo—a city with roughly the same GDP as South Korea, which is only surpassed by around nine countries worldwide. *(Photo by Kakidai, shared on the Creative Commons.)*

rapidly, it is simply bypassed with staggering local consequences (Castells, 2010, p. 147). Hence, it is vital to ensure that a city stays connected to national and global networks and maintains effective and efficient flows of people, capital, goods and services within the city itself. These cities comprise not only the public organisations often borrowing their names but all other actors partaking in the overall functioning of society, representing the public and private spheres, as well as civil society.

Globalisation has indeed positive effects, such as accelerated research and development, capital investment and technological progress. However, only a limited group of affluent and upcoming countries are fully experiencing them, as there are many losers too (e.g., Hashai & Buckley, 2021; Yusuf, 2003). However, its adverse effects are much more universally felt in the impact of changes, trends, disturbances, disruptions and disasters that spread wider and further than ever before. For instance, the 2010 volcano eruption of Eyjafjallajökull in Iceland was just a fraction of its 1821–23 eruption (Sigmundsson et al., 2010). Yet, the impact on our globalised world reached as far as Kenyan flower pickers losing their jobs as the flowers could not be flown through the volcanic ash into Europe for sale (Harcourt, 2010). Food prices in Africa are today largely determined by distant factors, such as the increasing demand for meat from the rapidly growing Chinese middle class (Abbott & Borot de Battisti, 2011) or the 2022 Russian attack on Ukraine (Osendarp et al., 2022). The global price of specific technological components spikes when local events hit particular areas, like the 2011 floods in Thailand that decimated a significant proportion of the global hard drive production there (Goodman & Polycarpou, 2013). Also, local measures to manage the COVID-19 pandemic had global consequences on many supply chains (Guan et al., 2020). We are increasingly connected, for better and for worse.

Increasing complexity

In addition to the increasing complexity of our globalised world, most domains of modern society witness technological development at a staggering speed, growing scale of industrial installations, increasing degree of integration of various systems, and a progressively more aggressive and competitive environment (Rasmussen & Svedung, 2000, p. 10). In this dynamic context, multiple actors from the state, market and civil society converge to develop and ensure the daily functioning of society. This functioning of society depends largely on effective and efficient flows of people, capital, goods and services within and between cities, regions and countries.

There are ample examples of recent events which, in various ways, have affected different critical flows and, thus, the functioning of society, for example, floods, terrorist threat, pandemics, droughts and severe winter conditions. In addition to these dramatic events, countless examples of more mundane incidents also impede sustainable development by disturbing and disrupting critical flows and functions every day, for example, recurring gridlock traffic and frequent power cuts. Regardless of type or scale, it is increasingly difficult for actors to overview and manage risk and the consequences of actual events and decisions since their effects can spread rapidly and widely throughout society (Rasmussen & Svedung, 2000, p. 10). At the heart of this increasing challenge for sustainability lay, in other words, dependencies through which effects cascade (Little, 2002). Consequently, the risk of disturbances or disruptions of one critical flow is contingent on the risk of disturbances or disruptions in all other flows it depends upon (Box 6.1).

Understanding and managing the complexity of society entails understanding and managing dependencies. A dependency is a connection between two entities in society (flows, sectors, infrastructures, actors, etc.) through which the state of one influences the state of the other. These dependencies do not only allow the effects of an unwanted event to cascade throughout society (Little, 2002). They also transmit the effects of human decisions and actions (Rinaldi et al., 2001), for good and bad, making it difficult to foresee the actual effects of our policies and practice.

It has been suggested for decades that society is getting increasingly complex (Perrow, 1991). This process is nothing that has happened overnight, nor will it be in the future. It is a gradual, continuous process increasing both the number and intensity of dependencies in society. Such slowly evolving increase of complexity—sometimes called creeping dependencies (Hills, 2005)—accumulate and eventually reach a threshold over which we lose overview and much of our ability to maintain our critical flows and societal functions (Box 6.2). Growing complexity also increases the likelihood that two or more failures interact in ways that are difficult to anticipate, as well as increases their consequences as they cascade wider and further throughout society (Perrow, 2008, p. 165).

One of the changes that drive these creeping dependencies is the process towards increasing effectiveness and efficiency. Such a drive for optimisation has been vital for the development of modern society, as its positive effect of increasing cost-effectiveness frees up resources to aim even higher or to utilise for other important things. However, optimisation implies the exploitation of the advantages of operating at the fringes of conventional practice, approaching the boundaries of safety and sustainability (Rasmussen & Svedung, 2000, p. 14). In addition to the risk of overstepping these

BOX 6.1 Dependable space

The loss of the Galaxy IV satellite in 1998 disrupted around 90% of all pagers in the United States, including communications within health care, and affected credit card purchases and ATM transactions all over the country (Rinaldi et al., 2001).

BOX 6.2 Fire in the hole

The 2002 Akalla tunnel fire in Stockholm, Sweden, likely started in a faulty cable splice on a high-voltage electric power cable. The fire destroyed several high-voltage electric power cables and many central telecommunications and IT cables, causing around 50,000 people and companies with a total of 30,000 employees to lose electric power. The fire also caused parts of the roof to collapse, complicating the ensuing response and recovery activities. Most customers had to wait more than two days for the electricity to return.

On top of the inconvenience for modern households to get by without electricity for that long, the resulting costs and potential danger of the event were substantial. Parts of the Metro system stopped working, demanding the complicated evacuation of passengers through tunnels and increasing the traffic on the streets above. Traffic lights and streetlights were not functioning, causing traffic flow to slow considerably. Telephone landlines, mobile phone networks and IT networks stopped working. Elevators had to be manually evacuated, and the increasingly common code locks were generally made useless. Water and sewage pumps stopped, and failing refrigerators and freezers caused foodstuff to spoil. In other words, most authorities and companies in and around the affected area had to disband normal activities, and several critical societal functions stopped working for a substantial period (Deverell, 2004).

boundaries, optimisation has the downside of increasing vulnerability by reducing buffers that could be used to maintain critical flows or functions during disturbances. Hence, optimisation means increased efficiency in everyday circumstances but also increased vulnerability to disturbances (Carlson & Doyle, 2000), as steadily fewer buffers make smaller and smaller disturbances potentially leading to disruptions of entire systems. It is, in other words, no coincidence that the concept of 'just-in-time' or 'lean production', applied in almost every segment of modern society, originally was called 'fragile production' (Oliver & Hunter, 1998, p. 90).

At the same time as the ever-increasing push for optimisation, there is also an ongoing process of diversification of actors responsible for maintaining and developing the most critical flows and functions in society. These two processes are closely related because the arguments for allowing more actors to be involved often focus on the expected increase in cost-effectiveness through competition. This process is generally called institutional fragmentation (De Bruijne & Van Eeten, 2007; Trein & Ansell, 2021) and increases complexity further as it adds dependencies between multiple actors, posing many new challenges for safety and sustainability (De Bruijne & Van Eeten, 2007).

Another closely related process is the increasingly aggressive and competitive environment in which most actors operate. This environment is a result of the processes of optimisation and institutional fragmentation. It has the effect of focusing the incentives of decision-makers on short-term financial gain rather than on safety and sustainability (Rasmussen & Svedung, 2000). Under these pressures of cost-effectiveness and competition, the defined boundaries of safety and sustainability are increasingly approached, making actors accustomed to a performance that earlier was considered risky and effectively recalibrating the boundaries of safety and sustainability (Kirwan, 2011, pp. 16—17). This process is not confined to the market environment of the private sector. It also influences the public sector due to increasing service demands and reduced resource allocations. Although eroding margins are normal as human beings gain experience, the challenge lies in knowing when we have gone too far (Kirwan, 2011).

The slow incremental increase of complexity through creeping dependencies, driven by the processes of optimisation, institutional fragmentation and an increasingly aggressive and competitive environment, make it progressively more difficult for actors to grasp and manage risk to their critical flows and functions. Without acknowledging these processes and their consequences, there is a grave risk that we allow society to slowly drift into danger.

Changing antagonistic threats

The last key process of change I want to mention briefly is the changing nature of antagonistic threats. The intense wars between states of our past have been increasingly transformed into low-intensity conflicts between more varied types of parties (UCDP, 2022). From being generally both spatially and temporally bounded, with notable exceptions such as the Napoleonic Wars and the two World Wars, modern conflicts are increasingly extending beyond borders and over time (Figure 6.10)—enough to suggest a distinction between old and new wars (Kaldor, 1999), even if interstate armed conflicts may be on the rise again with Russia's invasion of Ukraine and tensions brewing elsewhere (e.g., between Armenia and Azerbaijan, Kyrgyzstan and Tajikistan, China and Taiwan, etc.). Humankind still has the capacity to annihilate ourselves and the world many times, and weapons of mass destruction are proliferating again (Figure 6.11). Even a relatively localised nuclear war could have disastrous global consequences.

In addition to the transforming face of armed conflict, antagonistic threats are also changing in relation to various forms of terrorism, sometimes referred to as increasing asymmetrical threats (Kegley, 2003). The more geographically bounded organisations of the past, using terrorist activities to fight for a localised cause, have got company in loose global networks using terror to fight for global systemic change. Terrorism and warfare are also increasingly spreading to the digital dimension of our modern world, with cyber security now on everybody's lips. The distinction between armed conflict and the fight against terrorism is increasingly often blurred, frequently leading to what can only be referred to as attempts to crack a nut with a sledgehammer. The result is severe collateral damage, which could, at least in my mind, have counterproductive effects as injustice and suffering may further spur support and involvement in terrorist organisations. The blurring of warfare and the fight against terrorism is also, at times, undermining

Figure 6.10 Terrorism protection in Maastricht. *(Kim Willems | Dreamstime.com.)*

Figure 6.11 There are no winners in a nuclear war. *(Street art by Klisterpeter, published with permission.)*

international law, such as the detention of individuals without trial or increasingly frequent drone attacks against targets within the borders of sovereign states.

Crime is also changing, from individual or small bands of thieves or bandits operating in particular geographical areas, to increasingly organised and international crime cartels. Although organised crime has been around for centuries, with examples such as the Chinese Triads or Italian Mafia, what is changing is the increasing geographic spread and reach of such organisations (Mallory, 2012). Today we have various forms of organised crime cartels on all continents—many with regional and even global tentacles reaching and competing with each other. Well organised, well equipped and with what is seemingly the upper hand against our law enforcing agencies.

Conclusion

Managing risk for sustainable development is complicated by ongoing processes of change that continuously transform our risk landscape. Our future can no longer be expected to be an extrapolation of our past, as our world is increasingly complex and dynamic. It demands us to integrate multiple and sometimes escalating change processes in our assessments of our future. We know what we need to do and are increasingly figuring out how to do it. However, we also need to get over the old truth so eloquently described by George Orwell: 'People can foresee the future only when it coincides with their own wishes, and the most grossly obvious facts can be ignored when they are unwelcome' (Orwell, 1945). The transformations needed for the sustainability of humankind are unfortunately unwelcome for many, who like to fly on weekend trips to lovely cities elsewhere, have their large steaks, and drive around in their large cars. So, it is high time to wake up and realise that we are on a dangerous path, and we cannot afford to be delayed by climate deniers and others who are sowing doubt in our minds about uncertainty in the expectations of our future. The future will always be uncertain, but the best thing we can do is act the best way we can with the best information we have at hand. We now have more or less unanimous information pointing towards a not-at-all-desirable future if we continue as we have done in the past. We must act on it. Now.

References

Abbott, P., & Borot de Battisti, A. (2011). Recent global food price shocks: Causes, consequences and lessons for African Governments and donors. *Journal of African Economies, 20*, i12–i62. https://doi.org/10.1093/jae/ejr007

Agnew, J. A. (2009). *Globalization and sovereignty*. Rowman & Littlefield Publishers.

Ahmed, N., Marriott, A., Dabi, N., Lowthers, M., Lawson, M., & Mugehera, L. (2022). *Inequality kills: The unparalleled action needed to combat unprecedented inequality in the wake of COVID-19*. Oxfam. https://doi.org/10.21201/2022.8465

Armstrong, J. A. (2001). Postcommunism and nationalism. In M. Guibernau, & J. Hutchinson (Eds.), *Understanding nationalism* (pp. 182–206). Polity Press.

Bairoch, P. (1991). *Cities and economic development: From the dawn of history to the present*. University of Chicago Press.
Billig, M. (1995). *Banal nationalism*. Sage.
Boardman, J., Shepheard, M. L., Walker, E., & Foster, I. D. L. (2009). Soil erosion and risk-assessment for on- and off-farm impacts: A test case using the Midhurst area, West Sussex, UK. *Journal of Environmental Management, 90*, 2578–2588. https://doi.org/10.1016/j.jenvman.2009.01.018
Borrelli, P., Robinson, D. A., Panagos, P., Lugato, E., Yang, J. E., Alewell, C., Wuepper, D., Montanarella, L., & Ballabio, C. (2020). Land use and climate change impacts on global soil erosion by water (2015-2070). *Proceedings of the National Academy of Sciences of the U S A, 117*(36), 21994–22001. https://doi.org/10.1073/pnas.2001403117
Burke, E. J., Brown, S. J., & Christidis, N. (2006). Modeling the recent evolution of global drought and projections for the twenty-first century with the Hadley Centre climate model. *Journal of Hydrometeorology, 7*, 1113–1125.
Carlson, J. M., & Doyle, J. (2000). Highly optimized tolerance: Robustness and design in complex systems. *Physical Review Letters, 84*, 2529–2532.
Castells, M. (2010). *The rise of the network society*. Wiley-Blackwell.
Chandler, T. (1987). *Four thousand years of urban growth: An historical census*. St. David's University Press.
Conway, E. (2010). *NASA. Is Antarctica melting*.
Costello, A., Abbas, M., Allen, A., Ball, S., Bell, S., Bellamy, R., Friel, S., Groce, N., Johnson, A., Kett, M., Lee, M., Levy, C., Maslin, M. A., McCoy, D., McGuire, B., Montgomery, H., Napier, D., Pagel, C., Patel, J., ... Patterson, C. (2009). Managing the health effects of climate change. *Lancet, 373*, 1693–1733.
Crush, J., Frayne, B., & Pendleton, W. (2012). The crisis of food insecurity in African cities. *Journal of Hunger & Environmental Nutrition, 7*, 271–292. https://doi.org/10.1080/19320248.2012.702448
De Bruijne, M., & Van Eeten, M. (2007). Systems that should have failed: Critical infrastructure protection in an institutionally fragmented environment. *Journal of Contingencies and Crisis Management, 15*, 18–29.
Deverell, E. (2004). *Elavbrottet i Kista den 29–31 maj 2002: Organisatorisk och interorganisatorisk inlärning i kris*.
Dunne, J. P., Stouffer, R. J., & John, J. G. (2013). Reductions in labour capacity from heat stress under climate warming. *Nature Climate Change, 3*, 563–566. https://doi.org/10.1038/NCLIMATE1827
Elsner, J. B., Kossin, J. P., & Jagger, T. H. (2008). The increasing intensity of the strongest tropical cyclones. *Nature, 455*, 92–95. https://doi.org/10.1038/nature07234
Eriksen, T. H. (1994). *Kulturelle veikryss: Essays om kreolisering*. Universitetsforlaget.
Eriksson, A., Betti, L., Friend, A. D., Lycett, S. J., Singarayer, J. S., von Cramon-Taubadel, N., Valdes, P. J., Balloux, F., & Manica, A. (2012). Late Pleistocene climate change and the global expansion of anatomically modern humans. *Proceedings of the National Academy of Sciences, 109*, 16089–16094. https://doi.org/10.1073/pnas.1209494109
Espenshade, T. J., Guzman, J. C., & Westoff, C. F. (2003). The surprising global variation in replacement fertility. *Population Research and Policy Review, 22*, 575–583.
Fainstein, S. S., & Campbell, S. (2002). Theories of urban development and their implications for policy and planning. In S. S. Fainstein, & S. Campbell (Eds.), *Readings in urban theory* (pp. 1–15). Blackwell.
Folke, C., Jansson, Å., Larsson, J., & Costanza, R. (1997). Ecosystem appropriation by cities. *AMBIO: A Journal of the Human Environment, 26*, 167–172.
FRED. (2022). *Distribution of total net worth in USA*. https://fred.stlouisfed.org/graph/?id=WFRBST01134,WFRBSN09161,WFRBSN40188,WFRBSB50215,#.
Freeman, P. K. (2003). Natural hazard risk and privatization. In A. Kreimer, M. Arnold, & A. Carlin (Eds.), *Building safer cities: The future of disaster risk* (pp. 33–44). The World Bank.
Gapminder. (2019). *Gini coefficient*. www.gapminder.org/data.
Gellner, E. (1989). *Plough, sword and book: The structure of human history*. University of Chicago Press.
Gellner, E. (1997). *Nationalism*. Phoenix.
Glaeser, E. L. (1998). Are cities dying. *The Journal of Economic Perspectives, 12*, 139–160.
Gombert, P., Charmoille, A., Christophe, D., & D'hotelans, R. (2010). Impact of the expected climate change on the stability of underground cavities in France. *Advances in Research in Karst Media*, 521–526.

Goodman, A., & Polycarpou, L. (2013). The sustainability-social networking Nexus. *Sustainability, 6*, 26–32.
Guan, D., Wang, D., Hallegatte, S., Davis, S. J., Huo, J., Li, S., Bai, Y., Lei, T., Xue, Q., Coffman, D., Cheng, D., Chen, P., Liang, X., Xu, B., Lu, X., Wang, S., Hubacek, K., & Gong, P. (2020). Global supply-chain effects of COVID-19 control measures. *Nature Human Behaviour, 4*(6), 577–587. https://doi.org/10.1038/s41562-020-0896-8
Hallegatte, S., Ranger, N., Mestre, O., Dumas, P., Corfee-Morlot, J., Herweijer, C., & Wood, R. M. (2011). Assessing climate change impacts, sea level rise and storm surge risk in port cities: A case study on Copenhagen. *Climatic Change, 104*, 113–137. https://doi.org/10.1007/s10584-010-9978-3?LI=-true#page-1
Harcourt, W. (2010). Where did all the flowers go?: Contradictions in world economies. *Development, 53*, 301–303.
Hashai, N., & Buckley, P. J. (2021). The effect of within-country inequality on international trade and investment agreements. *International Business Review, 30*(6), 101862. https://doi.org/10.1016/j.ibusrev.2021.101862
Held, D., & McGrew, A. (2001). Globalization. In J. Krieger, & M. E. Crahan (Eds.), *The Oxford companion to politics of the world* (pp. 324–327). Oxford University Press.
Held, D., McGrew, A., Goldblatt, D., & Perraton, J. (1999). *Global transformations: Politics, economics and culture*. Stanford University Press.
Hettne, B., Sörlin, S., & Östergård, U. (1998). *Den globala nationalismen: Nationalstatens historia och framtid*. SNS Förlag.
Hills, A. (2005). Insidious environments: Creeping dependencies and urban vulnerabilities. *Journal of Contingencies and Crisis Management, 13*, 12–20. https://doi.org/10.1111/j.0966-0879.2005.00450.x
Hoegh-Guldberg, O., Jacob, D., Taylor, M., Guillén Bolaños, T., Bindi, M., Brown, S., Camilloni, I. A., Diedhiou, A., Djalante, R., Ebi, K., Engelbrecht, F., Guiot, J., Hijioka, Y., Mehrotra, S., Hope, C. W., Payne, A. J., Pörtner, H. O., Seneviratne, S. I., Thomas, A., … Zhou, G. (2019). The human imperative of stabilizing global climate change at 1.5 °C. *Science, 365*(6459), eaaw6974. https://doi.org/10.1126/science.aaw6974
HPTCC. (2004). *A more secure world: Our shared responsibility*.
IPCC. (2007). *Climate change 2007: The physical science basis*. Cambridge University Press.
IPCC. (2021). *Climate change 2021: The physical science basis*. Cambridge University Press.
IPCC. (2022). *Climate change 2022: Impacts, adaptation and vulnerability*. Cambridge University Press.
Jakob, M., & Lambert, S. (2009). Climate change effects on landslides along the southwest coast of British Columbia. *Geomorphology, 107*, 275–284. https://doi.org/10.1016/j.geomorph.2008.12.009
Johnson, D. L., Ambrose, S. H., Bassett, T. J., Bowen, M. L., Crummey, D. E., Isaacson, J. S., Johnson, D. N., Lamb, P., Saul, M., & Winter-Nelson, A. E. (1997). Meanings of environmental terms. *Journal of Environmental Quality, 26*, 581–589. https://doi.org/10.2134/jeq1997.00472425002600030002x
Kaldor, M. (1999). *New and old wars: Organized violence in a global era*. Polity Press.
Kegley, C. W. (2003). *The new global terrorism: Characteristics, causes, controls*. Prentice-Hall.
Kirwan, B. (2011). Incident reduction and risk migration. *Safety Science, 49*, 11–20.
Knutson, T., Camargo, S. J., Chan, J. C. L., Emanuel, K., Ho, C.-H., Kossin, J., Mohapatra, M., Satoh, M., Sugi, M., Walsh, K., & Wu, L. (2020). Tropical cyclones and climate change assessment: Part II: Projected response to Anthropogenic warming. *Bulletin of the American Meteorological Society, 101*(3), E303–E322. https://doi.org/10.1175/bams-d-18-0194.1
Knutson, T. R., McBride, J. L., Chan, J. C. L., Emanuel, K., Holland, G. J., Landsea, C., Held, I. M., Kossin, J. P., Srivastava, A. K., & Sugi, M. (2010). Tropical cyclones and climate change. *Nature Geoscience, 3*, 157–163. https://doi.org/10.1038/ngeo779
Kohl, U. (2019). Territoriality and globalization. In S. Allen, D. Costelloe, M. Fitzmaurice, P. Gragl, & E. Guntrip (Eds.), *The Oxford handbook of jurisdiction in international law* (pp. 299–329). Oxford University Press. https://doi.org/10.1093/law/9780198786146.003.0013
Kremer, M. (1993). Population growth and technological change: One million B.C. To 1990. *Quarterly Journal of Economics, 108*, 681–716. https://doi.org/10.2307/2118405

Kundzewicz, Z. W., Hirabayashi, Y., & Kanae, S. (2010). River floods in the changing climate—observations and projections. *Water Resources Management, 24*, 2633—2646. https://doi.org/10.1007/s11269-009-9571-6?LI=true#page-1

Lacis, A. A., Schmidt, G. A., Rind, D., & Ruedy, R. A. (2010). Atmospheric CO2: Principal control knob governing Earth's temperature. *Science, 330*, 356—359. https://doi.org/10.1126/science.1190653

Lewis, J. (1990). The vulnerability of small island states to sea level rise: The need for holistic strategies. *Disasters, 14*, 241—249. https://doi.org/10.1111/j.1467-7717.1990.tb01066.x

Little, R. G. (2002). Controlling cascading failure: Understanding the vulnerabilities of interconnected infrastructures. *Journal of Urban Technology, 9*, 109—123.

Luhmann, N. (1998). *Observations on modernity*. Stanford University Press.

Mallory, S. L. (2012). *Understanding organized crime*. Jones & Bartlett.

Malthus, T. (1798). An essay on the principle of population. J. Johnson.

Mann, M. (2013). *The sources of social power: Volume 4, Globalizations 1945—2011*. Cambridge University Press.

Marotzke, J., Milinski, S., & Jones, C. D. (2022). How close are we to 1.5 °C or 2 °C of global warming. *Weather, 77*(4), 147—148. https://doi.org/10.1002/wea.4174

McGranahan, G., Balk, D., & Anderson, B. (2007). The rising tide: Assessing the risks of climate change and human settlements in low elevation coastal zones. *Environment and Urbanization, 19*, 17—37. https://doi.org/10.1177/0956247807076960

Nordhaus, W. D. (2006). *The economics of Hurricanes in the United States* (p. W12813). NBER Working Paper Series.

O'Flaherty, B. (2005). *City economics*. Harvard University Press.

Oliver, N., & Hunter, G. (1998). The financial impact of 'Japanese' manufacturing methods. In R. Delbridge, & J. Lowe (Eds.), *Manufacturing in transition*. Routledge.

Orwell, G. (1945). London letter. *Partisan Review, 12*, 77—82.

Osendarp, S., Verburg, G., Bhutta, Z., Black, R. E., de Pee, S., Fabrizio, C., Headey, D., Heidkamp, R., Laborde, D., & Ruel, M. T. (2022). Act now before Ukraine war plunges millions into malnutrition. *Nature, 604*(7907), 620—624. https://doi.org/10.1038/d41586-022-01076-5

Perrow, C. B. (1991). A society of organizations. *Theory and Society, 20*, 725—762. https://doi.org/10.1007/BF00678095

Perrow, C. B. (2008). Complexity, catastrophe, and modularity. *Sociological Inquiry, 78*, 162—173.

Preston, S. H. (1995). Human mortality throughout history and prehistory. In J. L. Simon (Ed.), *The state of humanity* (pp. 30—36). Blackwell.

Quarantelli, E. L. (2003). Urban vulnerability to disasters in developing countries: Managing risks. In A. Kreimer, M. Arnold, & A. Carlin (Eds.), *Building safer cities: The future of disaster risk* (pp. 33—44). The World Bank.

Rahmstorf, S. (2012). Modeling sea level rise. *Nature Education Knowledge, 3*, 4.

Rasmussen, J., & Svedung, I. (2000). *Proactive risk management in a dynamic society*. SRSA.

Raworth, K. (2012). *A safe and just space for humanity: Can we live within the Doughnut.*

Rees, W. E., & Wackernagel, M. (2008). Urban ecological footprints: Why cities cannot be sustainable—and why they are a key to sustainability. *Urban Ecology*, 537—555. https://doi.org/10.1007/978-0-387-73412-5_35

Rinaldi, S. M., Peerenboom, J. P., & Kelly, T. K. (2001). Identifying, understanding, and analyzing critical infrastructure interdependencies. *IEEE Control Systems Magazine, 21*, 11—25.

Ritzer, G., & Dean, P. (2015). *Globalization: A basic text* (2 ed.). John Wiley & Sons.

Sassen, S. (1995). On concentration and centrality in the global city. In P. L. Knox, & P. J. Taylor (Eds.), *World cities in a world-system* (pp. 63—78). Cambridge University Press.

Satterthwaite, D. (1997). Urban poverty: Reconsidering its scale and nature. *IDS Bulletin, 28*, 9—23. https://doi.org/10.1111/j.1759-5436.1997.mp28002002.x

Satterthwaite, D., Huq, S., Reid, H., Pelling, M., & Romero Lankao, P. (2009). In J. Bicknell, D. Dodman, & D. Satterthwaite (Eds.), *Adapting to climate change in urban areas: The possibilities and constraints in low- and middle-income Nations* (pp. 3—47). Earthscan.

Shaver, P. (2011). How did life evolve. In P. Shaver (Ed.), *Cosmic heritage: Evolution from the big bang to conscious life* (pp. 117–136). Springer. http://link.springer.com/10.1007/978-3-642-20261-2_12.
Shrider, E. A., Kollar, M., Chen, F., & Semega, J. (2021). *Income and poverty in the United States: 2020*. US Census Bureau.
Sigmundsson, F., Hreinsdóttir, S., Hooper, A., Árnadóttir, T., Pedersen, R., Roberts, M. J., Óskarsson, N., Auriac, A., Decriem, J., Einarsson, P., Geirsson, H., Hensch, M., Ófeigsson, B. G., Sturkell, E., Sveinbjörnsson, H., & Feigl, K. L. (2010). Intrusion triggering of the 2010 Eyjafjallajökull explosive eruption. *Nature, 468*, 426–430. https://doi.org/10.1038/nature09558
Simon, J. L. (1995). Introduction. In J. L. Simon (Ed.), *The state of humanity* (pp. 1–28). Blackwell.
Small, C., & Nicholls, R. J. (2003). A global analysis of human settlement in coastal zones. *Journal of Coastal Research, 19*, 584–599.
Stewart, F. (2000). The root causes of humanitarian emergencies. In E. W. Nafziger, F. Stewart, & R. Väyrynen (Eds.), *War, hunger and displacements: The origins of humanitarian emergencies*. Oxford University Press.
Stewart, F. (2002). *Horizontal inequalities: A neglected dimension of development* (Vol. 81, pp. 1–40). Queen Elizabeth House Working Paper.
Stiglitz, J. E. (2012). *The price of inequality*. W.W. Norton & Company.
von Storch, H., & Woth, K. (2008). Storm surges: Perspectives and options. *Sustainability Science, 3*, 33–43. https://doi.org/10.1007/s11625-008-0044-2
Tacoli, C. (Ed.). (2006). *The Earthscan reader in rural-urban linkages*. Earthscan.
Tauger, M. B. (2021). *Agriculture in world history*. Routledge. http://books.google.se/books?id=o_WSzQEACAAJ&hl=&source=gbs_api.
Temmerman, S., Vries, M. B. D., & Bouma, T. J. (2012). Coastal marsh die-off and reduced attenuation of coastal floods: A model analysis. *Global and Planetary Change, 92–93*, 267–274. https://doi.org/10.1016/j.gloplacha.2012.06.001
Tomlinson, J. (1999). *Globalization and culture*. University of Chicago Press.
Trein, P., & Ansell, C. K. (2021). Countering fragmentation, taking back the state, or partisan agenda-setting? Explaining policy integration and administrative coordination reforms. *Governance, 34*(4), 1143–1166. https://doi.org/10.1111/gove.12550
UCDP. (2022). *Uppsala conflict data program*. https://ucdp.uu.se.
United Nations. (2019a). *World population prospects: The 2019 revision*. https://doi.org/10.1017/CBO9781107415324.004
United Nations. (2019b). *World urbanization prospects: The 2018 revision. United Nations.*
Van Evera, S. (1994). Hypotheses on nationalism and war. *International Security, 18*, 5–39.
Wang, Q., Wang, X., & Li, R. (2022). Does urbanization redefine the environmental Kuznets curve? An empirical analysis of 134 countries. *Sustainable Cities and Society, 76*, 103382. https://doi.org/10.1016/j.scs.2021.103382
Webster, P. J., Holland, G. J., Curry, J. A., & Chang, H.-R. (2005). Changes in tropical cyclone number, duration, and intensity in a warming environment. *Science, 309*, 1844–1846. https://doi.org/10.1126/science.1116448
Wilkinson, R. (2005). *The impact of inequality: How to make sick societies healthier*. Routledge.
Wilkinson, R., & Pickett, K. (2009). *The spirit level: Why greater equality makes societies stronger*. Bloomsbury Press.
Wolf, E. (1997). *Europe and the people without history*. University of California Press.
World Bank. (2022). *World development indicators database*. https://data.worldbank.org.
Yusuf, S. (2003). Globalisation and the challenge for developing countries. *Journal of African Economics, 12*, 35–72.
Zhang, G. J., Cai, M., & Hu, A. (2013). Energy consumption and the unexplained winter warming over northern Asia and North America. *Nature Climate Change*.
Ziv, J.-C., & Cox, W. (2007). *Megacities and affluence: Transport & land use considerations*.

PART II

Approaching the world

CHAPTER 7

Conceptual frames for risk, resilience and sustainable development

Introduction

With the current state of the world on our minds, it is time to lay the foundation for understanding and addressing our sustainability challenges. Our future is uncertain, and we must find ways to manage this uncertainty while steering the development of communities and societies towards sustainability. However, our world is also complex, ambiguous and dynamic, rendering much of our efforts ineffective at best and counterproductive at worst. Sustainability Science has much to offer to grasp the fundamental interactions between nature and society—between us, human beings, and our environment (Kates et al., 2001, p. 642). I believe Immanuel Kant (1968, p. 75) was correct when stating that '[t]houghts without content are empty, ideas without concepts are blind' (my translation from German). Therefore, I am focussing this chapter on presenting conceptual frames for approaching our world, drawing on Sustainability Science in addressing the challenges of uncertainty, complexity, ambiguity and the dynamic character of our world.

Before presenting a range of key concepts, which are all central to managing risk and resilience for sustainable development, I feel obliged to present the fundamental philosophical assumptions on which they rest—building a coherent framework for understanding and addressing our world that can bridge the various functional sectors and academic disciplines necessary for building a safe and sustainable society. The chapter introduces a comprehensive approach to the concepts of development, sustainability, risk, value, hazard, vulnerability, capacity and resilience and lays the foundation for managing risk in this context.

Philosophical assumptions about our world

Mentioning philosophical assumptions in a book like this may cause the eyes of both students and practitioners to glaze over. Even researchers from some scientific disciplines may feel somewhat uncomfortable, as there are many diverse ways to approach our world. However, as I have pointed out earlier, managing risk and resilience for sustainable development are intrinsically transdisciplinary issues that demand a common foundation. A foundation that facilitates the contribution of a wide range of functional sectors and scientific disciplines into describing and addressing the core challenges of humankind, which in turn demands clarity in four central philosophical assumptions.

Sustainability Science, Second Edition
ISBN 978-0-323-95640-6, https://doi.org/10.1016/B978-0-323-95640-6.00011-7

What constitutes our world?

The world is dynamic and complex (Dewey, 1922; Kates et al., 1990). Although parts of the world are determined by processes over which human activity has little influence—for example, tidal cycles or the movement of tectonic plates—human activity has increasingly become the most important determining factor of our future (Simon, 1996, pp. 2–3; Steffen et al., 2015). However, regardless of what our world is determined by, this book rests on the assumption that the world does exist whether I am around to observe it or not (Blaikie, 1991, p. 121; Keat & Urry, 1975). The alternative seems rather egocentric to me. This assumption may be considered evident concerning a mountain, river or other physical entities that have been around long before any human beings were around to observe them. However, I assume that a particular social norm and other more intangible entities would continue to exist even if I suddenly die. At least as long as the social norm is shared by people (which they are by definition) and not all of us go extinct (which this book tries to avoid).

Assuming that the world exists independently of the observer—commonly referred to as realist ontology—does not automatically entail that it is directly accessible to the observer. We do not necessarily have direct access to reality through our senses, and the entities we perceive are not necessarily the real entities—as they are. There may be several reasons for that. It could be that our imperfect sensory and intellectual capabilities make it impossible for us to perceive reality accurately and unambiguously (Blaikie, 2007, p. 15) or that our cultural assumptions block direct access to it (Hammersley, 1992). It could also be that we only experience reality partially (empirical domain), and there may be a lot of other things going on regardless of whether we observe them or not (actual domain), which in turn are produced by underlying processes and mechanisms (real domain) that are out of our perceptual reach (Bhaskar, 2008, pp. 46–48). In other words, the world is there, but not directly accessible to us (Hammersley, 1992, p. 69) for whatever reason. I explain the implications of that in the following sections, but I want you to imagine a landscape with rolling hills and assume that it is still there when you are not (Figure 7.1).

What constitutes knowledge about our world?

In a world that is not accessible through direct observation, where the empirical rests on our experiences, there can be no objective search for truth (Crotty, 1998, pp. 8–9). In other words, meaning cannot be discovered but must be constructed through social processes in which there is a constant struggle over what is considered true or false (Barnes et al., 1996; Winther Jørgensen & Phillips, 1999, pp. 11–12). Imagine the landscape again. What do you see? I argue that what you see likely depends on who you are, what you have on your mind, who else you are with and so on. Greider and Garkovich (2010, p. 1) illustrate this with brilliant simplicity:

Figure 7.1 The world is there, even if I am not there to observe. *(Photo by Stuart Brabbs, shared on the Creative Commons.)*

> *Every river is more than just one river. Every rock is more than just one rock. Why does a real estate developer look across an open field and see comfortable suburban ranch homes nestled in quiet cul-de-sacs, while a farmer envisions endless rows of waving wheat and a hunter sees a five-point buck cautiously grazing in preparation for the coming winter?*

What the real estate agent, farmer and hunter know about the same landscape varies substantially because the philosophical base for knowledge rests on social practice, on the practical knowledge of people acting and utilising artefacts in specific social contexts (Tanesini, 1999, pp. 11–14)—even if they are the same person. This argument follows John Dewey's pragmatic philosophy in which thought and action never can be separated (Dewey, 2012), as meaning and knowledge are forged in action (Dewey, 1906, pp. 306–307). His philosophy of knowledge is also inclusive in that everyone is a capable participant in generating knowledge (Greenwood & Levin, 2007, p. 61). But what, then, is knowledge?

Frank P. Ramsey, another pragmatic philosopher, distinguishes knowledge from belief by stating that belief is knowledge only if it is formed in a reliable process and never leads to mistakes (Ramsey, 1931, p. 258). It is thus inadequate to believe something, regardless of its empirical support, if that belief leads to errors (Sahlin, 1990, p. 4). Hence, knowledge comprises beliefs in whose validity we are reasonably confident (Hammersley, 1992, p. 50).

Science proposes theories to help us make sense of our complex world. Ramsey looks upon theory as being divided into existing entities (here, α, β, γ), axioms and a dictionary, which can be expressed as '($\exists\alpha$, β, γ): dictionary. axioms' (Ramsey, 1931, p. 231) and is referred to as the 'Ramsey Sentence' (e.g., Mellor, 1980; Sahlin, 1990). Here, the entities are the building blocks of theory. The axioms are the rules for how the entities function and interact with each other. The dictionary is our ability to find the entities and axioms in the empirical domain of the world. This 'Ramsey Sentence' is not only helping us to understand the relationship between the theoretical and the empirical (Sahlin, 1990, pp. 140—158), but it also gives us a philosophical framework for managing the complexity in what we perceive when observing the real world. Ramsey (1931, pp. 212—215) distinguishes between the world, which is the home of what we try to explain and understand, and our theoretical construction, which is a tool for making sense of the world. In other words, to grasp the complexity of reality, we need to implicitly or explicitly create models of it (Conant & Ashby, 1970), or human-environment systems as they are referred to in the context of this book (Figure 7.2). The vital link between reality and our models is our capability to identify what is relevant to include for what we attempt to explain, understand or improve, as well as our capability to test our hypotheses for that particular explanation, understanding and improvement. Returning to the landscape, the real estate agent, farmer and hunter would include very different things in their mental models of the rolling hills. For instance, the real estate agent may include the urban areas within commuting distance, the farmer may include soil nutrients, and the hunter may estimate a safe shooting angle towards the hills. The landscape is the same, but our knowledge of it is socially constructed in context.

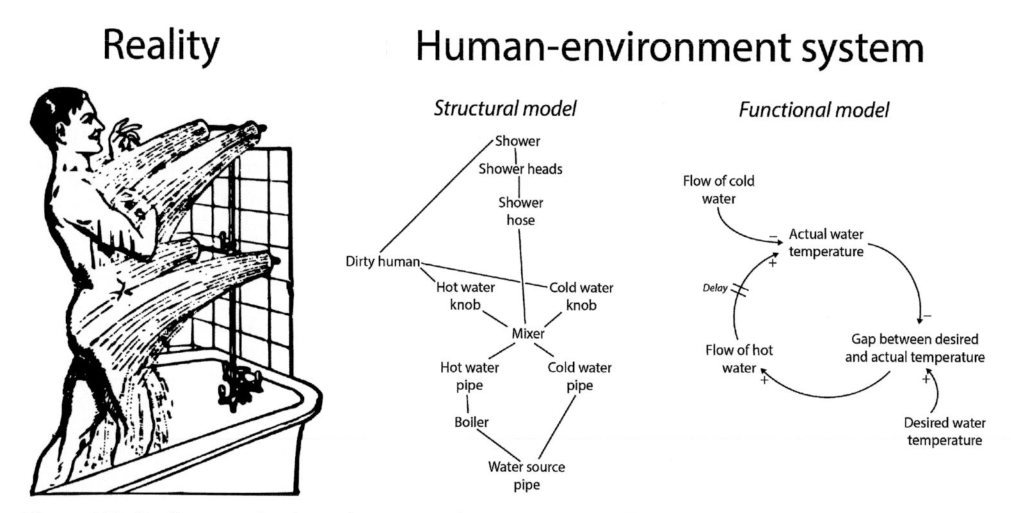

Figure 7.2 Reality conceived as a human-environment system for attempting to explain, understand and improve the discomfort of initially fluctuating shower temperatures in old buildings (by the way, it is the delay's fault).

What is the role of values?

It is impossible to be objective and value-free with the epistemological assumption above. However, when abandoning the timeworn axiological assumption of value-freedom in the pursuit of knowledge, a question arises concerning how to distinguish between the legitimate and illegitimate influence of values (Bueter, 2022, p. 307). As soon as we assume both the process and resulting knowledge to be value-laden, we take on a responsibility to address this problem. Regardless of directly embracing the influence of values—revealing our value-commitments and taking advantage of them in our pursuit of knowledge (Ellis, 2019, p. 987; Kourany, 2010, p. 58)—or still pursuing the unreachable ideal of value-neutrality—attempting to minimise the impact of our values—the influence of values is generally considered legitimate in (1) choosing and framing the phenomenon or problem, as well as the purpose of the research, (2) choosing a theoretical perspective and substantive theory to guide the research, (3) selecting the study context, (4) choosing methods for data collection and analysis, and when (5) deciding how to frame and communicate the results (Elliott, 2017, p. 166; Lincoln et al., 2017). However, reflexivity is vital for recognising our values and their consequences—also when assuming the process not to be outright political (Atkinson & Hammersley, 2007, p. 15)—and all values, as well as the knowledge they engender, must be subjected to critique (Kourany, 2010, p. 58). Reflexivity and being explicit about the values that influence our knowledge open up for a more constructive dialogue and thorough critique concerning commonly recognised standards for the pursuit of knowledge (e.g., Ellis, 2019, p. 987; Rolin, 2012).

I have so far only elaborated on the role of values in the pursuit of knowledge, which is the primary objective of traditional knowledge-oriented science (Hammersley, 2000; Weber, 1949). However, science can also be solution-oriented—here referred to as design science (Chapter 12)—with fundamental axiological consequences. Irrespective of your assumption concerning the role of values in the pursuit of knowledge, all solution-oriented activities are intrinsically value-driven. This difference is because all solutions entail, without exception, some claim about how things ought to be, but is often presented as the only legitimate alternative validated by empirical research (Hammersley, 2017, pp. 135–137). This practice is, however, fundamentally wrong since such claims can only be derived from values (Simon, 1996). It includes all recommendations based on knowledge-oriented studies, most often made with little or no reflection whatsoever.

Traditional science is well equipped to deal with how things are in the world (Checkland & Holwell, 2007, pp. 3–5), but less so in dealing with how things ought to be (Simon, 1996, p. 5). This distinction between 'is' and 'ought to be'—between the descriptive and the prescriptive—has been problematic for centuries, as it is easy to stray from the former to the latter without proper care. I believe the Scottish

Enlightenment philosopher David Hume was the first to point out this problem. Some scholars claim that Hume advocates a complete division between 'is' and 'ought to be' (Figure 7.3), which is rather theatrically illustrated by the principle's common epithet 'Hume's Guillotine' (Black, 1964, p. 166). Nevertheless, it is essential to note that values also have a grounding in reality—in the myriad experiences of human interaction (Hearn, 2012, pp. 215–216)—but I do not believe that undermines Hume's argument. Statements about 'how things ought to be' can simply not be empirically inferred from statements about 'how things are', as these two are entirely different from each other (Hume, 1896, p. 469). For instance, the prescriptive statement 'the authorities should lower the speed limit on all 90 km/h roads to 70 km/h' cannot be empirically inferred from the descriptive statement 'the number of fatalities per car in accidents is 38% lower on 70 km/h roads compared to 90 km/h roads'. Descriptive statements should be allowed to inform prescriptive statements, but the statement itself will always be essentially inferred from value preference—that is, if we value the potentially saved human lives higher than the costs and inconvenience of longer travel times. Hence, it is vital to be transparent about what values prescriptive statements rest upon.

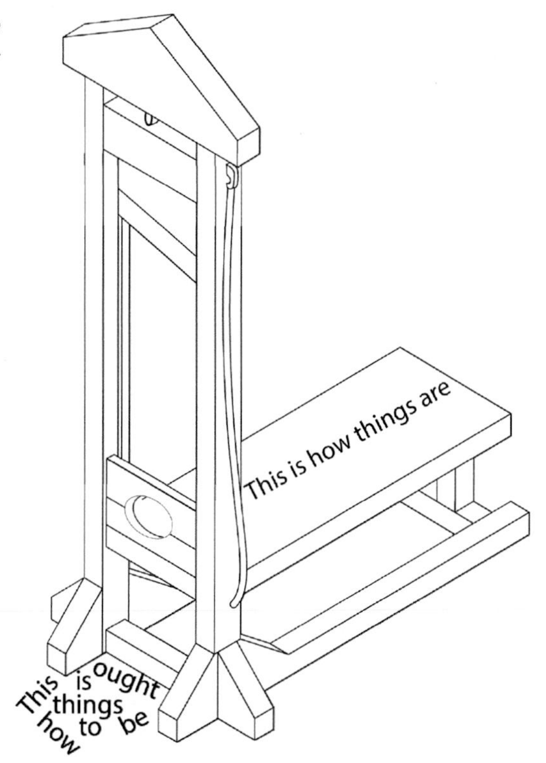

Figure 7.3 Hume's guillotine states the impossibility of inferring prescriptive statements from descriptive statements.

To what extent is complexity considered?

The fourth and final philosophical assumption to reflect on—before focussing the rest of this chapter on constructing a coherent conceptual framework—concerns the consequences of complexity on research. Or, more precisely, the extent to which we assume or allow complexity to play a role in what we set out to do. This assumption relates to perennial debates about how to approach phenomena or problems best—in parts or as wholes.

As I show in the first part of this book, Sustainability Science is concerned with pursuing knowledge of a wide range of phenomena and solving an equally wide range of real-world problems. How these phenomena and problems should be approached has been debated for decades, mainly in the form of more or less fierce advocacy for two seemingly disparate standpoints—reductionism and holism (e.g., Malanson, 1999, pp. 746–747).

Reductionism approaches a phenomenon or problem by reducing it to its constituent parts and studying or addressing the parts and their interactions one by one. If still complex, the parts are, in turn, reduced into their constituent parts, and so on (Heylighen et al., 2007, p. 3). Reductionism is central in Newtonian Science—the dominant and immensely successful paradigm in science for centuries. Reductionism simply works for many phenomena and problems, ranging from the simple—with apparent and stable cause-and-effect relations—to the complicated—with more intricate chains of cause-and-effect relations that are difficult but not impossible to grasp (Kurtz & Snowden, 2003; Snowden & Boone, 2007). However, it does not work for complex phenomena and problems—with large numbers of dynamically and nonlinearly interacting parts (Cilliers, 1998, pp. 3–5)—regardless of how hard we try. However, the vastness and sheer complexity of our universe make it impossible to research anything without reducing it to some extent (Churchman, 1970, pp. B43–44).

On the other hand, holism assumes that complex systems often have properties that cannot be reduced to their parts and interactions (Figure 7.4)—commonly referred to as emergent properties (Chapter 9). Although this idea was first suggested by Aristotle (1801, p. 199), Smuts (1926) coined the term *holism,* and the notion of emergent properties is visible in Comte's (2000) and Durkheim's (1993, p. 110) social scientific work before it was later introduced into the natural sciences (e.g., Smuts, 1926; von Bertalanffy, 1950). Holism is a necessary philosophical assumption for complex problems (cf. Laughlin, 2003)—including many, if not most, phenomena or problems of Sustainability Science—since reducing the system under study to its constituent parts and their pairwise interactions results, invariably, in emergent properties being ignored or at least inexplicable. However, if not explicitly focussing on a closed system in the sense of being completely isolated from its surroundings (no exchange of energy, matter, information),

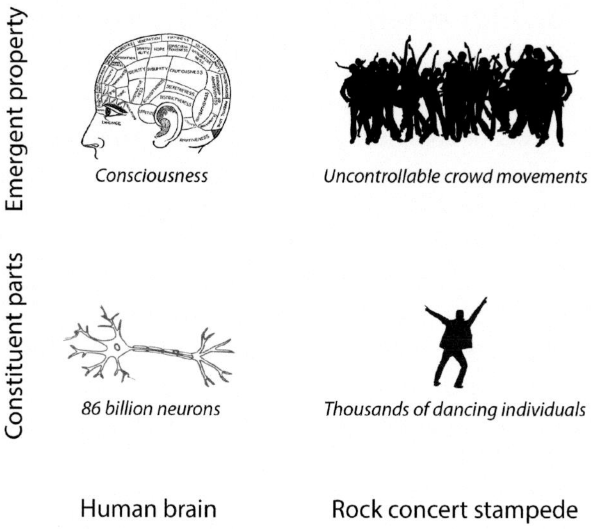

Figure 7.4 Examples of emergent properties.

Emergent property

Consciousness

Uncontrollable crowd movements

Constituent parts

86 billion neurons

Thousands of dancing individuals

Human brain

Rock concert stampede

studying it always requires some reduction (Cilliers, 2005, p. 258). Otherwise, we end up studying the entire universe all the time.

There is no simple solution for the problem of including the right amount of complexity. Formally, it is about balancing Ashby's (Ashby, 1957, pp. 202–268) Law of Requisite Variety—that the system we construct of reality can only model reality if it has sufficient complexity to represent it—and Ockham's Razor (Checkland, 1981, pp. 35–36)—that this complexity should be limited to only include what is relevant for the particular phenomenon or problem. In practice, it is about setting the boundary of the system we are interested in, which is not as straightforward as often expected and simply an analytical choice (cf. Heylighen et al., 2007, p. 16). Thus, we cannot know any complex system entirely, and any knowledge we think we have is provisional and contingent on the boundaries we apply (Cilliers, 2005). More on complexity in Chapter 9.

Development, sustainability and risk

In addition to the philosophical assumptions above, I also need to introduce some concepts that are helpful in Sustainability Science approaches to risk and resilience for sustainable development. Three of them are elaborated on in this section, that is, the concepts of development, sustainable development and risk.

The concept of development

Although the word development has been used for at least 250 years (Harper Dictionary, 2022), it was not until the end of World War II that it became an important concept (Thomas, 2000b, p. 3). Ideas about development have changed back and forth since then, for example, the Soviet model of development (Smekal, 1991, pp. 32–39), modernisation theories (Organski, 1965; Rostow, 1960), dependency theory (Dos Santos, 1970; Frank, 2004), world systems theory (Wallerstein, 1974), another development (Hettne, 1995, pp. 160–206), and human development (ul Haq, 1995). This diversity of ideas has spurred numerous and often competing definitions of development, making it difficult to communicate about it. It has been suggested that this Babylonian confusion (cf. Hagelsteen & Becker, 2014), to a great extent, is the result of the concept being used in three different ways (Figure 7.5): (1) as a description of a desired future state of society; (2) as a process of change over time or (3) as deliberate efforts of various actors aimed at improvement (Thomas, 2000a, p. 29).

Development may, in other words, refer to a desired state (goal), the process of getting there (change), as well as our efforts to get there (activities). Presenting a desired state of society implies some variables that human beings value and aspire to change from their current state. Development is, in other words, inherently normative (Seers, 1989). In this context, the change is the transformation of the set of variables over time, and our efforts refer to purposeful activities we carry out to drive or steer this change towards the desired state. The three parts are, thus, fundamentally related to each other (Thomas, 2000a, p. 29), enabling us to look at development as having five components:

1. A set of variables that human beings value and aspire to change.
2. A descriptive statement about the current state of the set of variables.
3. A prescriptive statement about the desired state of the set of variables.
4. A prescriptive description of a preferred expected scenario of change over time.
5. A set of purposeful activities aimed at driving or steering the change.

Figure 7.5 Our preferred development. *(Source: Slavoljub Pantelic/Shutterstock.com.)*

Most definitions of development only include one or a few of these components explicitly (e.g., Chambers, 1997; Rist, 2006, p. 13; Seers, 1989, p. 481; South Commission, 1990, pp. 10–11; Todaro, 1989, p. 620; UNDP, 1990, pp. 10–11). However, these components can be seen as incremental because it is unfeasible to focus a definition of development on one without at least the implicit involvement of the others before it on the list above. For example, it is neither possible to define a desired state of some variable in the world without first defining that variable and having some idea of its current state that requires development. Nor to define development activities without expressing this desired state and the required change that the activities are designed to bring about (Örtengren, 2003, pp. 9–15).

Although Hettne (1995, p. 15) argues that development is contextual and therefore eludes any fixed and final definition, the concept of development in this book includes all five components since sustainable development requires intentional human activity. Development is, thus, viewed as a preferred expected scenario of change in a set of variables over time, from a current to a desired state, and includes purposeful activities to drive or steer this change (Figure 7.6). It is important to note that this preferred expected scenario is not fixed over time but changes as we go along. For instance, the invention of irrigation revolutionised what farmers could expect in terms of harvests, and a political decision to build a new tramline changes what homeowners along the line expect in terms of property prices. There may also be several scenarios competing for precedence. Still, at any point in time, there is one preferred expected scenario that either implicitly or explicitly comprises our desires for the future.

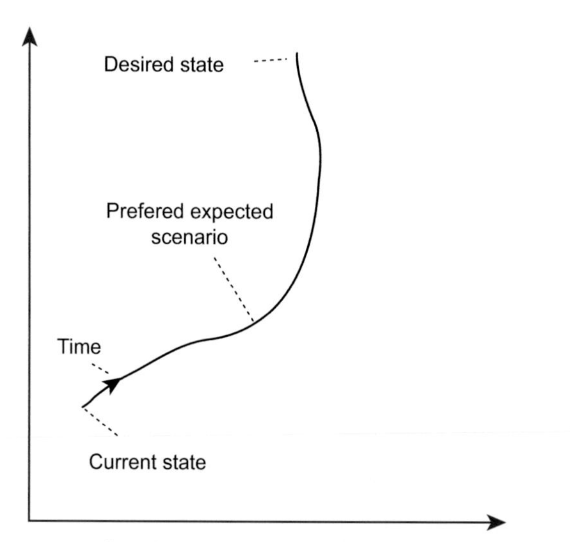

Figure 7.6 Development as a preferred expected scenario of change in a set of variables over time, from a current to a desired state, including purposeful activity.

The concept of sustainable development

Sustainable development is commonly defined as 'development that meets the needs of the present without compromising the ability of future generations to meet their own needs' (WCED, 1987, p. 43). As mentioned in Chapter 3, it is generally conceived as resting on three pillars: economic development, social development and environmental protection. Looking more closely into the term 'sustainable', it is defined as something that is 'able to be upheld or defended' (Oxford English Dictionary, 2020). The first part of this definition indicates that sustainable development is development that can be maintained over time, while the second part suggests that sustainable development is development that can be safeguarded from the impact of adverse events and processes. These two parts are closely related, as it is not only events and processes that may impact development, but the means for development may also increase or create new events and underlying processes that, in turn, make it challenging to maintain development over time. For instance, our dependency on fossil fuels for energy has allowed great developmental leaps in many societies since the Industrial Revolution. Still, it is, at the same time, the main cause of climate change and ocean acidification that are now threatening the existence of all societies.

Whether sudden and dramatic or gradual and obscure, adverse events and their underlying processes may cause deviations from our preferred expected development scenario (Figure 7.7), limiting the sustainability of our development. Hence, sustainable development is development that can be maintained over time and safeguarded from the impact of adverse events and their underlying processes. It is important to note

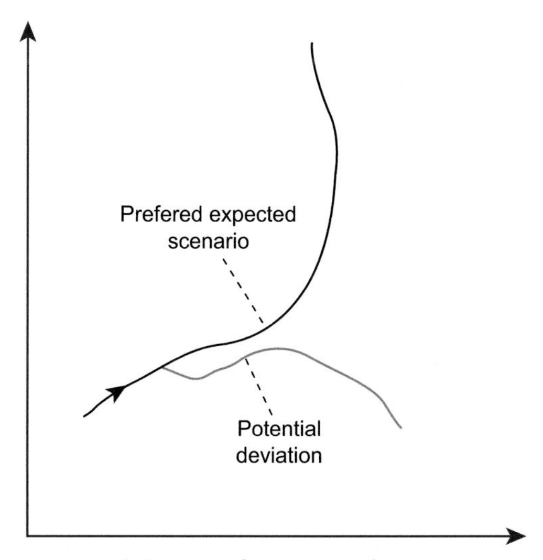

Figure 7.7 Potential deviation from the preferred expected scenario.

that any events or processes resulting in a more positive development of the defined set of variables than initially preferred and expected would automatically lead to the preferred expected scenario being updated from there and not be considered a deviation.

As the future is uncertain (Japp & Kusche, 2008, p. 80) and human beings are fundamentally incapable of predicting it (Simon, 1990, pp. 7–8; Taleb, 2007), there is not only one but myriad possible scenarios that deviate to various degrees from our preferred expected scenario (Figure 7.8). However, human beings can influence their future by structuring these uncertain scenarios and using them as mental tools to anticipate the consequences of different courses of action and then select activities that appear to lead to our desired state or goal (Renn, 2008, p. 1; Simon, 1990, p. 11). In this context, sustainable development requires the ability to manage risk (UNDP, 2004, pp. 9–27; United Nations, 2015).

The concept of risk

Risk is a contested concept with numerous definitions, creating the potential for miscommunication and misunderstandings (Aven & Renn, 2009a; Fischhoff et al., 1984; Rosa, 1998). In everyday language, the term 'risk' often means the likelihood or probability of an event, or sometimes a destructive incident that may or may not occur (Sjöberg & Thedéen, 2003, p. 16). In science, it is used more precisely, although the definition of risk varies significantly (Aven & Renn, 2009a, pp. 1–2; Lupton, 2013; Renn, 2008, pp. 12–45; Zinn & Taylor-Gooby, 2006). Nevertheless, most of the definitions

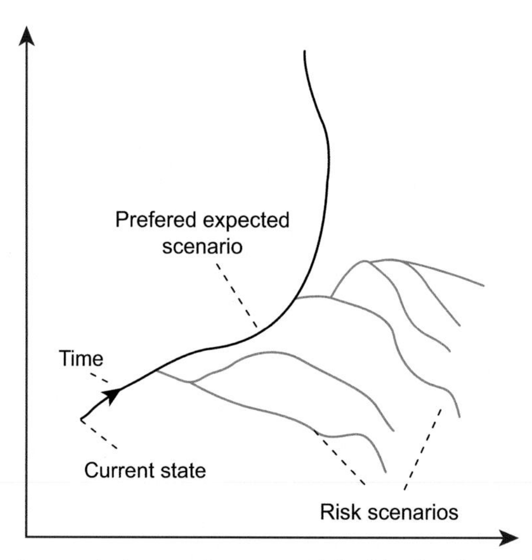

Figure 7.8 The preferred expected scenario as a sustainable development trajectory of the human-environment system under study, with risk scenarios as deviations from it.

have three aspects in common. First, they all distinguish between reality and possibility, as the concept of risk makes no sense if the future is predetermined or independent of present human activity (Renn, 1992, p. 56; Luhmann, 1993; Zinn, 2008, pp. 3–4). The future must, in other words, be uncertain (Japp & Kusche, 2008, p. 80; Renn, 1998b, p. 51) and any future event must at least be perceived as being possible to influence (Zinn, 2008, p. 4). That is to say; there would be no risk in gambling if the game were 100% rigged. Second, all definitions of risk explicitly or implicitly entail that these uncertain futures must have the potential to impact what human beings value (Renn, 1998a, p. 51; Renn, 2008, p. 2), or at least be so perceived (Slovic, 1987; Slovic et al., 1982). In other words, there would be no risk in gambling—even if the game were not rigged—if the stake is a grain of sand and takes place in a desert. Finally, and closely related to the previous aspect, risk must be defined in relation to a preferred expected outcome (Kaplan & Garrick, 1981; Luhmann, 1995, pp. 307–310; Zinn, 2008, p. 4). It means there would be no risk in gambling—even if the game were not rigged and the stakes were high—if the participant has no preference for winning or losing. Taking these three aspects of risk together means that risk is a representation of potential negative deviations in any variable or set of variables representing what human beings value from its preferred expected development over time (Figure 7.8).

Risk may be defined as uncertainty about what could happen and what the consequences would be (Aven, 2007, p. 747; Aven & Renn, 2009a). Early notions of risk could entail both positive and negative outcomes (Bernstein, 1996), which is still common in conceptualisations of financial risk (e.g., Greiner, 2013) and is a central aspect of theories about risk-taking (Zinn, 2020). However, the concept of risk here focuses on negative outcomes (Renn, 2008, p. 2). It is evident that any outcomes of decisions, actions or events can be either positive or negative. It is, therefore, necessary for clarity to follow the conventional contemporary notion of risk as negative and leave potential positive outcomes to be considered opportunities (cf. Japp & Kusche, 2008; Luhmann, 1993). Moreover, if risk is defined in relation to a preferred expected future scenario, any hints that there are alternative and even more positive scenarios that could be expected would result in that scenario quickly becoming the new preferred expected scenario against which we define risk scenarios.

It is important to note that there is nothing objective about risk (Wynne, 1982, 1998) since any notion of it is based on perceptions (Slovic, 1987), is culturally mediated (Douglas & Wildavsky, 1983), and can be socially amplified (Kasperson & Kasperson, 1996). It is, therefore, socially constructed and does not exist ontologically (Aven & Renn, 2009a, pp. 8–10; Renn, 2008, pp. 2–3; Slovic, 1992, p. 119). However, what does exist are the actual events that produce consequences that human beings experience, interpret and consider when making sense of the present and envisaging the future (Renn, 2008, p. 2). Few could dispute that people are actually harmed or even killed in accidents, disasters or by pollution (Shrader-Frechette, 1991, p. 30). These direct or indirect experiences create a link between risk, as a social construction, and reality (Renn, 2008, p. 2), meaning that it is vital not to mix ontology and epistemology in this context (Rosa, 2010).

The realist ontology presented earlier in this chapter does not require epistemological realism of viewing risk as real and objective (Aven & Renn, 2009a; Kunreuther & Slovic, 1996, p. 119; Renn, 2008, pp. 2–3). Nor does social constructivist epistemology require an ontology that reduces risk to only subjective issues of power and interest (Aven & Renn, 2009a, p. 9; Renn, 2008, p. 3). Instead, the ontology and epistemology presented in this book form a philosophical foundation for highlighting this link between risk, as socially constructed, and the real world. It is important to note that all human beings take part in experiencing and interpreting the world, making the social construction of risk rooted equally in science and public values and preferences (Aven & Renn, 2009a, pp. 8–9; MacGregor & Slovic, 2000, p. 49; Renn, 2008, pp. 3–4).

Fitting them together

The conceptual approach to development, sustainable development and risk presented here indicates that all three concepts are essentially connected to each other. Facilitating sustainable development entails purposeful human activity to make sure that any potential deviation from the preferred expected development scenario is avoided or minimised. Therefore, anticipating potential deviations, or risk scenarios, is vital for managing risk, making analysing risk a requisite for sustainable development (Haimes, 2004, pp. 101–106; Santos et al., 2020).

Managing risk for sustainable development

Simon (2002, p. 604) states that '[t]he reading of history persuades me that the most dangerous villains we will encounter along the way will rarely be the forces of nature'. Although some hazards are mainly outside the control of intentional human activity, such as earthquakes and cyclones, what we do on Earth determines our future to a much greater degree than ever before in human history (Chapter 2). This section focuses on managing risk and introduces additional key concepts, such as value, hazard, vulnerability and capacity, which are vital for managing risk and sustainability.

The essence of managing risk

Analysing risk is, here, the practice of structuring risk scenarios and comparing them with the preferred expected scenario, making a risk analysis into the answer to three questions (Kaplan & Garrick, 1981, p. 13): (1) what can happen?; (2) how likely is it to happen? and (3) if it happens, what are the consequences? However, it is essential to note that there is uncertainty in any possible answer to all three questions, which may be better captured by what Aven (2011, p. 518) refers to as the (A,C,U)-risk perspective, where there is uncertainty (U) in both whether or not an event (A) occurs and what consequences (C) it may have (Figure 7.9). Put in different words, analysing risk in relation to sustainable development entails considering what events (A) can divert us away from our preferred expected development trajectory, what consequences (C) they may have on what we value, and what uncertainties (U) there are concerning both events and consequences (A and C).

Wiek et al. (2011, p. 205) present an interesting approach to sustainable development in their integrated sustainability research and problem-solving framework. This framework is structured in four modules: (1) analysing the current problem constellation(s); (2) creating and crafting sustainability visions; (3) exploring less desirable future scenarios that might become reality without interventions towards sustainability, and (4) developing and testing strategies to transition from the current state to sustainable states without getting deflected towards undesirable pathways (Wiek et al., 2011). Translating this pragmatic approach into the language of this book tells us that sustainable development requires us to (1) analyse the current situation; (2) define our preferred expected scenario; (3) analyse potential deviations from our preferred expected scenario (risk scenarios), and (4) design and implement sets of activities that maintain our development trajectory along the preferred expected scenario (Figure 7.8). However, this is nothing new and can be found in the heart of Vickers (1968, p. 468) early writing on society as a complex adaptive system:

Figure 7.9 Uncertainty in both events and their consequences. *(Source: Andy Dean Photography/ Shutterstock.com)*

> *The problem for R [the regulation process] is to choose a way of behaving which will neutralise the disturbance threatening the maintenance of E [the expectations, preferred system]. Success means initiating behaviour which will reduce the deviation between the actual course of affairs and the course which would be consonant with E; or at least preventing its nearer approach to the limit of the unacceptable or the disastrous.*

This process is the essence of managing risk for sustainable development and entails a systematic analysis of what human beings value, the events that can have a negative impact on that, how vulnerable they are to the impact of these events, and what capacity we have for protecting them.

Values and what is expressed as valuable

Values may be seen as 'desirable trans-situational goals, varying in importance, that serve as guiding principles in a person's life or other social entity' (Schwartz, 1994, p. 21). In other words, values are what people care about (Keeney, 1992, p. 2). To grasp what human beings value, it is vital to understand how values come to be ascribed to whatever is declared to have value.

'No man is an island, entire or itself' (Donne, 1624, p. 415). This timeworn quote by a 17th-century English poet indicates that human beings are social beings functioning together in society. Giddens (1984) takes this idea further by stating that how human beings experience their social context influences how they perceive and understand it and, therefore, also how they will act in that social context. These actions, in turn, produce and reproduce social structures, which guide and restrict the actions that human beings may take (Giddens, 1984, pp. 25–26). Human actions are, thus, fundamentally linked to social structures, which are representations of established patterns of behaviour and have the purpose of keeping order while coordinating stable activities (Hardcastle et al., 2005, p. 224). What human beings value is, in other words, socially constructed in context, where prolonged human action creates social structures that direct human beings in what value to ascribe. It is, however, rare that society is homogenous, granting room for individual variation, as several social structures may compete for dominance. The more heterogeneous the society, the more individual variation is possible in what human beings value. Values may thus be seen as acquired 'both through socialisation to dominant group values and through the unique learning experience of individuals' (Schwartz, 1994, p. 21).

Values are notoriously challenging to measure (Slovic, 1995, p. 369), and the methods used are predisposed to biases (Payne et al., 1992, pp. 121–122) regardless of the assumptions upon which the value elicitation is based (Fischhoff, 1991). However, to manage risk for sustainable development, it is not the values we need to elicit but what human beings express as valuable and important to protect.

To understand what people express in particular contexts, it is crucial to consider that we can know more than we can tell (Polanyi, 1966b, p. 22). What people know can be divided into explicit and tacit knowledge (Figure 7.10) (Nonaka, 1994, p. 16). Explicit knowledge consists of concepts, information and insights that can be specified, stored and transmitted to others (Connell et al., 2003, p. 141). Tacit knowledge, on the other hand, is not directly transmittable and consists of knowledge that makes up our mental models for creating meaning to our experiences, as well as our know-how and skills to apply in specific contexts (Nonaka, 1994; Polanyi, 1967). However, explicit and tacit knowledge are closely connected, as 'explicit knowledge must rely on being tacitly understood and applied' (Polanyi, 1966a, p. 7).

Tacit knowledge comprises subsidiary awareness and focal awareness (Figure 7.10), where the phenomenon in our focal awareness is made identifiable by subconsciously assembled clues in our subsidiary awareness that are not identifiable in isolation (Polanyi, 1966a, pp. 2–7). An example is the psychiatrist showing his students a patient having a seizure. After letting the students discuss if it was an epileptic or a hystero-epileptic seizure, he settles the argument by stating, 'you have seen a true epileptic seizure. I cannot tell you how to recognise it; you will learn this by more extensive experience' (Polanyi, 1961, p. 458). The statement that the seizure was an actual epileptic seizure is possible to transmit across the classroom and is an example of explicit knowledge. However, the psychiatrist's knowledge of diagnosing the patient is tacit and less straightforward to share with the students. Only the diagnosis itself is in his focal awareness and accessible to him, as he is only subsidiary aware of each of the many clues and indicators he more or less subconsciously observed.

What is in our focal awareness is not only determined by individual characteristics, knowledge and so on but is constantly changing depending on context. Each situation gives us a sense of what is relevant to what we are doing. Our experience of similar situations, our idea of what the situation calls for or demands, and our sense of aim or direction all combine to supply us with this 'relevance structure' (Marton & Booth, 1997, p. 143). What we have talked about recently, what roles the people around us have, what goals we think they have and so on are thus crucial for how we understand, interpret and

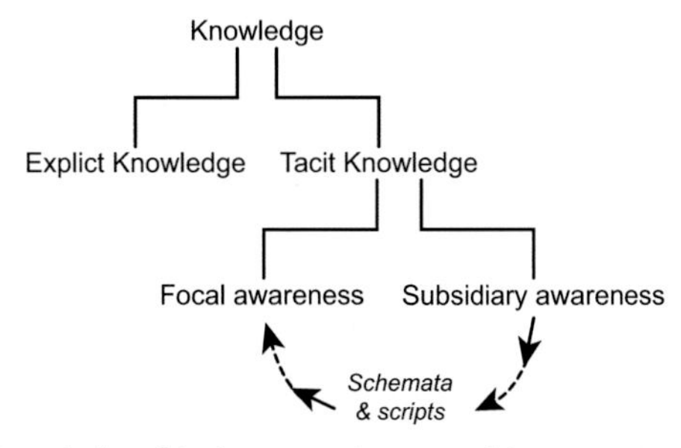

Figure 7.10 The relationship between key cognitive concepts of knowledge.

remember incoming information. These mental structures or processes are referred to in cognitive science as 'schemata' and are constantly amendable (Bartlett, 1995, p. 208). The current schemata of an individual guide her interpretation of the incoming situation, as well as her expectations of and attention to it (Boland et al., 2001, p. 394). Our tacit knowledge comprises, in other words, a part of our schemata (Nonaka, 1994, p. 16). Another closely related cognitive tool we utilise to get by in our complex world is called 'script'. Scripts are cognitive chains of expectations of actions and effects in particular situations (Abelson, 1981; Schank & Abelson, 1977), which assist individuals in their reactions in those situations without focussing much of their focal awareness on their actions. The primary function of schemata and scripts is to facilitate coherence in our perception and experience of a situation by filling in gaps in the available information (Figure 7.10).

What actors consider valuable and important to protect is usually not explicitly stated when managing risk for sustainable development but instead relies on an implicit assumption that all actors agree on this issue (Nilsson & Becker, 2009). However, studies indicate that such an assumption is dubious at best and flawed at worst (e.g., Becker, 2012). In short, what individuals express as valuable in any given situation is socially constructed in context and is determined by their values and what they have in their focal awareness at that time. This approach takes into account the functions of their current schemata and scripts and indicates that human beings construct their mental models of reality through active selection and interpretation of information around them (Vennix, 2001, p. 14). It may then be argued that discussing what is valuable and important to protect is useless, as each account is destined to be subjective and fragmented. However, this is not at all the case. Explicit dialogue may facilitate the integration of individual mental models—each giving a limited perspective on the world—into one shared model, which is vital for creating a common understanding of the challenge at hand (Becker & Tehler, 2013; Vennix, 2001). Each individual account is indeed unlikely to give a complete picture of what is considered valuable on its own. Still, it is likely that what is mentioned in the dialogue between several individuals triggers additional scripts and amends schemata. Thus, activating additional knowledge by moving it from their subsidiary awareness to their focal awareness. What the group comes up with is also highly contextual, but it is still likely to be a richer picture than the sum of each individual account (Becker & Tehler, 2013). And more importantly, it is their common picture of what is valuable, making it achievable for the actors involved to formulate and pursue common goals when managing risk (Becker & Tehler, 2013). Without such an explicit common picture, there is a grave danger that the actors might unwittingly impede each other's work by focussing on protecting different things. For instance, the Ministry of Agriculture may focus on securing state revenues by promoting the production of cash crops, while the Red Cross Society focuses on reducing the risk of famine by advocating for increased diversity of food crops.

Hazards that trigger destructive events

Explicit consideration of what is valuable and important to protect is often forgotten when analysing risk. People instead jump directly to identifying and selecting a set of hazards (e.g., Coppola, 2015, p. 41). However, deciding what hazards to include depends, essentially, on how we, at least implicitly, think they can impact what we value. How can we otherwise explain why the public health department identifies cholera while the public works department identifies landslides? In addition to what the involved actors consider valuable and important to protect, what hazards are identified also depends on their risk perception. While various demographical factors have been identified as influencing risk perception (e.g., Armaş, 2006; Bontempo et al., 1997; Chauvin et al., 2007; Flynn et al., 1994; Lam, 2005; Sjöberg, 1998; Slovic, 1987), not all such differences lead to differences in how hazards are ranked internally (Becker, 2011).

Selecting a set of hazards to include is a crucial step and is sometimes referred to as hazard analysis (e.g., Coppola, 2015, pp. 46–50) and is aimed at establishing necessary spatial, temporal and magnitudinal aspects of each hazard included when managing risk for sustainable development (Box 7.1). A clear definition of each hazard's location and spatial extent, its speed of onset and duration, frequency or likelihood, and magnitude or intensity are requisites for this part of the analysis (Coppola, 2015, pp. 49–50). The more specifically each hazard event is defined, the easier it is to construct risk scenarios. That said, it is impossible to include all possible events in a risk analysis, which calls for categorising such initiating events and allowing one hazard event to represent a number of them. This approach is called partitioning the risk scenario space (Kaplan et al., 2001, pp. 810–811).

Having identified a relevant hazard, it is crucial to analyse the factors that contribute to it, as these may be connected to, and amplified by, processes related to human activity (Hewitt, 1983, p. 25; Kates et al., 1990; Renn, 2008, p. 5). Examples of such connections are mining and pollution, logging and flash floods, irrigation for agriculture and sinkholes and so on. It is important to note that a specific hazard can also impact contributing factors for other hazards. For example, earthquakes or heavy rain may trigger landslides. When managing risk for sustainable development, it is vital to expand the risk scenario

BOX 7.1 Key aspects of hazard analysis

Location—Where is the source of the hazard event located?
Spatial extent—What geographical area is directly affected by the hazard event?
Speed of onset—How fast is the hazard event evolving?
Duration—For how long time is the hazard event taking place?
Frequency or likelihood—How often is the hazard event happening?
Magnitude or intensity—What is the destructive force of the hazard event?

space by not only including the sudden and dramatic but also the gradual and mundane. For example, as presented in Chapter 5, not only earthquakes and cyclones may trigger deviations away from our preferred expected development trajectory, but also soil erosion and freshwater availability.

As also presented in Chapter 5, human activity significantly influences most hazards. They are thus also possible for us to prevent or mitigate before they happen by adapting our human-environment system. It is obviously impossible to prevent the tremor of earthquakes or torrential rain in a cyclone. Yet, we can prevent a range of hazards, such as floods, landslides, erosion, epidemics, transport accidents, conflict and crime. I am here not referring to our capacity to mitigate disturbances, disruptions and disasters by reducing our vulnerability to the impact of these events, which I will come back to in a moment, but our capacity to reduce the speed of onset, spatial extent, duration, magnitude and likelihood of the actual hazards themselves.

Finally, I would like to stress again the importance of not skipping explicit analysis of what is considered valuable and important to protect before, or at least in connection to, analysing what hazards can trigger deviations away from our preferred expected development trajectory. Without such explicit analysis or dialogue, it may, at best, be more difficult to identify relevant hazards and, at worst, undermine the entire foundation for collaboration. For instance, without a shared understanding of what is valuable and important to protect, drought may be a top priority for the Ministry of Agriculture but not at all interesting for the Ministry of Transport.

Vulnerability to impact

Regardless of whether an event is triggered by a hydrometeorological, geological, biological, accidental or antagonistic hazard (Chapter 5), it will not result in unwanted consequences unless it occurs in a conducive setting (Wisner et al., 2004, pp. 3—16; Santos et al., 2020). Such a setting is determined by factors from all spheres of society (Hearn Morrow, 1999; Wisner et al., 2004, pp. 49—84) and is primarily a result of human activity (Hewitt, 1983, pp. 24—29; Oliver-Smith, 1999). In short, most disasters stem from unresolved development issues (Figure 7.11) (Wijkman & Timberlake, 1984). Destructive events are, therefore, not discrete, unfortunate and detached from ordinary societal processes but are intrinsic products of everyday human-environment relations over time (Fordham, 2007, pp. 338—339; Hewitt, 1983, p. 25; Oliver-Smith, 1999). The consistent failure of most professionals and academics to grasp and address this undisputable fact has spurred a rather vocal movement among disaster scholars who suggest that society mainly engages in Disaster Risk Creation. It is by acknowledging and changing all those practices that we have a chance ever to enjoy a safe and sustainable society (e.g., Bankoff & Hilhorst, 2022; Davis & Alexander, 2022; Lewis & Kelman, 2012). We are still not heard, unfortunately.

Figure 7.11 Inequality drives vulnerability: high-risk Rocinha favela on the steep slopes above luxury properties in Rio, Brazil. *(Source: Dabldy | Dreamstime.com.)*

The susceptibility to harm is referred to as vulnerability. It is never a general attribute but must always be defined in relation to the impact of a specific hazard event (e.g., Aven, 2007, p. 747; Aven & Renn, 2009b, pp. 588–589; Blaikie et al., 1994, pp. 9–10; Cannon, 2008, p. 351; Dilley & Boudreau, 2001, p. 232; Gallopín, 2006, p. 294; Hollenstein et al., 2002; Twigg, 2004, p. 13; Wisner et al., 2004, pp. 11–13). Vulnerability analysis can, as such, be conceived as the answer to three questions—analogous to the risk analysis questions presented earlier in the chapter—(1) what can happen, given a specific hazard event?, (2) how likely is that to happen, given that hazard event? and (3) if it happens, what are the consequences (Hassel, 2010, p. 35). However, analysing vulnerability is also associated with the same uncertainty concerning both events and consequences (Aven, 2007, p. 747; Aven & Renn, 2009b, p. 589). If risk is defined as uncertainty about what could happen and what the consequences would be (Aven, 2007, p. 747; Aven & Renn, 2009a), vulnerability is defined as uncertainty about what could happen and what the consequences would be, given a specific hazard event (Aven, 2007, p. 747).

Since vulnerability is determined by physical, environmental, social, cultural, political and economic factors, which we have the potential to address, we can reduce our vulnerability to all possible types of hazards by adapting our human-environment system. For instance, the Bam earthquake in Iran and the San Simeon earthquake in California in 2003 measured both around 6.6 on the Richter scale, but the former killed more than

25,000 people while the latter killed two people. It is rarely entirely fair to compare two earthquakes like that, as there are factors other than magnitude on the Richter scale that determine the destructive potential of the tremor itself. However, such a comparison is still indicating vast differences in vulnerability. Remember that the 2010 earthquake in Haiti had a magnitude of 7.1 and killed up to 300,000 people, while the massive 8.8 magnitude earthquake in Chile the same year released 350 times more energy yet killed 525 people. Such differences are easy to find for most types of events between different places on our planet—generally connected to differences in development. However, it is not so that the poor always are the most vulnerable, even if that is commonly stated within the international community. For example, poor people living in makeshift shacks may be less likely to die in an earthquake than wealthier people living in unreinforced brick buildings. That said, it is rarely so that the really well-off people are worse affected than any other group living in the same location. In other words, although the level of development is a fundamental determinant of vulnerability, I urge caution in generalisations not rooted in a specific hazard event.

The only general thing is that vulnerability is highly contextual, and different individuals and households differ in vulnerability depending on various factors (Buckle, 1998) and differences in access to various resources (Hearn Morrow, 1999). For instance, gender and age are often decisive factors in determining the consequences of an event, although being regularly ignored, often resulting in women and older people being disproportionately affected. However, to say categorically that women and older people are more vulnerable, even in relation to a specific event, may overlook variations within these groups that may lead us astray. Remember the 1995 Chicago heat wave referred to in Chapter 5, which indicated not only the need to disaggregate our population according to gender, age, ethnicity and so on, but also that vulnerability is about what is referred to as intersectionality (Crenshaw, 1989). It is about the interaction of several factors that together result in marginalisation in terms of access to vital resources (Figure 7.12), which in the Chicago heat wave case left elderly African-American men more vulnerable than any other demographic group (Klinenberg, 2002). Understanding intersectionality is, in other words, central to understanding vulnerability (Figure 7.13). Therefore, the only way forward is to be mindful of the likelihood of variations in vulnerability when managing risk for sustainable development, to disaggregate the data we collect, and to target our activities on the most vulnerable we find—whoever that is in each specific context. Finally, it is by reducing vulnerability that we have the greatest potential to reduce risk.

Capacity to react or to proact

When analysing vulnerability, it becomes clear that it is not only structural issues, in the sense of a complex combination of physical, environmental, social, cultural, political and economic factors, that determine vulnerability. Human agency also plays a vital role

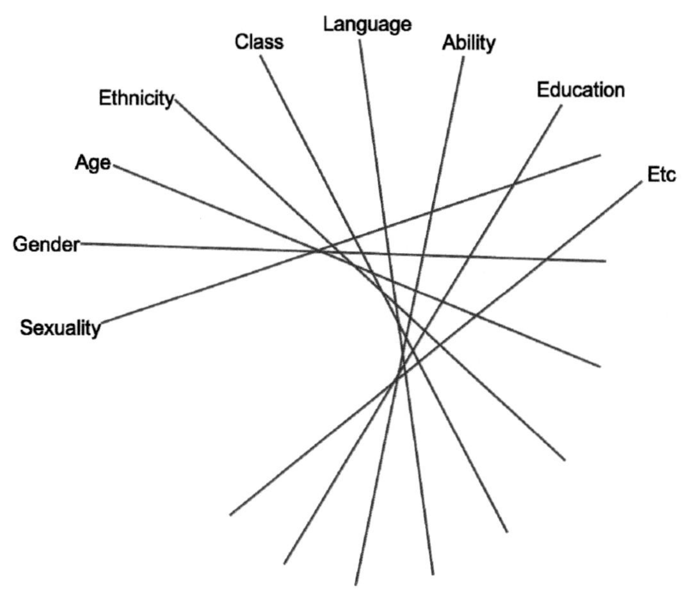

Figure 7.12 Intersectionality: The notion that different social categorisations intersect to create complex systems of disadvantage and domination.

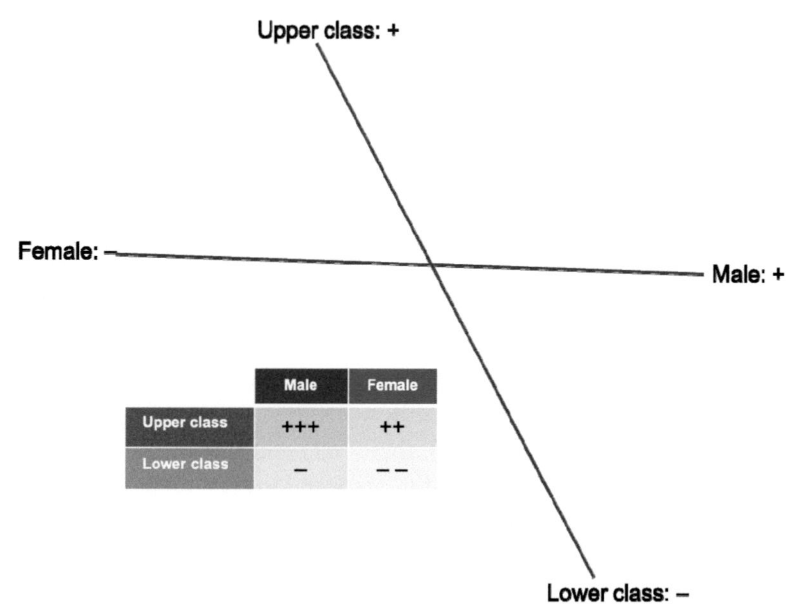

	Male	Female
Upper class	+++	++
Lower class	–	– –

Figure 7.13 Example of two intersecting social categorisations resulting in compounded advantages and disadvantages.

(Renn, 2008, p. xiii), as purposeful human activity influences what could happen and what the consequences would be. This influence may be direct through reactive activities that affect the course of events in a specific risk scenario—for instance, recognising the need to evacuate before flood waters reach a critical level, the actual evacuation to safer grounds, rapid salvage of damaged food crops and so on. It may also be indirectly, by proactive activities influencing what risk scenarios are feasible altogether. For example, constructing permanent levees to protect settlements from flood water, awareness-raising campaigns clarifying when and where to evacuate in a flood, and so on. Hence, it is essential to include, when analysing risk or vulnerability, the capacities of individuals, communities and organisations to take actions to limit the impact in each risk scenario.

Proactive activities, with more indirect influence on risk scenarios, are vital for managing risk to facilitate sustainable development. However, they are, to a large extent, already set at the beginning of the period we want to analyse and are thus less feasible to incorporate when analysing risk. It is not at all to say that analysing and reducing risk and preparing for effective response and recovery only happen before a hazard has triggered some destructive event. On the contrary, these activities must continue even amid calamity to protect what human beings value.

Developing capacity is, in other words, a third way to adapt our human-environment system to reduce risk for sustainable development, in addition to preventing or mitigating hazards and reducing vulnerability. Both in the sense of reactive capacities to act in actual disturbances, disruptions or disasters and proactive capacities to manage risk that, of course, also include our capacity to address hazards and vulnerabilities. I come back to this when operationalising the concept of resilience in Chapter 8.

The concept of resilience

Our world is complex and dynamic, continuously transforming due to ongoing processes of change (Chapter 6). As it is in this dynamic environment that communities and societies must develop, it has been increasingly recognised that they have to be resilient to be sustainable (Becker, 2021; Berkes, 2007; Boyd et al., 2013; Levin et al., 1998; Manyena, 2006; Perrings, 2006; van Niekerk, 2015). This idea is primarily based on arguments that sustainable development demands societies to be resilient to safeguard what they value over time (cf. Becker & Tehler, 2013). Although the exact meaning of the concept of resilience differs (see Pendall et al., 2010), there has been a proliferation of publications focussing on the importance of resilience for safety and sustainability (e.g., Becker, 2021; Becker et al., 2011; Becker et al., 2014; Coaffee, 2008; Duit et al., 2010; Handmer & Dovers, 1996; Hollnagel, 2006; Mazur, 2013; Rockström et al., 2009; Tu et al., 2019).

The concept of resilience has been used for different purposes in various scientific disciplines, from psychology to engineering, resulting in equally diverse definitions as for the concepts of development and risk presented earlier in this chapter. Many of these definitions describe resilience as either (1) ability to 'bounce back' to a single equilibrium (e.g., Cohen et al., 2011; Pimm, 1984), as (2) a measure of robustness or buffering capacity before a disturbance forces a system from one stable equilibrium to another (e.g., Berkes & Folke, 1998; Holling, 1973) or as (3) ability to adapt in reaction to a disturbance (Pendall et al., 2010, p. 76). Although these definitions are useful for their intended purposes (Figure 7.14), organisations, communities or societies are human-environment systems that adapt to changes over time (Branlat & Woods, 2010, pp. 11–13; Harvey, 1968), making single- or multi-equilibrium approaches to resilience less suitable in this context (Woods, 2015)—at least in relation to sustainable development that requires long-term perspectives by definition. Moreover, organisations, communities or societies entail human beings with the ability not only to react to disturbances but also to anticipate and learn from them (cf. Hollnagel, 2009). The resilience of human-environment systems includes not only reactive qualities but also proactive qualities (Becker et al., 2016; Hollnagel, 2017). This conclusion calls for modifications of conventional reactive notions of resilience in complex adaptive systems when applying them in relation to sustainable development (Figure 7.14).

Sustainable development requires the ability of a human-environment system to achieve its objectives in an as high proportion of cases as possible under varying conditions (cf. Hollnagel, 2014). Sustainability cannot be considered independently of its core activities since resilience is not only about addressing various adverse outcomes but also about how the system performs every day. Resilience can thus be seen as the 'intrinsic ability of a system to adjust its functioning before, during, or following changes and disturbances, so that it can sustain required operations under both expected and unexpected conditions' (Hollnagel, 2011, p. xxxvi). It is important to note that, in the context of this book, both the human-environment system we focus on and the environment it exists in are changing repeatedly, if not continuously, over time. Hence, altering what adjustments are feasible, what functioning is intended, what operations are required, what conditions are unexpected, and so on. Although this perspective on resilience holds, considering its inherent temporality, resilience is better conceived as the

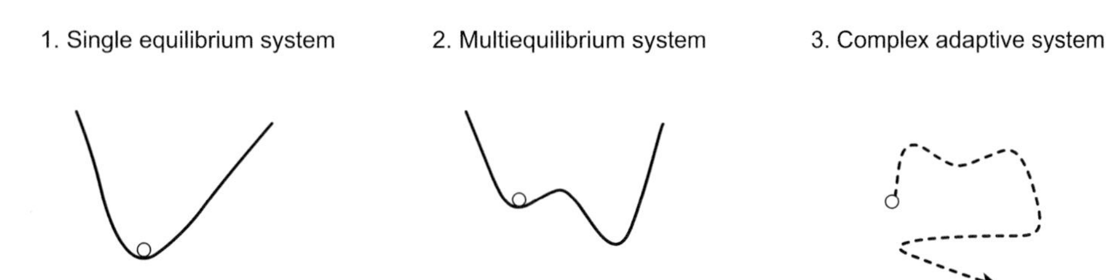

Figure 7.14 Three common approaches to resilience represented as system states in state space, which is the set of all possible states of a system regardless of variables.

ability to develop along a preferred development trajectory through the state space of the system (Figure 7.8).

Some of the highly influential scientists suggesting the planetary boundaries presented in Chapter 4 defines resilience as 'the capacity of a system to continually change and adapt yet remain within critical thresholds' (Stockholm Resilience Center, 2012). When adding Raworth's (2012) social boundaries to define her safe and just space for humanity (Figure 4.8), resilience can be seen as the capacity of a human-environment system to continuously develop while remaining within human and environmental boundaries (Figure 7.15). However, the critical question is what capacities we need to be able to develop and maintain our communities and society along our preferred expected development trajectory and within these boundaries. It is important to note that the human and environmental boundaries and our preferred expected development trajectory within them are all social constructions and equally based on our empirical understanding of the world and our value preferences. Resilience is, in other words, about protecting and maintaining what human beings value, now and for the future.

For any human-environment system to be resilient in our dynamic world, it must have the ability to adapt. That is to say; it must be able to transform itself to continue to develop along its preferred expected trajectory within human and environmental boundaries, regardless of what happens. I elaborate in more detail on how human-environment systems adapt in the next chapter, but what is often forgotten is the inputs these systems require to start adapting. This is why Hollnagel (2009) calls the ability to adapt responding, as the system responds by adapting to input from other key abilities for resilience. However, as the term responding connotes, in this broader societal context, only the reactive response to adverse events, the name of this ability is maintained as adaptation throughout this book.

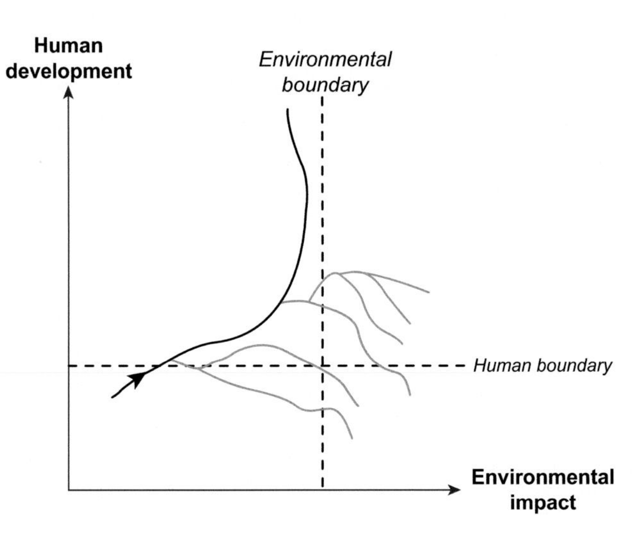

Figure 7.15 A preferred expected development trajectory, with potential deviations crossing human and environmental boundaries.

As I mentioned earlier, human-environment systems involve human beings with the ability to anticipate events before they happen. Such anticipation can be explicitly documented—for example, forecast reports, development plans and so on—but is more often just an implicit part of our everyday human experience. We routinely anticipate what could happen and use that as input for our decisions and actions. For instance, several decisions about my travel plan for an important meeting in Brussels are informed by implicit appraisal of what can happen on the way that may delay my arrival. When booking the train tickets, I look at the time available for changing trains in Copenhagen and make sure to book an earlier departure of the local train from Lund than suggested since experience indicates enough likelihood of delays to warrant a bit more margin. I also look at the time available at the train stations in Hamburg and Cologne and if there are later departures that would still get me to the meeting even if something happens on the way. On the day of departure, I look at the weather and time of day and consider how much time I need to walk to the train station, which takes longer if the roads are icy in winter. Then, when there are no contingencies, I have time for a coffee, work and perhaps even a stroll in these lovely European cities and arrive well before the meeting. However, when a delay causes me to miss a connection, I can still get to the meeting in time, even if it entails some stress. In other words, we routinely assess the likelihood and consequence of multiple risk scenarios and adapt our behaviour and activities based on that assessment and our underlying value preference.

Although I go into more detail on how human-environment systems anticipate their many potential futures in the next chapter, it is essential to note that such anticipation may be both formal and informal. For instance, in the form of formalised technical procedures, such as risk assessments or weather forecasts, and in the form of informal mechanisms, such as indigenous knowledge of foreboding signs in the environment. Explicit or implicit, formal or informal, the ability of human-environment systems to anticipate is central to its resilience as it provides vital input to decisions and actions for adapting the system in our dynamic world (Figure 7.16). Although using the more contentious term prediction, Vickers (1968, p. 463) clarifies the need for anticipation by stating that:

> *Clearly, a modern society would not hold together for a year, if R [the regulation process] did not begin to work until it received signals of incipient breakdown. These signals are supplemented in varying degrees by prediction. [...] The ability of R [the regulation process] to act on merely predictive signals is, however, limited by several factors. Such signals never compel action, though they may invite it. They make less impact on the mind. And they are no more reliable than prediction itself.*

Another input for adapting human-environment systems comes from recognising what 'is or could be a threat in the near term' (Hollnagel, 2009, p. 120), as well as what the impact is on the human-environment system in an ongoing disturbance, disruption or disaster (Becker et al., 2016). Hollnagel (2009, pp. 124–125) focuses on the

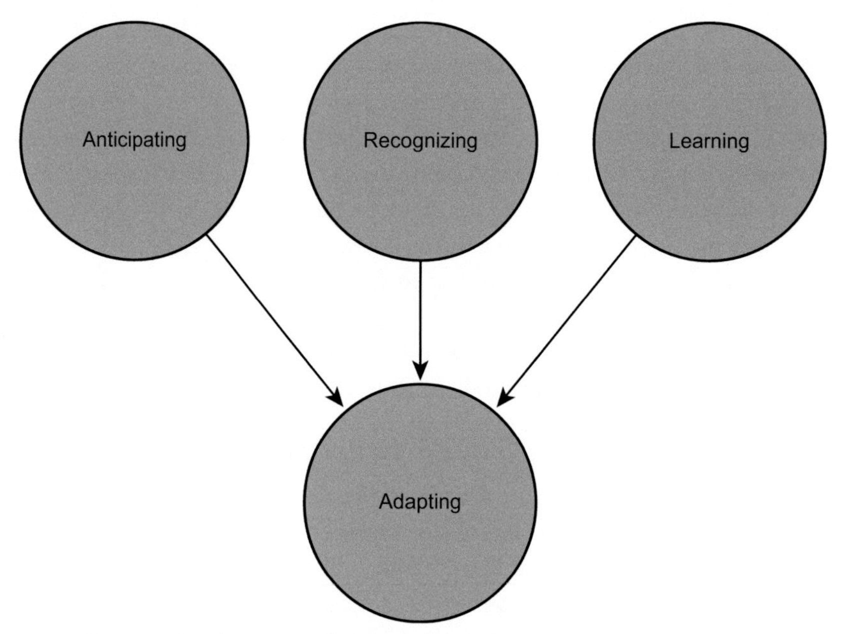

Figure 7.16 Three central abilities for resilience as inputs to a fourth.

former by emphasising the need to monitor specific predefined indicators of potential problems, for example, actual river flow, number of cholera cases in the area and so on. He calls this monitoring, which constitutes the proactive part of recognising critical states in human-environment systems that require action. The other part of this ability is reactive, referred to as impact assessment of the situation in actual adverse events—what Woods (1988) calls detection. In such situations, the human-environment system must first recognise the impact before it can do something to address it. Both parts of the ability to recognise critical states in human-environment systems generate vital input to decisions and actions for adapting the system to protect and maintain what human beings value (Figure 7.16).

Finally, human beings can also learn from experiences—both their own and others. Hollnagel (2009, p. 127) states that a 'resilient system must be able to learn from experience'. The focus here is neither on what failed in a specific event nor on who is to blame for it. Learning is instead a continuous planned process focussing on how the human-environment system functions, its interdependencies, links between causes and effects, and so on (Hollnagel, 2009, pp. 129–130)—a process that not only generates input for adapting the human-environment systems to manage future events better but also feeds back signals about the abilities of the system to anticipate, recognise, adapt and learn in general.

In short, the ability to adapt requires input concerning what is anticipated to have the potential to become a problem in the future, what is recognised as critical or soon to be critical in the current situation, or what is learnt to be a problem from experience (Becker et al., 2016). However, the connections between these four fundamental abilities for resilience are neither restricted to these inputs nor the learning feedback just mentioned. For instance, what we anticipate as a potential future problem may inform what variables we monitor to identify when that problem situation arises, and recognising a critical situation may trigger the anticipation of new potential future scenarios. Similarly, the ability to learn not only feeds back signals about the other abilities but also demands signals about their performance to facilitate learning in the first place. The four abilities for resilience are, in other words, locked in a nexus in which they are all interdependent (Figure 7.17).

Suppose resilience is the capacity of a human-environment system to continuously develop along a preferred expected trajectory while remaining within human and environmental boundaries. In that case, resilience is an emergent property determined by the ability of the human-environment system to anticipate, recognise, adapt to and learn from variations, changes, disturbances, disruptions and disasters that may cause harm to what human beings value. Resilience is, in other words, a means to reach the ends of safety and sustainability (Becker et al., 2014).

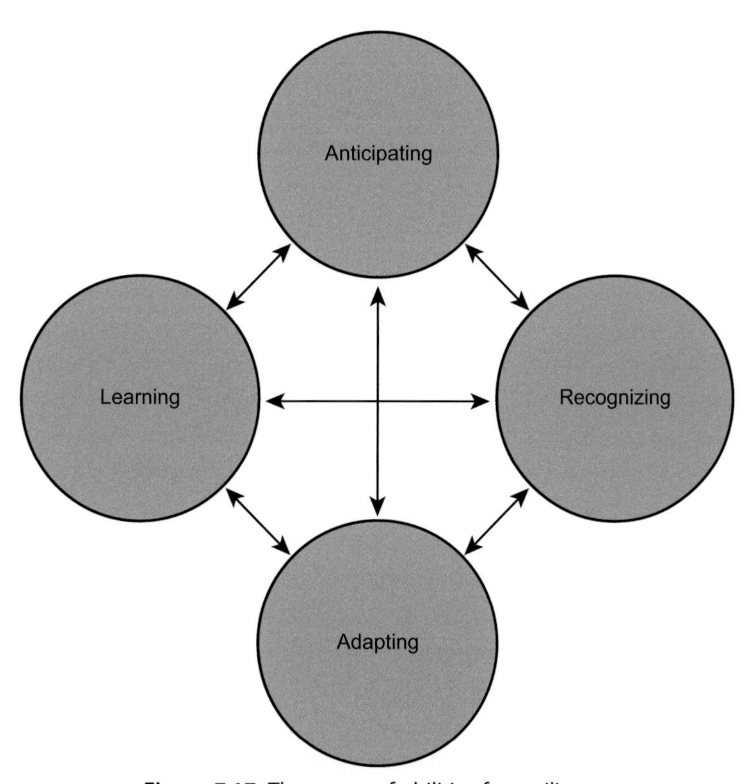

Figure 7.17 The nexus of abilities for resilience.

Conclusion

Our future is uncertain, and our world is complex, ambiguous and dynamic. To manage uncertainty in what the future holds for us, we utilise the concept of risk and structure various risk scenarios that could divert us away from our preferred expected future. This preferred expected future is a description of a sustainable development trajectory on which we protect and maintain what we as human beings value, now and for the future. To assess and address risk scenarios, we must first establish what is valuable to us and important to protect. Then, we need to identify what hazards could initiate events that have the potential to disturb, disrupt or destroy these valuable aspects. However, what really determines the consequences of such events is their vulnerability to the impact of each specific event, as well as what capacities we have to actively limit the consequences.

To grasp the complexity of our world, we construct explicit or implicit models of it. Models that I refer to in this book as human-environment systems to highlight the intrinsic links between nature and society on all levels. I elaborate more on human-environment systems in Chapter 11 and the importance of making such models explicit to manage risk and resilience for sustainable development.

Regardless of how much we think we know about our world, our knowledge is always ambiguous in the sense of including multiple ways of explaining and understanding observed phenomena and solving experienced problems. To cope with this ambiguity, we need to involve a wide range of actors, including experts, policymakers and the public, as well as state, market and civil society, and allow for dialogue and mutual understanding of various standpoints.

Finally, the dynamic character of our world makes static approaches to safety and sustainability futile. Sustainable development entails communities, societies or any other human-environment systems following their preferred expected development trajectory while remaining within human and environmental boundaries. Human-environment systems must, in other words, be resilient to be sustainable. It means that to follow that development trajectory, human-environment systems must be able to anticipate, recognise, adapt to and learn from variations, changes, disturbances, disruptions and disasters that may cause harm to what human beings value. I devote Chapter 8 to further operationalise the concept of resilience, maintaining a clear link to managing risk and suggesting connections to other common approaches to addressing risk and disasters, such as disaster risk reduction, climate change adaptation and disaster management.

References

Abelson, R. P. (1981). Psychological status of the script concept. *American Psychologist, 36*, 715–729.
Aristotle. (1801). *The metaphysics of Aristotle*. Davis, Wilks: Taylor.
Armaş, I. (2006). Earthquake risk perception in Bucharest, Romania. *Risk Analysis, 26*, 1223–1234.

Ashby, W. R. (1957). *An introduction to cybernetics*. Chapman & Hall Ltd.
Atkinson, P., & Hammersley, M. (2007). *Ethnography: Principles in practice*. Routledge.
Aven, T. (2007). A unified framework for risk and vulnerability analysis covering both safety and security. *Reliability Engineering & System Safety, 92*, 745–754.
Aven, T. (2011). On some recent definitions and analysis frameworks for risk, vulnerability, and resilience. *Risk Analysis, 31*, 515–522.
Aven, T., & Renn, O. (2009a). On risk defined as an event where the outcome is uncertain. *Journal of Risk Research, 12*, 1–11. https://doi.org/10.1080/13669870802488883
Aven, T., & Renn, O. (2009b). The role of quantitative risk assessments for characterizing risk and uncertainty and delineating appropriate risk management options, with special emphasis on terrorism risk. *Risk Analysis, 29*, 587–600.
Bankoff, G., & Hilhorst, D. (Eds.). (2022). *Why vulnerability still matters: The politics of disaster risk creation*. Routledge.
Barnes, B., Bloor, D., & Henry, J. (1996). *Scientific knowledge: A sociological analysis*. Athlone.
Bartlett, F. C. (1995). *Remembering: A study in experimental and social psychology*. Cambridge University Press.
Becker, P. (2011). Whose risks? -Gender and the ranking of hazards. *Disaster Prevention and Management, 20*, 423–433.
Becker, P. (2012). The importance of explicit discussions of what is valuable in efforts to reduce disaster risk: A case study from Fiji. *Asian Journal of Environment and Disaster Management, 4*, 477–492.
Becker, P. (2021). Advancing resilience for sustainable development: A capacity development approach. In W. Leal Filho, R. Pretorius, & L. O. de Sousa (Eds.), *Sustainable development in Africa: Fostering sustainability in one of the world's most promising continents* (pp. 525–540). Springer. https://doi.org/10.1007/978-3-030-74693-3_29
Becker, P., Abrahamsson, M., & Tehler, H. (2011). An emergent means to assurgent ends: Community resilience for safety and sustainability. In E. Hollnagel, E. Rigaud, & D. Besnard (Eds.), *Proceedings of the fourth resilience engineering symposium, June 8-10, 2011* (pp. 29–35). MINES ParisTech.
Becker, P., Abrahamsson, M., & Tehler, H. (2014). An emergent means to assurgent ends: Societal resilience for safety and sustainability. In C. P. Nemeth, & E. Hollnagel (Eds.), *Resilience engineering perspectives* (Vol. 3). Ashgate.
Becker, P., Abrahamsson, M., & Tehler, H. (2016). An emergent means to assurgent ends: Societal resilience for safety and sustainability. In *Resilience engineering in practice* (Vol. 2, pp. 29–40). CRC Press. https://doi.org/10.1201/9781315605708
Becker, P., & Tehler, H. (2013). Constructing a common holistic description of what is valuable and important to protect: A possible requisite for disaster risk management. *International Journal of Disaster Risk Reduction, 6*, 18–27. https://doi.org/10.1016/j.ijdrr.2013.03.005
Berkes, F. (2007). Understanding uncertainty and reducing vulnerability: Lessons from resilience thinking. *Natural Hazards, 41*, 283–295.
Berkes, F., & Folke, C. (1998). Linking social and ecological systems for resilience and sustainability. In F. Berkes, & C. Folke (Eds.), *Linking social and ecological systems: Management practices and social mechanisms for building resilience* (pp. 1–25). Cambridge University Press.
Bernstein, P. L. (1996). *Against the gods: The remarkable story of risk*. John Wiley & Sons.
von Bertalanffy, L. (1950). An outline of general system theory. *The British Journal for the Philosophy of Science, 1*, 134–165.
Bhaskar, R. (2008). *A realist theory of science* (2 ed.). Routledge.
Black, M. (1964). The gap between "is" and "should". *Philosophical Review, 73*, 165–181.
Blaikie, N. W. H. (1991). A critique of the use of triangulation in social research. *Quality and Quantity, 25*, 115–136. https://doi.org/10.1007/BF00145701
Blaikie, N. W. H. (2007). *Approaches to social enquiry* (2 ed.). Polity Press.
Blaikie, P. M., Cannon, T., Davis, I. R., & Wisner, B. (1994). *At risk: Natural hazards, people's vulnerability, and disasters*. Routledge.
Boland, R. J. J., Singh, J., Salipante, P., Aram, J. D., Fay, S. Y., & Kanawattanachai, P. (2001). Knowledge representations and knowledge transfer. *Academy of Management Journal, 44*, 393–417. https://doi.org/10.2307/3069463

Bontempo, R. N., Bottom, W. P., & Weber, E. U. (1997). Cross-cultural differences in risk perception: A model-based approach. *Risk Analysis, 17*, 479–488.
Boyd, E., Cornforth, R. J., Lamb, P. J., Tarhule, A., Lélé, M. I., & Brouder, A. (2013). Building resilience to face recurring environmental crisis in African Sahel. *Nature Climate Change, 3*, 631–638. https://doi.org/10.1038/NCLIMATE1856
Branlat, M., & Woods, D. D. (2010). *How do systems manage their adaptive capacity to successfully handle disruptions? A resilience engineering perspective*. Arlington: Complex Adaptive Systems —Resilience, Robustness, and Evolvability, Proceedings from the Association for the Advancement of Artificial Intelligence conference. *November 2010*.
Buckle, P. (1998). Re-defining community and vulnerability in the context of emergency management. *Australian Journal of Emergency Management, 13*, 21–26.
Bueter, A. (2022). Bias as an epistemic notion. *Studies in History and Philosophy of Science, 91*, 307–315. https://doi.org/10.1016/j.shpsa.2021.12.002
Cannon, T. (2008). Vulnerability, "innocent" disasters and the imperative of cultural understanding. *Disaster Prevention and Management, 17*, 350–357. https://doi.org/10.1108/09653560810887275
Chambers, R. (1997). *Whose reality counts? Putting the first last*. Intermediate Technology Publications.
Chauvin, B., Hermand, D., & Mullet, E. (2007). Risk perception and personality facets. *Risk Analysis, 27*, 171–185. https://doi.org/10.1111/j.1539-6924.2006.00867.x
Checkland, P. (1981). *Systems thinking, systems practice*. John Wiley & Sons.
Checkland, P., & Holwell, S. (2007). Action research: Its nature and validity. In N. Kock (Ed.), *Information systems action research: An applied view of emerging concepts and methods* (pp. 3–16). Springer Science.
Churchman, C. W. (1970). Operations research as a profession. *Management Science, 17*, B37–B53.
Cilliers, P. (1998). *Complexity & postmodernism: Understanding complex systems*. Routledge.
Cilliers, P. (2005). Complexity, deconstruction and relativism. *Theory, Culture & Society, 22*(5), 255–267. https://doi.org/10.1177/0263276405058052
Coaffee, J. (2008). Risk, resilience, and environmentally sustainable cities. *Energy Policy, 36*, 4633–4638.
Cohen, L., Pooley, J. A., Ferguson, C., & Harms, C. (2011). Psychologists understandings of resilience: Implications for the discipline of psychology and psychology practice. *Australian Community Psychologist, 23*, 7–22.
Comte, A. (2000). *The positive philosophy, vol. II (II)*. Batoche Books.
Conant, R. C., & Ashby, W. R. (1970). Every good regulator of a system must be a model of that system. *International Journal of Systems Science, 1*, 89–97.
Connell, N. A. D., Klein, J. H., & Powell, P. L. (2003). It's tacit knowledge but not as we know it: Redirecting the search for knowledge. *Journal of the Operational Research Society, 54*, 140–152. https://doi.org/10.2307/4101605
Coppola, D. P. (2015). *Introduction to international disaster management* (3 ed.). Butterworth-Heinemann http://books.google.se/books?id=xyrDoQEACAAJ&hl=&source=gbs_api.
Crenshaw, K. (1989). *Demarginalizing the intersection of race and sex: A black feminist critique of antidiscrimination doctrine, feminist theory and antiracist politics* (Vol. 1989, pp. 139–167). University of Chicago Legal Forum.
Crotty, M. (1998). *The foundations of social research: Meaning and perspective in the research process*. Sage Publications.
Davis, I. R., & Alexander, D. E. (2022). *A glass half-full or half-empty? A dialogue on progress in disaster risk reduction (working paper 2022-1)*. University College London.
Dewey, J. (1906). The experimental theory of knowledge. *Mind, New Series, 15*, 293–307.
Dewey, J. (1922). *Human nature and conduct: An introduction to social psychology*. Henry Holt & Co.
Dewey, J. (2012). *The public and its problems*. Pennsylvania State University Press.
Dilley, M., & Boudreau, T. E. (2001). Coming to terms with vulnerability: A critique of the food security definition. *Food Policy, 26*, 229–247.
Donne, J. (1624). *Devotions vpon emergent occasions, and Seuerall steps in my sicknes*. Printed by AM for Thomas Iones.
Dos Santos, T. (1970). The structure of dependence. *The American Economic Review, 60*, 231–236.
Douglas, M., & Wildavsky, A. (1983). *Risk and culture: An essay on the selection of technological and environmental dangers*. University of California Press.

Duit, A., Galaz, V., Eckerberg, K., & Ebbesson, J. (2010). Governance, complexity, and resilience. *Global Environmental Change, 20*, 363–368. https://doi.org/10.1016/j.gloenvcha.2010.04.006
Durkheim, É. (1993). *Ethics and the sociology of morals*. Prometheus Books.
Elliott, K. C. (2017). *A tapestry of values: An introduction to values in science*. Oxford University Press.
Ellis, R. D. (2019). The role of values in scientific theory selection and why it matters to medical education. *Bioethics, 33*(9), 984–991. https://doi.org/10.1111/bioe.12612
Fischhoff, B. (1991). Value elicitation: Is there anything in there. *American Psychologist, 46*, 835–847.
Fischhoff, B., Watson, S. R., & Hope, C. (1984). Defining risk. *Policy Sciences, 17*, 123–139.
Flynn, J., Slovic, P., & Mertz, C. K. (1994). Gender, race, and perception of environmental health risks. *Risk Analysis, 14*, 1101–1108.
Fordham, M. H. (2007). Disaster and development research and practice: A necessary eclecticism. In H. Rodríguez, E. L. Quarantelli, & R. R. Dynes (Eds.), *Handbook of disaster research* (pp. 335–346). Springer.
Frank, A. G. (2004). The development of underdevelopment. In S. M. Wheeler, & T. Beatley (Eds.), *The sustainable urban development reader* (pp. 38–41). Routledge.
Gallopín, G. C. (2006). Linkages between vulnerability, resilience, and adaptive capacity. *Global Environmental Change, 16*, 293–303. https://doi.org/10.1016/j.gloenvcha.2006.02.004
Giddens, A. (1984). *The constitution of society: Outline of the theory of structuration*. University of California Press.
Greenwood, D., & Levin, M. (2007). *Introduction to action research: Social research for social change*. Sage Publications.
Greider, T., & Garkovich, L. (2010). Landscapes: The social construction of nature and the environment. *Rural Sociology, 59*, 1–24. https://doi.org/10.1111/j.1549-0831.1994.tb00519.x
Greiner, S. P. (2013). *Investment risk and uncertainty: Advanced risk awareness techniques for the intelligent investor*. John Wiley & Sons.
Hagelsteen, M., & Becker, P. (2014). A great Babylonian confusion: Terminological ambiguity in capacity development for disaster risk reduction in the international community. In *Proceedings of the 5th international disaster risk conference* (pp. 298–300). Switzerland: Davos, 24-28/08/2014.
Haimes, Y. Y. (2004). *Risk modeling, assessment, and management*. John Wiley & Sons.
Hammersley, M. (1992). *What's wrong with ethnography?: Methodological explorations*. Routledge.
Hammersley, M. (2000). *Taking sides in social research*. Routledge.
Hammersley, M. (2017). On the role of values in social research: Weber vindicated? *Sociological Research Online, 22*(1), 130–141. https://doi.org/10.5153/sro.4197
Handmer, J. W., & Dovers, S. R. (1996). *A typology of resilience: Rethinking institutions for sustainable development* (Vol. 9, pp. 482–511). Organization & Environment. https://doi.org/10.1177/108602669600900403
ul Haq, M. (1995). *Reflections on human development*. Oxford University Press.
Hardcastle, M.-A. R., Usher, K. J., & Holmes, C. A. (2005). An overview of structuration theory and its usefulness for nursing research. *Nursing Philosophy, 6*, 223–234. https://doi.org/10.1111/j.1466-769X.2005.00230.x
Harper, Dictionary (2022). *Development*. https://www.etymonline.com/word/development.
Harvey, E. (1968). Technology and the structure of organizations. *American Sociological Review, 33*, 247–259.
Hassel, H. (2010). *Risk and vulnerability analysis in society's proactive emergency management: Developing methods and improving practices*. Lund University.
Hearn, J. (2012). *Theorizing power*. Palgrave Macmillan.
Hearn Morrow, B. (1999). Identifying and mapping community vulnerability. *Disasters, 23*, 1–18. https://doi.org/10.1111/1467-7717.00102
Hettne, B. (1995). *Development theory and the three worlds: Towards an international political economy of development*. Longman.
Hewitt, K. (1983). The idea of calamity in a technocratic age. In K. Hewitt (Ed.), *Interpretations of calamity* (pp. 3–32). Allen & Unwin.
Heylighen, F., Cilliers, P., & Gershenson, C. (2007). Complexity and philosophy. In J. Bogg, & R. Geyer (Eds.), *Complexity, science and society* (pp. 117–134). Radcliffe Publishing.
Hollenstein, K., Bieri, O., & Stückelberger, J. (2002). *Modellierung der Vulnerability vonSchadenobjekten gegenüber Naturgefahrenprozessen*.

Holling, C. S. (1973). Resilience and stability of ecological systems. *Annual Review of Ecology and Systematics, 4*, 1–23.
Hollnagel, E. (2006). Resilience - the challenge of the unstable. In E. Hollnagel, D. D. Woods, & N. G. Leveson (Eds.), *Resilience engineering: Concepts and precepts* (pp. 9–17). Ashgate.
Hollnagel, E. (2009). The four cornerstones of resilience engineering. In C. P. Nemeth, & E. Hollnagel (Eds.), *Resilience engineering perspectives, volume 2: Preparation and restoration* (pp. 117–133). Ashgate.
Hollnagel, E. (2011). Prologue: The scope of resilience engineering. In E. Hollnagel, J. Pariès, D. D. Woods, & J. Wreathall (Eds.), *Resilience engineering in practice: A guidebook*. CRC Press.
Hollnagel, E. (2014). *Safety-I and safety-II: The past and future of safety management*. Ashgate.
Hollnagel, E. (2017). *Safety-II in practice: Developing the resilience potentials*. Routledge.
Hume, D. (1896). *A treatise of human nature*. Clarendon Press.
Japp, K. P., & Kusche, I. (2008). Systems theory and risk. In J. O. Zinn (Ed.), *Social theories of risk and uncertainty: An introduction* (pp. 76–105). Blackwell Publishing.
Kant, I. (1968). *Kritik Der Reiner vernunft*. Walter de Gruyter.
Kaplan, S., & Garrick, B. J. (1981). On the quantitative definition of risk. *Risk Analysis, 1*, 11–27. https://doi.org/10.1111/j.1539-6924.1981.tb01350.x
Kaplan, S., Haimes, Y. Y., & Garrick, B. J. (2001). Fitting hierarchical holographic modeling into the theory of scenario structuring and a resulting refinement to the quantitative definition of risk. *Risk Analysis, 21*, 807–819. https://doi.org/10.1111/0272-4332.215153
Kasperson, R. E., & Kasperson, J. X. (1996). The social amplification and attenuation of risk. *The Annals of the American Academy of Political and Social Science, 545*, 95–105.
Kates, R. W., Clarke, W. C., Corell, R. W., Hall, J. M., Jaeger, C. C., Lowe, I., McCarthy, J. J., Schellnhuber, H. J., Bolin, B., Dickson, N. M., Faucheux, S., Gallopín, G. C., Huntley, B., Jodha, N. S., Kasperson, R. E., Mabogunje, A., Matson, P. A., Mooney, H., Moore, B., … Svedin, U. (2001). Sustainability science. *Science, 292*, 641–642.
Kates, R. W., Turner, B. L., II, & Clarke, W. C. (1990). The great transformation. In B. L. Turner, II, W. C. Clarke, R. W. Kates, J. F. Richards, J. T. Mathews, & W. B. Meyer (Eds.), *The great transformation* (pp. 1–17). Beacon Press.
Keat, R., & Urry, J. (1975). *Social theory as science*. Routledge.
Keeney, R. L. (1992). *Value-focused thinking: A path to creative decisionmaking*. Harvard University Press.
Klinenberg, E. (2002). *Heat wave: A social autopsy of disaster in Chicago*. University of Chicago Press.
Kourany, J. A. (2010). *Philosophy of science after feminism*. OUP USA.
Kunreuther, H. C., & Slovic, P. (1996). Science, values, and risk. *The Annals of the American Academy of Political and Social Science, 545*, 116–125.
Kurtz, C. F., & Snowden, D. J. (2003). The new dynamics of strategy: Sense-making in a complex and complicated world. *IBM Systems Journal, 42*(3), 462–483. https://doi.org/10.1147/sj.423.0462
Lam, L. T. (2005). Parental risk perceptions of childhood pedestrian road safety: A cross cultural comparison. *Journal of Safety Research, 36*, 181–187. https://doi.org/10.1016/j.jsr.2005.03.003
Laughlin, R. B. (2003). Fractional quantization: Nobel lecture in physics 1998. In G. Ekspong (Ed.), *Nobel lectures in physics 1996–2000* (Vol. 8, pp. 264–286). World Scientfic. https://doi.org/10.1142/4973
Levin, S. A., Barrett, S., Aniyar, S., Baumol, W., Bliss, C., Bolin, B., Dasgupta, P., Ehrlich, P., Folke, C., Gren, I.-M., Holling, C. S., Jansson, A., Jansson, B.-O., Mäler, K.-G., Martin, D., Perrings, C., & Sheshinski, E. (1998). Resilience in natural and socioeconomic systems. *Environment and Development Economics, 3*, 221–262.
Lewis, J., & Kelman, I. (2012). The good, the bad and the ugly: Disaster risk reduction (DRR) versus disaster risk creation (DRC). *PLoS Currents: Disasters, 4*, Article e4f8d4eaec6af8. https://doi.org/10.1371/4f8d4eaec6af8
Lincoln, Y. S., Lynham, S. A., & Guba, E. G. (2017). Paradigmatic controversies, contradictions, and emerging confluences, revisited. In N. K. Denzin, & Y. S. Lincoln (Eds.), *The Sage handbook of qualitative research* (pp. 213–263). Sage Publications.
Luhmann, N. (1993). *Risk: A sociological theory*. Walter de Gruyter.
Luhmann, N. (1995). *Social systems*. Stanford University Press.
Lupton, D. (2013). *Risk*. Routledge.

MacGregor, D. G., & Slovic, P. (2000). Perceived risk and driving behavior: Lessons for improving traffic safety in emerging market countries. In H. von Holst, Å. Nygren, & Å. E. Andersson (Eds.), *Transportation, traffic safety, and health: Human behavior* (pp. 35–54). Springer.

Malanson, G. P. (1999). Considering complexity. *Annals of the Association of American Geographers, 89*, 746–753.

Manyena, S. B. (2006). The concept of resilience revisited. *Disasters, 30*, 433–450.

Marton, F., & Booth, S. (1997). *Learning and awareness*. Routledge.

Mazur, L. (2013). Cultivating resilience in a dangerous world. In I. Worldwatch (Ed.), *State of the world 2013: Is sustainability still possible* (pp. 353–362). Island Press.

Mellor, D. H. (Ed.). (1980). *Prospects for pragmatism: Essays in memory of F. P. Ramsey*. Cambridge University Press.

van Niekerk, D. (2015). Disaster risk governance in Africa. *Disaster Prevention and Management, 24*(3), 397–416. https://doi.org/10.1108/DPM-08-2014-0168

Nilsson, J., & Becker, P. (2009). What's important? Making what is valuable and worth protecting explicit when performing risk and vulnerability analyses. *International Journal of Risk Assessment and Management, 13*, 345–363.

Nonaka, I. (1994). *A dynamic theory of organizational knowledge creation* (Vol. 5, pp. 14–37). Organization Science. https://doi.org/10.2307/2635068

Oliver-Smith, A. (1999). Peru's five-hundred-year earthquake: Vulnerability in historical context. In A. Oliver-Smith, & S. M. Hoffman (Eds.), *The angry earth: Disaster in anthropological perspective* (pp. 74–88). Routledge.

Organski, A. F. K. (1965). *The stages of political development*. Knopf.

Örtengren, K. (2003). *Logical framework approach - a summary of the theory behind the LFA method*. Sida.

Oxford English Dictionary. (2020). *Oxford dictionary of English*. Oxford University Press.

Payne, J. W., Bettman, J. R., & Johnson, E. J. (1992). Behavioral decision research: A constructive processing perspective. *Annual Review of Psychology, 43*, 87–131.

Pendall, R., Foster, K. A., & Cowell, M. (2010). Resilience and regions: Building understanding of the metaphor. *Cambridge Journal of Regions, Economy and Society, 3*, 71–84.

Perrings, C. (2006). Resilience and sustainable development. *Environment and Development Economics, 11*, 417–427.

Pimm, S. L. (1984). The complexity and stability of ecosystems. *Nature, 307*, 321–326.

Polanyi, M. (1961). Knowing and being. *Mind, 70*, 458–470. https://doi.org/10.2307/2251603

Polanyi, M. (1966a). The logic of tacit inference. *Philosophy, 41*, 1–18. https://doi.org/10.2307/3749034

Polanyi, M. (1966b). *The tacit dimension*. Doubleday & Company.

Polanyi, M. (1967). Sense-giving and sense-reading. *Philosophy, 42*, 301–325. https://doi.org/10.2307/3748494

Ramsey, F. P. (1931). *The Foundations of Mathematics, and other logical essays*. Kegan Paul, Trench, Trubner & Co.

Raworth, K. (2012). *A safe and just space for humanity: Can we live within the Doughnut*.

Renn, O. (1992). Concepts of risk: A classification. In S. Krimsky, & D. Golding (Eds.), *Social theories of risk*. Praeger.

Renn, O. (1998a). The role of risk perception for risk management. *Reliability Engineering & System Safety, 59*, 49–62. https://doi.org/10.1016/S0951-8320(97)00119-1

Renn, O. (1998b). Three decades of risk research: Accomplishments and new challenges. *Journal of Risk Research, 1*, 49–71. https://doi.org/10.1080/136698798377321

Renn, O. (2008). *Risk governance: Coping with uncertainty in a complex world*. Earthscan.

Resilience Center, Stockholm (2012). *What is resilience*. http://www.stockholmresilience.org/research/whatisresilience.4.aeea46911a3127427980004249.html.

Rist, G. (2006). *The history of development: From western origins to global faith*. Zed Books.

Rockström, J., Steffen, W., Noone, K., Persson, Å., Chapin, F. S., III, Lambin, E. F., Lenton, T. M., Scheffer, M., Folke, C., Schellnhuber, H. J., Nykvist, B., de Wit, C. A., Hughes, T. P., van der Leeuw, S., Rodhe, H., Sörlin, S., Snyder, P. K., Costanza, R., Svedin, U., … Foley, J. A. (2009). Planetary boundaries: Exploring the safe operating space for humanity. *Ecology and Society, 14*, 32.

Rolin, K. (2012). A feminist approach to values in science. *Perspectives on Science, 20*(3), 320–330. https://doi.org/10.1162/posc_a_00068
Rosa, E. A. (1998). Metatheoretical foundations for post-normal risk. *Journal of Risk Research, 1*, 15–44. https://doi.org/10.1080/136698798377303
Rosa, E. A. (2010). The logical status of risk - to burnish or to dull. *Journal of Risk Research, 13*, 239–253. https://doi.org/10.1080/13669870903484351
Rostow, W. W. (1960). *The stages of economic growth: A non-communist manifesto*. Cambridge University Press.
Sahlin, N.-E. (1990). *The philosophy of F.P. Ramsey*. Cambridge University Press.
Santos, P. P., Chmutina, K., Meding, J. V., & Raju, E. (Eds.). (2020). *Understanding disaster risk*. Elsevier.
Schank, R. C., & Abelson, R. P. (1977). *Scripts, plans, goals, and understanding: An inquiry into human knowledge structures*. L. Erlbaum Associates.
Schwartz, S. H. (1994). Are there universal aspects in the structure and contents of human values. *Journal of Social Issues, 50*, 19–45.
Seers, D. (1989). The meaning of development. In C. Cooper, & E. V. K. Fitzgerald (Eds.), *Development studies revisited: Twenty-five years of the Journal of development studies* (pp. 480–497). Frank Cass & Company.
Shrader-Frechette, K. S. (1991). *Risk and rationality: Philosophical foundations for populist reforms*. University of California Press.
Simon, H. A. (1990). Prediction and prescription in systems modeling. *Operations Research, 38*, 7–14.
Simon, H. A. (1996). *The sciences of the artificial*. MIT Press.
Simon, H. A. (2002). Forecasting the future or shaping it. *Industrial and Corporate Change, 11*, 601–605.
Sjöberg, L. (1998). Worry and risk perception. *Risk Analysis, 18*, 85–93.
Sjöberg, L., & Thedéen, T. (2003). Att reflektera over risker och teknik. In G. Grimvall, P. Jacobsson, & T. Thedéen (Eds.), *Risker i tekniska system*. Studentlitteratur.
Slovic, P. (1987). Perception of risk. *Science, 236*, 280–285.
Slovic, P. (1992). Perceptions of risk: Reflections on the psychometric paradigm. In S. Krimsky, & D. Golding (Eds.), *Social theories of risk*. Praeger.
Slovic, P. (1995). The construction of preference. *American Psychologist, 50*, 364–371.
Slovic, P., Fischhoff, B., & Lichtenstein, S. (1982). Why study risk perception. *Risk Analysis, 2*, 83–93. https://doi.org/10.1111/j.1539-6924.1982.tb01369.x
Smekal, P. (1991). *Teorier om utveckling & underutveckling*. Uppsala University.
Smuts, J. C. (1926). *Holism and evolution*. Macmillan.
Snowden, D. J., & Boone, M. E. (2007). A leader's framework for decision making. A leader's framework for decision making. *Harvard Business Review, 85*(11), 68–76, 149 https://pubmed.ncbi.nlm.nih.gov/18159787.
South Commission. (1990). *The challenge to the South*. Oxford University Press.
Steffen, W., Broadgate, W., Deutsch, L., Gaffney, O., & Ludwig, C. (2015). The trajectory of the anthropocene: The great acceleration. *The Anthropocene Review, 2*, 81–98. https://doi.org/10.1177/2053019614564785
Taleb, N. N. (2007). *The black swan: The impact of the highly improbable*. Random House.
Tanesini, A. (1999). *An introduction to feminist epistemologies*. Blackwell Publishing.
Thomas, A. (2000a). Meanings and views of development. In T. Allen, & A. Thomas (Eds.), *Poverty and development into the 21st century*. Oxford University Press.
Thomas, A. (2000b). Poverty and the 'end of development. In T. Allen, & A. Thomas (Eds.), *Poverty and development into the 21st century*. Oxford University Press.
Todaro, M. P. (1989). *Economic development in the third world*. Longman.
Tu, C., Suweis, S., & D'Odorico, P. (2019). Impact of globalization on the resilience and sustainability of natural resources. *Nature Sustainability, 2*(4), 283–289. https://doi.org/10.1038/s41893-019-0260-z
Twigg, J. (2004). *Disaster risk reduction: Mitigation and preparedness in development and emergency programming*. Overseas Development Institute.
UNDP. (1990). *Human development report 1990*. Oxford University Press.
UNDP. (2004). *Reducing disaster risk: A challenge for development*. John Swift Print.

United Nations. (2015). *Transforming our world: The 2030 Agenda for sustainable development (A/RES/70/1)*. United Nations.

Vennix, J. A. M. (2001). *Group model building: Facilitating team learning using system dynamics*. John Wiley.

Vickers, G. (1968). Is adaptability enough. In W. F. Buckley (Ed.), *Modern systems research for the behavioral scientist: A sourcebook* (pp. 460–473). Aldline Publishing.

Wallerstein, I. M. (1974). *The modern world-system*. Academic Press.

WCED. (1987). *Our common future*. Oxford University Press.

Weber, M. (1949). *The methodology of the social sciences*. The Free Press.

Wiek, A., Withycombe, L., & Redman, C. L. (2011). Key competencies in sustainability: A reference framework for academic program development. *Sustainability Science, 6*, 203–218. https://doi.org/10.1007/s11625-011-0132-6

Wijkman, A., & Timberlake, L. (1984). *Natural disasters: Acts of god or acts of man*. Earthscan.

Winther Jørgensen, M., & Phillips, L. (1999). *Diskursanalys som teori och metod*. Studentlitteratur.

Wisner, B., Blaikie, P. M., Cannon, T., & Davis, I. R. (2004). *At risk: Natural hazards, people's vulnerability and disasters*. Routledge.

Woods, D. D. (1988). Coping with complexity: The psychology of human behaviour in complex systems. In L. P. Goodstein, H. B. Andersen, & S. E. Olsen (Eds.), *Tasks, errors, and mental models* (pp. 128–148). Taylor & Francis.

Woods, D. D. (2015). Four concepts for resilience and the implications for the future of resilience engineering. *Reliability Engineering & System Safety, 141*, 5–9. https://doi.org/10.1016/j.ress.2015.03.018

Wynne, B. (1982). Institutional mythologies and dual societies in the management of risk. In H. C. Kunreuther, & E. V. Ley (Eds.), *The risk analysis controversy: An institutional perspective*. Springer.

Wynne, B. (1998). May the sheep safely graze? A reflexive view of the expert–lay knowledge divide. In S. Lash, B. Szerszynski, & B. Wynne (Eds.), *Risk, environment and modernity: Towards a new ecology* (pp. 44–83). SAGE Publications.

Zinn, J. O. (2008). Introduction. In J. O. Zinn (Ed.), *Social theories of risk and uncertainty: An introduction* (pp. 1–17). Blackwell Publishing.

Zinn, J. O. (2020). *Understanding risk-taking*. Palgrave Macmillan.

Zinn, J. O., & Taylor-Gooby, P. (2006). Risk as an interdisciplinary research area. In P. Taylor-Gooby, & J. O. Zinn (Eds.), *Risk in social science* (pp. 20–53). Oxford University Press.

CHAPTER 8

Resilience—from panacean to pragmatic

Introduction

Our world is truly dynamic. Our challenges constantly change at seemingly ever-increasing speed, demanding greater forethought and flexibility than ever in human history. Static protective and preserving measures are no longer feasible, if ever, as both human-environment systems and their sustainability challenges are continuously transforming. Here, resilience comes in as a means for safety and sustainability.

The term resilience is not in any way a new term. It is much older than commonly believed and neither originally invented by engineers to describe the performance of springs and other objects they bend, stretch or compress (Rankine, 1858, p. 273), nor by ecologists to describe stability and complexity of ecosystems (Holling, 1973). The term resilience was used as early as a few centuries AD and found its way into scientific English at least as early as the first decades of the seventeenth century when Sir Francis Bacon used the concept to describe the strength of echoes (Alexander, 2013, pp. 2707–2709). However, the use of the term resilience has literally exploded in recent years (Figure 8.1), being applied for various purposes in a broad range of functional sectors and academic disciplines.

Although many approaches to resilience are useful for their intended purposes, most are purely descriptive and not designed in relation to risk, safety, sustainability and other inherently normative concepts where values and preferences are central. Whilst such descriptive approaches may be useful when focusing on purely technical systems or ecosystems, they fail miserably to capture the momentous importance of agency, knowledge, power and conflict in systems that entail human beings (Cannon & Müller-Mahn, 2010; Olsson et al., 2015). They do not help guide how to make communities, society or any type of human-environment systems resilient in practice—given that is what we aspire to do. Moreover, resilience is often used as a mere buzzword, without conceptual clarity or apparent utility (Alexander, 2013, p. 2713). Some scholars even argue that 'the resilience discourse of the UNDP, World Bank and others is a tool in a bigger game [...] in order to get [states] to reform their institutions in the interests of global capital' (Joseph, 2013, p. 51). This neoliberal resilience agenda corresponds to observable changes in resilience on the local level all across the world (e.g., Apgar et al., 2015; Becker, 2017; Lama et al., 2017), and may be an important reason for the unprecedented proliferation of the

Sustainability Science, Second Edition
ISBN 978-0-323-95640-6, https://doi.org/10.1016/B978-0-323-95640-6.00007-5

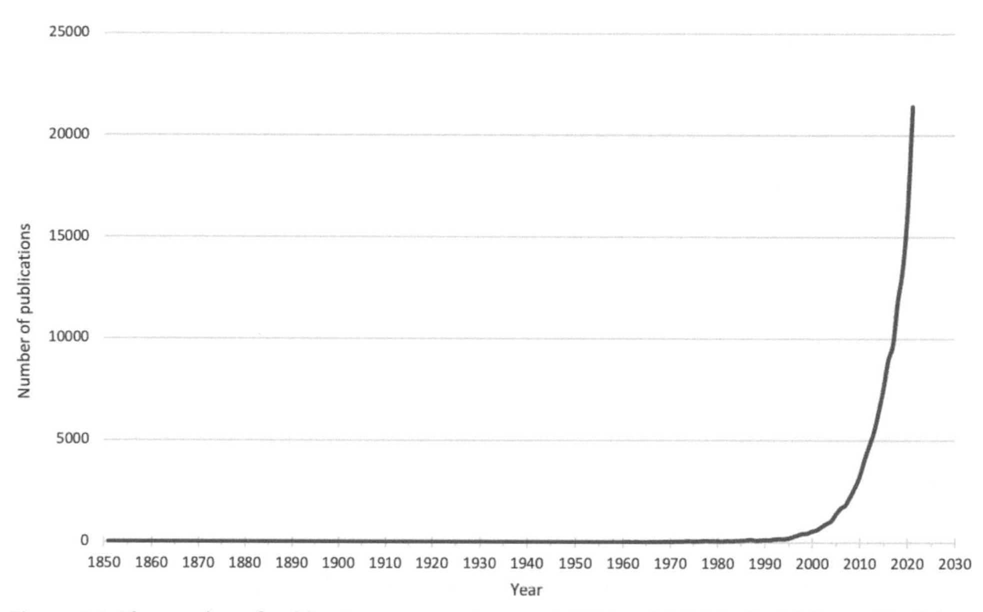

Figure 8.1 The number of publications per year between 1851 and 2021 in the bibliographical database Scopus mentioning 'resilience' in title, abstract or keyword (as of 24 August 2022).

concept. The problematic applications of various conceptualisations of resilience on human-environment systems have been met by wholesale criticism of all applications (e.g., Cannon & Müller-Mahn, 2010; Joseph, 2013; Olsson et al., 2015). However, abandoning the concept is impossible and unconstructive considering its ubiquity—like throwing out the baby with the bathwater. I argue that it is more productive to operationalise resilience in such a way it becomes useful also in relation to human-environment systems. For that, it must assist us in identifying what is needed for a human-environment system to be resilient in context.

Resilience can be a helpful concept to grasp and address pressing safety and sustainability challenges, but it is not a panacea. As just mentioned, resilience is as liable to insufficient, inappropriate or unsuccessful definitions and activities as any other approach. Just introducing a new concept does not suffice. To make a difference, we must link the conceptual with the actual. We must operationalise the concept to provide a clear link between theory and practice. This chapter focuses on just that—on attempting to make it understandable and possible to identify in actual contexts (Rist, 2006, pp. 9–10). Furthermore, the chapter suggests how this approach to resilience links to other preceding approaches to the safety and sustainability of society in relation to risk and disruptive events. But first, a word of caution concerning the prospect of measuring and comparing resilience between contexts and over time.

Inherent problems for measuring and comparing resilience

Resilience is defined here as a human-environment system's capacity to continuously develop along a preferred expected trajectory whilst remaining within human and environmental boundaries (Chapter 7). Such resilience is then an emergent property determined by the ability of the human-environment system to anticipate, recognise, adapt to and learn from variations, changes, disturbances, disruptions and disasters that may cause harm to what human beings value. When considering this approach to resilience, it is evident that it is relative and context-dependent, just like the concepts of sustainable development and risk. It is, therefore, impossible to compare the resilience of different human-environment systems without setting the same preferences and defining the same variables for measuring deviations from these preferences. This limitation may be straightforward, such as not tolerating any deaths from diarrhoeal diseases and looking at mortality statistics. However, it is often more complicated in the context of sustainable development when there are multiple values to include and numerous risks to consider. Moreover, it is equally difficult to compare the resilience of one system over time, as the preferred expected development trajectory is likely to change from the first instance to the other—at least over extended periods. Again, the only way to compare the resilience of a system over time is to compare the same variables and to keep the preferences constant, artificially or not.

This limitation may disappoint policymakers and practitioners who prefer a set of straightforward indicators to measure. However, for the concept of resilience to have any meaning in relation to the sustainability of society, it must be able to capture the normative and context-dependent qualities of the concepts of sustainable development and risk. In other words, any attempt to devise an objective and universal measurement of resilience is destined to fail as long as the concept of resilience is defined in relation to our continuously transforming world.

It is also important to remember that all knowledge of the world is socially constructed in context and that the human-environment systems I refer to throughout this book are models we construct to attempt to make sense of an overwhelmingly intricate reality. As soon as any endeavour to explain, understand or improve something includes more than one person, power takes centre stage as it determines who is allowed to be involved, contribute, decide, etc. Hence, power influences what is included and what is excluded in the human-environment system in the first place, making explicit boundary critique vital (Ulrich, 2002). Power also influences the definition of such a system's preferred expected development trajectory, what risk scenarios are deemed relevant, what options are available to maintain the preferred trajectory, etc. Power is, however, not only operating in the sense of direct decision-making and coercion but also through nondecisions and control of the agenda and through shaping social structures that regulate what thoughts and actions are thought possible in the first place (Lukes, 2005). In addition to these different dimensions of domination, power operates through its less contentious siblings of authority and legitimacy (Hearn, 2012). Power is, in other words, not only about one having power over another but also about the power of the collective to get things done (Hearn, 2012)—two sides of power that are intrinsically linked, much like the chicken and the egg. Although I agree with Russell's (Russell, 2004, p. 4) timeless statement that power is as a fundamental concept for social science as energy is for physics, I have no room for further elaborating on it here but provide slightly more detail in Chapter 14.

Resilience is not absolute and objective but relative and contextual. Although this approach is not radical or unexpected, it is vital to make the resulting restrictions clear for all actors increasingly interested in measuring the resilience of communities and society. Regardless of these inherent restrictions, identifying and assessing resilience is still possible by operationalising the concept to guide what to look for in context.

Operationalising resilience

Resilience is, as stated earlier, an emergent property of human-environment systems (cf. Pariès, 2006, p. 48). This approach to resilience entails, in other words, that the level of safety and sustainability is determined by the internal attributes of such systems. I elaborate more on emergence in Chapter 9, but it is traditionally explained by contrasting *resultants* and *emergents* (Lewes, 1875, p. 413). A resultant property of a system is a property that is traceable in the properties of the various components that constitute it, whilst an emergent property is not and emerges only in the interactions of the components (Figure 7.4). Resilience is, therefore, not an additive result but emerges in the complex interactions of the components of the human-environment system in question. To not succumb to the sheer complexity of this matter, Rasmussen (1985) suggests approaching such a complex system as a functional hierarchy from purpose, through increasingly concrete levels of functions, to the perceivable forms of the system contributing in the real world to meet its purpose (Figure 8.2).

Hollnagel (2009, 2017) suggests that resilience emerges in the performance of interdependent and dynamically coupled functions, which in this book are categorised under the four required generalised functions presented in Chapter 7: anticipating, recognising, adapting, and learning (Becker et al., 2011, 2016). The performance of each of these generalised functions is inherently dependent on the performance of the three others. However, it is equally important to understand that each function's performance also depends on the performance of numerous constituent functions further down the levels of abstraction of Rasmussen's (1985) functional hierarchy (Figure 8.2). In other words, resilience is not only an emergent property based on the performance of the four required

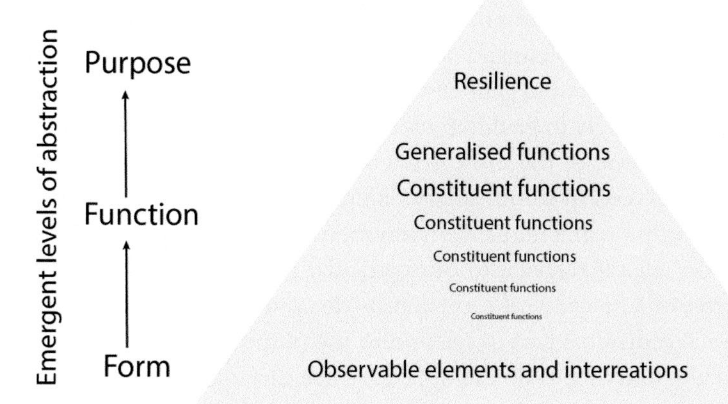

Figure 8.2 An abstraction hierarchy of resilience. *(Adapted from Rasmussen (1985).)*

generalised functions but the performance of any function is also emergent in itself. For instance, the performance of anticipation emerges in the performance of different forward-looking constituent functions—such as risk analysis and forecasting—and the performance of recognition emerges in the complex interaction of various functions for monitoring and impact assessment (Becker & Hagelsteen, 2016). Therefore, the performances of functions are not resultants but emergent properties of the performance of functions below on the functional hierarchy, irrespective of how they are conceptualised or categorised. This hierarchy continues all the way down to the assemblages of perceivable elements and their interrelations that constitute the actual performance of the functions. For instance, the tsunami detection buoys, their two redundant communications systems, the tsunami warning centre (with all its equipment and personnel), the World Meteorological Organization's dedicated Global Telecommunication System, etc., interact dynamically to perform the function of tsunami monitoring. Yet, it would all be in vain if not connected to a functioning tsunami warning system (e.g., sirens, direct messaging), if people in harm's way do not understand and trust the warning, if they think it is a false alarm, if they do not know what to do, if they are not able to do it (e.g., disabled, incarcerated), etc. Identifying which elements and interrelations can be strengthened in such assemblages is, therefore, key to developing resilience in this context (Chapter 14).

In short, to operationalise the resilience of human-environment systems, it helps to look at them as having purpose, function and form.

Purpose

The dictionary definition of purpose is 'the reason for which something is done or created or for which something exists' (Oxford English Dictionary, 2020) and it is the most abstract level in Rasmussen's (1985) systems hierarchy. It provides the overall rationale for constructing human-environment systems to understand, explain and address a particular phenomenon or problem and determines the main boundaries for that system. For instance, if we are interested in gender roles or the economy of modern Sweden, the systems we construct are likely to be different. Nevertheless, they would most likely share some elements, as even one of the most gender-equal countries in the world still has a substantial gendered division of labour, salary differences, etc. (Elwér et al., 2013). Moreover, considering a purpose of a human-environment system helps us to think about the system and focus on what is relevant to our particular purpose. Rubin's (1949) deliberation on the shutter of a mechanical camera provides a suitable analogy, as he perceives the function of each individual part in relation to the purpose of the shutter as a whole. Imagine the difficulty of trying to do the opposite since each observable physical element may have myriad functions for various purposes (Figure 8.3). For instance, a spring may also be used in a watch, and a sturdy leaf shutter may perhaps be used to cut cigars. However, the links between the purposes we ascribe and what we observe go both ways, which Kant (1892, p. 281) points out in his classic example of early scientists dissecting plants and animals:

> In fact, they can as little free themselves from this teleological proposition as from the universal physical proposition; for as without the latter we should have no experience at all, so without the former we should have no guiding thread for the observation of a species of natural things which we have thought teleologically under the concept of natural purposes.

Figure 8.3 What is the purpose of this? *(Photo by Michel Villeneuve, shared on the Creative Commons.)*

Although human-environment systems may have a range of purposes, depending on the phenomena or problems focused on, this book focuses specifically on risk, resilience and sustainability. Hence, the purpose of the human-environment systems referred to here is directly deducible from the definition of resilience presented above—to continuously develop along a preferred expected trajectory whilst remaining within human and environmental boundaries. To do that, they must protect what human beings value, now and for the future.

Without an explicit purpose for the human-environment system we are constructing, it is very difficult to explain, understand or improve its resilience. Having this purpose continuously in mind assists us in grasping or addressing challenges on the lower levels of abstraction, as 'solutions of subproblems immediately have their place in the whole picture, and it is immediately possible to judge whether a solution is correct or not' (Rasmussen, 1985, pp. 239—240).

Function

For a human-environment system to meet the stated purpose, it must be able to perform a set of functions. These functions can be conceptualised on different levels of abstraction, and the number of such levels between purpose and form depends on the type of system and the aim of the study (Rasmussen, 1985, p. 235). Although there are infinite ways to describe the functions necessary for a human-environment system to continuously develop along a preferred expected trajectory whilst remaining within human and environmental boundaries, I present one way that hopefully provides structure and rigour (Figure 8.4).

Returning to the definition of resilience presented in Chapter 7, resilience is an emergent property determined by the ability of the human-environment system to anticipate, recognise, adapt to and learn from variations, changes, disturbances, disruptions and disasters that may cause harm to what human beings value. These four requisite abilities are, in other words, the four required generalised functions that together contribute to meeting the purpose of human-environment systems in this context. They are operating continuously and simultaneously, though with different intensities over time.

Anticipating

The function of anticipating potential deviations away from a human-environment system's preferred expected development trajectory is a crucial function contributing to its resilience. It is a fundamentally proactive function in the sense of focusing on what has not yet happened (Figure 8.5). Even amid calamity, anticipation continuously provides input regarding what may occur in the human-environment system at various future points, although the time to these points tends to be shorter and shorter, the more precarious the immediate situation is. For instance, it is difficult to imagine a flooded coastal community focusing on anticipating the location of the coastline in 50 years due to

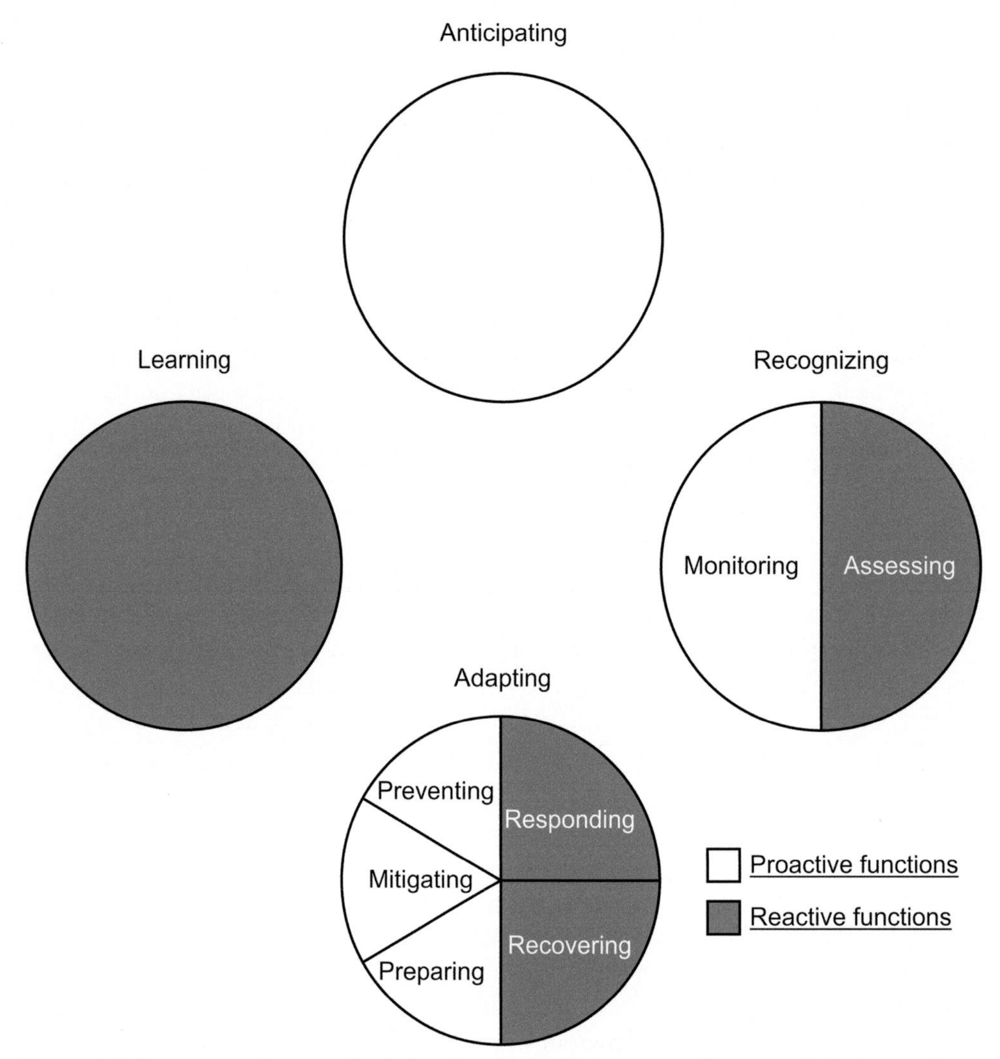

Figure 8.4 Four required generalised functions contributing to resilience and their constituent functions.

intensified erosion in an altered climate, whilst that would not at all be exceptional when planning the development of the community in everyday circumstances. There is no doubt that the flooded community in this example makes the right priorities when focusing on more pressing issues—such as anticipating the level of floodwater tomorrow, the number of evacuated people to care for at that time, etc.—but it is not always the most urgent that is the most important. However, the urgent has an unfortunate way of claiming precedence over everything else on our minds, diverting our attention away from what might be more important, with the notion of attending to it as soon

Figure 8.5 Can we anticipate trouble?. *(Source: Lightspring/Shutterstock.com.)*

as the first urgency has been addressed. This situation is referred to as the 'tyranny of the urgent' (Hummel, 1967), as there are often new urgencies taking up our time and attention, leaving the really important unattended. Just consider how little actual efforts are focusing on addressing climate change, as people are busy thinking about pandemics, refugee influx, spikes in energy prices, etc. It is as if we are stuck in some kind of permacrisis (Borges de Castro et al., 2021), where one urgency is seamlessly followed by the next, effectively diverting attention away from dealing with absolute existential peril.

In addition to being explicit or implicit, formal or informal, different inputs may also inform anticipation. Looking more closely into this, it seems like anticipation is based on two different archetypes of input—experiential or analytical (Marx et al., 2007; Slovic et al., 2002; Vervoort et al., 2012). Experiential input to the function of anticipation is characterised by direct reliance on past experiences when looking ahead into the future. Anticipated potential deviations using such input are, in other words, recurrences of past events, often relatively recent, that have affected the human-environment system itself. On the other hand, analytical input is characterised as the result of systematic analysis of the functioning of the human-environment system over a defined period ahead. In other words, anticipated potential deviations based on analytical input are descriptions of scenarios based on available knowledge in a more comprehensive sense. Whilst analytical input also relies on observations of past events, it is not using them as direct representations of potential deviations and not restricting them to events that have affected the specific human-environment system. The observations are instead used to explain and understand the structures and functions of different human-environment systems, which in turn are used to anticipate what may happen in the specific human-environment system under various future circumstances.

Both of these archetypes of input have their benefits and drawbacks, as they result from two qualitatively different mental processes (Marx et al., 2007; Slovic et al., 2002). On the one side, the process entails concrete and often personal memories, focusing on specific causal chains and evoking vivid images and strong emotions. In contrast, the other mental process entails intangible and detached theories, focusing on logical rules and abstract symbols and evoking deliberation (Marx et al., 2007, pp. 48–49). Whereas experiential input is self-evidently valid since the anticipated potential deviations have already been experienced, and is oriented towards immediate action, analytical input requires justification and evidence and is oriented towards delayed action (Slovic et al., 2002, pp. 330–331). Experiential input facilitates participation and engagement in decisions and actions (Vervoort et al., 2012), whilst analytical input facilitates a shared understanding of the challenges at hand (Vervoort et al., 2012) and directs decisions and actions towards the most efficient use of the available resources.

Human beings have many fundamental biases constraining several crucial functions contributing to resilience (e.g., Johnson & Levin, 2009; Meyer, 2006). Although these are addressed in more detail in Chapter 13, it is essential to note that anticipation is never completely informed by one type of input or the other. It is always a mix of both (Marx et al., 2007, p. 48). However, the tendency for human beings to be biased by recent experiences (Kahneman & Tversky, 1982) that evoke emotions (Slovic et al., 2002) often causes experiential input to take precedence over analytical input when informing adaptation decisions and activities (Vervoort et al., 2012). A central challenge for the resilience of human-environment systems is, thus, to balance the need for analytical input—to base anticipation on available state-of-the-art knowledge—with the need for experiential input—to stimulate engagement in actual decisions and actions.

Finally, it is important to emphasise that there are limits to our anticipation, not only due to our predisposition towards experiential input but also because of limitations or flaws in the analytical input. Even when not trying to predict the future, which is impossible in this context, but instead to anticipate a wide range of potential future scenarios, there will always be discrepancies between the anticipated scenarios and what actually happens. Westrum (1992) stresses the need to have requisite imagination when considering potential scenarios, but that is not helping us to identify the scenarios that we do not know and cannot imagine (Taleb, 2007). Also, our models of the human-environment system may be flawed and result in counterproductive output to the other functions contributing to resilience. However, we must do the best we can with the knowledge we have.

Recognising

Recognising imminent or actual deviations away from the preferred expected development trajectory of the human-environment system is also a crucial function for resilience. It has both a proactive and a reactive part, constituting functions on a lower level of abstraction. Although these constituent functions are relatively straightforward, they are unfortunately also complicated by fundamental biases presented in Chapter 13.

The first constituent function for recognition is monitoring (Box 8.1). This function focuses on continuously monitoring critical parameters in the human-environment system that might, in the near term, be causing deviations from the preferred expected development trajectory. These parameters are generally selected based on the output from the functions of anticipation or learning and are monitored through various means. Regardless of parameter or means, monitoring requires predefined thresholds at which the adaptation function is activated. For example, a developing country monitoring the number of cases of meningococcal meningitis may define 10 new cases per 100,000 people per week as the threshold indicating an outbreak (The Sphere Project, 2004, p. 282), whilst a watershed authority may define a specific water level at a point upstream as the threshold indicating the coming of a particular flood scenario downstream (Figure 8.6). Rivers have also been monitored for centuries to get indications of severe drought, with the so-called hunger stones in Germany and the Czech Republic as epitomic examples. The most famous one is in Děčín, with markings all around for the water levels of most years with severe drought and famine since the 15th century and an inscription saying, 'When you see me, weep'[1] (Figure 8.7). Again, monitoring thresholds are highly contextual and may not only focus on parameters related to specific hazards but also on parameters indicating vulnerability. For instance, the population density of an informal settlement in a flood-prone area or the illiteracy rate amongst a population.

In addition to the proactive constituent monitoring function, recognition also has a reactive part. This other constituent function focuses on assessing the consequences of actual deviations from the preferred expected development trajectory (Box 8.1). The utility of such assessment is, in other words, to describe the current state of the human-environment system during or right after a specific disturbance, disruption or disaster, to inform reactive adaptation efforts addressing the immediate consequences and restoring the system to the preferred expected development trajectory. In contrast to monitoring, the function of impact assessment focuses on parameters that are more

BOX 8.1 Two functions for recognising

Monitoring—monitoring critical parameters that are or could in the near term be causing deviations from the preferred expected scenario.

Assessing—assessing the consequences of actual deviations from the preferred expected scenario.

[1] My translation from German: 'Wenn Du mich siehst, dann weine'.

Figure 8.6 Monitoring the River Stour in the United Kingdom. *(Photo by Lewis Clarke, shared on the Creative Commons.)*

Figure 8.7 The hunger stone of Děčín, the Czech Republic. *(Photo by Dr Bernd Gross, shared in the Creative Commons.)*

directly related to what is considered valuable and important to protect in the human-environment system. Such parameters are often associated with the life, health and well-being of people and the functioning of society but could include whatever is assigned to have value. Hence, the number, severity and location of injured people are common parameters, as well as the quality of available drinking water, state of transport infrastructure, households without electricity, etc. It is important to note that such impact assessment is done in relation to the preferred expected development trajectory, making the output highly contextual concerning many parameters. For instance, thousands of people without clean drinking water is an awful normality in many parts of Mozambique but creates public outcry when happening in urban Australia.

Adapting

As our world is uncertain, complex, ambiguous and dynamic, a human-environment system must have the ability to adapt to continuously develop along its preferred expected scenario whilst remaining within human and environmental boundaries. Otherwise, the systems eventually end up on entirely different developmental paths, which are undesirable at best and cataclysmic at worst. It is important to note that some scholars distinguish explicitly between coping—immediate short-term responses to adversity— and adapting—long-term adjustments or even transformation of the system (e.g., Berkes & Jolly, 2001; Kirk & Rifkin, 2020), but I chose to keep the entire width of such activities under the overall generalised function of adaptation. It could be disaggregated in many ways, but the approach to resilience presented in this book concretises the generalised function of adaptation by defining five constituent functions on a lower level of abstraction (Box 8.2).

BOX 8.2 Five constituent functions for adapting human-environment systems

Preventing—reducing the likelihood of anticipated potential deviations from the preferred expected scenario.

Mitigating—reducing beforehand the consequences of anticipated potential deviations from the preferred expected scenario.

Preparing—preparing beforehand to respond to and recover from anticipated potential deviations from the preferred expected scenario.

Responding—addressing the immediate consequences of actual deviations from the preferred expected scenario.

Recovering—restoring the consequences of actual deviations to the preferred expected scenario.

First, human-environment systems adapt to anticipated potential deviations from the preferred expected development trajectory by reducing the likelihood of the deviations in the first place. It is referred to as *preventing* and is the preferred adaptation function, if possible concerning the characteristics of the anticipated event, and feasible with regard to social, technical, administrative, political, legal, economic and environmental constraints (Coppola, 2007, pp. 202—205). Prevention is, thus, an inherently proactive function and requires input from anticipation, monitoring or learning. As suggested in Chapter 5, prevention is possible for all but a few types of events. The associated costs are often relatively low compared to the potential or expected costs if waiting until the event happens before reactively adapting the system. However, different potential events may require different prevention activities, often overwhelming not only different constraints but also the capacity for anticipation.

Secondly, human-environment systems adapt to anticipated potential deviations from the preferred expected development trajectory by reducing the consequences of such deviations beforehand. It is referred to as *mitigating* and is the next proactive adaptation choice after prevention. In contrast to preventing, there is not a single event that cannot be mitigated as long as it can be anticipated and the activities are socially, technically, administratively, politically, legally, economically and environmentally feasible (Coppola, 2007, pp. 202—205). Again, the associated costs are relatively low compared to the potential or expected costs if the events actually occur, though the various possible events may require diverse activities. It has, therefore, been suggested not only to focus mitigation on traditional activities that focus on specific events but also on reducing vulnerability through more general poverty reduction and development. That way, potential consequences may be mitigated regardless of the actual event and our limits to anticipation, as there seems to be a clear link between human development and disaster (Fordham, 2007). Unfortunately, a substantial part of the climate change community uses the term mitigation differently and in relation to interventions for reducing the human impact on the climate system by reducing greenhouse gas emissions and improving greenhouse gas sinks (IPCC, 2022, p. 2915). However, this approach to climate change mitigation is about reducing the likelihood of potential deviations away from the preferred expected trajectory and falls, in this framework, under the constituent function of prevention.

Thirdly, human-environment systems adapt to anticipated potential deviations from the preferred expected development trajectory by preparing beforehand to respond to and recover from them when occurring. It is referred to as *preparing* and is the last resort regarding proactive adaptation choices after preventing and mitigating. All possible events can be prepared for, both in the long-term, as informed by anticipation or learning, and in the short-term, as informed by either anticipation or monitoring. However, activities reducing the likelihoods or consequences of potential events beforehand may target only parts of the risk scenario space due to limits to anticipation and the constraints listed above. Therefore, finding ways to prepare for the rest is vital—not only by preparing for specific high-risk events but also by addressing the needs that may arise in destructive events more generally (Abrahamsson et al., 2007). For instance, it is possible to conceive that people need clean drinking water regardless if they have been affected by a cyclone, drought, contamination or terrorist attack and to prepare the human-environment system to provide that regardless. However, such activities are also liable to the same categories of constraints as for preventing and mitigating.

This approach suggests that human-environment systems adapt to their challenges holistically to ensure that all potential deviations away from the preferred expected development trajectory are addressed by at least one of the three proactive functions for adaptation. These three constituent functions may, in other words, be seen as incremental in the sense of mitigating what is not possible and feasible to prevent entirely and preparing for what is not possible and feasible to fully mitigate. Considering the relatively low costs associated with prevention, mitigation and preparedness, compared to the potential and expected costs associated with actual deviations away from the preferred expected development trajectory, they should be an easy sell. They really would if it were not for the fundamental biases and challenges presented in Chapter 13.

In addition to the three proactive constituent functions for adaptation, there are two inherently reactive functions. The fourth constituent function for adapting human-environment systems is referred to as *responding*, connoting that human-environment systems adapt to actual deviations from the preferred expected development trajectory to address their immediate consequences. Responding requires, in other words, a situation that is markedly deviating from the preferred and expected state of the human-environment system at that particular time. For instance, a situation with 18 cases of meningitis spotted the last week in a town of 45,000 inhabitants, a situation in which a series of bombs have detonated in the public transport system of London, or a situation in which a tropical reef has lost 50% of its coral. Although responding is as vital as any other constituent function for adaptation—saving lives, alleviating suffering and maintaining critical societal functions in actual disrupting events—it is by definition too late to save the lives that have already been lost and to protect the buildings that have already collapsed (Figure 8.8).

Finally, human-environment systems adapt to actual deviations from the preferred expected scenario by restoring the system to the preferred expected development trajectory. It is referred to as *recovering* and comprises all possible aspects of human-environment systems, such as reconstructing damaged or destroyed buildings and infrastructure, restoring livelihood opportunities, rehabilitating contaminated wells, restoring eroded agricultural soil, etc. Recovering is by far the most resource-intensive function for adaptation and is, by definition, the last link of the chain of constituent functions of adaptation. That said, it could, in some circumstances, be a central part of an elaborate strategy for resilience. For instance, an old pub on the bank of River Ouse in York, England, gets

Figure 8.8 Most lives were already lost when the response to the 2010 Haiti earthquake started. *(Photo by Magnus Hagelsteen.)*

flooded on average four times per year (and seemingly increasing) but can, through intense focus on preparing and recovering, get the pub ready for a coming flood in half an hour and open it up again about 4 h after the water has subsided (Figure 8.9).

It is often suggested that situations after major disturbances, disruptions or disasters provide a 'window of opportunity' for proactive adaptation (e.g., Berkes & Ross, 2012; Christoplos, 2006). It is, in other words, important to have several functions ongoing at the same time. This notion is indicated in the common slogan 'build back better' (Kennedy et al., 2008; United Nations, 2015), which unfortunately often fails to address the question 'better for whom' (Bosher et al., 2021, p. 529; Cheek &

Figure 8.9 The Kings Arms Pub in York on 22 February 2022, with floodwater 458 cm above normal (the record is 540 cm, recorded on 4 November 2000). *(Photo by Malcolmxl5, shared on the Creative Commons.)*

Chmutina, 2022). It is important to note, though, that it is not the constituent function of recovery that pushes the system further along the preferred expected development trajectory than at the start of the deviation. It is the combination of recovery and one or more of the proactive adaptation functions of prevention, mitigation and preparedness that can deliver such output. Whilst the actually implemented activities can fulfil several functions simultaneously, it is important not to blur the distinction between the functions if we are to understand resilience and how to develop it further.

To summarise, the performance of adaptation emerges in the performance of the constituent functions for preventing, mitigating, preparing for, responding to and recovering from changes, trends, or shocks that undermine sustainable development (Becker, 2021).

Learning

Last but not in any way least is the function of learning from experience. Learning is fundamental both in evaluating what happened in an actual event to inform anticipation, recognition and adaptation concerning potential future events and in the sense of learning how the human-environment system works in general. Learning is, in other words, mainly a reactive function, although learning from deliberating counterfactual scenarios may be considered a borderline case. Such scenarios have, per definition, not actually occurred. Yet, carefully considering what could have happened if they did can offer valuable analytical input to the anticipation discussed above, as well as to our understanding of the ability of the human-environment system to perform vital functions for resilience. The experience is, in such cases, not directly lived but imagined. Our imagination is, as such, important for learning, as it allows us to draw real-life lessons from hypothetical scenarios (Currie, 2016; Skolnick Weisberg, 2016, p. 307). Moreover, we can also learn from other peoples' experiences (Manski, 1993), even if we unfortunately often seem reluctant to do so (e.g., Johnson & Levin, 2009). In short, we learn from experience—our own, others and imagined.

Although there are countless approaches to learning, it is here seen as driven by a continuous cycle of action and reflection (Dewey, 1933; Gibbs, 1988; Kolb, 1984). In other words, actors in a human-environment system experience a concrete event—ordinary or exceptional—make observations of what happened and reflect on it, form abstract ideas explaining and understanding what happened, and utilise these ideas when anticipating, recognising and adapting for future events.

However, learning is not only occurring on the individual level, with each person acting and reflecting in isolation. Wenger (1998) suggests learning as also inherently social and taking place in a nexus of four interconnected components. First, it is not just that an individual can learn what to do and how. Doing it together with others can become a practice and provide a framework for communicating about shared perspectives, resources and routines that facilitates everyone's sustained engagement in the activity. Wenger (1998, p. 5) refers to this as 'learning as doing'. Wenger (1998, p. 5) calls the second component 'learning as experience', which concerns our human ability to experience the world and our engagement with it as meaningful. In other words, we do not only learn *what* to do and *how* but also *why*. Whilst this is fundamental for learning, such meaning is created not only individually but also collectively. It provides a framework for communicating about our changing ability to create meaning as we learn. The next component concerns how learning changes our identity and creates personal histories of becoming who we are in a particular moment. Similar to the previous two components, it provides a framework for communicating about these issues and connects to available theories (Wenger, 1998, pp. 11—15). Wenger (1998, p. 5) calls this component 'learning as becoming' and explains how learning transforms who we are and what we do. For instance, my decision to stop flying privately, cease to eat lamb, beef and pork, and not own a car is not only based on my understanding that these activities comprise a substantial part of our individual carbon footprints in the affluent parts of the world (*practice* and *meaning*). I do it also because I identify as an environmentalist, a humanitarian, a Professor of Risk and Sustainability, a Quaker, an outdoorsy person, etc. (*identity*). It is, simply, what I do, being me. In addition, I also refrain from these activities because I belong to various more or less easily identifiable groups of people who share my environmental concern and passion for saving the world—for example environmentalists, humanitarians, like-minded academics and the Religious Society of Friends. That is the final component of Wenger's (1998, p. 5) nexus and concerns 'learning as belonging' to a community of people who engage in dialogue and joint activities, share information and help each other. It is not about having the same job title, reading the same books or being a member of the same organisation. It is about building relationships through which members of the community of practice can interact and learn from each other. Not necessarily daily, but enough to build trust, inspire each other, and renew each member's energy and direction before going back to their home context and doing their work—with or without other members of that community.

Form

Form is the most concrete level of abstraction in Rasmussen's (1985) systems hierarchy and entails the myriad of interconnected elements that make up the perceivable dimension of human-environment systems. In addition to this structural aspect of such systems, there is also a dynamic side in the sense of various processes that are central to human-environment systems as we observe reality 'not as a rigid quasi-object, but as a continuous, unending stream of events' (Sztompka, 1993, p. 9). However, these processes are shaped by the elements and their relationships just as much as they shape the processes. It is, in other words, vital to grasp the complexity of human-environment systems in any attempts to explain, understand or improve resilience, as it is in the assemblages of elements and interrelations that the ability emerges to perform the necessary functions in this context. It is important to note that many elements and interrelations are often necessary for one particular function and that one particular element or interrelation may contribute to several functions.

There are numerous ways to categorise form (e.g., Becker et al., 2011). Still, to facilitate identifying what constitutes the functions presented above, I suggest a general typology of the interconnected elements making up the core structure of human-environment systems. It is vital to remember that human-environment systems include, per definition, various other elements that compose the contexts in which the functions are performed but do not contribute directly to the functions. The structural typology I suggest only includes elements and interrelations that do and consists of four main categories (Box 8.3).

First, a human-environment system's ability to perform functions contributing to resilience is partly determined by the involved people and their knowledge, skills, methods, tools, funding, etc. Purposeful activity is not conceivable without such essential human and material resources. Luckily, human beings are incredibly resourceful, and although there are always resource constraints, we have proven capable of great deeds. Such human and material resources are often seen as including formal education, which in itself is increasingly acknowledged as an important factor for adaptation (e.g., Wamsler et al., 2012), as well as technical skills and equipment, financial capital, etc. These are all essential resources for the resilience of human-environment systems, especially in higher-level conceptualisations of such systems in modern industrial society, such as cities and countries. However, human beings are also endowed with other resources that may be less formalised and technical. Knowledge, skills, tools and various other resources are inherent parts of culture (Tylor, 1920, p. 1), regardless if you view culture as a tool (e.g., Rappaport, 1984, p. 233) or as a cognitive aspect (e.g., Geertz, 1973, p. 89). Culture endows, in other words, human-environment systems with exceptional flexibility in coping with boundless diversity of social and natural settings (Halstead & O'Shea, 1989, p. 1), irrespective of access to the more formalised resources of modern industrial society. Remember how human beings conquered more or less all corners of the planet long before such resources were available (Chapter 2). However, the increasingly dynamic character of our world, compared to the generally slow cultural changes, demands further integration of these two sets of resources. Such integration may also be necessary

BOX 8.3 Four categories of elements constituting from

1. Knowledge, skills, tools and other resources;
2. Organisation on all levels;
3. Rules, regulations and other formal institutions and
4. Norms, values and other informal institutions.

because the solutions to our current challenges may not only come from innovation—even if that seems to be what most people expect—but also from rediscovering old solutions that have served us well in the past before being forgotten (Becker & Lama, 2020, p. 51).

Regardless of the amount of knowledge, skills, tools and other resources available in a human-environment system, its ability to perform the functions contributing to resilience is also partly determined by the organisation of the resources on appropriate levels. It is, in other words, not enough for a community to have hundreds of well-trained and equipped individuals if they cannot work together in any way. Such an organisation comes in the form of formal organisations in legal terms—such as authorities, companies or NGOs—as well as other forms of social organisation along kinship, community, ethnicity, faith, interest, etc.

It is important to note that not only individuals organise into groups of various kinds. Also, multiple groups organise amongst each other, and so on upward in the organisational hierarchy of human-environment systems. Although all social organisation is fluid and requires continuous reconstruction (Sztompka, 1993, pp. 9–10), many formal organisations and forms of social organisation persist over time—some for years, some for millennia. However, there is also organisation that takes place to address a specific situation at a particular time and then most often dissolves again. This organisation is referred to as emergent and can be categorised as having new tasks and new structure in the specific situation, in contrast to organisations with old tasks and old structure (established organisation), with old tasks but new structure (extending organisation), or with new tasks but old structure (expanding organisation) (Dynes, 1970; Quarantelli, 1994, p. 206). However, organisations are in themselves manifestations of sets of underlying formal and informal institutions (Handmer & Dovers, 2007, p. 30).

The ability of a human-environment system to perform the functions contributing to its resilience is also partly determined by its existing rules, regulations, legislation, policies, etc. (Figure 8.10). Such formal institutions can be codified or customary and provide persistent and predictable guidelines for behaviour and interaction amongst individuals and organisations that facilitate coexistence and collective activities by reducing the need for constant negotiation (Handmer & Dovers, 2007, p. 30). Formal institutions are formally sanctioned by some level of organisation and are rooted in the norms, values and other informal institutions of the human-environment system.

Formal institutions are actively implemented through coercion, cooperation, exhortation, or a combination of more than one of these three (Handmer & Dovers, 2007, pp. 110–120). The two former implementation styles are tightly linked to the old story of the stubborn donkey, where coercion is the stick and cooperation is the carrot. Coercion is, in other words, driven by credible threats of punitive measures if not complying—like the stick being wielded by the hurried farmer. Examples of this style of implementation

Figure 8.10 A legal framework (also available as an app). *(Shared by Wolters Kluwer on the Creative Commons.)*

are found in building codes, legislation concerning pollution, urban planning, policing, etc. On the other hand, cooperation is driven by incentives and offers of support and collaboration—like the carrot hanging in front of the donkey's head. For instance, tax deductions for households investing in double-glazed windows to reduce energy needs (e.g., Boardman, 2004), or resettlement grants for households living in high-risk areas (e.g., Thatte, 2012). However, human beings are not donkeys and can also be motivated to comply with formal institutions through education and moral suasion to raise their awareness and understanding of the reasons for doing so. Carlzon (2008) summarises the core idea of such exhortation well when stating that 'a person who does not have access to information cannot take responsibility. A person who has information cannot resist from taking responsibility' (my translation from Swedish). However, this may not actually be the case, which we will return to in Chapter 13.

In addition to formal institutions, the ability of a human-environment system to perform the functions contributing to its resilience is also partly determined by the norms, values and other informal institutions of that system more directly. Although informal institutions have the same purpose as formal institutions and are underlying their existence, they are without formal sanctioning and implementation. However, informal institutions discourage deviant behaviour through various social sanctions (Helmke & Levitsky, 2004), such as criticism, ridicule, shame and disregard. These sanctions may be addressed directly to the deviant individual or group or to a larger social entity they belong to—adding another dimension of social control. For instance, an individual who breaks a social norm by not removing his hat when entering a church or not washing before entering a mosque may bring shame to his whole family, which further motivates him to stay in line.

Norms, values and other informal institutions are inherent parts of culture (e.g., Geertz, 1973; Tylor, 1920) and comprise a central part of the social structure that is both directing and being directed by human agency (Giddens, 1984). They underlie formal institutions, set boundaries for and shape organisation, and influence what knowledge, skills, tools and other resources are deemed practical, preferable and even possible. Although entirely fundamental for the resilience of a human-environment system, informal institutions are notoriously challenging and slow to alter and too often overlooked or ignored in efforts to build resilience.

Informal institutions must be distinguished from mere behavioural regularities (Brinks, 2003; Helmke & Levitsky, 2004). When revisiting the example of the man not removing his hat when entering a church or not washing before entering a mosque, it is obvious that such deviations away from social norms would bring social sanctions in devout Christian or Muslim communities. However, even if most people take their hats off when entering a restaurant and wash their hands before eating, it is unlikely that deviating from such behavioural regularities would cause any reaction in many parts of the world. The core distinction between an informal institution and a behavioural regularity is, in other words, that the former is attached to a prescriptive behavioural rule that triggers social sanctions if broken, whilst the latter may be connected to practical incentives, habits, etc. (Brinks, 2003). Whilst it is undoubtedly so that behavioural regularities influence risk—for example, bad hygiene practices, not wearing seat belts, ignoring building codes, etc.—they are more the dynamic outcome of the complex combination of the four categories of elements described here, than elements in themselves.

As a complement to the framework of structural categories of form just presented, it may also be helpful to consider observable processes that, in various ways, combine and interact to perform different functions for resilience. There are many such processes, and several typologies suggested in the literature, especially in relation to adaptation (e.g., Agrawal, 2010; Halstead & O'Shea, 1989; McCay, 1978; Thornton & Manasfi, 2010). For instance, Becker and Lama (2020) suggest a framework of 13 processes, out of which 10 focus directly on adapting to reduce risk (first-order), two on adapting the means of adaptation (second-order), and one focuses on adapting the ends of adaptation (third-order) (Figure 8.11). First, risk is reduced by moving away from its source or towards better opportunities (mobility) or by changing the environment (design). Few things are more symptomatic of the human condition than designing and applying technology in the broadest possible sense (Elias, 1995), and risk is regularly reduced through this process (design). Risk is also reduced by adjusting the production or consumption of vital resources. It is reduced by increasing the output of essential resources and distributing their production across space (extensification), by increasing the output of vital resources

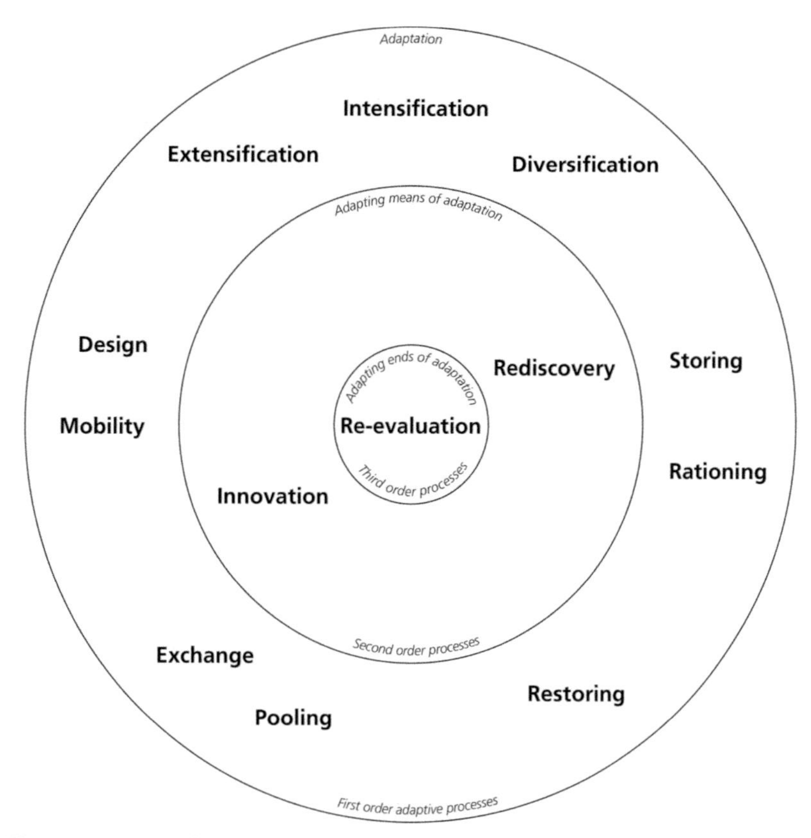

Figure 8.11 An ordered process-oriented framework for adaptation to reduce risk.

within the same space (intensification), or by diffusing it across different resources (diversification). Risk is also reduced by diffusing the consumption of vital resources across time (storing) and regulating their consumption (rationing). Regulating consumption in combination with reducing risk by restoring previously consumed resources (restoring) is what conservation is all about. Finally, risk is also reduced by sharing or trading resources. It is done by pooling vital resources (pooling) or exchanging resources across social groups (exchange). The two second-order processes in Becker and Lama's (2020) framework have already been alluded to earlier in this section, with risk being reduced indirectly either by coming up with new means for adaptation (innovation) or by reintroducing old means for adaptation (rediscovery). Last but not least, risk can also be reduced indirectly by adjusting our aspirations and expectations for the future (reevaluation). This third-order process is often forgotten or ignored, but it seems increasingly likely that modifying our preferred expected development trajectory will become necessary for the survival of humankind.

Linking resilience to other frameworks

It is evident in this approach to resilience that most of what is operationalised under the concept has roots in a relatively widespread academic and professional pedigree. It is not at all a coincidence, as I consider the alternative unfeasible at best. We have introduced enough new concepts by aggressively redefining their predecessors to claim novelty and precedence. For instance, even early definitions of disaster management or emergency management included prevention/mitigation and preparedness, in addition to response and recovery (e.g., McLoughlin, 1985, p. 166; Mileti et al., 1975; Stallings & Quarantelli, 1985, p. 94), whilst the rhetoric advocating for disaster reduction or disaster risk reduction decades later redefined these predecessors to merely involve reactive response and recovery (e.g., Jeggle, 2001). The reason for this may have been the profound imbalance in favour of reactive over proactive activities in practice. However, if we consider that an acceptable reason for such a terminological slant, then I believe the successors are in trouble too, considering the continuous focus on disaster response. I have nothing against the inventions of first disaster reduction (e.g., IDNDR, 1994; Jeggle, 2001) and then disaster risk reduction (e.g., Twigg, 2004; Vermaak & van Niekerk, 2004; White et al., 2004) as attempts to increase the focus on proactive activities. However, advocating for these new concepts by consciously stating that their conceptual predecessors ignored proactive activities is simply unfair.

Similarly, climate change adaptation was at least initially introduced as something novel and distinct. However, the immense focus on adaptation to moderate actual or potential harm, over adaptation to exploit beneficial opportunities, has spurred intense debate over the extent to which disaster risk reduction and climate change adaptation overlap (e.g., Kelman & Gaillard, 2008; Mercer, 2010; Mitchell & van Aalst, 2008, pp. 1–22; Schipper, 2009; Shea, 2003)—especially since many definitions of climate change adaptation only include adverse effects. For example, as the '[a]ctions to reduce the vulnerability of a system (e.g., a city), population (e.g., a vulnerable population in a city) or individual to the adverse impacts of anticipated climate change due to emission of greenhouse gases' (Satterthwaite et al., 2009, p. 9). This particular definition is also interesting because it is linked to categories of actual adaptation activities, including protection (disaster avoidance), preparedness, response and rebuilding (Satterthwaite et al., 2009, p. 36). These are described identically as the prevention/mitigation, preparedness, response and recovery of traditional definitions of emergency management or disaster management referred to earlier in this section.

Instead of introducing resilience as a new concept that overrides all previous approaches to building a safe and sustainable society by managing risk and disaster situations, I intend to suggest how the most common conceptual predecessors fit this new resilience framework—including emergency management and disaster management from 1970,

1980 and onwards, disaster reduction from the 1990s, and disaster risk reduction and climate change adaptation from the last decades.

Approaches to emergency management and disaster management that include prevention, mitigation, preparedness, response and recovery are de facto comprising all overall ways a human-environment system adapts—as presented earlier in this chapter. However, most users of these approaches implicitly limit these functions to a few central functional sectors—such as emergency- or disaster management, buildings and infrastructure—and focus mainly on what is here referred to as adaptation. It does not mean they completely ignore functions for anticipating, recognising or learning, but they are secondary at best and more or less implicit. Risk assessment and hazard monitoring are, for instance, commonly mentioned as a requisite for proactive activities, and lessons learnt are so more often mentioned than actually done that it more or less has become a cliché.

The Sendai Framework for Disaster Risk Reduction, as well as the preceding Hyogo Framework for Action and Yokohama Strategy, indicate what disaster reduction and disaster risk reduction cover. Two dimensions emerge when analysing these frameworks. First, they specify disaster reduction and disaster risk reduction to comprise risk assessment, prevention/mitigation, and preparedness for response and recovery. Secondly, they emphasise the effective implementation of disaster reduction and disaster risk reduction as requiring effective legal and institutional frameworks, close collaboration between organisations, well-functioning organisations, and sufficient human and material resources. Disaster reduction and disaster risk reduction are, in other words, explicitly focusing on the proactive ways in which a human-environment system can adapt and on ways for anticipation. Although not included in the list of priorities, as such, the frameworks also mention functions for recognising and learning. Moreover, they also indicate what is needed in human-environment systems to perform these functions.

One of the more influential early definitions of disaster risk reduction states that it is the 'conceptual framework of elements considered with the possibilities to minimise vulnerabilities and disaster risks throughout a society, to avoid (prevention) or to limit (mitigation and preparedness) the adverse impacts of hazards, within the broad context of sustainable development' (ISDR, 2004, p. 17). After being initially endorsed and advocated for by the UN, for some reason the official UN definition of disaster risk reduction was later changed to the 'concept and practice of reducing disaster risks through systematic efforts to analyse and manage the causal factors of disasters, including through reduced exposure to hazards, lessened vulnerability of people and property, wise management of land and the environment, and improved preparedness for adverse events' (UNISDR, 2009, pp. 10–11). Then, after the adoption of the Sendai Framework, the definition of disaster risk reduction was again rephrased and stated that it 'is aimed at preventing new and reducing existing disaster risk and managing residual risk, all of which contribute to strengthening resilience and therefore to the achievement of sustainable development' (United Nations, 2016, p. 16). Disaster risk management is then seen as the application of disaster risk reduction policies and strategies.

IPCC defines climate change adaptation in human systems as 'the process of adjustment to actual or expected climate and its effects, to moderate harm or exploit beneficial opportunities' (IPCC, 2022, p. 2898). It is the most influential definition and includes both negative and positive effects and actual and potential effects. Although other definitions, as well as the main focus of policy and practice, more or less only include adverse effects (remember Satterthwaite et al., 2009), both potential and actual effects are generally included. Adaptation covers, in other words, both proactive and reactive activities to manage risks related to a changing climate. Since adapting to the potential negative impacts of climate change is by far the main focus, it has been suggested that it is more or less a part of disaster risk reduction in practice (Mercer, 2010; Mitchell & van Aalst, 2008, p. 4).

Indeed, the disaster risk reduction community often fails to define and address the problem holistically enough (Bankoff & Hilhorst, 2022; Bosher et al., 2021; Lewis & Kelman, 2012), but it is better to correct that than to just invent new concepts. It is not only confusing for policymakers and practitioners, as new words are invented that mean the same thing or at least very similar things. For instance, every national authority and county administrative board in Sweden have been legally obliged for decades to regularly perform a risk and vulnerability analysis, with the Swedish Civil Contingencies Agency (MSB) providing guidelines for how to do that. Then, a few years ago, there was a new legal requirement for them to also perform a so-called climate and vulnerability analysis, with the Swedish Meteorological and Hydrological Institute (SMHI) providing guidelines. There are striking resemblances in the descriptions of these requirements and an almost complete overlap between what these analyses should include. Yet, the processes are largely separate and often involve different people. Such parallel structures for disaster risk reduction and climate change adaptation are found all over the world (e.g., Becker et al., 2013) and have been suggested to be the result of incentives and pressures from global and international actors having such parallel structures themselves and the response-oriented legacy of most disaster risk reduction institutions (Becker et al., 2021). Regardless of the reasons, the effects of parallel structures are overwhelmingly negative and include unclear mandate and leadership, uncoordinated or duplication of efforts, inefficient use of resources, and competition for resources and control (Becker et al., 2021). Whilst some actors benefit from keeping these essentially overlapping policy areas apart, few human-environment systems do.

To summarise, the approach to resilience presented in this book is not intended to undermine existing approaches to building a safe and sustainable society. On the contrary, it consciously connects to and between them whilst suggesting a coherent framework for what we must do to manage the core challenges of humankind.

Conclusion

Resilience is a means to reach the ends of safety and sustainability in our rapidly and continuously transforming world. It is nothing objective and universal but relative and context-dependent—just as the concepts of development and risk. Assessing and comparing resilience requires, in other words, proper care concerning keeping variables and preferences constant over space and time, artificially or not. Resilience is not a simple panacea for all possible evils in our world. To have any purpose, the concept of resilience must be operationalised and used to guide appropriate practical initiatives. Otherwise, it becomes merely rhetorical and loses its meaning.

I have devoted this chapter to suggesting one way of operationalising the resilience of human-environment systems by approaching them as having purpose, function and form and presenting a dictionary for finding these in the empirical world. This approach provides, thus, a link between the conceptual and the actual—between the definition of the concept of resilience, as presented in Chapter 7, and what human-environment systems need to be resilient in practice. Although the required generalised functions presented in this chapter are complete, it is always possible to reorganise, add and remove functions and forms on lower levels of abstraction. However, I argue that an observer or agent aspiring to grasp or develop the resilience of any human-environment system is better equipped to do so with a comprehensive guide for making this leap from theory to practice and back again. Not only focusing on the formal and technical but also on the informal and social. Culture is key. Do not forget it.

References

Abrahamsson, M., Eriksson, K., Fredholm, L., & Johansson, H. (2007). Analytical input to emergency preparedness at the municipal level–a case study. In *Proceedings of disaster recovery and relief: Current & future approaches (TIEMS 2007).*

Agrawal, A. (2010). Local institutions and adaptation to climate change. In R. Mearns, & A. Norton (Eds.), *Social dimensions of climate change: Equity and vulnerability in a warming world* (pp. 173–198). World Bank.

Alexander, D. E. (2013). Resilience and disaster risk reduction: An etymological journey. *Natural Hazards and Earth System Sciences, 13*, 2707–2716.

Apgar, M. J., Allen, W., Moore, K., & Ataria, J. (2015). Understanding adaptation and transformation through indigenous practice: The case of the Guna of Panama. *Ecology and Society, 20*(1). https://doi.org/10.5751/es-07314-200145

Bankoff, G., & Hilhorst, D. (Eds.). (2022). *Why vulnerability still matters: The politics of disaster risk creation.* Routledge.

Becker, P. (2017). Dark side of development: Modernity, disaster risk and sustainable livelihoods in two coastal communities in Fiji. *Sustainability, 9*, 1–23. https://doi.org/10.3390/su9122315

Becker, P. (2021). Advancing resilience for sustainable development: A capacity development approach. In W. Leal Filho, R. Pretorius, & L. O. de Sousa (Eds.), *Sustainable development in Africa: Fostering sustainability in one of the world's most promising continents* (pp. 525–540). Springer. https://doi.org/10.1007/978-3-030-74693-3_29

Becker, P., Abrahamsson, M., & Hagelsteen, M. (2013). Parallel structures for disaster risk reduction and climate change adaptation in Southern Africa. *Jàmbá: Journal of Disaster Risk Studies, 5*, 1–5. https://doi.org/10.4102/jamba.v5i2.68

Becker, P., Abrahamsson, M., & Tehler, H. (2011). An emergent means to assurgent ends: Community resilience for safety and sustainability. In E. Hollnagel, E. Rigaud, & D. Besnard (Eds.), *Proceedings of the fourth resilience engineering symposium, June 8-10, 2011* (pp. 29–35). MINES ParisTech.

Becker, P., Abrahamsson, M., & Tehler, H. (2016). An emergent means to assurgent ends: Societal resilience for safety and sustainability. *In Resilience Engineering in Practice, 2*, 29–40. https://doi.org/10.1201/9781315605708. CRC Press.

Becker, P., & Hagelsteen, M. (2016). Kapacitetsutveckling för katastrofriskreducering. In S. Baez Ullberg, & P. Becker (Eds.), *Katastrofriskreducering: Perspektiv, praktik, potential* (pp. 265–291). Studentlitteratur.

Becker, P., Hagelsteen, M., & Abrahamsson, M. (2021). 'Too many mice make no lining for their nest' – reasons and effects of parallel governmental structures for disaster risk reduction and climate change adaptation in Southern Africa. *Jàmbá Journal of Disaster Risk Studies, 13*(1). https://doi.org/10.4102/jamba.v13i1.1041

Becker, P., & Lama, P. D. (2020). Narratives of change: First-, second-, and third-order adaptive processes in Nepal and the Maldives. *Southeast Asia, 7*, 35–56.

Berkes, F., & Jolly, D. (2001). Adapting to climate change: Social-ecological resilience in a Canadian western arctic community. *Conservation Ecology, 5*(2), 18.

Berkes, F., & Ross, H. (2012). Community resilience: Toward an integrated approach. *Society & Natural Resources, 26*, 5–20. https://doi.org/10.1080/08941920.2012.736605

Boardman, B. (2004). New directions for household energy efficiency: Evidence from the UK. *Energy Policy, 32*, 1921–1933. https://doi.org/10.1016/j.enpol.2004.03.021

Borges de Castro, R., Emmanouilidis, J. A., & Zuleeg, F. (2021). *Europe in the age of permacrisis*. https://www.epc.eu/en/Publications/Europe-in-the-age-of-permacrisis~3c8a0c.

Bosher, L., Chmutina, K., & van Niekerk, D. (2021). Stop going around in circles: Towards a reconceptualisation of disaster risk management phases. *Disaster Prevention and Management: An International Journal, 30*(4/5), 525–537. https://doi.org/10.1108/dpm-03-2021-0071

Brinks, D. M. (2003). Informal institutions and the rule of law: The judicial response to state killings in Buenos Aires and São Paulo in the 1990s. *Comparative Politics, 36*, 1–19. https://doi.org/10.2307/4150157

Cannon, T., & Müller-Mahn, D. (2010). Vulnerability, resilience and development discourses in context of climate change. *Natural Hazards, 55*(3), 621–635. https://doi.org/10.1007/s11069-010-9499-4

Carlzon, J. (2008). *Riv pyramiderna!: en bok om den nya människan, chefen och ledaren.* Natur & Kultur.

Cheek, W., & Chmutina, K. (2022). 'Building back better' is neoliberal post-disaster reconstruction. *Disasters, 46*(3), 589–609. https://doi.org/10.1111/disa.12502

Christoplos, I. (2006). *The elusive 'Window of Opportunity' for risk reduction in post disaster recovery* (pp. 2–3), 2006/02/02-03.

Coppola, D. P. (2007). *Introduction to international disaster management*. Butterworth-Heinemann (Elsevier).

Currie, G. (2016). Imagination and learning. In A. Kind (Ed.), *The Routledge handbook of philosophy of imagination* (pp. 407–419). Routledge.

Dewey, J. (1933). *How we think: A restatement of the relation of reflective thinking to the educative process*. Heath and company.

Dynes, R. R. (1970). *Organized behavior in disaster*. Heath Lexington Books. https://archive.org/details/organizedbehavio0000dyne.

Elias, N. (1995). Technization and civilization. *Theory, Culture & Society, 12*, 7–42.

Elwér, S., Harryson, L., Bolin, M., & Hammarström, A. (2013). Patterns of gender equality at workplaces and psychological distress. *PLoS One, 8*, e53246. https://doi.org/10.1371/journal.pone.0053246

Fordham, M. H. (2007). Disaster and development research and practice: A necessary eclecticism. In H. Rodríguez, E. L. Quarantelli, & R. R. Dynes (Eds.), *Handbook of disaster research* (pp. 335–346). Springer.

Geertz, C. (1973). *The interpretation of cultures: Selected essays*. Basic Books.

Gibbs, G. (1988). *Learning by doing: A guide to teaching and learning methods. Oxford Polytechnic.*

Giddens, A. (1984). *The constitution of society: Outline of the theory of structuration*. University of California Press.

Halstead, P., & O'Shea, J. (1989). Introduction: Cultural responses to risk and uncertainty. In P. Halstead, & J. O'Shea (Eds.), *Bad year economics*. Cambridge University Press.
Handmer, J. W., & Dovers, S. (2007). *The handbook of disaster and emergency policies and institutions*. Earthscan.
Hearn, J. (2012). *Theorizing power*. Palgrave Macmillan.
Helmke, G., & Levitsky, S. (2004). Informal institutions and comparative politics: A research agenda. *Perspectives on Politics, 2*, 725–740. https://doi.org/10.1017/S1537592704040472
Holling, C. S. (1973). Resilience and stability of ecological systems. *Annual Review of Ecology and Systematics, 4*, 1–23.
Hollnagel, E. (2009). The four cornerstones of resilience engineering. In C. P. Nemeth, & E. Hollnagel (Eds.), *Resilience engineering perspectives, volume 2: Preparation and restoration* (pp. 117–133). Ashgate.
Hollnagel, E. (2017). *Safety-II in practice: Developing the resilience potentials*. Routledge.
Hummel, C. E. (1967). *Tyranny of the urgent*. InterVarsity Press.
IDNDR. (1994). *Yokohama strategy and plan of action for a safer world: Guidelines for natural disaster prevention, preparedness and mitigation*.
IPCC. (2022). *Climate change 2022: Impacts, adaptation and vulnerability*. Cambridge University Press.
ISDR. (2004). *Living with risk: A global review of disaster reduction initiatives*.
Jeggle, T. (2001). The evolution of disaster reduction as an international strategy: Policy implications for the future. In U. Rosenthal, A. Boin, & L. K. Comfort (Eds.), *Managing crises: Threats, Dilemmas, opportunities* (pp. 316–341). Charles C Thomas.
Johnson, D., & Levin, S. A. (2009). The tragedy of cognition: Psychological biases and environmental inaction. *Current Science, 97*, 1593–1603.
Joseph, J. (2013). Resilience as embedded neoliberalism: A governmentality approach. *Resilience, 1*, 38–52.
Kahneman, D., & Tversky, A. (1982). The psychology of preferences. *Scientific American, 246*, 160–173.
Kant, I. (1892). *Kant's Kritik of judgement*. Macmillan & Co.
Kelman, I., & Gaillard, J. (2008). Placing climate change within disaster risk reduction. *Disaster Advances, 1*, 3–5.
Kennedy, J., Ashmore, J., Babister, E., & Kelman, I. (2008). The meaning of 'build back better': Evidence from post-tsunami Aceh and Sri Lanka. *Journal of Contingencies and Crisis Management, 16*, 24–36. https://doi.org/10.1111/j.1468-5973.2008.00529.x
Kirk, C. P., & Rifkin, L. S. (2020). I'll trade you diamonds for toilet paper: Consumer reacting, coping and adapting behaviors in the COVID-19 pandemic. *Journal of Business Research, 117*, 124–131. https://doi.org/10.1016/j.jbusres.2020.05.028
Kolb, D. A. (1984). *Experiential learning: Experience as the source of learning and development* (Vol 1). Prentice-Hall.
Lama, P. D., Becker, P., & Bergström, J. (2017). Scrutinizing the relationship between adaptation and resilience: Longitudinal comparative case studies across shocks in two Nepalese villages. *International Journal of Disaster Risk Reduction, 23*, 193–203. https://doi.org/10.1016/j.ijdrr.2017.04.010
Lewes, G. H. (1875). *Problems of life and mind: First series — the foundation of a creed (2)*. Trübner & Co.
Lewis, J., & Kelman, I. (2012). The good, the bad and the Ugly: Disaster risk reduction (DRR) versus disaster risk creation (DRC). *PLoS Currents: Disasters, 4*. https://doi.org/10.1371/4f8d4eaec6af8
Lukes, S. (2005). *Power: A radical view*. Palgrave Macmillan.
Manski, C. F. (1993). Dynamic choice in social settings: Learning from the experiences of others. *Journal of Econometrics, 58*(1–2), 121–136. https://doi.org/10.1016/0304-4076(93)90115-l
Marx, S. M., Weber, E. U., Orlove, B. S., Leiserowitz, A., Krantz, D. H., Roncoli, C., & Phillips, J. (2007). Communication and mental processes: Experiential and analytic processing of uncertain climate information. *Global Environmental Change, 17*, 47–58. https://doi.org/10.1016/j.gloenvcha.2006.10.004
McCay, B. J. (1978). Systems ecology, people ecology, and the anthropology of fishing communities. *Human Ecology, 6*, 397–422. https://doi.org/10.1007/BF00889417
McLoughlin, D. (1985). A framework for integrated emergency management. *Public Administration Review, 45*, 165–172.
Mercer, J. (2010). Disaster risk reduction or climate change adaptation: Are we reinventing the wheel. *Journal of International Development, 22*, 247–264. https://doi.org/10.1002/jid.1677

Meyer, R. (2006). Why we under-prepare for hazards. In R. J. Daniels, & D. F. Kettl (Eds.), *On risk and disaster: Lessons from Hurricane Katrina* (pp. 153–174). University of Pennsylvania Press.
Mileti, D. S., Drabek, T. E., & Haas, J. E. (1975). *Human systems in extreme environments: A sociological perspective*. Institute of Behavioral Science, University of Colorado.
Mitchell, T., & van Aalst, M. (2008). *Convergence of disaster risk reduction and climate change adaptation: A review for DFID*.
Olsson, L., Jerneck, A., Thoren, H., Persson, J., & O'Byrne, D. (2015). Why resilience is unappealing to social science: Theoretical and empirical investigations of the scientific use of resilience. *Science Advances, 1*, 1–12.
Oxford English Dictionary. (2020). *Oxford dictionary of English*. Oxford University Press.
Pariès, J. (2006). Complexity, emergence, resilience. In E. Hollnagel, D. D. Woods, & N. G. Leveson (Eds.), *Resilience engineering: Concepts and precepts* (pp. 43–53). Ashgate.
Quarantelli, E. L. (1994). *Emergent behaviors and groups in the crisis time periods of disaster*. DRC Preliminary Paper.
Rankine, W. J. M. (1858). *A manuel of applied mechanics*. Richard Griffin and Company.
Rappaport, R. A. (1984). *Pigs for the ancestors: Ritual in the ecology of a new Guinea people*. Yale University Press.
Rasmussen, J. (1985). The role of hierarchical knowledge representation in decisionmaking and system management. *IEEE Transactions on Systems, Man and Cybernetics, 15*, 234–243.
Rist, G. (2006). *The history of development: From western origins to global faith*. Zed Books.
Rubin, E. (1949). Vorteile der Zweckbetrachtung für die Erkentnis. In E. Rubin (Ed.), *Experimenta psychologica: Collected scientific papers in German* (p. 66). Munksgaard: English & French.
Russell, B. (2004). *Power: A new social analysis*. Routledge.
Satterthwaite, D., Huq, S., Reid, H., Pelling, M., & Romero Lankao, P. (2009). In J. Bicknell, D. Dodman, & D. Satterthwaite (Eds.), *Adapting to climate change in urban areas: The possibilities and constraints in low- and middle-income Nations* (pp. 3–47). Earthscan.
Schipper, E. L. F. (2009). Meeting at the crossroads?: Exploring the linkages between climate change adaptation and disaster risk reduction. *Climate & Development, 1*, 16–30.
Shea, E. L. (2003). Living with a climate in transition: Pacific communities plan for today and tomorrow. *Asia Pacific Issues, 66*, 1–8.
Skolnick Weisberg, D. (2016). Imaginatin and child development. In A. Kind (Ed.), *The Routledge handbook of philosophy of imagination* (pp. 300–313). Routledge.
Slovic, P., Finucane, M. L., Peters, E., & MacGregor, D. G. (2002). Rational actors or rational fools: Implications of the affect heuristic for behavioral economics. *The Journal of Socio-Economics, 31*, 329–342. https://doi.org/10.1016/S1053-5357(02)00174-9
Stallings, R. A., & Quarantelli, E. L. (1985). Emergent citizen groups and emergency management. *Public Administration Review, 45*, 93–100.
Sztompka, P. (1993). *The sociology of social change*. Blackwell.
Taleb, N. N. (2007). *The black swan: The impact of the highly improbable*. Random House.
Thatte, C. D. (2012). Resettlement due to Sardar Sarovar Dam, India. In C. Tortajada, D. Altinbilek, & A. K. Biswas (Eds.), *Impacts of large Dams: A global assessment* (pp. 259–276). Springer. http://link.springer.com/10.1007/978-3-642-23571-9_12.
The Sphere Project. (2004). *Humanitarian charter and minimum standards in disaster response*. The Sphere Project.
Thornton, T. F., & Manasfi, N. (2010). Adaptation—genuine and Spurious: Demystifying adaptation processes in relation to climate change. *Environment and Society: Advances in Research, 1*, 132–155. https://doi.org/10.3167/ares.2010.010107
Twigg, J. (2004). *Disaster risk reduction: Mitigation and preparedness in development and emergency programming*. Overseas Development Institute.
Tylor, E. B. (1920). *Primitive culture: Researches into the development of mythology, philosophy, religion, language, art and custom*. John Murray.
Ulrich, W. (2002). Bondary critique. In H. G. Daellenbach, & R. L. Flood (Eds.), *The informed student guide to management science* (pp. 41–42). Thomson.
UNISDR. (2009). *UNISDR terminology on disaster risk reduction*.
United Nations. (2015). *Sendai framework for disaster risk reduction 2015-2030. United Nations*.

United Nations. (2016). *Report of the open-ended intergovernmental expert working group on indicators and terminology relating to disaster risk reduction (A/71/644). United Nations.*

Vermaak, J., & van Niekerk, D. (2004). Disaster risk reduction initiatives in South Africa. *Development Southern Africa, 21*, 555–574. https://doi.org/10.1080/0376835042000265487

Vervoort, J. M., Kok, K., Beers, P. J., van Lammeren, R., & Janssen, R. (2012). Combining analytic and experiential communication in participatory scenario development. *Landscape and Urban Planning, 107*, 203–213. https://doi.org/10.1016/j.landurbplan.2012.06.011

Wamsler, C., Brink, E., & Rantala, O. (2012). Climate change, adaptation, and formal education: The role of Schooling for increasing Societies' adaptive capacities in El Salvador and Brazil. *Ecology and Society, 17*(2). https://doi.org/10.5751/ES-04645-170202

Wenger, E. (1998). *Communities of practice: Learning, meaning, and identity*. Cambridge University Press.

Westrum, R. (1992). Cultures with requisite imagination. In J. A. Wise, D. Hopkin, & P. Stager (Eds.), *Verification and validation of complex systems: Human factors issues* (pp. 401–416). Springer.

White, P., Pelling, M., Sen, K., Seddon, D., Russel, S., & Few, R. (2004). *Disaster risk reduction: A development concern.*

CHAPTER 9

Grasping complexity

Introduction

I restate several times throughout this book that our world is uncertain, complex, ambiguous and dynamic. It is not merely rhetorical since these four constitute fundamental challenges for us to understand, explain and improve our sustainability challenges and the world. Although each is essential in its own right, one can be seen as more fundamental than the others because it influences them. I am here referring to complexity. Whilst a simple system can entail uncertainty or dynamic change—such as a set of dice (uncertain outcome) or a pendulum (dynamic position)—the uncertainty and dynamic character escalate with increasing complexity. Similarly, the potential for ambiguity in our knowledge about a system increases the more complex it is since it becomes more and more difficult to ascertain how it works. The underlying reasons for these connections to complexity will soon become apparent when we elaborate on complexity. For now, it is enough to appreciate the particular importance of complexity and that it deserves a chapter of its own.

It is of utmost importance to overcome the common conflation of everyday notions of being *complex* and *complicated* (Figure 9.1). Whilst these two words sometimes are used synonymously in ordinary language, being complicated denotes intricate chains of cause-and-effect relations that are difficult but not impossible to fully understand (Snowden & Boone, 2007). On the other hand, complexity is not a euphemism for ignorance (Yates, 1978, p. R201). It is not our lack of understanding that makes something complex, transforming it into something simple when we finally understand it. As introduced in Chapter 7, complexity entails large numbers of elements interacting dynamically and non-linearly, so it is impossible to ever fully grasp the system (Cilliers, 1998, pp. 3—5). This chapter is about such complexity.

However, it is essential to remember that the human-environment systems in this book are models constructed to attempt to make sense of an overwhelmingly intricate reality. We can never *know* if what we try to understand, explain or improve is complex, only perceive it as such when fitting our observations. Moreover, the purpose we have in mind when constructing a human-environment system fundamentally determines how we perceive the complexity of that piece of reality. Ashby's analogy of the brain demonstrates this with amusing clarity and simplicity:

> *To the neurophysiologist, the brain, as a feltwork of fibers and a soup of enzymes, is certainly complex […]. To a butcher the brain is simple, for he has to distinguish it from only about thirty other 'meats'*
> **(Ashby, 1973, p. 1).**

Sustainability Science, Second Edition
ISBN 978-0-323-95640-6, https://doi.org/10.1016/B978-0-323-95640-6.00004-X

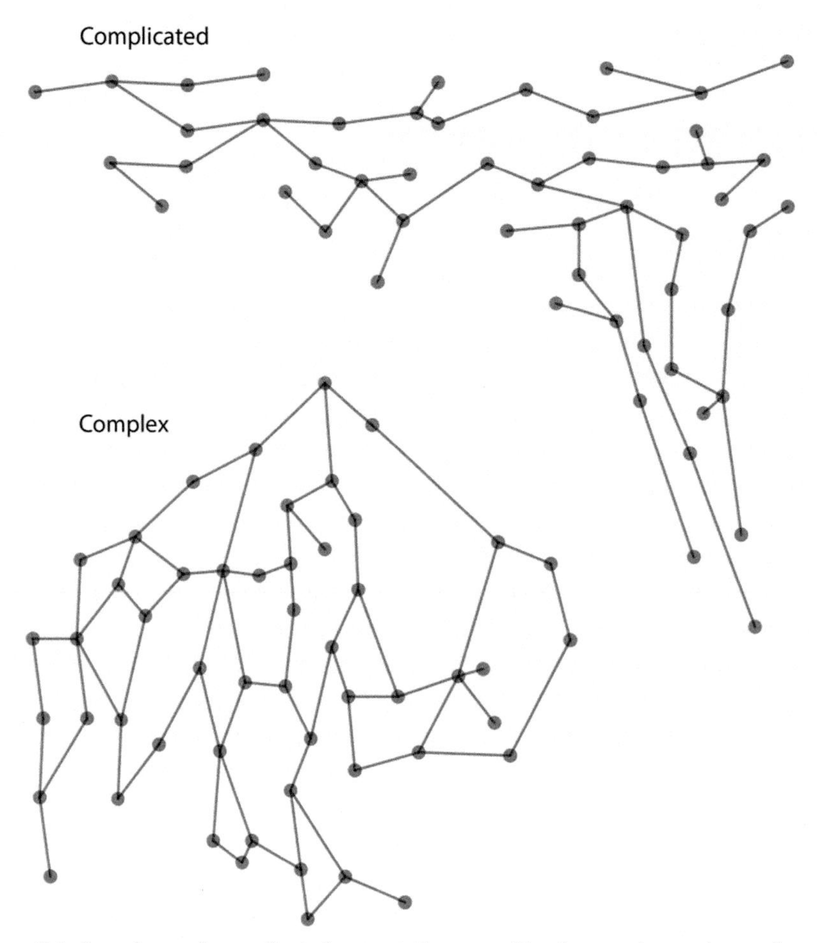

Figure 9.1 Complex and complicated are not the same. The former demands non-linearity.

I have structured this chapter in three parts. First, it elaborates on the constitution of complexity in this context—what complexity is or is required for something to be complex. Secondly, the chapter elaborates on the manifestations of complexity—what complexity does in terms of complex systems' general structure and function. It is about feedback, disproportionality between cause and effect, hierarchical aggregation, emergence, path dependence, self-organisation and far-from-equilibrium dynamics. Thirdly, the chapter introduces significant consequences of complexity on our ability to understand, explain and improve phenomena and problems concerning risk and resilience, which are sometimes difficult to distinguish from the manifestations of complexity. It concerns requisite relationality, surprise, intrinsic volatility, cascading effects and tipping points. The chapter ends with concluding remarks on complexity and the importance of doing our utmost to be mindful of complexity and its manifestations and cope with its consequences.

The constitution of complexity

Complexity is a contested concept with many definitions (see Backlund, 2002; Horgan, 1995; Richardson & Cilliers, 2001), which all may be useful for their intended purposes. However, I find it immensely difficult to either select or suggest a concise definition that captures everything needed to guide us in our work in actual contexts. It should perhaps not be surprising that complexity cannot be given a simple definition (Cilliers, 1998, p. 2), but there are still a set of characteristics that together describe what complexity is. First, there are four fundamental characteristics of complex systems, each of which is necessary but not sufficient for constituting complexity on its own (Cilliers, 1998, pp. 3–4).

1. A complex system invariably consists of a relatively *large number of elements*. Without such a bulk of parts, their individual behaviour can often be formally described in conventional terms. The higher degree of freedom for each element, the smaller number necessary for complexity to arise.[1]
2. A complex system's *elements must interact dynamically*. The interactions can transfer something physical or information, but without interaction, a change in one element does not affect other elements. Moreover, without the interaction changing dynamically or having the capacity to change dynamically, the transfer between elements is constant and cannot transmit change.
3. The *interaction between elements must be fairly rich* for a system to be complex. It means that the overall element in the system influences and is influenced by a reasonably large number of other elements. The more elements in the system—out of which only some are redundant—the sparser the interaction can be in relative terms for complexity to still arise.
4. The *interaction must be non-linear*. It means that there are loops of interactions, allowing a change in an element to exert influence on that same element after it has been transmitted through parts of the system and frequently altered during the transmission.[2] In a linear system, a change in an element only disperses away from it along branching chains of interacting elements (Figure 9.1).

These four fundamental characteristics are present in all complex systems and give rise to consequential manifestations of complexity presented later.

Principle of locality

In addition to the constitution of complexity itself, other common characteristics of complex systems are vital for the kinds we consider in this book. The first concerns that each element is always only acting in response to the inputs available to it locally (Figure 9.2)

[1] This means it takes vastly more neurons either firing or not for a system to be complex than human beings with tremendous behavioural variability.

[2] That is, intensify, dampen or change direction (increase to decrease or decrease to increase).

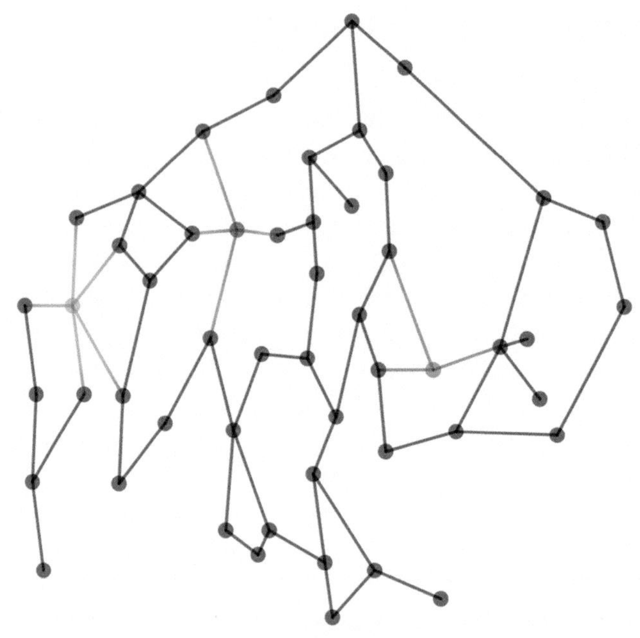

Figure 9.2 Sketch demonstrating the principle of locality, where the blue, red and green elements only receive input from their direct connections.

whilst being oblivious to the overall state or behaviour of the complex system as a whole (Cilliers, 1998, pp. 4—5). This is vital and referred to as the *principle of locality* (Heylighen et al., 2007). For an element to know everything happening to the system as a whole, it must be as complex as the system (Cilliers, 1998, pp. 4—5), which is impossible for the systems of interest in this book. The locality principle entails, in other words, that all actions are local and in response to locally changing conditions, without any involved element being aware of the full effects of their actions throughout the system. Knowing this should help us develop our humility as actors in complex human-environment systems and our tolerance and patience with other actors whose decisions and actions may make sense under the local conditions under which they operate.

Although the interactions between elements can be long-range in spatial terms, most usually have a relatively short direct reach and involve, for practical reasons, primarily inputs from neighbouring elements (Cilliers, 1998, p. 4). However, that does not preclude extensive influence since the route from one element to any other often involves only a few steps in systems with rich interaction. For instance, Milgram (1967) suggested any pair of individuals in the U.S. in the 1960s to be, on average, six people away from each other. More recent studies have shown that anybody in the world is between five and seven steps away (Dodds et al., 2003). The influence may, of course, be altered, suppressed or enhanced along the way (Cilliers, 1998, p. 4), but this means that 'each actor in a complex system controls little, but influences everything' (Bergström & Dekker, 2014, p. 2). In short, complexity results from rich, dynamic and non-linear interaction between many elements that only respond to the limited input each receives.

Openness and boundaries

The second additional characteristic concerns if a system is *open* or *closed* in the sense of its elements interacting with elements outside the system itself or not (von Bertalanffy, 1968). Cilliers (1998, p. 4) states that complex systems are *usually* open systems, and closed systems are *usually* merely complicated. Whilst it is often wise not to be categorical in matters of complexity, I claim that all complex systems of interest concerning the safety and sustainability of humankind are open systems. All living organs and organisms, ecosystems, hydrological systems, climate systems, organisations, business systems, the economy, societies, social-ecological systems, and the aid system are open systems (e.g., von Bertalanffy, 1968; Maturana & Varela, 1980, p. 134; Kauffman, 1993; Capra, 2005, p. 37; Freeze & Cherry, 1979; IPCC, 2021; Checkland, 1981, p. 256; Scott, 1981; Jackson, 2003, p. 6; Clegg et al., 2006; Sundström & Hollnagel, 2006, p. 242; Arthur et al., 1997b; Nicolis & Prigogine, 1989; Byrne & Callaghan, 2014; Donati, 2012, p. 104; Berkes et al., 2003, pp. 22–23; Ramalingam, 2013). I cannot imagine any closed complex system that is relevant in the context of this book, and all human-environment systems are certainly open. This has an important consequence that has already been alluded to in Chapter 7 when introducing assumptions concerning the extent to which we allow complexity to play a role in our work.

Open systems have, per definition, not an unambiguous *system boundary* for us to adopt when constructing them to help us make sense of our often overwhelmingly intricate reality. Therefore, the boundaries we set are more a feature of our need for boundaries than the complex reality we want to understand, explain or improve (Cilliers, 1998, p. 4). What to include and exclude is simply an analytical choice (Heylighen et al., 2007, p. 16; Richardson et al., 2001, pp. 8–10) and all boundaries are transient given a sufficiently long timeframe (Richardson et al., 2001, p. 8). However, they are not arbitrary but demand serious consideration. The deliberations of what to include and exclude are considered *boundary judgements* to underline their inherently subjective nature (Ulrich, 1996, pp. 156–158) and making them always entails analytical sacrifice (Bergström & Dekker, 2014). Such boundary judgements are connected in such a way to both descriptive statements about what we know about the phenomenon or problem and prescriptive statements about our purpose and objectives for addressing them that changing one automatically causes changes in the others (Figure 9.3) (Ulrich, 2000, pp. 251–252). Thus, it is crucial to systematically scrutinise what is included and excluded when addressing any phenomenon or problem and the descriptive and prescriptive statements on which those boundary judgements are based. Such scrutiny is critical when multiple actors with

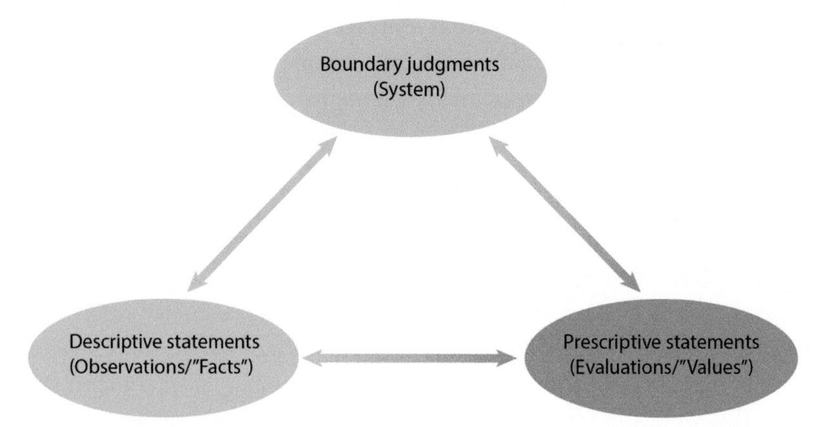

Figure 9.3 The nexus of boundary judgements, descriptive statements and prescriptive statements (Ulrich, 2000, pp. 251–252).

different perspectives are involved in the process who may agree on neither (Ashby, 1973, p. 1; Flood, 1987, pp. 177–178).

The systematic scrutiny of what to include and exclude is called *boundary critique* (Ulrich, 1996, pp. 171–176; Ulrich, 2000, pp. 254–266) and must be practised carefully when managing risk and resilience for sustainable development. The notion of 'something is better than nothing' is misleading concerning the construction of system boundaries, for which 'take nothing for granted' may be more appropriate (Richardson et al., 2001, p. 9). Chapter 11 provides guidelines for approaching this boundary challenge in practice. For now, it is enough to appreciate the need for humility about our limited ability to represent reality and for blatant openness about the incompleteness and provisionality of the results.

Manifestations of complexity

Whilst the constitution of complexity itself is arguably relatively straightforward—that is, many elements with rich, dynamic and non-linear interactions—the manifestations of such complexity are indeed striking, with far-reaching significance for all of us. I do not have room to elaborate on all the exciting contributions of complexity science. Still, I devote this section to introducing the most central manifestations of complexity for understanding, explaining and improving our challenges for sustainability and the world.

Feedback

Most fundamentally, the non-linearity of dynamic interactions in complex systems provides the necessary looped structures for changes in elements to come back as input to the same elements—sometimes directly and sometimes after several intervening

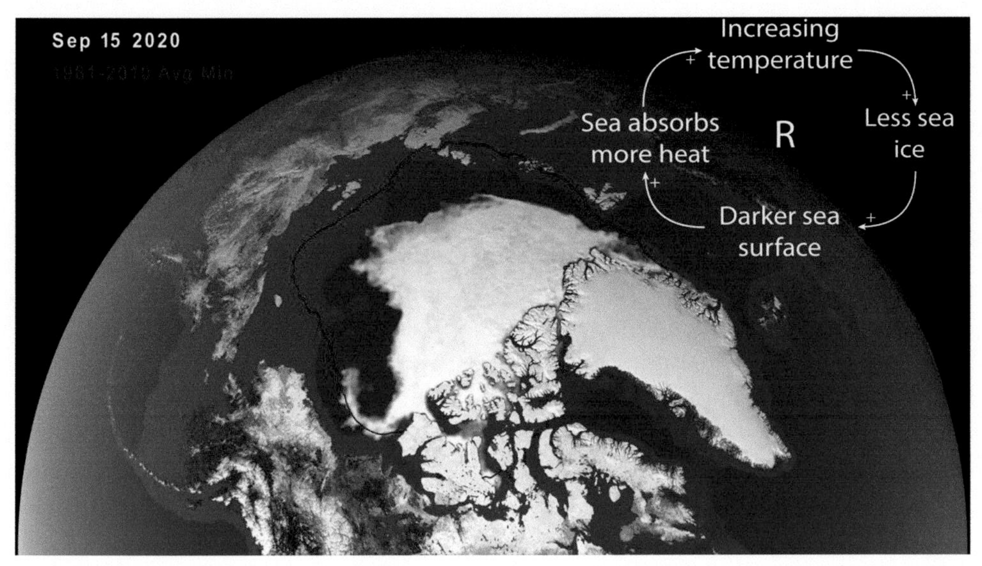

Figure 9.4 A reinforcing feedback loop diminishing the Arctic Sea ice cover. Satellite image from NASA of the minimum ice cover in 2020, compared to a 30-year average minimum cover (red line).

steps (Cilliers, 1998, p. 4). This returning input is called *feedback* and can result in immediate or delayed effects. It can either reinforce the initial change in the element—by returning input that increases the change further (Figure 9.4)—or balance the initial change—by returning input that counteracts it.[3] These two types of feedback are sometimes called positive feedback and negative feedback (e.g., Deutsch, 1948; IPCC, 2021; Meadows et al., 1972), but it is important to note that these names have nothing to do with the normative meaning of positive and negative. They should be understood in relation to the initial change, with positive feedback denoting the same direction of the incoming input as the outgoing change (initial increase feeds back further increase, or initial decrease feeds back further decrease) and negative feedback denoting the opposite direction of the incoming input as the outgoing change (initial increase feeds back decrease, or initial decrease feeds back increase) (Heylighen, 2001, p. 10). However, such positive feedback provides the foundation for so-called vicious and virtuous cycles (Meadows & Wright, 2009, p. 187), which indeed carry normative meaning. For clarity, I suggest following the convention of naming the two types of feedback loops reinforcing and balancing (Meadows & Wright, 2009; Senge, 2006), but there are additional suggestions found in the literature (e.g., Coetzee et al., 2016). Although feedback is again relatively straightforward to comprehend, it is fundamental for other manifestations of complexity and a range of vital processes for resilience—for example, learning

[3] The technical term for this manifestation of complexity is recurrency (Cilliers, 1998, p. 4).

and adaptation (e.g., Coetzee et al., 2016; Luhmann, 1995, p. 110; McMillan, 2003, p. 149). Realising actual feedback is, therefore, of utmost importance to grasp and address the phenomena and problems of interest in this book, and Chapter 11 introduces established tools to help us.

Disproportionality between cause and effect

The non-linearity of dynamic interactions in complex systems has another fundamental effect, closely linked to feedback. It results in *disproportionality between cause and effect*, between input and output (Byrne & Callaghan, 2014, p. 18; Cetina, 2005). This disproportionality means that small changes sometimes cause massive effects, whilst big changes sometimes fail to generate much effect at all (Heylighen, 2001, p. 10; Homer-Dixon, 2014, p. 126). The relationship between cause and effect may also change drastically over some input levels. It is referred to as threshold effects or tipping points and is further addressed in a later section on the consequences of complexity. Remember the terrifying potential tipping point of the Amazon rainforest in Chapter 4. In addition, the relationship between cause and effect may change over time as the complex system evolves. This decoupling of input and output in terms of proportionality and permanence is a central feature of complexity and cannot be found in simple or complicated systems.

The disproportionality between cause and effect is closely related to the *butterfly effect*. This effect has to do with *sensitive dependence on initial conditions* but with the same implications. It means that slight differences in the initial conditions of a complex system may result in significant differences at a later stage. Lorenz (1963) accidently discovered this effect when manually entering numbers in his complex weather model based on printed results that provided three decimals and getting vastly different outcomes than when using the numbers stored in the computer with six decimals (a difference of less than 0.1%). He became famous for describing conditions under which a system behaves chaotically, known as the Lorenz attractor, producing a butterfly-looking pattern when plotted (Figure 9.5). The effect was later popularised as the butterfly effect after Lorenz, in a presentation at a conference, used the analogy of a butterfly flapping its wings to illustrate the minuscule changes that, under certain conditions, could either trigger or thwart a tornado (Lorenz, 2000).

Whether we refer to it as the butterfly effect or disproportionality between cause and effect, this manifestation of complexity has essential consequences on our ability to grasp and address complex systems and the issues we may have in them. For instance, it exacerbates the earlier discussed problem of deciding what to include and exclude in the system, as minor differences in our assumptions concerning the system boundary may have significant effects on the appropriateness of the system in relation to our purposes (Richardson et al., 2001, p. 9). I elaborate on other consequences of complexity in later sections.

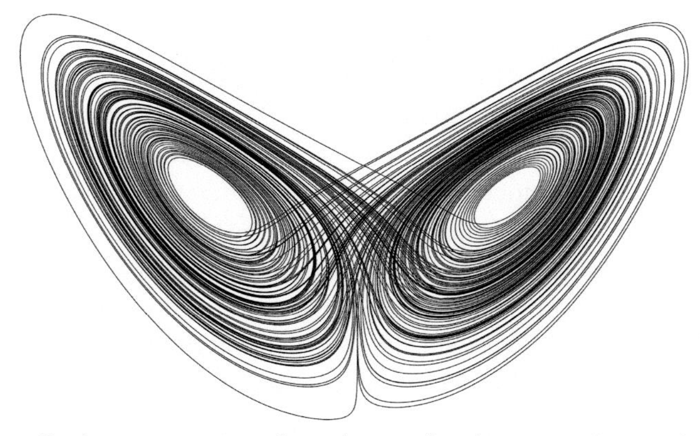

Figure 9.5 A butterfly shape appearing when plotting the chaotic oscillation of system states between the two attractors of a Lorenz system.

Hierarchical aggregation

The next manifestation of complexity that is crucial for us to understand concerns aggregation and hierarchy. It has to do with both our human inclination to categorise what we observe into groups—and these groups into larger groups of groups, and so on—and the actual organisation of elements in complex systems (Holland, 1995, pp. 10—12; Levin, 1998, pp. 432—433). These two sides are closely related since our categorisations are both enabled and restricted by what we can observe. Furthermore, all complex systems develop observable inhomogeneities in the organisation of their elements (Levin, 1998, p. 432). It means elements tend to aggregate into groups by interacting with some elements more than others. These groups are, in turn, interacting more with certain other groups and so on.

Consider any element of a complex system you can imagine. Then zoom in to the level of complexity below, and you will find that it comprises elements and interactions of its own. For instance, an organism consists of interacting organs, which in turn comprise interacting tissues, interacting cells and so on for as far down the quantum levels as the physicists have endeavoured. If you, instead, zoom out, you will find that interacting organisms constitute interacting populations, which in turn constitute interacting ecosystems and so on up to the biosphere and perhaps beyond. In short, complex systems are hierarchical (Cilliers, 1998; Heylighen et al., 2007). They exist, function and interact across levels of complexity, which Gunderson and Holling (2002) refer to as a 'panarchy', and the interactions across levels are fundamentally important regardless of focal level. Moreover, this ubiquitous process of *hierarchical aggregation* is not imposed on complex systems but emerges in the local interactions between elements that are always subsystems of constituent elements on the level of complexity immediately below (Levin, 1998, p. 432).

Emergence

The different levels of complexity we observe in the hierarchical aggregation of interacting elements are not arbitrary but guided by the emergence of observable properties that cannot be reduced to the constituent elements and interactions on the level below. In other words, it is not a coincidence that we identify the ant colony as a level of complexity above the myriad of interacting ants or the brain as above 86 billion interacting neurons. We recognise the ant colony because it exhibits remarkable ingenuity in solving the problems it encounters (Figure 9.6) at the same time as each ant has a minimal behavioural portfolio, with significant randomness in terms of activities (see Bonabeau, 1998; Hofstadter, 1979, pp. 311–336). The resilience of the ant colony is, in other words, an emergent property allowing the ant colony to survive over long periods and regular exposure to various hazards, even when each ant almost always dies when circumstances do not fit its behaviour (Holland, 1995, p. 11). That said, there may be other levels of complexity that could be interesting for other purposes. Regardless of which, the complex behaviour of the whole emerges in the simple actions and interactions of the parts. Similarly, we identify the brain because human consciousness emerges in the electric impulses between all those neurons, and a rock concert stampede because the dangerous and incontrollable crowd movements emerge in the benign dancing of thousands of individuals (Figure 7.4). This *emergence* of properties of the whole that cannot be reduced to its parts is perhaps *the* paramount manifestation of complexity (Schneider, 2012, p. 138).

Figure 9.6 A tree-hanging ant nest (*Crematogaster Castanea*) in Krantzkloof nature reserve, South Africa. *(Shared by JMK on the creative commons.)*

It is helpful to briefly revisit how elements aggregate in the evolution of complex systems to understand emergence and emergent properties. Emergence is impossible in symmetric systems. The key is, therefore, the inhomogeneities or asymmetries in the organisation of the interacting elements, sometimes referred to as *broken symmetry* (Flood, 1987, p. 180). Consider, for instance, how a human egg cell develops into a baby. After fertilisation, the cell divides into two identical cells, each of which divides again and again. If this symmetry continued for nine months, it would result in a blob of around 1.25 trillion identical cells (cf. Hirsch, 1977). Luckily, after 16 cells, they start to differentiate into outer layer cells and inner cell mass. The new symmetry is 'now called broken symmetry because the original symmetry is no longer evident' (Anderson, 1972, p. 395). This symmetry breaking happens repeatedly throughout the pregnancy, resulting in the 1.25 trillion cells coming out as a human being. This broken symmetry is fundamental for how complexity can arise out of simplicity (Anderson & Stein, 1987). It provides the foundation for emergence since it breaks the symmetry between levels of complexity (Anderson, 1972).

There are essential properties of complex systems that are not emergent but resultant in Lewes (1875, p. 413) original vocabulary (Chapter 8). For instance, we tend to be somewhat obsessed with the weight of the newborn baby just discussed, which is simply the aggregated weight of all elements constituting her body. Similarly, we can report the number of staff and volunteers of a Red Cross Society by counting everybody in each department and volunteer group and adding them together upwards along the organisational hierarchy. However, most properties that matter to us are emergent properties (Heylighen et al., 2007, p. 6), regardless of the level of complexity we are interested in—such as the life, health and consciousness of the baby and the organisational culture, efficiency and resilience of the Red Cross Society. The most important takeaway is that it is not feasible to explain, understand or improve any of these emergent properties by studying or addressing their elements in isolation and aggregating what we learn or achieve. Any potential success based on such a hit-and-miss approach to complex systems would be dumb luck. The devil is not in the details but in the relations.

Emergent properties of complex systems are not only emerging in the rich, dynamic and non-linear interaction between elements (*upward causation*) (Heylighen, 2001, p. 10). They also constrain those elements' behaviour (Campbell, 1974; Heylighen et al., 2007, p. 122; Macdonald & Macdonald, 2010a). This notion of the macro-level influencing the micro-level is a central tenet in most social science, regardless of whether the argument for such *downward causation* is explicit (e.g., Popper & Eccles, 1977, p. 20; Luhmann, 1981; Newman, 1996, p. 248; van den Bergh & Gowdy, 2003; Hedström, 2005; Sawyer, 2005; Manzo, 2014), or phrased in other ways (e.g., Alexander & Giesen, 1987; Durkheim, 2005; Giddens, 1984; Marx, 1976, p. 3; Weber, 2005). Emergent properties can also influence the elements of complex systems that do not comprise human beings (e.g., Campbell, 1974; Kauffman, 1993, p. 376). Heylighen (2007, p. 122) explains how the interactions between elements constrain them since 'they can no longer act as if they are independent from the others'. In other words, the whole imposes a certain coherence or coordination on its parts.

The remaining issue regarding emergent properties of complex systems concerns the circumstances of direct causal links between them. It is perhaps obvious that one emergent property can influence another directly if the two are emergent properties of two different systems. They are then simply properties of two interacting elements on a higher level of complexity. It is less straightforward when the two emergent properties emerge in the same complex system or in the interactions of the same elements. Under such circumstances, the possibility of a direct causal link depends on whether the emergent property itself can be considered endowed with causal powers or not. It is not easy to imagine relevant emergent properties with their own causal powers in the complex systems of primary interest in this book. Still, there are arguments both for and against regarding the brain (cf. Bogg & Geyer, 2007; Macdonald & Macdonald, 2010b; Popper & Eccles, 1977). Without their own causal powers, any observed association between emergent properties of the same complex system must be indirect and explained with reference to its interacting elements (cf. Hedström, 2005, pp. 115–116). Coleman's (1986, pp. 1321–1324) now classic example of Weber's suggested causal link between protestant religious doctrine and the capitalist economy is an informative illustration of this. Whilst an association between these two emergent properties can be observed in some 19th-century societies in Europe (link 1 in Figure 9.7), explaining it requires us to understand how the protestant doctrine influenced individual values (link 2: downward causation), how these values influenced the myriad economic activities of protestant individuals (link 3) and how the capitalist economy emerged in the interaction between these economic activities (link 4: emergence). Even if all three steps are equally important for explaining the observed association, the last step has been the most challenging (Hedström, 2005, pp.

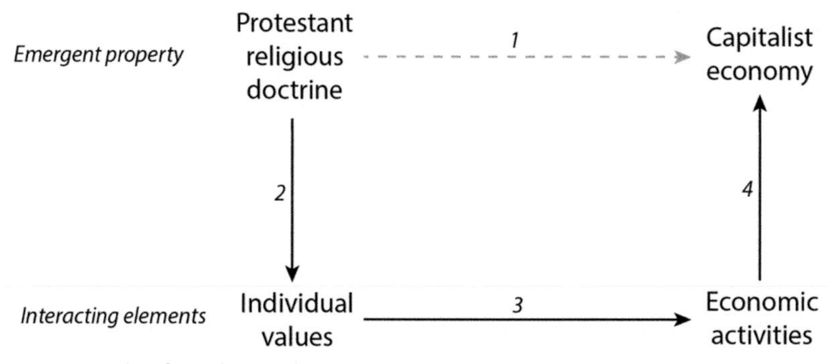

Figure 9.7 An example of an observed association between two emergent properties explained as a chain of causal links across levels of complexity. *(based on Coleman, 1986.)*

115–116). Emergence may be not only the paramount manifestation of complexity but also the most demanding for us to grasp.

Path dependence

It is essential not to forget that all the complex systems we are interested in here have a history. They evolve over time, and their past is co-responsible for their present (Cilliers, 1998, p. 4). In other words, the current state of any complex system depends on the path it has taken to reach it. This manifestation of complexity is referred to as *path dependence*. It captures how past events, regardless of how small, can result in lasting effects on the complex system that are difficult, if not impossible, to undo or change (Arthur, 1988, 1994). It is as if complex systems have a memory not located in any specific part but distributed throughout the systems (Cilliers, 2016b, p. 68). This intuition is informative since the rich, dynamic and non-linear interaction between the elements of complex systems gives rise to path dependence. It means that a change in one element constrains its future behaviour and the behaviour of other elements as the change propagates through the system. However, the interacting elements are not only constraining each other directly but also through emergent properties that may arise or change (remember emergence and downward causation), resulting in a 'strong dynamical fixation of historical constraints' (Hooker, 2004, p. 476). Reinforcing feedback is a significant cause of path dependence (Arthur et al., 1997a, p. 2), regardless of whether we call it increasing returns, self-reinforcement or positive feedback (cf. Hämäläinen & Lahtinen, 2016, p. 16; Page, 2006, p. 88). Symmetry breaking is also an important cause of path dependence, as constraints may push a complex system over a critical threshold, referred to as a *bifurcation point* (Prigogine & Stengers, 1984), and into system states that are qualitatively different than before the bifurcation (Nicolis, 1995, p. 8). A slight initial fluctuation may effectively result in a wide-ranging and permanent consequence (Brinsmead & Hooker, 2011, p. 825).

Path dependence is, as such, closely related to the idea of a *lock-in effect*, in which the path dependence results in such a strong anchor point that it is onerous to move forward (Hämäläinen & Lahtinen, 2016, p. 14). There are numerous examples of such lock-in effects, out of which the prevailing dominance of the QWERTY keyboard (Figure 9.8), regardless of superior alternatives, is the archetypical example (Arthur, 1989, p. 126; David, 1985). There are, of course, many other lock-in effects that are of more direct relevance to the context of this book. For instance, IPCC presents lock-in effects of fossil fuel-related infrastructure as critical for long-term mitigation strategies (IPCC, 2021, pp. 1–112). It highlights how inappropriate responses to climate change lock in high levels of vulnerability and risk (IPCC, 2022, p. 85). Lock-in effects are identified in the continuing production of chemicals despite strong imperatives for reduction (Persson et al., 2022, p. 1514) and as essential to anticipate and avoid when adapting infrastructure

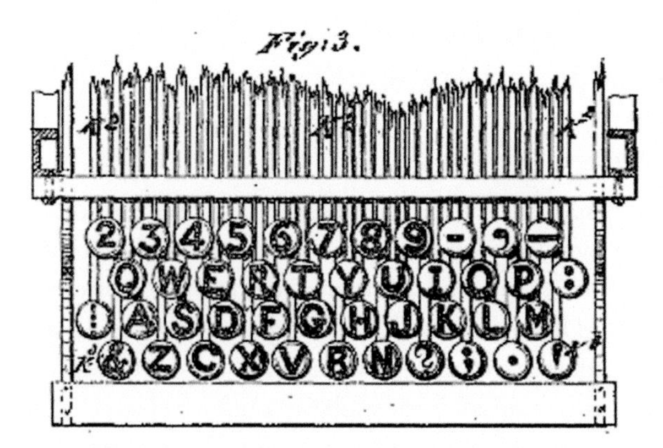

Figure 9.8 The QWERTY typewriter key layout from 1878.

systems (Payo et al., 2015). In short, it is not enough to just state that history matters (Page, 2006). We need to consider *how* it matters.

Self-organisation

The next crucial manifestation of complexity for us to understand is *self-organisation*. Remember that the structure of complex systems is not imposed on them but emerges in the local interactions between their elements.[4] 'Structure is the *result* of action in the system, not something that has to exist in an a priori fashion' (Cilliers, 2016a, p. 89, original emphases). However, this fits poorly into our intuitive worldview, which predisposes us to assume that someone must have arranged the elements in a particular order when confronted with something organised (Heylighen, 2001, p. 2). We are generally inclined to attribute it to some mysterious intelligence if we cannot identify anyone responsible for the seemingly peculiar organisation. Nevertheless, the organisation of complex systems is not the making of some external or centralised authority (Centeno et al., 2015, p. 67; Cilliers, 1998; Macy & Willer, 2002)—commander, controller, coordinator or creator. It is the result of self-organisation (Figure 9.9), which is 'the appearance of structure or pattern without an external agent imposing it' (Heylighen, 2001, p. 2). The elements may have varying influence on the structure and functioning of the system, and there may even be observable subsystems for coordination or control, for example, the brain and central nervous system of an organism and the CEO and line management of an organisation. Yet, the actual organisation of activities and events in complex systems are overwhelmingly self-organised (e.g., Camazine et al., 2003;

[4] Although both organisation and emergent properties emerge in the local interactions between elements, self-organisation results in patterns of interacting elements and emergence results in properties that cannot be reduced to the elements (cf. Page, 2015, p. 32).

Figure 9.9 A self-organising shoal of fish moving as one.

McMillan, 2003; Taylor & Van Every, 2000). It is 'order for free', in Kauffman's words (Lewin, 1992, p. 25). Centralised control may have some advantages, but 'at some level it must itself be based on distributed control' (Heylighen, 2001, p. 9) and fully centralised control means the system loses its adaptability and survives only as long as the environment is stable (Cilliers, 1998, p. 108).

Far-from-equilibrium dynamics

Self-organisation is vital for complex systems' ability to adapt to changes in their environment. It allows them to transform themselves in response to the changes, which, in turn, often affect the environment (Cilliers, 1998, p. 108). However, not all self-organising systems can adapt continuously. Heylighen (2001, p. 2) distinguishes between self-organisation towards equilibrium and self-organisation *far from equilibrium*. This distinction is essential for explaining and understanding other manifestations of complexity. Equilibrium is, here, a state of maximum order where the elements and interactions are locked, and there is no change or activity. Some self-organising systems can only strive towards their equilibrium and are, therefore, not adaptive—like the spontaneous alignment of atomic spins during the magnetisation of a ferromagnet (Haken, 1983, p. 3) or the arrangement of molecules during crystallisation (Heylighen, 2001). As Schrödinger (1992) pointed out some 80-odd years ago, equilibrium is death! *Adaptation,* thus, requires complex systems to operate far from equilibrium—'in a region intermediate between total order and complete disorder' (Gell-Mann, 1995, p. 17)—where they have enough structure to maintain the coherence of the systems at the same time as enough freedom to adapt. This region is poetically referred to as *the edge of chaos* (e.g., Kauffman, 1993; Lewin, 1992; Ramalingam, 2013, p. 146).

Figure 9.10 The combined entropy (disorder) of a system and its environment can only increase over time.

In technical terms, the second law of thermodynamics stipulates that disorder (entropy) can only increase in any system left to itself (Heylighen, 2001, p. 2). Although this is a fundamental law of our universe that is impossible to break (Figure 9.10), it does not preclude increasing local order as long as the total disorder is increasing (Cilliers, 1998; Gell-Mann, 1995, p. 19; Prigogine & Stengers, 1984). This prospect for local order is relatively straightforward to grasp in the case of self-organisation towards equilibrium. If looking at the example of crystallisation, the increase in the order of the crystal (decreasing entropy) is compensated by increasing disorder of the liquid the crystalising molecules were dissolved in (increasing entropy), resulting in a total increase in disorder of the crystal and liquid together until equilibrium is reached and everything stops (Heylighen, 2001). Death!

The case of self-organisation far from equilibrium is somewhat less straightforward, as these systems seek to avoid equilibrium death. Continuous adaptation means continuous generation of aggregated surplus disorder (entropy), which they must discard. These complex systems accomplish that by actively exporting or dissipating disorder (entropy) from the systems into their environments.[5] This process can only be achieved by importing more ordered resources and exporting more disordered waste, which explains why all the complex systems of interest in this book are open systems. For example, an organism consumes food (higher order) and excretes waste (lower order). Not to mention the immense resource consumption and waste production of modern societies. In short, these complex systems 'manages to increase its own organisation at the expense of the order in the environment' (Heylighen, 2001, p. 2). Waste becomes a source of order (Prigogine & Stengers, 1984, p. 143).

Complex systems operating far from equilibrium require constant flows of resources to maintain their organisation and to ensure their survival (Cilliers, 1998, p. 4). Although they continuously adapt to changes in their environment, they retain their coherence and viability in relation to it in the sense that the systems are continuously observable. The

[5] The arrangement for this is referred to as a *dissipative structure* (Prigogine & Stengers, 1984).

systems appear stable, regardless of the dynamic processes underlying them. However, this stability must not be confused with equilibrium, which Parsons (1951) did for social systems. It is dynamic stability in which the complex system can remain in reproducible steady states (Prigogine & Stengers, 1984). For instance, I have been around for half a century, but all the muscle and fat cells that make up my body have been replaced between one and four times since birth, and there are cells in my blood and organs that are replaced every three days or weekly (Fischetti, 2021). Regardless of this constant flux, I have constituted some kind of stable whole ever since I drew my first breath. The same goes for all complex systems that keep operating far from equilibrium and neither drift too far towards the debilitating constraints of equilibrium nor into the chaos of removing all constraints on their elements—that is, all complex systems that remain on the edge of chaos.

Consequences of complexity

The manifestations of complexity introduced above have real consequences on our ability to understand, explain and improve phenomena and problems concerning risk and resilience. As already alluded to in the introduction to this chapter, complexity is not only a fundamental challenge in itself but also exacerbates the other three specified in this book. Complexity escalates uncertainty, as feedback, disproportionality between cause and effect, emergence, and downward causation introduce inherent unpredictability and undermine our ability to grasp how our systems of interest work. Complexity is also the root cause of their dynamic character since far-from-equilibrium dynamics and the resulting self-organisation and adaptation, as well as path dependence and lock-in effects, both cause and constrain change. Finally, complexity engenders ambiguity, as the sheer unknowability of complex systems opens up for multiple interpretations. However, this does not mean that we cannot know anything about these systems, nor that their complexity completely incapacitates us. 'Outcomes are still bounded even though they are not predictable' (Etkin, 2016, p. 155). For example, nobody can predict the weather six months from now, but I know there will be no category 5 cyclone in Chad or blizzard in Singapore (Figure 9.11). That said, there are other consequences of complexity—often closely linked to the four fundamental challenges just mentioned—that require attention if we are to move forward. Especially, since complexity requires and enables profound changes in our policymaking and practices (cf. Brinsmead & Hooker, 2011, p. 809).

Requisite relationality

The most fundamental consequence of complexity is *requisite relationality*. If the system of interest is complex, it matters little how hard we try to understand, explain or improve it by dividing it into parts and studying or addressing each part in isolation. Such a

Figure 9.11 Singapore has a typically tropical climate, with the lowest recorded temperature of 19.4°C/66.9°F.

reductionist approach may work for simple or complicated systems (Chapter 7), but it should be clear now that it is inappropriate for complex systems. There are three primary reasons for this. The first is obviously related to emergence, resulting in observed properties of the system that simply cannot be reduced to properties of its elements (Holland, 1998, p. 225; Lewes, 1875). Secondly, the state of anything in a complex system depends on many other things, hampering all attempts to grasp or address even a single element or issue in isolation. The third reason for the inappropriateness of reductionist approaches to complex systems is related to the assumption of *ceteris paribus*—that all elements that are not explicitly included stay the same over time. This assumption clearly does not hold for complex systems since the rich, dynamic and non-linear interactions between elements preclude attempts to keep them constant (Anderson, 1999, p. 217; Ashby, 1957, p. 5). It all leads to the absolute necessity of relationality (e.g., Emirbayer, 1997; Rika, 2016; Urry, 2003)—the perspective that nothing important in the world exists in isolation—and of paying serious attention to the effects of relations on the effectiveness of policies and practices for safety and sustainability. Unfortunately, there are countless examples of policymakers and practitioners ignoring interactions, focussing on particular and often somewhat arbitrarily cherrypicked elements or issues, and still expecting results without ever considering what else is needed (e.g., Becker, 2021a). We also tend to ignore complexity when we attempt to explain unexpected events (Figure 9.12):

> *When we are faced with unexpected occurrences, especially when they have catastrophic results, we tend to ascribe their cause to a rare combination of unlikely circumstances. When we have to explain the crash of the stock-market, an earthquake, or the sudden outbreak of political violence, we try to find a number of factors that combined to cause it, often with the hope of showing that the chances of the same combination of factors occurring again are slim. This kind of analysis,*

Figure 9.12 A sign commemorating the three mile Island nuclear accident in 1979. *(Shared by Z22 on the creative commons)*

> *however, is the result of trying to explain the behaviour of large, complex systems by extrapolating from the behaviour of small, simple systems. Unfortunately this extrapolation fails*
>
> ***(Cilliers, 1998, p. 96).***

We can never solve these problems entirely since we must always reduce the actual complexity of the phenomena or problems we set out to study or address (Chapter 7). We must, nevertheless, not resign ourselves to being overwhelmed by their complexity and only focus on some isolated parts that simply seem important for apprehending the phenomenon or generating the desired outcome without careful and systematic consideration of how the system works. Even if we can never fully understand how it works, we must relentlessly seek to recognise the nexus of causes and effects of the phenomenon or what else is needed to generate that outcome. We must apply a systems perspective (Chapter 11), if you like. We must also realise the value of broad participation and different perspectives and interpretations—referred to as epistemological pluralism (Healy, 2003; Miller et al., 2008)—when seeking to understand phenomena or address problems in complex systems. Then, armed with a richer understanding of the system, we are better equipped to design activities that are more likely to result in what we want to achieve, regardless of whether that is what we initially desired or an updated objective based on what we have learnt. We may not be able to address everything in the system, but it is at least more likely that what we end up doing will have an actual impact and not just result in wasted resources. Chapter 11 provides further guidelines for how to do this in practice.

Surprise

The next important consequence of complexity is *surprise* (Duit & Galaz, 2008). Surprise can also emanate from insufficient information or knowledge (McDaniel et al., 2003, p. 266). However, complexity engenders contingency in the sense of absence of certainty in what may happen. Luhmann (2014, p. 25) even states that by 'complexity we would like to understand that there are always more possibilities than can be actualised'. It means there can only be one event happening at a particular moment out of a myriad of possible events. Several interlocking reinforcing and balancing feedback loops in complex systems regularly result in convoluted recurrences that amplify specific changes and suppress others, often with significant delays that further challenge human cognition. Add emergence and downward causation, and you have unpredictability in what will happen, even with a reasonably good understanding of how the complex system works (Heylighen, 2001, p. 10).

However, for the actual event to constitute a surprise, it must fall outside the normal range of variation for that system (Woods, 2015, p. 6). For instance, just because you do not know what the weather would be like before waking up in the morning, it would not be a surprise if it was either sunny or snowing in Sweden in December. On the other hand, if it snowed in Singapore, it would definitely constitute a surprise. Although the dynamic stability of complex systems operating far from equilibrium keeps them within their normal variability most of the time, changes in their elements, interactions or environment can suddenly push them into unexpected states (Figure 9.13). *Surprise!* Furthermore, it also constitutes a surprise when such complex systems unexpectedly remain within their normal variability regardless of intense activity to change their current states. Even combinations of changes that are insignificant on their own can interact to produce surprise or even catastrophe (Perrow, 1984). More on that in a moment.

Traditionally, we consider surprises as unwelcome and, often, disruptive events, prompting attempts to avert or cope with them through more knowledge, tighter planning, more quality control and standardisation (McDaniel et al., 2003, p. 266). Whilst more knowledge about how the system works and better preparedness to deal with surprise before it happens are essential for resilience (Woods, 2015, p. 6), it is not possible to ever preclude surprise since it is a consequence of complexity itself (McDaniel et al., 2003, p. 266). Moreover, control and standardisation may even be counterproductive, as it directs perception, tapers attention and reduces flexibility (Olsen et al., 2020). So, rather than erroneously act as if complex systems are characterised by disconnection and stability, we must acknowledge their complexity and dynamic character and design policies and practices accordingly (Duit et al., 2010, p. 364). Remember that surprise is not a function of ignorance, or at least not only a function of ignorance. Accepting this fact lets us refocus attention away from trying to predict what will happen and towards understanding the phenomenon or problem (McDaniel et al., 2003, p. 269).

Figure 9.13 The massive seagrass die-off in florida constitutes a surprise (Gunderson, 2001). *(Shared by FWRI (CC BY-NC-ND 2.0).)*

Surprise is unpredictable by definition, but there are patterns to surprise, and it is often possible to characterise recurring classes of it (Woods, 2015, pp. 6–7). However, we do not need to predict an event or even a class of events to do something about it (cf. Clearfield & Tilcsik, 2018, p. 84). Instead of desperately attempting to predict what *will happen,* we can try to anticipate myriad potential events that *could happen*—remember Westrum's (1992) requisite imagination from Chapter 8. We should, then, do what we can to prevent destructive events and mitigate potential consequences beforehand. However, in an unpredictable world that is rapidly changing, we must do everything we can to remain flexible (Duit & Galaz, 2008).

Intrinsic volatility

Perrow (1984) takes the consequences of complexity further. It is not just that combinations of insignificant changes can interact to produce surprise, as described in the previous section. They can also generate disaster. He suggests that complex systems are subject to destructive events inherent to the systems themselves and, therefore, unavoidable. This *intrinsic volatility* of complex systems is a consequence of complexity and makes some disasters simultaneously unpredictable and inevitable. It is essential to remember that most disasters are preventable (Chapter 8) when complexity is just hampering understanding and effective risk reduction. However, rich, dynamic and non-linear interaction can, under certain conditions, cause disaster in itself. These are the disasters Perrow (1984) describes, with the Three Mile Island nuclear accident being his archetypical case (Figure 9.12):

That accident had its cause in the interactive nature of the world for us that morning and in its tight coupling—not in the discrete failures, which are to be expected and which are guarded against with backup systems. Most of the time we don't notice the inherent coupling in our world, because most of the time there are no failures, or the failures that occur do not interact. But all of a sudden, things that we did not realise could be linked (buses and generators, coffee and a loaned key) became linked. The system is suddenly more tightly coupled than we had realised. When we have interactive systems that are also tightly coupled, it is 'normal' for them to have this kind of an accident, even though it is infrequent. It is normal not in the sense of being frequent or being expected—indeed, neither is true, which is why we were so baffled by what went wrong. It is normal in the sense that it is an inherent property of the system to occasionally experience this interaction

(Perrow, 1984, p. 8).

There are numerous other examples of such *normal accidents* in which small errors that are inevitable in complex systems combine and spread between parts that are dependent on each other in various ways[6] (see Rinaldi et al., 2001) to produce baffling effects (Clearfield & Tilcsik, 2018, p. 32). For example, the 2020 stock market crash was triggered by growing uncertainty concerning what impact the Covid-19 pandemic *would* have, making the Dow Jones Industrial Average fall 36% from 20 February to 23 March 2020, even before the real onset of the pandemic (Figure 9.14). It had nothing to do with any actual impact of the virus. The slump ended about a week after the global death toll started to escalate, with unprecedented recovery and continuous growth throughout the rest of the pandemic until the war in Ukraine began to loom. Seemingly random spikes in food prices in Africa (Gaupp, 2020) and rapid transformation in dominant plant species in freshwater marshes of the Everglades (Gunderson, 2001) are other examples.

No matter how hard we try to grasp and address this kind of destructive event, we often make the wrong diagnosis and may even aggravate the problem by trying to solve it (Clearfield & Tilcsik, 2018, p. 32). When we encounter a problem in a complex system, we regularly rely on active and often automatic technological devices to address it. For instance, when we want a more aesthetically pleasing grand lecture hall for 1200 people with combustible wood on the walls and ceiling, we install a sprinkler system and fire alarm and count on these installations to work (Figure 9.15). If not, it would most likely lead to disaster. Similarly, the Boeing 737 Max flight stability problem was addressed by installing a Maneuvering Characteristics Augmentation System, which led to two fatal crashes when malfunctioning. Relying on such technological solutions is not only demanding that everything works all the time. It also increases the system's complexity, which can increase its 'catastrophic potential' through additional combinations of failures (Perrow, 1984). Regardless of how improbable these disasters are, the additional

[6] Perrow (1984) refers to this as 'tight coupling' and to the systems of interest as 'complex and tightly coupled systems'.

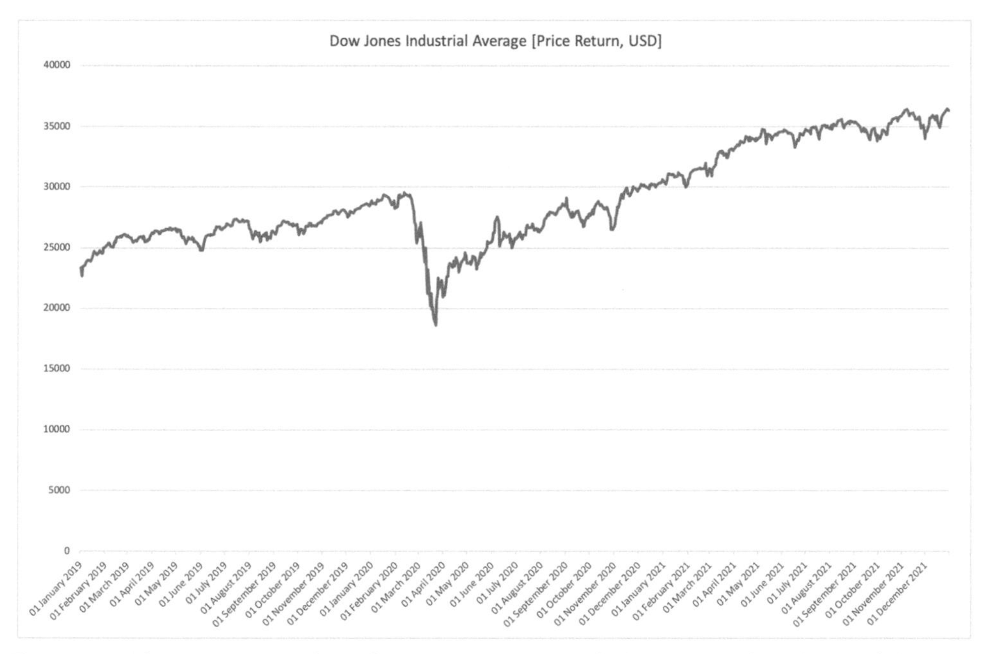

Figure 9.14 The Dow Jones industrial average 2019–21, with the 2020 stock market crash between 20 February and 23 March 2020 visible.

Figure 9.15 The Aula magna at Stockholm University—an architectural masterpiece or disaster waiting to happen? *(Shared by Arild Vågen on the creative commons.)*

combinations of failures increase their probabilities of occurring ('t Hart, 2013). This relationship between complexity and disaster explains why the overall process of increasing complexity is problematic (Chapter 6).

Instead of further increasing the complexity of already complex systems, Perrow (2008) argues for reducing their complexity through increased modularity. Modularisation is about removing or, at least, relaxing dependencies between parts of a system.[7] It is important to note that this may only work if removing the actual dependence, not merely ignoring it, which unfortunately is rather common when governing complex challenges (e.g., Becker, 2020; Becker, 2021b; Gilissen et al., 2016; Metz et al., 2020).

Cascading effects

Complexity is not only generating unpredictable and unpreventable failures and disasters or hampering our understanding and handling of preventable ones. It also makes them harder to contain as their effects are transmitted through the rich interactions between elements (Clearfield & Tilcsik, 2018, p. 32). It brings us to the next consequence of complexity: *cascading effects* (Duit & Galaz, 2008). This consequence is closely related to the intrinsic volatility of complex systems described above. It is based on the same interactions or dependencies, but now our focus is not on how failure or disaster can emerge but how their consequences can propagate throughout systems—immediately or to different places at various times due to the delays mentioned earlier. The same effect is sometimes described as falling dominos (Clearfield & Tilcsik, 2018, p. 32), but that analogy connotes linearity that is rarely appropriate for neither causes nor effects of accidents and disasters in complex systems (Hollnagel, 2014, pp. 64–66). Cascading effects are pervasive and found in diverse contexts ranging from interconnected infrastructures (Little, 2002) and power blackouts (Anderson et al., 2007; Kinney et al., 2005) to invasive species (Linders et al., 2019) and marine plastic pollution (Li et al., 2021, p. 4). They are particularly interesting concerning climate change, which is expected to have colossal cascading effects (IPCC, 2022). They have long been an important feature of understanding disasters (e.g., Di Baldassarre et al., 2018; Little, 2004; Månsson, 2019; Naqvi & Monasterolo, 2021; Ongkowijoyo & Doloi, 2017; Santos et al., 2020; United Nations, 2015). This means it is rarely enough to only consider the direct effects of variations, changes, disturbances, disruptions or disasters without considering how they cascade throughout the system of interest.

Tipping points

Complex systems do not only exhibit disproportionality between cause and effect, but the relationship between cause and effect may also change drastically over a certain level

[7] Making the couplings less tight, in Perrow's own language.

Figure 9.16 Tipping point just ahead.

as a result of the rebalancing of interlocking reinforcing and balancing feedback loops. It means that complex systems can have thresholds over which they suddenly shift to a completely different regime of states[8] (Scheffer et al., 2009, p. 53). This consequence of complexity is called *tipping points*[9] (Gladwell, 2000) and constitutes a challenge for understanding and explaining phenomena and improving problems in complex systems (Figure 9.16). Tipping points are especially central to the multi-equilibrium approaches to resilience mentioned in Chapter 7, where resilience is defined as the system's capacity to avoid tipping (e.g., Holling & Gunderson, 2002). However, tipping points are also consequential to the approach of this book since they bring human-environment systems into fundamentally different regimes of states—altering their capacities to continuously develop along their preferred expected trajectories whilst remaining within human and environmental boundaries.

The discovery of the Antarctic ozone hole in the mid-1980s was an early paradigm-shattering example of such a tipping point in environmental systems (Homer-Dixon, 1991, p. 80), and there are countless other examples since then. We have already heard of the potential tipping point of the Amazon rainforest that we seem to approach dreadfully fast (Lovejoy & Nobre, 2019). IPCC (2021) forewarns us of expected tipping points in our climate system over which the rate of change escalates, and changes become irreversible. However, tipping points were first identified in the social parts of human-environment systems (Grodzins, 1957), although sometimes conceptualised as a critical mass over which a change becomes self-sustaining and often escalating

[8] Also known as an attractor (Lenton, 2013).

[9] Also known as threshold effects (Duit & Galaz, 2008).

(Granovetter, 1978; Schelling, 1971, 1978). Tipping points are prevalent there too. They are crucial for understanding collective action (Macy, 1991), financial crashes (May et al., 2008), social movements (Hedström, 1994), political transformation (Walby, 2009) and many other social phenomena and processes (Hedström, 2005, pp. 90–96)—also for stabilising Earth's climate (Otto et al., 2020).

Regardless of the critical importance of tipping points, most policymaking and practice overlook or ignore them. The reason for this is simple. Complexity is impossible to grasp fully, and tipping points are difficult to anticipate (Lenton, 2013; Moore, 2018). Even more so in the social parts than natural parts of human-environment systems (Stadelmann-Steffen et al., 2021, p. 3). Identifying a tipping point and describing its causes with hindsight is one thing, but anticipating it is something else (Moore, 2018, p. 635). However, we must beware of their ubiquity, and there are generic indicators of systems approaching tipping points that have the potential to assist us in anticipating them (Scheffer et al., 2009).

There are three general types of tipping points that can be called bifurcation-type, noise-induced, and rate-dependent tipping points (Ashwin et al., 2012) with different early indicators. However, actual tipping points are usually combinations (Lenton, 2013, p. 6).

The first type concerns tipping points in which a small, steady change forces the system the last bit past a critical threshold, causing a significant change in its regime of states (Lenton, 2011, p. 202). Such tipping points can be either reversible or irreversible but are indicated beforehand by slower and slower recovery after small short-term fluctuations the closer the system is to the tipping point (Lenton, 2013, p. 6; Scheffer et al., 2009, p. 53; Wissel, 1984). This *critical slowing down* can also assist anticipation through associated increases in variance and autocorrelation in the fluctuations (Scheffer et al., 2009, pp. 53–54).

On the other hand, noise-induced tipping points are caused by short-term internal variability without any steady change (Lenton, 2011, p. 202)—such as combinations of rare disturbances and disruptions (Harley & Paine, 2009). Whilst these tipping points can be just as rapid and profound as the first type, they are not expected to show the same early warning indications (Lenton, 2013, p. 6). However, noise-induced tipping points are sometimes preceded by a period of 'flickering' in which the noise pushes the system back and forth between alternative states before settling on the other side of the tipping point (Lenton, 2013, p. 7). Although it is a weaker early indication than critical slowing down and related indicators, it has proven informative in various contexts spanning from food security (Krishnamurthy et al., 2020) to financial markets (Gatfaoui & de Peretti, 2019).

The last general type of tipping points is commonly referred to as rate-dependent. It comprises tipping points driven neither by steady change (bifurcation-type) nor short-term variability (noise-induced). They are instead driven by the rate of change exceeding

Figure 9.17 A compost approaching the rate-dependent tipping point of catching fire due to escalating internal heat release by organisms respiring organic matter. *(Shared by SB Johnny on the creative commons.)*

some threshold rate over which runaway change occurs (Figure 9.17) (Lenton, 2013, p. 7). In other words, the rate of change is so fast the system cannot absorb it to stay in its current regime of states (Ritchie & Sieber, 2016). Rate-dependent tipping points were initially believed not to have the same early indicators as the bifurcation-type (Ashwin et al., 2012). However, research now suggests that increases in both variance and autocorrelation may be observed as the system approaches the threshold rate before tipping (Ritchie & Sieber, 2017, p. 1)—although with substantially different delays for different levels of noise (Ritchie & Sieber, 2016).

In addition to early indications of tipping points by paying attention to different changes over time, system-specific spatial patterns may also arise before tipping and, thus, serve as early indications (Scheffer et al., 2009, p. 56). For instance, the spontaneous formation of regular vegetation patterns through spatial self-organisation in ecosystems has been suggested as a prominent early indication of a tipping point towards desertification, although spatial self-organisation may also help evade tipping altogether (Rietkerk et al., 2021, p. 1). In other words, it is essential to consider the kind of system when interpreting spatial patterns (Scheffer et al., 2009, p. 56).

Finally, whilst tipping points in complex systems imply risk, they also comprise opportunities for positive change (Scheffer et al., 2012, p. 344). Tipping points towards carbon-neutral societies may even be required for humankind to avoid climate catastrophe (Otto et al., 2020). Here, complexity may act in our favour. Without tipping points, more than 50% of the population may be required to adopt environmentally friendly behaviour before the majority can sway the rest. Yet, research suggests that a coordinated 10% is enough for their behaviour to predominate in a structurally undifferentiated setting (Hedström, 2005, p. 93). However, actual social interaction is inherently differentiated with people belonging to various groups. Hedström (2005, pp. 96–98) demonstrates that under such more realistic circumstances, even one person belonging to a relatively small group can set in motion a sequential process of tipping points, with no coordination whatsoever, eventually involving the entire population. The probability of such successful dissemination increases sharply with the first additional people enacting the behaviour from the start. Still, it is enough if they constitute 10% of their group to sway everybody in it. Even if each individual in that first group has little influence over those in another group, once everyone has adopted the behaviour, their collective influence can push the other group over the tipping point. Then, these two groups exert collective influence on a third, and so on, until the behaviour predominates in the entire population (Hedström, 2005, pp. 96–98).

Conclusion

The phenomena and problems we are concerned with are complex, regardless of whether we ignore their complexity or not. Whilst we cannot know the complexity of the parts of reality we focus on, the human-environment systems we construct to make sense of them must fit our observations. Continuing to approach complex phenomena or problems as if they are merely simple or complicated may be the principal blunder of humankind in the 21st century.

Although it may be relatively straightforward to grasp what constitutes complexity on the most fundamental level—that is, the rich, dynamic and non-linear interaction between many elements—its manifestations are indeed astonishing and consequences seemingly overwhelming. Undoubtedly, complexity is a fundamental challenge for us to understand, explain and improve our sustainability challenges and the world. However, this challenge does not provide a blank cheque to disregard complexity—quite the contrary. The realisation that the systems we are concerned with are complex, open and operating far from equilibrium prompts us to overcome our linear intuitions and unlearn much of what we have been taught in school concerning how to approach phenomena and problems—or at least learning when it is not appropriate, if ever.

Approaching such complex systems as if they are complex is essential for any endeavour to manage risk and resilience for sustainable development. Such an approach entails recognising the importance of feedback and the disproportionality between cause and effect it may give rise to. It involves understanding how interacting elements aggregate to form hierarchies of interacting systems within systems, with emergent properties on each level that cannot be explained by studying the elements on the level below in isolation (e.g., resilience). Such emergence is rooted in the interactions between the elements and the emergent property may influence them. Approaching a complex system as if it is complex entails recognising path dependence as past events both constrain and enable current events, which in turn constrain or enable future events. It involves understanding that the interactions between elements can result in elaborate organisation without the involvement of a commander, controller, coordinator or creator and that such self-organisation exceeds organisation based on centralised control in all but very few complex systems. Such an approach entails appreciating that only complex, open systems operating far from equilibrium can be adaptive. Any observed stability is dynamic in the sense of being continuously maintained—never resulting from reaching equilibrium. Finally, it requires humility concerning our limited ability to represent reality and blatant openness about the incompleteness and provisionality of our knowledge of it.

Complexity requires us to focus not only on the parts but on the interactions between them. Without such requisite relationality, we can neither succeed in understanding and explaining phenomena nor in improving problems. Complexity requires us to do everything we can to remain flexible. Surprise is inherent to complexity, and we cannot predict what will happen. When also failing to anticipate what could happen—out of myriad potential events—we only have our flexibility left. Complexity requires us to reconsider many of our intuitive measures to reduce risk. Complexity is not only hampering understanding and effective risk reduction but is generative of disaster in itself, making some disasters simultaneously unpredictable and inevitable. Complexity requires us to consider how effects can cascade throughout complex systems, regularly ignoring sectorial, political, administrative, organisational or other boundaries. Finally, complexity requires us to abandon our intuitive expectations of continuous linear change, as tipping points are equally ubiquitous as unpredictable. We must attempt to anticipate tipping points. However, it is best to do everything in our power not to push complex systems we depend on for our survival too far away from their dynamic stability that has proven to be able to sustain us over time.

It is essential to understand that complexity is not only debilitating, making effective policymaking and practice so much more difficult. It also offers new ways to grasp and address the challenges at hand. Although we can never fully understand how a complex system works, we must relentlessly seek to recognise the nexus of causes and effects of the phenomenon of interest or everything needed to generate the sought-after outcome. A deeper understanding of its interlocking feedback loops may reveal where relatively small investments can generate large outcomes. Path dependence does not only pose a problem but provides a possibility for us to have a significant influence in shaping, although not determining future conditions.

Self-organisation is already essential for social change towards sustainability. Imagine what could happen if the social movement for sustainability reaches that sequence of tipping points just mentioned. When you think about how to improve the world, start making it a better place yourself. It is not a pipe dream of an old environmentalist and Quaker. It is science! If you are concerned about climate change, make the necessary changes to reduce your carbon footprint. Stop flying privately and minimise your business travel. Use public transport for as much of your transportation needs as possible, and do not own a car if not absolutely necessary. Stop eating lamb, mutton, beef and pork. Do not fall for the urge to buy things you do not need. These are the four primary sources of carbon emissions that you could reduce by yourself and save money whilst doing it. Do it! Engage! Others are bound to follow. Perhaps enough people will be mobilised also to create the necessary political pressure for meso- and macro-level changes. Perhaps enough to save the world.

References

Alexander, J. C., & Giesen, B. (1987). From reduction to linkage: The long view of the micro-macro link. In J. C. Alexander, B. Giesen, R. Munch, & N. J. Smelser (Eds.), *The micro-macro link* (pp. 1–42). University of California Press.

Anderson, P. W. (1972). More is different. *Science, 177*, 393–396.

Anderson, P. W. (1999). Complexity theory and organization science. *Organization Science, 10*, 216–232. https://doi.org/10.2307/2640328

Anderson, C. W., Santos, J. R., & Haimes, Y. Y. (2007). A risk-based input–output methodology for measuring the effects of the august 2003 northeast blackout. *Economic Systems Research, 19*, 183–204. https://doi.org/10.1080/09535310701330233

Anderson, P. W., & Stein, D. L. (1987). Broken symmetry, emergent properties, dissipative structures, life: Are they related? In F. E. Yates, A. Garfinkel, D. O. Walter, & G. B. Yates (Eds.), *Self-organizing systems* (pp. 445–457). Plenum Press.

Arthur, W. B. (1988). Urban systems and historical path dependence. In J. H. H. Ausubel, & Robert (Eds.), *Cities and their vital systems: Infrastructure past, present, and future* (pp. 85–97). National Academies Press. https://doi.org/10.17226/1093

Arthur, W. B. (1989). Competing technologies, increasing returns, and lock-in by historical events. *The Economic Journal, 99*(394), 116–131.

Arthur, W. B. (1994). *Increasing returns and path dependency in the economy*. The University of Michigan Press.

Arthur, W. B., Durlauf, S. N., & Lane, D. (1997). Introduction. In W. B. Arthur, S. N. Durlauf, & D. Lane (Eds.), *The economy as an evolving complex system II* (pp. 1–14). ABP.

Arthur, W. B., Durlauf, S. N., & Lane, D. (Eds.). (1997b). *The economy as an evolving complex system II*. ABP.

Ashby, W. R. (1957). *An introduction to cybernetics*. Chapman & Hall Ltd.

Ashby, W. R. (1973). Some peculiarities of complex systems. *Cybernetic Medicine, 9*, 1–7.

Ashwin, P., Wieczorek, S., Vitolo, R., & Cox, P. (2012). Tipping points in open systems: Bifurcation, noise-induced and rate-dependent examples in the climate system. *Philosophical Transactions of the Royal Society, 370*(1962), 1166–1184. https://doi.org/10.1098/rsta.2011.0306

Backlund, A. (2002). The concept of complexity in organisations and information systems. *Kybernetes, 31*, 30–43. https://doi.org/10.1108/03684920210414907

Becker, P. (2020). The problem of fit in flood risk governance: Regulative, normative, and cultural-cognitive deliberations. *Politics and Governance, 8*(4), 281–293. https://doi.org/10.17645/pag.v8i4.3059

Becker, P. (2021a). Fragmentation, commodification and responsibilisation in the governing of flood risk mitigation in Sweden. *Environment and Planning C: Politics and Space, 39*(2), 393–413. https://doi.org/10.1177/2399654420940727

Becker, P. (2021b). Tightly coupled policies and loosely coupled networks in the governing of flood risk mitigation in municipal administrations. *Ecology and Society, 26*(2), 34. https://doi.org/10.5751/es-12441-260234

Bergström, J., & Dekker, S. W. A. (2014). Bridging the macro and the micro by considering the meso: Reflections on the fractal nature of resilience. *Ecology and Society, 19*, 22.

Berkes, F., Colding, J., & Folke, C. (Eds.). (2003). *Navigating social-ecological systems: Building resilience for complexity and change*. Cambridge University Press.

Bogg, J., & Geyer, R. (Eds.). (2007). *Complexity, science and society*. Radcliffe Publishing.

Bonabeau, E. (1998). Social insect colonies as complex adaptive systems. *Ecosystems, 1*(5), 437–443. https://doi.org/10.1007/s100219900038

Brinsmead, T. S., & Hooker, C. (2011). Complex systems dynamics and sustainability. In C. Hooker (Ed.), *Philosophy of complex systems* (pp. 809–838). Elsevier. https://doi.org/10.1016/b978-0-444-52076-0.50026-2

Byrne, D., & Callaghan, G. (2014). *Complexity theory and the social sciences: The state of the art*. Routledge.

Camazine, S., Deneubourg, J.-L., Franks, N. R., Sneyd, J., Theraula, G., & Bonabeau, E. (2003). *Self-organization in biological systems*. Princeton University Press.

Campbell, D. T. (1974). Downward causation in hierarchically organised biological systems. In *Studies in the philosophy of biology: Reduction and related problems* (pp. 179–186). Macmillan Press.

Capra, F. (2005). Complexity and life. *Theory, Culture & Society, 22*, 33–44.

Centeno, M. A., Nag, M., Patterson, T. S., Shaver, A., & Windawi, A. J. (2015). The emergence of global systemic risk. *Annual Review of Sociology, 41*, 65–85. https://doi.org/10.1146/annurev-soc-073014-112317

Cetina, K. K. (2005). Complex global microstructures. *Theory, Culture & Society, 22*(5), 213–234. https://doi.org/10.1177/0263276405057200

Checkland, P. (1981). *Systems thinking, systems practice*. John Wiley & Sons.

Cilliers, P. (1998). *Complexity & postmodernism: Understanding complex systems*. Routledge.

Cilliers, P. (2016a). Boundaries, hierarchies and networks in complex systems. In P. Rika (Ed.), *Critical complexity* (pp. 85–96). De Gruyter.

Cilliers, P. (2016b). What can we learn from a theory of complexity? In P. Rika (Ed.), *Critical complexity* (pp. 67–76). De Gruyter.

Clearfield, C., & Tilcsik, A. (2018). *Meltdown: Why our systems fail and what we can do about it*. Penguin Press.

Clegg, S., Courpasson, D., & Phillips, N. (2006). *Power and organizations*. SAGE Publications.

Coetzee, C., Van Niekerk, D., & Raju, E. (2016). Disaster resilience and complex adaptive systems theory. *Disaster Prevention and Management, 25*(2), 196–211. https://doi.org/10.1108/dpm-07-2015-0153

Coleman, J. S. (1986). Social theory, social research, and a theory of action. *American Journal of Sociology, 91*(6), 1309–1335. https://doi.org/10.1086/228423

David, P. A. (1985). Clio and the economics of QWERTY. *The American Economic Review, 75*(2), 332–337. https://doi.org/10.2307/1805621

Deutsch, K. W. (1948). Some notes on research on the role of models in the natural and social sciences. *Synthese, 7*, 506–533.

Di Baldassarre, G., Nohrstedt, D., Mård, J., Burchardt, S., Albin, C., Bondesson, S., Breinl, K., Deegan, F. M., Fuentes, D., Lopez, M. G., Granberg, M., Nyberg, L., Nyman, M. R., Rhodes, E., Troll, V., Young, S., Walch, C., & Parker, C. F. (2018). An integrative research framework to unravel the interplay of natural hazards and vulnerabilities. *Earth's Future, 6*(3), 305–310. https://doi.org/10.1002/2017EF000764

Dodds, P. S., Muhamad, R., & Watts, D. J. (2003). An experimental study of search in global social networks. *Science, 301*(5634), 827–829. https://doi.org/10.1126/science.1081058

Donati, P. (2012). *Relational sociology: A new paradigm for the social sciences*. Routledge.

Duit, A., & Galaz, V. (2008). Governance and complexity: Emerging issues for governance theory. *Governance, 21*(3), 311–335. https://doi.org/10.1111/j.1468-0491.2008.00402.x

Duit, A., Galaz, V., Eckerberg, K., & Ebbesson, J. (2010). Governance, complexity, and resilience. *Global Environmental Change, 20*, 363–368. https://doi.org/10.1016/j.gloenvcha.2010.04.006

Durkheim, E. (2005). *Suicide: A study in sociology*. Routledge.

Emirbayer, M. (1997). Manifesto for a relational sociology. *American Journal of Sociology, 103*, 281–317.

Etkin, D. (2016). *Disaster theory: An interdisciplinary approach to concepts and causes*. Butterworth-Heinemann (Elsevier).

Fischetti, M. (2021). A new you in 80 days: Cell turnover is vast and swift. *Scientific American, 324*, 76.
Flood, R. L. (1987). Complexity: A definition by construction of a conceptual framework. *Systems Research, 4*, 177–185.
Freeze, A., & Cherry, J. A. (1979). *Groundwater*. Prentice-Hall.
Gatfaoui, H., & de Peretti, P. (2019). Flickering in information spreading precedes critical transitions in financial markets. *Scientific Reports, 9*(1), 5671. https://doi.org/10.1038/s41598-019-42223-9
Gaupp, F. (2020). Extreme events in a globalized food system. *One Earth, 2*(6), 518–521. https://doi.org/10.1016/j.oneear.2020.06.001
Gell-Mann, M. (1995). What is complexity? Remarks on simpicity and complexity by the nobel prizewinning author of the quark and the jaguar. *Complexity, 1*, 16–19. http://scholar.google.comjavascript:void(0).
Giddens, A. (1984). *The constitution of society: Outline of the theory of structuration*. University of California Press.
Gilissen, H. K., Alexander, M., Beyers, J.-C., Chmielewski, P., Matczak, P., Schellenberger, T., & Suykens, C. (2016). Bridges over troubled waters: An interdisciplinary framework for evaluating the interconnectedness within fragmented flood risk management systems. *Journal of Water Law, 25*(1), 12–26.
Gladwell, M. (2000). *The tipping point: How little things can make a big difference*. Little, Brown and Company.
Granovetter, M. (1978). Threshold models of collective behavior. *American Journal of Sociology, 83*(6), 1420–1443. http://www.jstor.org/stable/10.2307/2778111.
Grodzins, M. (1957). Metropolitan segregation. *Scientific American, 197*(4), 33–41. http://www.jstor.org/stable/10.2307/24941940.
Gunderson, L. H. (2001). Managing surprising ecosystems in southern Florida. *Ecological Economics, 37*(3), 371–378. https://doi.org/10.1016/s0921-8009(01)00179-3
Gunderson, L. H., & Holling, C. S. (Eds.). (2002). *Panarchy: Understanding transformations in human and natural systems*. Island Press.
Haken, H. (1983). *Synergetics: An introduction* (3 ed.). Springer.
Hämäläinen, R. P., & Lahtinen, T. J. (2016). Path dependence in operational research—how the modeling process can influence the results. *Operations Research Perspectives, 3*, 14–20. https://doi.org/10.1016/j.orp.2016.03.001
Harley, C. D. G., & Paine, R. T. (2009). Contingencies and compounded rare perturbations dictate sudden distributional shifts during periods of gradual climate change. *Proceedings of the National Academy of Sciences, 106*(27), 11172–11176. https://doi.org/10.1073/pnas.0904946106
Healy, S. (2003). Epistemological pluralism and the 'politics of choice. *Futures, 35*(7), 689–701. https://doi.org/10.1016/s0016-3287(03)00022-3
Hedström, P. (1994). Contagious collectivities: On the spatial diffusion of Swedish trade unions, 1890-1940. *American Journal of Sociology, 99*(5), 1157–1179. https://doi.org/10.1086/230408
Hedström, P. (2005). *Dissecting the social: On the principles of analytical sociology*. Cambridge University Press.
Heylighen, F. (2001). The science of self-organization and adaptivity. In L. D. Kiel (Ed.), *Knowledge management, organizational intelligence and learning, and complexity, in encyclopedia of life support systems* (pp. 1–26). EOLSS Publishers.
Heylighen, F., Cilliers, P., & Gershenson, C. (2007). Complexity and philosophy. In J. Bogg, & R. Geyer (Eds.), *Complexity, science and society* (pp. 117–134). Radcliffe Publishing.
Hirsch, H. R. (1977). The dynamics of repetitive asymmetric cell division. *Mechanism of Ageing and Development, 6*(5), 319–332. https://doi.org/10.1016/0047-6374(77)90033-1
Hofstadter, D. R. (1979). *Godel, Escher, Bach: An eternal golden braid*. Basic Books.
Holland, J. H. (1995). *Hidden order: How adaptation builds complexity*. Perseus Books.
Holland, J. H. (1998). *Emergence: From chaos to order*. Oxford University Press.
Holling, C. S., & Gunderson, L. H. (2002). Resilience and adaptive cycles. In L. H. Gunderson, & C. S. Holling (Eds.), *Panarchy: Understanding transformations in human and natural systems* (pp. 25–62). Island Press.
Hollnagel, E. (2014). *Safety-I and safety-II: The past and future of safety management*. Ashgate.
Homer-Dixon, T. F. (1991). On the threshold: Environmental changes as causes of acute conflict. *International Security, 16*, 76–116.

Homer-Dixon, T. F. (2014). Complexity: Shock, innovation and resilience. In T. Webb, & S. Novkovic (Eds.), *Co-Operatives in a post-growth era: Creating Co-operative economics* (pp. 115–133). Zed Books.
Hooker, C. A. (2004). Asymptotics, reduction and emergence. *The British Journal for the Philosophy of Science, 55*(3), 435–479. https://doi.org/10.1093/bjps/55.3.435
Horgan, J. (1995). From complexity to perplexity. *Scientific American, 272*(6), 104–109. https://doi.org/10.1038/scientificamerican0695-104
IPCC. (2021). *Climate change 2021: The physical science basis*. Cambridge University Press.
IPCC. (2022). *Climate change 2022: Impacts, adaptation and vulnerability*. Cambridge University Press.
Jackson, M. C. (2003). *Systems thinking: Creative holism for managers*. John Wiley & Sons.
Kauffman, S. A. (1993). *The origins of order: Self organization and selection in evolution*. Oxford University Press. http://books.google.se/books?id=lZcSpRJz0dgC&printsec=frontcover&dq=intitle:the+origins+of+order+inauthor:kauffman&hl=&cd=1&source=gbs_api.
Kinney, R., Crucitti, P., Albert, R., & Latora, V. (2005). Modeling cascading failures in the north American power grid. *European Physical Journal B: Condensed Matter and Complex Systems, 46*, 101–107.
Krishnamurthy, R. P. K., Fisher, J. B., Schimel, D. S., & Kareiva, P. M. (2020). Applying tipping point theory to remote sensing science to improve early warning drought signals for food security. *Earth's Future, 8*(3), 1–14. https://doi.org/10.1029/2019ef001456
Lenton, T. M. (2011). Early warning of climate tipping points. *Nature Climate Change, 1*(4), 201–209. https://doi.org/10.1038/nclimate1143
Lenton, T. M. (2013). Environmental tipping points. *Annual Review of Environment and Resources, 38*(1), 1–29. https://doi.org/10.1146/annurev-environ-102511-084654
Levin, S. A. (1998). Ecosystems and the biosphere as complex adaptive systems. *Ecosystems, 1*(5), 431–436. https://doi.org/10.1007/s100219900037
Lewes, G. H. (1875). Problems of life and mind: First series – the foundation of a creed *(2)*. Trübner & Co.
Lewin, R. (1992). *Complexity: Life at the edge of chaos*. Collier Books.
Linders, T. E. W., Schaffner, U., Eschen, R., Abebe, A., Choge, S. K., Nigatu, L., Mbaabu, P. R., Shiferaw, H., & Allan, E. (2019). Direct and indirect effects of invasive species: Biodiversity loss is a major mechanism by which an invasive tree affects ecosystem functioning. *Journal of Ecology, 107*(6), 2660–2672. https://doi.org/10.1111/1365-2745.13268
Little, R. G. (2002). Controlling cascading failure: Understanding the vulnerabilities of interconnected infrastructures. *Journal of Urban Technology, 9*, 109–123.
Little, R. G. (2004). A socio-technical systems approach to understanding and enhancing the reliability of interdependent infrastructure systems. *International Journal of Emergency Management, 2*, 98–110.
Li, M., Wiedmann, T., Fang, K., & Hadjikakou, M. (2021). The role of planetary boundaries in assessing absolute environmental sustainability across scales. *Environment International, 152*, 106475. https://doi.org/10.1016/j.envint.2021.106475
Lorenz, E. N. (1963). Deterministic nonperiodic flow. *Journal of the Atmospheric Sciences, 20*(2), 130–141.
Lorenz, E. N. (2000). The butterfly effect. In R. Abraham, & Y. Ueda (Eds.), *The chaos avant-garde: Memories of the early days of chaos theory* (pp. 91–94). World Scientific.
Lovejoy, T. E., & Nobre, C. (2019). Amazon tipping point: Last chance for action. *Science Advances, 5*(12), eaba2949. https://doi.org/10.1126/sciadv.aba2949
Luhmann, N. (1981). Communication about law in interaction systems. In K. D. Knorr-Cetina, & A. V. Cicourel (Eds.), *Advances in social theory and methodology: Toward an integration of micro- and macro-sociologies* (pp. 234–256). Routledge.
Luhmann, N. (1995). *Social systems*. Stanford University Press.
Luhmann, N. (2014). *A sociological theory of law (E. King-utz & M. Albrow, trans.)*. Routledge.
Macdonald, C., & Macdonald, G. (2010a). Emergence and downward causation. In C. Macdonald, & G. Macdonald (Eds.), *Emergence in mind* (pp. 139–168). Oxford University Press. https://doi.org/10.1093/acprof:oso/9780199583621.001.0001
Macdonald, C., & Macdonald, G. (Eds.). (2010b). *Emergence in mind*. Oxford University Press. https://doi.org/10.1093/acprof:oso/9780199583621.001.0001
Macy, M. W. (1991). Chains of cooperation: Threshold effects in collective action. *American Sociological Review, 56*(6), 730. https://doi.org/10.2307/2096252

Macy, M. W., & Willer, R. (2002). From factors to actors: Computational sociology and agent-based modeling. *Annual Review of Sociology, 28*(1), 143–166. https://doi.org/10.1146/annurev.soc.28.110601.141117

Månsson, P. (2019). Uncommon sense: A review of challenges and opportunities for aggregating disaster risk information. *International Journal of Disaster Risk Reduction, 40*, 101149. https://doi.org/10.1016/j.ijdrr.2019.101149

Manzo, G. (Ed.). (2014). *Analytical sociology: Actors and networks*. John Wiley & Sons.

Marx, K. (1976). *A contribution to the critique of political economy*. Foreign Languages Press.

Maturana, H. R., & Varela, F. J. (1980). *Autopoiesis and Cognition: The realization of the living*. D. Reidel Publishing.

May, R. M., Levin, S. A., & Sugihara, G. (2008). Complex systems: Ecology for bankers. *Nature, 451*(7181), 893–895. https://doi.org/10.1038/451893a

McDaniel, R. R., Jordan, M. E., & Fleeman, B. F. (2003). Surprise, surprise, surprise! A complexity science view of the unexpected. *Health Care Management Review, 28*(3), 266–278.

McMillan, E. (2003). *Complexity, organizations and change*. Routledge.

Meadows, D. H., Randers, J., & Behrens, W. W., III (1972). *The limits to growth: A report to the club of rome*. Universe Books.

Meadows, D. H., & Wright, D. (2009). *Thinking in systems: A primer*. Earthscan.

Metz, F., Angst, M., & Fischer, M. (2020). Policy integration: Do laws or actors integrate issues relevant to flood risk management in Switzerland. *Global Environmental Change, 61*, 101945. https://doi.org/10.1016/j.gloenvcha.2019.101945

Milgram, S. (1967). The small-world problem. *Psychology Today, 1*(1), 61–67.

Miller, T. R., Baird, T. D., Littlefield, C. M., Kofinas, G., Chapin, F. S. I. I. I., & Redman, C. L. (2008). Epistemological pluralism: Reorganizing interdisciplinary research. *Ecology and Society, 13*(2). http://www.jstor.org/stable/10.2307/26268006.

Moore, J. C. (2018). Predicting tipping points in complex environmental systems. *Proceedings of the National Academy of Sciences, 115*(4), 635–636. https://doi.org/10.1073/pnas.1721206115

Naqvi, A., & Monasterolo, I. (2021). Assessing the cascading impacts of natural disasters in a multi-layer behavioral network framework. *Scientific Reports, 11*(1), 20146. https://doi.org/10.1038/s41598-021-99343-4

Newman, D. V. (1996). Emergence and strange attractors. *Philosophy of Science, 63*(2), 245–261. https://doi.org/10.1086/289911

Nicolis, G. (1995). *Introduction to nonlinear science*. Cambridge University Press.

Nicolis, G., & Prigogine, I. (1989). *Exploring complexity: An introduction*. W.H. Freeman and Company. http://www.worldcat.org/title/exploring-complexity-an-introduction/oclc/898843316.

Olsen, O. E., Juhl, K., Lindøe, P. H., & Engen, O. A. (Eds.). (2020). *Standardization and risk governance: A multi-disciplinary approach*. Routledge.

Ongkowijoyo, C., & Doloi, H. (2017). Determining critical infrastructure risks using social network analysis. *International Journal of Disaster Resilience in the Built Environment, 8*, 5–26. https://doi.org/10.1108/IJDRBE-05-2016-0016

Otto, I. M., Donges, J. F., Cremades, R., Bhowmik, A., Hewitt, R. J., Lucht, W., Rockström, J., Allerberger, F., McCaffrey, M., Doe, S. S. P., Lenferna, A., Morán, N., van Vuuren, D. P., & Schellnhuber, H. J. (2020). Social tipping dynamics for stabilizing Earth's climate by 2050. *Proceedings of the National Academy of Sciences, 117*(5), 2354–2365. https://doi.org/10.1073/pnas.1900577117

Page, S. E. (2006). Path dependence. *Quarterly Journal of Political Science, 1*(1), 87–115. https://doi.org/10.1561/100.00000006

Page, S. E. (2015). What sociologists should know about complexity. *Annual Review of Sociology, 41*, 21–41. https://doi.org/10.1146/annurev-soc-073014-112230

Parsons, T. (1951). *The social system*. The Free Press of Glencoe.

Payo, A., Becker, P., Otto, A., Vervoort, J. M., & Kingsborough, A. (2015). Experiential lock-in: Characterizing avoidable maladaptation in infrastructure systems. *Journal of Infrastructure Systems, 22*, 2515001. https://doi.org/10.1061/(asce)is.1943-555x.0000268

Perrow, C. B. (1984). *Normal accidents: Living with high-risk technologies*. Basic Books.

Perrow, C. B. (2008). Complexity, catastrophe, and modularity. *Sociological Inquiry, 78*, 162–173.

Persson, L., Carney Almroth, B. M., Collins, C. D., Cornell, S., de Wit, C. A., Diamond, M. L., Fantke, P., Hassellöv, M., MacLeod, M., Ryberg, M. W., Søgaard Jørgensen, P., Villarrubia-Gómez, P., Wang, Z., & Hauschild, M. Z. (2022). Outside the safe operating space of the planetary boundary for novel entities. *Environmental Science & Technology, 56*(3), 1510–1521. https://doi.org/10.1021/acs.est.1c04158

Popper, K. R., & Eccles, J. C. (1977). *The self and its brain*. Springer.

Prigogine, I., & Stengers, I. (1984). *Order out of Chaos: Man's new dialogue with nature*. New York: Bantam Books.

Ramalingam, B. (2013). *Aid on the edge of chaos*. Oxford University Press.

Richardson, K. A., & Cilliers, P. (2001). What is complexity science? A view from different directions. *Emergence: Complexity and Organization, 3*(1), 5–23. https://doi.org/10.1207/s15327000em0301_02

Richardson, K. A., Cilliers, P., & Lissack, M. (2001). Complexity science: A "gray" science for the "stuff in between". *Emergence: Complexity and Organization, 3*(2), 6–18. https://doi.org/10.1207/S15327000EM0302_02

Rietkerk, M., Bastiaansen, R., Banerjee, S., van de Koppel, J., Baudena, M., & Doelman, A. (2021). Evasion of tipping in complex systems through spatial pattern formation. *Science, 374*(6564), eabj0359. https://doi.org/10.1126/science.abj0359

Rika, P. (Ed.). (2016). *Critical complexity*. De Gruyter.

Rinaldi, S. M., Peerenboom, J. P., & Kelly, T. K. (2001). Identifying, understanding, and analyzing critical infrastructure interdependencies. *IEEE Control Systems Magazine, 21*, 11–25.

Ritchie, P., & Sieber, J. (2016). Early-warning indicators for rate-induced tipping. *Chaos, 26*(9), 093116. https://doi.org/10.1063/1.4963012

Ritchie, P., & Sieber, J. (2017). Probability of noise- and rate-induced tipping. *Physical Review E, 95*(5–1), 052209. https://doi.org/10.1103/PhysRevE.95.052209

Santos, P. P., Chmutina, K., Meding, J. V., & Raju, E. (Eds.). (2020). *Understanding disaster risk*. Elsevier.

Sawyer, R. K. (2005). *Social emergence: Societies as complex systems*. Cambridge University Press.

Scheffer, M., Bascompte, J., Brock, W. A., Brovkin, V., Carpenter, S. R., Dakos, V., Held, H., van Nes, E. H., Rietkerk, M., & Sugihara, G. (2009). Early-warning signals for critical transitions. *Nature, 461*(7260), 53–59. https://doi.org/10.1038/nature08227

Scheffer, M., Carpenter, S. R., Lenton, T. M., Bascompte, J., Brock, W., Dakos, V., van de Koppel, J., van de Leemput, I. A., Levin, S. A., van Nes, E. H., Pascual, M., & Vandermeer, J. (2012). Anticipating critical transitions. *Science, 338*, 344–348.

Schelling, T. C. (1971). Dynamic models of segregation. *Journal of Mathematical Sociology, 1*(2), 143–186. https://doi.org/10.1080/0022250x.1971.9989794

Schelling, T. C. (1978). *Micromotives and macrobehavior*. W. W. Norton.

Schneider, V. (2012). Governance and complexity. In D. Levi-Faur (Ed.), *The oxford handbook of governance*. Oxford University Press.

Schrödinger, E. (1992). *What is Life*. Cambridge University Press.

Scott, W. R. (1981). *Organizations: Rational, natural, and open systems*. Prentice-Hall.

Senge, P. (2006). *The fifth discipline: The art & practise of the learning organisation*. Currency & Doubleday.

Snowden, D. J., & Boone, M. E. (2007). A leader's framework for decision making. A leader's framework for decision making. *Harvard Business Review, 85*(11), 68–76, 149 https://pubmed.ncbi.nlm.nih.gov/18159787.

Stadelmann-Steffen, I., Eder, C., Harring, N., Spilker, G., & Katsanidou, A. (2021). A framework for social tipping in climate change mitigation: What we can learn about social tipping dynamics from the chlorofluorocarbons phase-out. *Energy Research & Social Science, 82*, 102307. https://doi.org/10.1016/j.erss.2021.102307

Sundström, G., & Hollnagel, E. (2006). Learning how to create resilience in business systems. In E. Hollnagel, D. D. Woods, & N. G. Leveson (Eds.), *Resilience engineering: Concepts and precepts* (pp. 235–252). Ashgate.

Taylor, J. R., & Van Every, E. J. (2000). *The emergent organization: Communication as its site and surface*. Lawrence Erlbaum Associates.

't Hart, P. (2013). After Fukushima: Reflections on risk and institutional learning in an era of mega-crises. *Public Administration, 91*(1), 101–113. https://doi.org/10.1111/padm.12021
Ulrich, W. (1996). Critical systems thinking for citizens. In R. L. Flood, & N. R. A. Romm (Eds.), *Critical systems thinking* (pp. 165–178). Plenum Press. http://link.springer.com/10.1007/b102400.
Ulrich, W. (2000). Reflective practice in the civil society: The contribution of critical systems thinking. *Reflective Practice, 1*, 247–268.
United Nations. (2015). *Proceedings third UN world conference on disaster risk reduction.*
Urry, J. (2003). *Global complexity*. Polity Press.
van den Bergh, J. C. J. M., & Gowdy, J. M. (2003). The microfoundations of macroeconomics: An evolutionary perspective. *Cambridge Journal of Economics, 27*, 65–84.
von Bertalanffy, L. (1968). *General systems theory: Foundations, development, applications.* George Braziller.
Walby, S. (2009). *Globalization and inequalities: Complexity and contested modernities.* Sage.
Weber, M. (2005). *The protestant ethic and the spirit of capitalism.* Routledge.
Westrum, R. (1992). Cultures with requisite imagination. In J. A. Wise, D. Hopkin, & P. Stager (Eds.), *Verification and validation of complex systems: Human factors issues* (pp. 401–416). Springer.
Wissel, C. (1984). A universal law of the characteristic return time near thresholds. *Oecologia, 65*(1), 101–107. http://www.jstor.org/stable/10.2307/4217501.
Woods, D. D. (2015). Four concepts for resilience and the implications for the future of resilience engineering. *Reliability Engineering & System Safety, 141*, 5–9. https://doi.org/10.1016/j.ress.2015.03.018
Yates, F. E. (1978). Complexity and the limits to knowledge. *American Journal of Physiology - Regulatory, Integrative and Comparative Physiology, 4*, R201–R204.

CHAPTER 10

Governing and governmentalisation

Introduction

It should be clear by now that contemporary society is confronted with grave sustainability challenges that require concerted action. It is paramount to ensure that humankind is not transgressing the fundamental planetary boundaries of Earth's life support systems (Chapter 4). Nevertheless, societies aspiring for sustainable development must also address the actual or potential impacts of a range of shocks and sudden, seasonal, or steady changes locally (Chapter 5). Some sustainability challenges are new, whilst others have been around since the dawn of civilisation (Chapter 2). Some actual sustainability challenges are not yet perceived as such by policymakers or the public but remain overlooked or ignored. Those that finally do catch our attention, however, may be transformed into issues requiring governing on the societal level. This chapter is about governing risk in relation to complex sustainability challenges and the processes through which they become something governable in the first place. It is about governing and governmentalisation.

Governing is not only exercised by the state, especially not in advanced liberal democracies (Rose & Miller, 1992, p. 174) where it has been redirected towards improving the well-being of their populations (Foucault, 2007). In such societies, and to various degrees in other kinds of societies imitating them, the governing of complex issues does not reside with the government in the sense of particular hierarchical political institutions with formal authority (Rhodes, 2007; Rosenau, 1992). Instead, governing is distributed and involves various actors across societal spheres and sectors (Dean, 2010; Miller & Rose, 2008). It is difficult to imagine a complex sustainability challenge that can be addressed by any single actor alone. They must be jointly governed by a web of actors (Folke et al., 2005; Renn, 2008) and the patterns of interactions amongst these actors are fundamental to society's capacity to achieve that (Becker & Bodin, 2022; Ingold et al., 2010). Moreover, governing sustainability challenges always involve something human beings value, a preferred future trajectory concerning its state, and a range of decisions and events with uncertain likelihoods and consequences. In other words, it is all about governing risk. Together.

Not a single complex sustainability challenge has always been a priority issue to govern, but some of them undergo a process of governmentalisation through which they become governable on the societal level. Regardless of all their differences, these sustainability challenges are, in other words, problematised and transformed into issues

Sustainability Science, Second Edition
ISBN 978-0-323-95640-6, https://doi.org/10.1016/B978-0-323-95640-6.00008-7

amenable to governing in this context. The framing of them as challenges is central to this process (e.g., Béland, 2009; Benford & Snow, 2000; Boström et al., 2017; Spence & Pidgeon, 2010). Yet, other processes at play deserve attention as they concurrently enable and constrain the resulting governing of risk.

This chapter has three main parts. First, it introduces a useful approach to understanding and explaining governing and governmentalisation, combining two influential perspectives. Then, applying this approach, the chapter elaborates on and tries to explain important common features of the governing of complex sustainability challenges. The third part elaborates on the governmentalisation of sustainability, suggesting a nexus of four constituent processes that simultaneously enables and constrains the governing of risk, which seem to be inherent to an underlying process of neoliberalisation. The chapter ends with some concluding remarks.

Perspectives on governing and governmentalisation

There are many ways to approach governing and governmentalisation. Whilst no perspective is perfect, and all have their strengths and weaknesses, I find more empirically oriented governmentality perspectives and new institutionalism particularly helpful to understand and explain what is going on in this context.

Governing and governmentality

Governing has been on human minds in one form or another since the dawn of civilisation. It traces its etymological roots to the Latin and Greek words for 'to steer' or 'to rule' (Oxford English Dictionary, 2020) and was a keen interest of famous mediaeval Arab and Renaissance European scholars (Khaldun, 1969; Machiavelli, 2014). However, it was not until the 1980s that the word 'governance' became ubiquitous (Bevir, 2012), denoting a qualitative shift away from Ibn Khaldun's and Machiavelli's focus on the government of the state. Instead, governing had permeated all spheres of society, with references to organisational governance, corporate governance, public governance, global governance, good governance, etc. (Bevir, 2012). Focus shifted from the state and its institutions to social practices and activities. From hierarchies to also involve networks. Around the turn of the millennium, these ideas found their way into safety and sustainability domains, and notions of risk governance started to emerge (van Asselt & Renn, 2011, pp. 432–433).

The governing of risk can be approached from various perspectives (e.g., Cedergren & Tehler, 2014; van Niekerk, 2015)—often referred to as risk governance (e.g., Aven & Renn, 2010; Renn, 2008) or sometimes the government of risk (Hood et al., 2001). Yet, *government* usually refers to particular hierarchical political institutions with formal authority, whilst *governance* denotes distributed, networked modes of governing

(Bevir, 2012; Rhodes, 2007; Rosenau, 1992). Modern liberal democracies have also seen a marked shift from government to governance (Blatter, 2003; Milward & Provan, 2000), indicated by the recent proliferation of the latter just described. However, this distinction between government and governance is confused by governmentality scholars using the word *government* to describe all efforts to govern people, also highlighting a fundamental transformation in the modes of governing in advanced liberal democracies (Dean, 1996; Foucault, 1982, 1991a; Rose, 1999). However, theories of governance and governmentality are complementary in this context—though appearing at first like strange bedfellows (Bevir, 2011). It is, therefore, essential to settle this conceptual impasse. Hence, I mainly use *governing* to describe action and process, with governance sometimes used synonymously and government used only concerning particular hierarchical political institutions (i.e. the government).

Foucault's contributions to the study of governing and power are so influential on our thoughts and practices (Figure 10.1) that his work is used in sociology (Dean, 2010), political science (Brass, 2000), human geography (Rutherford, 2016) and a range of other disciplines (Miller & Rose, 2008, p. 14). In a seminal series of lectures at the Collège de France in 1977–78, he elaborated a genealogy of the modern state—from ancient Greece to contemporary Western neoliberalism—focusing on shifts in the forms of power and the emergence of a new mode of governing (Foucault, 2007). Foucault argues

Figure 10.1 Michel Foucault passed away in 1984 but is still immensely influential. *(Photo shared by mike (CC BY-SA 2.0).)*

that the direct coercive power that dominated in earlier societies (Chapter 2) has not disappeared in advanced liberal democracies. Nor has the power of disciplinary institutions been relinquished. They are simply complemented by a new form of power penetrating further into our lives.

Foucault introduces the notion of *governmentality* to describe this new mode of governing and the process through which it emerged (Foucault, 1991a). It is how governing occurs in advanced liberal democracies (Joseph, 2010; Miller & Rose, 2008), where free individuals control, determine and delimit the liberty of others (Foucault et al., 1987, pp. 130—131). He uses this semantic unification of governing (*gouverner*) and modes of thinking (*mentalité*) to expound the emptiness of studying the technologies of governing without also considering their underpinning rationalities (Lemke, 2002, p. 50). These technologies refer to all people, techniques, tools, definitions, equipment and other resources that enable actors to envisage and act upon the conduct of others, individually and collectively, and often at a distance (Miller & Rose, 2008). Rationalities, on the other hand, refer to modes of thinking—ways of rendering reality thinkable in such a way that it becomes amenable to analysis and action (Dean, 2010).

Governing is, here, defined as the 'conduct of conduct' (Dean, 2010, p. 17). It is the situated activities undertaken by various actors, employing a range of technologies and rationalities, seeking to shape conduct by influencing the beliefs, interests, desires and aspirations of others, as well as themselves, for specific but shifting objectives and with relatively unpredictable outcomes (Dean, 2010; Miller & Rose, 2008). It involves governing both others and the self and is based on another rather eloquent play of words between the French verb *conduire*—to lead or drive—and its reflexive form *se conduire*—to behave or conduct oneself (see Foucault, 2007, pp. 257—258). Thus, understanding the governing of any issue requires attention to the regime of practices comprising the rationalities and technologies through which that is done (Dean, 2010, pp. 40—44)—the sets of relatively coherent, organised, routinised and ritualised ways of thinking and doing in a particular situation, at a certain time, and in a specific place (Dean, 2010, p. 31). In short, it requires considering the set of institutionalised ways of thinking and doing when governing risk.

Foucault's initial work on governmentality was further elaborated by his colleagues in Paris, who suggested risk, in itself, as a central technology and rationality for governing the self and society (Donzelot, 1979; Ewald, 1991). Risk is, in essence, a central rationality for imagining the world as governable in the sense of being able to intentionally direct it away from futures that, for different reasons, are deemed undesirable and towards some preferable future (cf. Luhmann, 1993; O'Malley, 2008). Risk also involves technologies for achieving that in practice—most explicitly visible in risk analyses where calculations provide the foundations for decision-making (Figure 10.2) (Donzelot, 1979; Ewald, 1991). The perspective of risk as governmentality is, in other words, fundamental for governing risk in general since the associated regimes of practices are inherently oriented

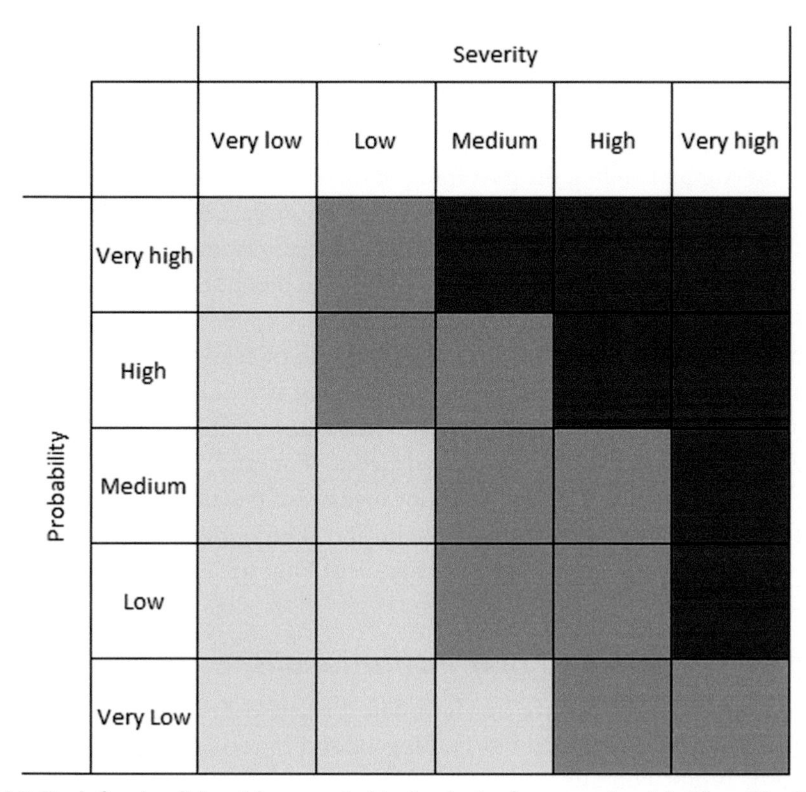

Figure 10.2 Tools for visualising risk are central technologies for governing risk. *(Shared by David Vose on the Creative Commons.)*

towards an uncertain future (cf. Miller & Rose, 2008, pp. 216–217). In other words, it appears impossible to study *the governmentality of risk* without considering *risk as governmentality*.

Regardless of the intrinsic importance of risk as governmentality, understanding, explaining and improving the governing of risk may demand a looser and more empirical interpretation of Foucault's work (cf. Rose, 1999, pp. 4–5)—an interpretation more aligned with the influential work of Dean (2010), Miller and Rose (2008). These scholars guide how to engage with the localised empirical context in which a particular risk is becoming governable and being governed (O'Malley, 2008).

In contrast to traditional risk management, the governing of risk concerns situations with many actors, multiple and often conflicting values and no single authority that can make binding decisions (Aven & Renn, 2010; Renn, 2008). These actors do not exist in isolation but depend on each other and are affected by the decisions and actions of others (Becker, 2018), forming the network of actors introduced above (Folke et al., 2005; Renn, 2008). However, the social relations making up the network are not only formed because actors depend on each other for some resource. Social relations can also form when actors convince each other that their problems or objectives are shared or linked and can be addressed by working together (Miller & Rose, 2008). Regardless of which, once established, any social relation denotes a kind of dependence (Luhmann, 1979). In other words, comprehending the governing of risk requires attention to networks of dependencies that enable governing practices to act upon the concerned places and actors (cf. Miller & Rose, 2008, p. 33). The patterns of social relations amongst actors in these 'networks of rule' (Rose & Miller, 1992, p. 189) are, as mentioned earlier, fundamental to society's capacity to govern risk (Ingold et al., 2010).

Since the governing of risk entails a 'complex web of actors, rules, conventions, processes and mechanisms' (Renn, 2008, p. 9), studying it requires an expansion of Foucault's first notion of governmentality as an assemblage of 'the institutions, procedures, analyses and reflections, the calculations and tactics' (Foucault, 1991a, p. 102). It requires an even more comprehensive conception of regimes of practices that also includes networks of involved actors (Miller & Rose, 2008, pp. 34—35). Including networks of actors also helps to overcome an institutionalist critique of Foucault's work that I will return to in a moment.

Institutionalisation and new institutionalism

The governmentality perspective just introduced suggests that regimes of practices are neither static nor predetermined but contingent and historically constituted (cf. Dean, 2010, p. 50). Whilst Foucault's analysis can be criticised for detaching governmentality from any particular institutional configuration (Friedland & Alford, 1991, p. 254), the governmentality perspective developed here emphasises the importance of actors in institutionalising practices. Institutionalisation is the process through which a practice becomes a convention, expectation, or even taken for granted amongst certain actors in specific situations (Meyer et al., 1987, p. 13). It is best understood as a dynamic, continuous process (Barley & Tolbert, 1997). To emphasise the role of actors in it, I draw on *new institutionalism*—another influential school of thought (Scott, 2014) that has been suggested as important for the study of human-environment interactions (Hotimsky et al., 2006).

New institutionalism grew out of a sociological critique of early influential organisational theorists' (e.g., Taylor, 1919; Weber, 1978) notion of organisations as closed, self-sufficient instruments for rational goal-oriented action (Scott, 2014). It built on another strand of critique, which showed that organisations are partly structured by external factors and adapt to environmental change actively, strategically, and in a goal-oriented manner (Hillman et al., 2009; Thompson, 1967). Early new institutionalists, then, effectively debunked the myth that organisations are structured and functioning only for rational goal-oriented efficiency (DiMaggio & Powell, 1983; Meyer & Rowan, 1977). Regardless of how common such ideas are in society, new institutionalism demonstrates that organisations are also structured by institutional rules (DiMaggio & Powell, 1991; Scott, 2014). These institutional rules can be regulative (e.g., legislation, policy), normative (e.g., norms, expectations) or cultural-cognitive (e.g., schema, frames) (Scott, 2014). Scholars place different weights on the importance of each of these elements, but they are most often combined in various ways (Figure 10.3). When all three align, their combined force is most formidable (Scott, 2014, pp. 70–71)—regardless of whether the resulting regime of practices is efficient or not (DiMaggio & Powell, 1983).

Institutions comprise 'regulative, normative, and cultural-cognitive elements that, together with associated activities and resources, provide stability and meaning to social life' (Scott, 2014, p. 56). They are socially constructed templates for action generated and maintained through ongoing social interaction (Meyer & Rowan, 1977, p. 346; Zucker, 1977, p. 728). Although generally persistent over time, institutions are simultaneously constituting and being constituted by the actions of actors who can innovate, act strategically and, therefore, contribute to institutional change (DiMaggio, 1991; Oliver, 1991; Scott, 2014). Institutions and actions are, in other words, inextricably linked.

Regardless of the fundamental importance of agency in institutionalisation, there are usually similarities in how actors organise within particular organisational fields (Becker et al., 2021; DiMaggio & Powell, 1983; Meyer & Rowan, 1977). Such isomorphism is driven by coercive, normative, and mimetic pressures (DiMaggio & Powell, 1983). Regulative elements of legislation and policy are enforced through formal authority and sanctions (Scott, 2014). Normative elements—such as expectations, obligations and identities—are imposed through norms and moral sanctions (Scott, 2014). Whilst cultural-cognitive elements—such as ideas and predispositions—are acquired through imitation and learning (Scott, 2014).

Although practices form the intrinsic basis for institutions (Barley & Tolbert, 1997), institutions are also conveyed by different types of 'carriers' (Jepperson, 1991, pp.

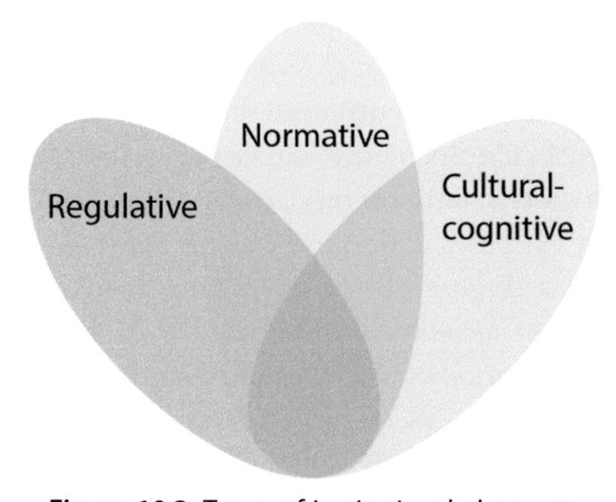

Figure 10.3 Types of institutional elements.

150–151). Different institutionalist scholars assign these carriers more or less weight (Scott, 2014, p. 95). Symbolic carriers encompass the legislation, policies, values, norms, frames and schemas that comprise the core of regulative, normative and cultural-cognitive elements. Relational carriers involve the patterns of social relations amongst actors (Strang & Meyer, 1993). These patterns are vital because they provide potential paths for transmitting institutional elements but are also observable indicators of institutions themselves (Scott, 2014, pp. 174–175). Artefacts comprise a third type of carrier of institutions (Scott, 2014, pp. 102–104), being objects intentionally made for specific purposes (Chapters 11 and 12). Artefacts embody technical and symbolic elements (Suchman, 2003, p. 99) and may have relational effects (Lupton, 2014). That is to say that artefacts may have fundamental material aspects at the same time as their meaning may vary and could influence how actors interact with each other. Thus, they are partly socially constructed through the different meanings actors ascribe to them. However, Orlikowski (1992) convincingly argues that artefacts tend to become reified and institutionalised, severing the link with their original purpose and meaning and becoming part of the institutional properties of the situation through their habitualised and routinised use. Artefacts do not determine institutions by themselves but provide the circumstances for structuring them (Barley, 1986; Scott, 2014, p. 177).

Many institutionalists have contributed to our knowledge of the foundations and direction of institutionalisation (Scott, 2014). Stinchcombe (2000) was one of the first to suggest the importance of initial conditions, and others point out the significance of a few decisive events at the national (Scott, 2014) or even global level (Drori et al., 2006). However, actors' agency influences the institutionalisation of regimes of practices more than suggested by these accounts (Johnson, 2007; King & Pearce, 2010; Migdal-Picker & Zilber, 2019; Schneiberg, 2007).

Instead of looking for an explanation based solely on a few macro-level conditions and events, van de Ven and Garud (1994) suggest paying attention to the many micro-level events in which actors who are faced with a new situation coinvent ways to deal with it. They argue that after a period of behavioural variation when actors test and adjust activities as they go along, some patterns of activities are increasingly preferred over others (rule-making events) until they dominate and become the convention (rule-following events). Consequently, interacting actors coinvent and update practices together, with high costs in terms of time, energy and resources. North (1990) refers to this as large setup costs and uses the notion of *increasing returns* to explain why flawed practices are not addressed (Figure 10.4)—even when apparent (Becker, 2021c). He highlights the importance of *incentives* and argues that flawed practices persist because further efforts in the same direction continue to be rewarded, whilst the costs of changing to an alternative increase over time (Scott, 2014). The status quo is, then, maintained due to a combination of three factors (North, 1990). First, actors are reluctant to consider alternatives after investing time and energy into learning the current practices (learning

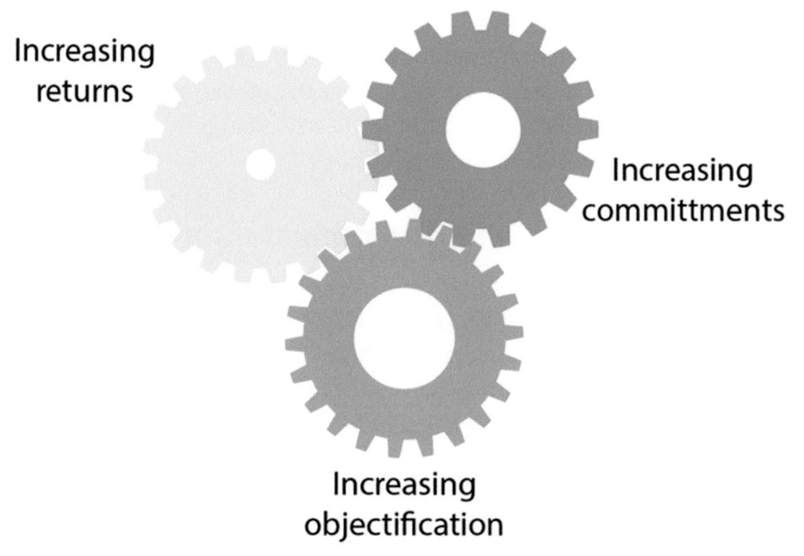

Figure 10.4 Mechanisms of institutionalisation (see Scott, 2014, pp. 144—151).

effects). Second, the contribution of each actor is facilitated by all actors following the same practices (coordination effects). Finally, new actors are motivated to adopt current practices as they appear to be commonly accepted (adaptive expectations). The resulting path-dependent persistence of even flawed practices is especially widespread in contexts with fuzzy feedback and subjective evaluations (North, 1990)—like in the governing of most sustainability challenges (Becker, 2020, 2021c; DeFries & Nagendra, 2017).

Selznick (1992, p. 232) argues that institutions hold actors hostage to their history not only through incentives but also through their normative order. This pattern is commonly associated with *increasing commitments* (Figure 10.4), and notions of *this is the way we do it*—often in relation to the *identity* and common practices of particular professional groups (Scott, 2014, pp. 145—148). Although closely related to coordination effects (North, 1990), such normative expectations are invaluable as they 'reduce the need for constant negotiation of expectations and behavioural contracts' (Handmer & Dovers, 2007, p. 30). However, they can also bind actors to flawed practices.

Finally, institutionalisation can also be driven by the *increasing objectification* (Figure 10.4) of the regime of practices (Berger & Luckmann, 1966). Such objectification is associated with notions of *this is how it is done*, which is a typical indicator of more cultural-cognitive elements of institutionalisation (Berger & Luckmann, 1966, p. 77; Scott, 2014, p. 148). Here, the underlying mechanism relates not to incentives or identity but to the objectification of shared ideas. Objectification involves the development and diffusion of a certain degree of consensus amongst actors concerning the meaning and value of an idea, where the diffusion shifts from imitation to routinisation. Although the underlying mechanism is separate, such objectification is often linked to an increasingly normative base that leaves less and less room for alternatives (Tolbert & Zucker, 1996, pp. 182–183). These shared ideas, thus, thicken and harden when diffused (Berger & Luckmann, 1966, p. 76)—not only amongst newly included actors but also for those already subscribing to the particular understanding.

Approaching governing and governmentalisation

This chapter is not only about governing risk but examines how a complex sustainability challenge becomes an issue that requires governing on the societal level in the first place. I refer to this as becoming governmentalised. Although the Oxford English Dictionary (2020) defines *governmentalisation* as the 'process of bringing something under the control or supervision of a government', it is used here in relation to the more distributed notion of governing outlined above. The concept of governmentalisation is central in Foucault's (1991a) original work on governmentality. He uses it to describe an escalation in the capacity for governing in advanced liberal democracies, which is less about the state taking over society (*étatisation*) and more about introducing regimes of practices that shape the conduct of conduct amongst and between various actors (Box 10.1). Governmentalisation is, therefore, understood as a particular process of institutionalisation that turns an issue into something governable on the societal level.

Focusing explicit attention on the institutionalisation of regimes of practices of governing risk motivates a careful combination of governmentality and new institutionalism (Lim, 2011). Governmentality has proven to be a helpful heuristic for grasping the complexities of governing risk in relation to sustainability (e.g., Hutter et al., 2014;

BOX 10.1 Foucault's governmentalisation and the state

Foucault suggested that at the centre of the construction of the modern state was not the bureaucracy or army, as others had suggested. Instead, he argued that 'what is really important for our modernity— that is, for our present—is not so much the *étatisation* of society, as the "governmentalisation" of the state' (Foucault, 1991a, p. 103). This governmentalisation meant that the state was simultaneously brought in and set aside to render governing more effective (Sawyer, 2015). Paradoxically, by drawing limits on the state and defining its relationship to society, the state's capacity to govern is increased through the new form of power relations that emerge and can penetrate wider and deeper than ever before. This is the essence of governmentality, and the modern state emerged by negotiating the limits between public and private, state and civil society, legislation and laissez-faire—not by limiting, boosting or banishing one side or the other (Foucault, 2007, pp. 144–145; Sawyer, 2015, p. 145). Foucault (2007) goes on to argue that although the problems and practices of governing have become the only real space of political struggle and contestation, the governmentalisation of the state has nevertheless made it into what it is today and allowed the state to survive.

Lövbrand et al., 2009; Rutherford, 2016). It emphasises the underlying rationalities behind why something becomes institutionalised but is less well suited to describe the mechanisms of institutionalisation. On the other hand, new institutionalism focuses explicitly on understanding these mechanisms but has been criticised for not analysing enough why something becomes institutionalised (Cooper et al., 2008). In short, governmentality perspectives mainly address how issues are constructed as governable (Dean, 2017, p. 2), whilst new institutionalism examines how governing is institutionalised (Scott, 2014).

Combining perspectives always risks provoking theoretical purists. However, O'Malley (2008, pp. 68–69) advocates pragmatism. He argues convincingly that governmentality perspectives are theoretically and methodologically flexible. They can be articulated with sociological analysis. He even suggests that such cross-fertilisation could facilitate overcoming common challenges. Johansson (2009) notes a similar eclecticism amongst institutionalist scholars. The approach to governing and governmentalisation presented in this chapter draws on the respective strengths of these two theoretical perspectives in an attempt to offset their weaknesses. There is agreement and controversy between new institutionalism and the original Foucauldian idea of governmentality (see Power, 2011). However, corresponding developments in the two theoretical perspectives over the years facilitate their combination.

The most persistent institutionalist critique of Foucault and his more orthodox followers concerns their lack of attention to actors (Friedland & Alford, 1991, pp. 253–254; Power, 2011). It is important to note that Foucault was never interested in action per se but rather in the conditions under which actors are constituted (Power, 2011, pp. 48–49). However, this critique is bypassed by the more empirically oriented governmentality scholars who explicitly include networks of actors in the regimes of practices (Miller & Rose, 2008, pp. 34–35). Moreover, many early institutionalists were open to a similar critique in their 'rhetorical defocalization of interest and agency' before DiMaggio (1988, p. 3) suggested the more explicit focus on agency that has become a hallmark of new institutionalism since then (e.g., Barley & Tolbert, 1997; DiMaggio, 1991; Johnson, 2007; Meyer, 2010; Migdal-Picker & Zilber, 2019). Understanding governing and governmentalisation requires giving agency a prominent position in the analysis, which is possible when combining governmentality and new institutionalism regardless of past critiques of the two theoretical perspectives.

In addition to the explicit focus on agency, DiMaggio (1988) also highlights the importance of interest and power (King & Pearce, 2010; Meyer, 2010). Institutionalisation is a profoundly political process that reflects the relative power of actors mobilising around organised interests (DiMaggio, 1988, p. 13). However, the intrinsic link between actions and institutions—agency and structure (Giddens, 1984)—introduces significant path dependency into institutionalisation (Schneiberg, 2007). It means that exercised agency and power in the past restrict and enable present agency and power (cf.

Emirbayer, 1997, p. 294) through symbolic (regulative, normative, cultural-cognitive), relational, and artefactual means (Figure 10.5).

Here, a link appears between new institutionalism and governmentality. Governmentality perspectives generally resist questions concerning the possession of power (Dean, 2010, pp. 16–17). They focus on the more distributed power that is productive of meaning, processes and objects (Miller & Rose, 2008, p. 9). This Foucauldian notion of power connects to DiMaggio and Powell's (1991) relaunch of new institutionalism as not only focusing on regulative- and normative elements but also on cultural-cognitive elements (Power, 2011, p. 50). This shift towards emphasising taken-for-granted assumptions, habits, and routines echoes governmentality scholars' notion of the intrinsic link between technologies and rationalities of governing. Hence, grasping governing and governmentalisation requires attention to both technologies and rationalities—to the rules and regulations, strategies and plans, relations and networks, borders and boundaries, procedures and tools, along with the norms and ideas that afford their meanings and motivations. Finally, DiMaggio (1991) suggests that agency, interests and power

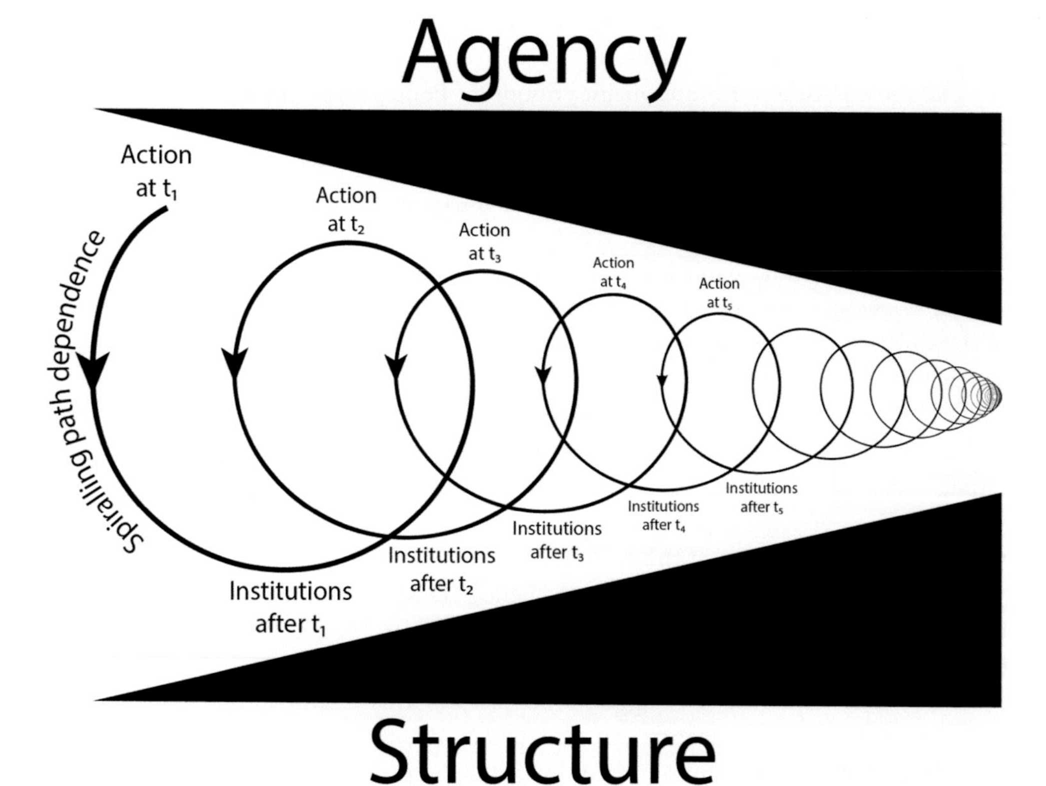

Figure 10.5 Spiralling path dependence of actions and institutions; between agency and structure.

are most apparent and amenable to study during the formation of a new institutional field. This idea is analogous to the suggestion from governmentality scholars to study problematisations (Dean, 2010).

Governing and institutionalisation can be considered on different levels (Schneider, 2012; Scott, 2014, pp. 104–107). There is an immensely rich literature focusing on the governing of various sustainability challenges (e.g., Bäckstrand, 2003; Bergsten et al., 2019; Bodin, 2017; Bodin & Tengö, 2012; Epstein et al., 2015; Guerrin et al., 2014; Lebel et al., 2013; Li et al., 2021; Lubell & Morrison, 2021; Pahl-Wostl et al., 2012; Sayles & Baggio, 2017; Widmer et al., 2019). However, this literature is overwhelmingly focused on the institutional (macro) level or the interaction between organisations (meso-level), whilst much less attention is paid to the level of the acting and interacting individuals who constitute organisations and reproduce institutions (micro-level). Although these levels are inseparable (Berger & Luckmann, 1966; Bourdieu, 1977; Elias, 1978; Foucault, 1995; Giddens, 1984; Latour, 2005), and governmentality perspectives rarely make an explicit issue of linking them, Miller and Rose (2008) suggest that any analysis of governing should begin with the practices of governing themselves.

All practices are done by individuals. However, even when data is collected by interviewing individuals, most studies miss the micro-level as a few participants are selected to represent their organisations and simply asked to recount what their organisations do and if their organisations interact. This approach disregards the social relations within organisations, which have proved crucial for grasping the governing of sustainability challenges (e.g., Becker, 2021c; Becker & Bodin, 2022). Moreover, it simplifies the social relations between organisations to the extent that the data may become thoroughly misleading. For instance, by ignoring the internal structural position of the actors linking two organisations (Becker, 2020). Ahrne (1994, p. 28) asserts that 'organizations cannot speak or move; they have no legs to walk with, and no eyes to see with. When organizations do something it is always individuals who act'.

Nevertheless, individuals in organisations do not act only for themselves but also on behalf of their organisations. Studying inter-organisational relations, therefore, presupposes studying interacting individuals who represent their organisations. However, as Callon and Latour (1981) suggest, individuals do not represent their organisations equally. They might be unequally successful in translating the interests, desires and forces of other individuals with whom they form alliances or argue.

Governing complex sustainability challenges

Armed with the approach suggested in the previous section, it is now time to engage with the governing of risk in relation to complex sustainability challenges. There are three particularly interesting common features of such governing that I want to highlight here: (1) multifarious fragmentation, (2) the location of responsibility, and (3) the role of the market. First, I discuss these features, then I turn your attention towards how they have become institutionalised. It is perhaps there the benefit of combining a more localised and empirically oriented governmentality perspective and new institutionalism becomes most visible.

Multifarious fragmentation

The most striking general feature of governing complex sustainability challenges is its fragmentation. Actors rationalise continuing greenhouse gas emissions by reporting them in a fragmented way that hides the actual contribution of their lifestyles (Chapter 15). Coastal erosion is often averted locally by constructing hard barriers, only to exacerbate erosion elsewhere. Water flows across the landscape, from upstream to downstream, regardless of whether human beings see the world as divided into countries, states, provinces, municipalities or by any other borders we construct to make sense of and govern social life. Vital activities to prevent desertification drop or even stop due to funding cuts during particular budget cycles, just to be revived later after large areas have been irrevocably lost. Many policies and the activities they entail impact biodiversity but are implemented and evaluated only against their own objectives. This multifarious fragmentation can be seen as spatial, temporal and functional.

Spatial fragmentation

Borders structure the regimes of practices governing risk in relation to most, if not all, sustainability challenges. National borders are seen, or at least portrayed, as delimiting natural social, political and economic entities (Wimmer & Glick Schiller, 2002) and heavily influence how we perceive and address any challenges (Carter et al., 2021). State or canton borders of federal countries are often also prominent in this regard (e.g., Bara & Doktor, 2010; Scanlan, 2019), as are many provincial or regional borders depending on the level of autonomy of the demarcated areas (e.g., Bond et al., 2012, pp. 57—59). Also, local-level borders structure many such regimes of practices (e.g., Becker, 2020), especially in decentralised political systems with an extended tradition of local self-governance—like in Scandinavia, reaching as far back as the Viking assemblies.

All these borders are generally taken for granted today but are virtually always historically contingent and could have looked very different. Just think about all the wars and associated changes in national borders over centuries or how the colonial powers negotiated the borders for the colonised (Figure 2.6). The same goes for most sub-national borders, which are at least as important for governing the lion's share of sustainability challenges. For instance, many North American state- or provincial borders were drawn using a ruler and compass (Figure 10.6)—as were many county or municipal borders there. Not to mention the creative manipulation of electoral district boundaries for political advantages (i.e. gerrymandering). The borders of Swedish municipalities are also

Figure 10.6 State and provincial borders of mainland North America. *(Adapted from Alex Covarrubias on the Creative Commons.)*

historically contingent. Most of them are identical to the outer borders of collections of parishes, which could have been drawn very differently as most parishes were formed many centuries ago to provide viable congregations to already constructed churches (Becker, 2021b). These borders are, as such, not only historically contingent but heavily path-dependent (Chapter 9).

Regardless of their contingency and path dependence, all these geopolitical- and administrative borders are central technologies in the sense of fundamentally structuring the governing of sustainability challenges today, resulting in actors consistently ignoring meaningful connections across them. Actors tend to focus on the problem within their own jurisdiction, with little or no attention to the effects their activities may have on circumstances elsewhere (e.g., Becker, 2020; Carter et al., 2021). Even in cases with explicit attention to cross-border links, there are many examples of geographically bounded entities ending up much more egocentric than intended, agreed upon or advantageous for themselves in the long run. Like many Western countries during the Covid-19 pandemic, hoarding medical supplies for their populations and not considering the impact of their policies on people in many less affluent countries, or many signatory parties to river basin commissions that only focus on maximising their access to water.

The prominence of national borders is generally explained with reference to nationalism—the patterns of belief and practice reproducing the world as a world of countries where we live as citizens (Billig, 1995, p. 15). This global ideology results in national borders being viewed as natural, self-evident and taken for granted in most places worldwide, despite their contingent character and all aspects connecting people and places across them. 'What holds humanity together today is the denial of what the human race has in common' (Hobsbawm, 1996, p. 265). Some sub-national borders carry similar meanings for the people within them, as salient regional or local identities may match or even trump the identity linked to the nation-state (cf. Billiet et al., 2021; Paasi, 2012).

National- and sub-national borders that are not imbued with the same ideological significance may still fragment the governing of sustainability challenges, demanding further explanation. It has been suggested that such spatial fragmentation may result from legal frameworks hindering cross-border linkages (Johannessen & Granit, 2015). However, the explanation becomes richer when considering this technology in the form of borders as intrinsically linked with a dominant rationality that reduces the spatial complexity of sustainability challenges to fit the legal *and* broader institutional environment. It means the borders themselves are not only institutionalised by regulative and even normative means but are embedded in cultural-cognitive patterns of ideas that people have come to simply take for granted (e.g., Becker, 2021b).

The same fragmenting rationality makes people liable to consider solutions to particular sustainability challenges that are locally rational but globally irrational. Becker (2020) provides an informative example concerning flood risk mitigation. Although all actors in a catchment area may agree that flood risk is a priority issue, some actors may consider flood risk mitigation in terms of increasing the retention of water upstream, whilst others consider flood risk mitigation by improving the local drainage capacity of areas they control (i.e. reducing the retention of water). Both alternatives reduce flood risk locally, but if the latter is implemented upstream, it aggravates flood risk downstream. In other words, both sides are locally rational in addressing their own experienced problem. On the one hand, increasing upstream retention can have various effects (Acreman et al., 2007), but can be implemented with acceptable trade-offs for other actors (Thaler, 2019). It is, in other words, not necessarily globally irrational, but it can be. On the other hand, improving local drainage always increases flow and downstream flood risk if no additional measures are implemented. It is thus locally rational but globally irrational.

Moreover, Becker (2021b) shows that the same reductionist rationality shapes how flood risk mitigation is also governed within municipal borders, with flood risk assessed and addressed for each planning area in isolation without taking into account the fact that their boundaries seldom coincide with hydrological boundaries. Implementing a detailed development plan may, thus, cause or aggravate flood risk in other planning areas and restrict future land use. In this context, the plans are key technologies (cf. Moisio & Luukkonen, 2015) that are intrinsically linked to the same spatially reductionist rationality as the borders. There are myriad similar examples, like the owner wanting to protect his beachfront property from coastal erosion by constructing hard erosion protection around it, multiplying the erosion of all his neighbours on the downstream side of the natural sediment movements along the shore (Figure 10.7).

This tension in rationality is alluded to in several seminal contributions to the social study of complex issues (e.g., Flyvbjerg, 1998; Wynne, 1982) and it is essential to note that there are no objective criteria to judge who is right. A potential way forward, however, is to ensure that what is valuable and worth protecting is made explicit (Nilsson & Becker, 2009). Informed dialogue has been suggested as a way to achieve this (Becker & Tehler, 2013). Even if actors may perceive risks differently (e.g., Armaş, 2006; Flynn et al., 1994), they may still share similar priorities (Becker, 2011).

Figure 10.7 Coastal erosion of Matunuck Beach, Rhode Island.

Temporal fragmentation

The rationality reducing the complexity of sustainability challenges in spatial terms is closely linked to a similar, temporal rationality often visible in regimes of practices governing risk in relation to complex sustainability challenges. Most measures to address any problem are designed based on a snapshot in time of what the problem context would look like, assuming that everything else remains the same (Hollnagel, 2014, p. 101). This rationality of *ceteris paribus* assumes that a range of essential conditions stay the same, at the same time as many of them cannot be regulated and are likely to change over time (Becker, 2021a). Although such assumptions are common in the application of any technology (Luhmann, 1993, p. 88), since they are required for forward-looking assessments of effects (Sayer, 1992, p. 216), they must be explicitly considered in the governing of sustainability challenges and not allowed to completely undermine our intentions.

Moreover, such temporal fragmentation is also prevalent in how the sustainability challenges we want to govern are perceived in the first place. The quintessential example is the determined set of time perspectives on climate change in IPCC's reports, which have become fundamental technologies thoroughly structuring the global dialogue without any real attention to what may happen after the year 2080 or 2100, except in particular academic milieux. Although it may be unlikely (but not at all impossible) that I am still alive in 2080, my children will be within the average life expectancy of their generation, and babies born today will be alive in 2100. We know this for a fact, yet it is consistently ignored in all attempts to govern climate change. A goal to keep global warming under a particular temperature increase—ideally below 1.5 °C (Chapter 6)—is not meaningful in relation to sustainability if only valid before a predetermined year, which has been remarkably constant since the start of IPCC and thus provides a shrinking timeframe for each passing year. We cannot claim to govern climate change successfully even if the projected temperatures are below the goal at that particular year if the climate projections still point upwards at that time (Figure 10.8). This oblivious pattern of thinking is based on a dominant temporally reductionist rationality that must be challenged.

There are countless similar examples in which time-related notions become central technologies in governing sustainability challenges with shocking effects. For instance, most urban drainage and stormwater systems across the world are designed to handle the water expected in particular rainfall events that are commonly defined in relation to their statistical return periods, with the 10-year rainfall being prevalent (see Haghighatafshar et al., 2020; Kim & Han, 2008; Shen et al., 2019). However, designing such systems for the 10-year rainfall rarely includes any considerations for what may happen during a heavier rainfall event than that—be it an 11-year rainfall or a 30-year rainfall—regardless of the consequences potentially being severe since the relationship between rainfall intensity and flood effects is nonlinear

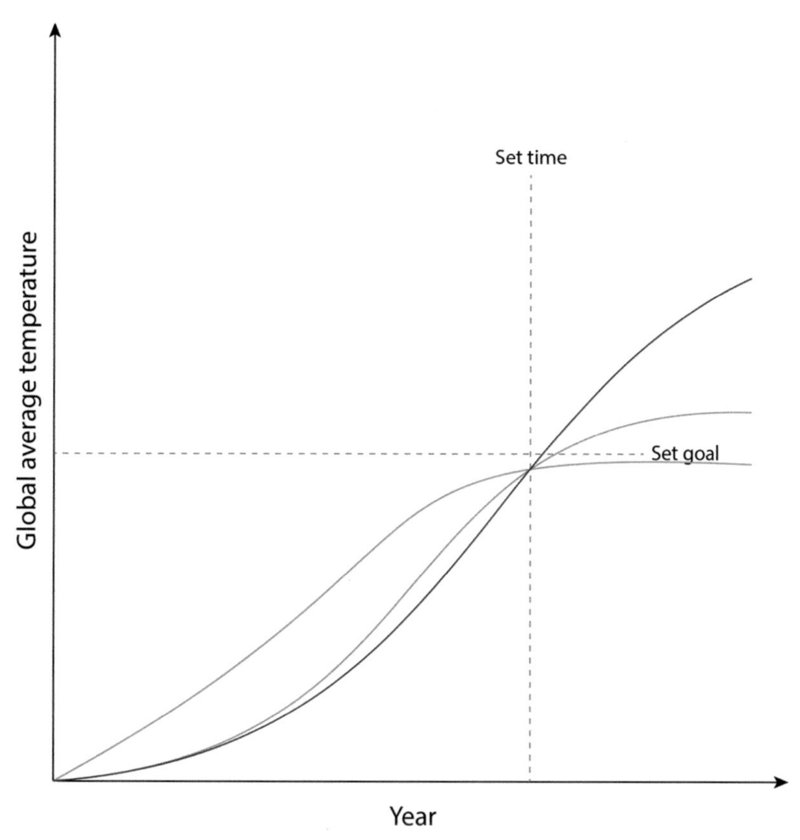

Figure 10.8 Three temperature projections all meeting the set goal at the set time, whilst only one provides a sustainable scenario below the threshold.

(Haghighatafshar et al., 2020; Leitão et al., 2013; Wilby et al., 2008). This problem is sometimes compounded by regulative constructions assigning responsibility for managing water. There are, for example, cases in Sweden in which the responsibility for water up to the 10-year rainfall is assigned to the water and sewage organisations, leaving a diffuse responsibility for the more significant flood scenarios to municipal administrations that rarely have the competence and resources available to assume it (Mobini et al., 2020). This kind of design rainfall is a key technology in governing flood risk mitigation that is not only fragmenting it. It is based on a flawed assumption of stationarity when rainfall patterns are indeed changing (Milly et al., 2008). What is now considered a 10-year rainfall is anticipated to occur more frequently in the future (IPCC, 2022). The future is not anymore, if ever, an extrapolation of the past (Becker & Payo, 2013). It demands an entirely new approach to designing urban drainage systems (Haghighatafshar et al., 2020)—as key technologies for governing flood risk mitigation (Boyd et al., 2014)—that can overcome the current temporal fragmentation and address our innate affinity for basing crucial decisions mainly on past experiences that may limit our options to govern the issue in the future (Payo et al., 2015).

The same temporally reductionist rationality is also often observable in the different groups of actors engaged at different points in time. Although the need for specialised knowledge and skills leads to a marked division of labour in governing most sustainability challenges, studies demonstrate unnecessary fragmentation (Becker, 2021b). It means there are regularly different groups of actors working together at different points in time, with the project documentation, the formal deliverables of earlier working groups, notes and minutes, and the individual memory of the remaining actors comprising the only link over time. Hence, there is a significant potential for fragmentation, as the recurrent or constant change of involved actors generally results in the erosion of institutional memory and the ability to learn (Carley, 1992; Hagelsteen & Becker, 2014b; Handmer & Dovers, 2007, p. 155; Raju & Becker, 2013, p. 89).

Functional fragmentation

The governing of complex sustainability challenges is also often functionally fragmented. Many policies and the activities they entail to address particular parts of a sustainability challenge impact other parts or other sustainability challenges—often unwittingly—but are implemented and evaluated only against their own objectives. This kind of fragmentation can be linked to the spatial- and temporal kinds just introduced since borders and timeframes regularly structure policies and activities. However, functional fragmentation also arises on its own when actors disregard the complexity of the world and the interconnectedness of the problems they focus on (Chapter 9). Countless examples of such functional fragmentation arise from actors prioritising different values. Remember, for instance, the example of the Ministry of Agriculture and the Red Cross Society providing incommensurable advice to farmers mentioned earlier (Chapter 7). This fragmentation can also arise when the actors fail to recognise important dependencies or simply ignore them to facilitate their own agenda.

The fit between the biophysical basis of complex sustainability challenges and the institutions attempting to address them is fundamental for effective governing (Folke et al., 2007). The same institutional fit is required for all problems, regardless of whether or not they are biophysical. It means that the regime of practices governing a particular problem must be able to capture and address all necessary parts of it. For example, banning the hunting of an endangered species matters little if its habitat is destroyed (CBD, 2020; Courchamp et al., 2003), or designing a new policy for disaster risk reduction matters little if the relevant actors are unaware of it or lack the human and material resources to implement it (GNCSODR, 2009). The purposes of the governing can, in other words, not be met without identifying and addressing entire assemblages of parts and their dependencies (cf. Becker, 2021a). For instance, criminalising homelessness is not addressing the problem of homelessness even if the issue may become less visible in public space, only exacerbating the vulnerability of the homeless (Murphy, 2019).

Early thinking about institutional fit was not only concerned with biophysical and institutional compatibility but also explicitly with the fit of legal frameworks regulating activities (Young & Underdal, 1997). Similar arguments for the importance of connecting and coordinating activities in the implementation of different policies governing sustainability challenges have been framed in terms of policy coherence (Nilsson et al., 2012), integration (Metz et al., 2020), and overcoming fragmentation (Becker, 2021c). Yet, fragmentation is often evident between actors implementing distinct but tightly coupled policies (e.g., Becker, 2021c). Such decoupling is a common theme in organisational theory (DiMaggio & Powell, 1983; Meyer & Rowan, 1977; Scott, 2014; Weick, 1976), and I will return to it in a moment when discussing viable explanations for the institutionalised fragmentation of the regime of practices.

Finally, functional fragmentation is not only prevalent within the governing of risk in relation to particular sustainability challenges but also between them. Addressing one problem may exacerbate another, which is further complicated when neither the problems nor solutions are evident, and there are many involved actors and conflicting values (cf. Alford & Head, 2017). For instance, short- and medium-term food security has so far mainly been tackled by increasing industrial agriculture based on monocultures that have devastating effects on biodiversity (Chappell & LaValle, 2011), which in turn is vital for long-term food security (Tscharntke et al., 2012). This issue is partly explained by most policymakers operating in silos (e.g., water, energy, agriculture, health) and lacking tools to grasp such dependencies, or at least which dependencies are most important (Nilsson et al., 2016, p. 321). However, the fragmentation is again best explained with reference to a dominant rationality that makes it self-evident and hidden in plain sight.

Locations of responsibility

Another interesting feature of the governing of sustainability challenges is the location of responsibility, which is multifaceted and involves various actors across societal spheres and at different levels depending on the particular sustainability challenge (Figure 10.9). At the global and regional levels, international organisations may play a role in setting standards and goals for sustainable development (Chapter 3). At the national level, governments are responsible for implementing and enforcing these standards and developing policies and activities to address sustainability challenges. This responsibility is sometimes also located with actors on the state or provincial level. At the local level, municipal administrations or other local authorities play essential roles in implementing sustainable practices and engaging citizens in actual efforts. It is also here we find the basis for most civil society organisations that provide space for people to mobilise and organise in relation to diverse issues and sometimes contradictory interests, although

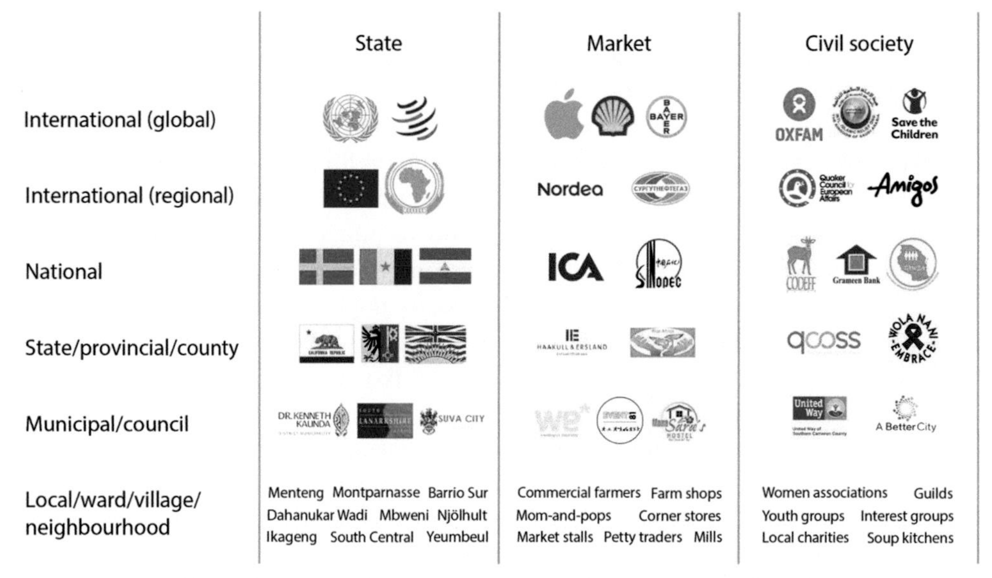

Figure 10.9 Examples of actors across societal spheres and levels.

some of them may also act on and influence the national and even international levels. Additionally, market actors, such as companies and corporations of various sizes, foci and reach, have the responsibilities to address sustainability challenges by implementing sustainable practices within their operations, providing sustainable goods and services, engaging in sustainable supply chain management, and collaborating with other actors to advocate for policy changes. They are also central actors in addressing complex sustainability challenges, which are elaborated on further in the next section. Finally, it is important not to forget that individuals also have important roles in addressing sustainability challenges through their consumption choices and voting patterns (if living in democracies). In short, addressing sustainability challenges requires action and cooperation across multiple spheres and levels of society. They require polycentric governance (Ostrom, 2010).

Although the concept of polycentricity was first used by Polanyi (1951), it was Vincent and Elinor Ostrom who developed and popularised it in relation to governing complex issues (Carlisle & Gruby, 2019). It refers to a complex form of governing with multiple decision-making centres operating with some degree of individual autonomy (Ostrom, 2005; Ostrom et al., 1961). The decision-making units in polycentric governing are generally overlapping and nested at multiple administrative levels (e.g., national, county, municipal), including particular governing mechanisms across levels (Ostrom, 2005). Because of this tiered structure, polycentric governing may achieve a balance between centralised and decentralised governing (Imperial, 1999). However, whilst the presence of numerous semiautonomous decision-making centres may seem sufficient to define a governing structure as polycentric, it does not ensure that these decision-making centres are sufficiently coordinated for actual polycentric governing to materialise (Carlisle & Gruby, 2019; Pahl-Wostl & Knieper, 2014). Polycentrism requires distributed power with no loss of coordination between actors across administrative boundaries, levels or societal spheres, concerning a spatially bounded issue (Andersson & Ostrom, 2008; Pahl-Wostl et al., 2012). In other words, the many decision-making centres must consider each other in collaborative or competitive interactions and be able to resolve disagreements for actual polycentric governing to materialise (Carlisle & Gruby, 2019; Ostrom et al., 1961). There are myriad examples of this latter, yet fundamental, aspect being flawed or outright missing. Instead of the involved actors contributing with parts harmonising with each other in a polyphonic manner, we regularly find a cacophony of fragmented and often clashing opinions, decisions and activities.

This problem of inadequate coordination can partly be explained by the responsibility for governing sustainability challenges often being pushed towards particular and less-powerful actors, whilst more powerful actors withdraw from responsibility, regardless of the number and diversity of contributing actors overall (Becker, 2021b). This process is clearly visible in the responsibilisation of citizens to shoulder all sorts of responsibilities when the state scales back its services or at least fails to meet growing needs (e.g., Bergström, 2018; Kelly, 2010; Soneryd & Uggla, 2015). It is also visible in municipal administrations being the main locus of responsibility for governing various sustainability challenges in many countries worldwide (e.g., Harjanne et al., 2016; Mancilla García et al., 2019).

This push of responsibility downwards has been significantly driven by notions of decentralisation and, sometimes, by a misconstrued idea of polycentrism. It is fundamental to note that decentralising responsibility for governing an issue to a lower level rarely leads to the polycentrism advocated by influential scholars concerning complex issues (Berardo & Lubell, 2019; Ostrom, 1990). First, decentralisation is commonly associated with a corresponding allocation of resources (Saito, 2011), which is often not the case (e.g., Becker, 2021b). Secondly, polycentrism is not an automatic result of decentralisation. At least not in the concept's original meaning, if not used as a mere synonym of decentralisation (e.g., Johannessen et al., 2019). The concentration of responsibility to a lower level can instead be seen, in combination with its fragmentation discussed earlier, as a clear sign of weak polycentrism (cf. Andersson & Ostrom, 2008). Governing can be more or less polycentric (Andersson & Ostrom, 2008), but passive or withdrawing actors on higher levels undermine governing as a lack of interaction with actors on higher administrative levels has been shown to have adverse effects on governing outcomes (Angst et al., 2018). Integrating multiple levels and spheres is, in other words, crucial in any attempt to understand or assess society's capacity to govern its challenges (Becker, 2012, 2020).

The role of the market

The final important feature of the governing of sustainability challenges elaborated on here is the role of the market—as one of the three main societal spheres, together with the state and civil society (Mol, 2010a,b). Here, the market is more than just 'an allocative mechanism but also an institutionally specific cultural system' (Friedland & Alford, 1991, p. 234). State organisations, like national authorities and municipal administrations, rely heavily on input from various private companies—the quintessential market actors. Not necessarily so much for coordination but for providing a broad range of goods and services perceived necessary for governing various issues. This pattern is perhaps not surprising considering the fundamental role of the market for production, and private companies have been implementing various activities decided and funded by the state on different levels. This reliance on private companies to provide public services is ubiquitous in most parts of the world. However, studies suggest that the role of the market has expanded to also provide crucial input to the decision-making processes per se—clearly visible in the dependence on input from various consultants. Numerous studies show how consultants are central to a range of important decisions on different administrative levels (e.g., Becker, 2021b; Öjehag-Pettersson & Granberg, 2019; Pedersen, 2004). For instance, in decision-making for climate change adaptation (Boyd et al., 2011; Jensen, Ørsted Nielsen, & Lilleøre Nielsen, 2016; Orderud & Naustdalslid, 2020). However, their influence does not stop there. They are not only bidding to win contracts but are also influencing perceptions of what needs to be done in the first place. The recent growth of consultancies is, in other words, not only driven by a growing need for input to different decision-making processes, as aptly suggested by Boyd et al. (2011), but also by their ability to create a need for services that the state actors do not even know exist yet (Figure 10.10). Their input drives demand for more input, motivated by further improvement in governing the particular challenge. Some studies suggest that consultants play a key role in formulating public policy itself (Jupe & Funnell, 2015). In other words, private companies are no longer only carrying out activities to implement particular decisions but also providing the foundations for the decisions themselves and influencing the experienced need for their services.

Finally, the role of the market is intrinsically linked to the fragmentation elaborated on earlier. Bevir (2011) notes the relationship between fragmentation and the increasing dependence of state actors on other actors. These 'other actors' are overwhelmingly market actors, although civil society organisations are also essential in providing basic services in some parts of the world. However, the expanding role of the market is connected to the concentration of responsibility to lower administrative levels (Becker, 2021b). The increasingly important role of consultants is a strategy these actors use to cope with this recently materialised responsibility within their existing resources. It is, in other words, in the vacuum of increasing responsibility, without a commensurate increase in

Figure 10.10 Consultants play key roles in governing most sustainability challenges. *(Source: Dusit/ Shutterstock.com.)*

the resources available to execute it, that the actors turn to consultants, who in turn provide whatever services they want to pay for (cf. Becker, 2021b; Jensen, Ørsted Nielsen, & Lilleøre Nielsen, 2016).

Interdependent mechanisms of institutionalisation

The multifarious fragmentation that is seen in the regime of practices governing risk in relation to sustainability challenges is a common theme across the world (Carter et al., 2021; Hegger et al., 2016; Marks, 2019; Rivera et al., 2015). Although there is 'no such thing as "good organization" in any absolute sense' (Ashby, 1962, p. 263), there is a broad understanding that fragmentation seriously undermines the governing of complex issues in general (see Alford & Head, 2017; Becker, 2009; Bouckaert et al., 2010; Cejudo & Michel, 2017; Folke et al., 2007). This link is perhaps best explained by Ashby's (1957) Law of Requisite Variety, which stipulates that any system governing another larger complex system must have a comparable degree of complexity to that system (Chapter 7). It means that the complexity of the governing system must match the complexity of the system being governed. This widespread notion is sometimes labelled the diversity hypothesis and assumes that organisational and institutional diversity is the most effective way to cope with complexity (Duit et al., 2010, p. 365). Others frame similar ideas concerning institutional fit (e.g., Bergsten et al., 2014; Hedlund et al., 2021; Mancilla García et al., 2019), or as the problem of fit (e.g., Folke et al., 2007; Widmer et al., 2019; Young & Underdal, 1997), highlighting the importance of fit between the system being governed and the system governing it. From this perspective, the governing of any sustainability challenge must be able to accommodate its complexity, not ignore it. So, how come fragmentation is so prevalent?

The literature is rife with attempts to explain fragmentation, often attributing it to a push for specialisation to improve efficiency (see Bouckaert et al., 2010; Cejudo & Michel, 2017; Hood & Dixon, 2015). Whilst these public administration-oriented contributions are persuasive, they fail to explain the range of fragmentation identified here. For that, it is helpful to focus on the institutionalisation of the regime of practices.

These regimes of practices—the sets of institutionalised ways of thinking and doing—are neither primordial nor do they appear out of thin air. Rather, they emerge as contingent reactions to path-dependent problems. Whilst particular events are commonly considered decisive (Drori et al., 2006), a more complex mix of factors is usually found behind any regimes of practices. For instance, a particular flood may often be considered the initiating event for flood risk to become governmentalised, but the regime of practices itself is not only determined by that event (Becker, 2021c). Instead, regimes of practices tend to emerge from established practices for governing something else—generally something as closely related to the problematised sustainability challenge as available. Like, the regime of practices governing flood risk carrying over a lot from established practices for managing urban drainage of more everyday rainfall (Becker, 2021c). The emerging regime of practices, thus, carries with it a symbolic (regulative, normative, cultural-cognitive), relational, and artefactual legacy (cf. Scott, 2014, pp. 95−104). This legacy provides the foundations for, and the initial direction of, the institutionalisation of the governing of the newly problematised sustainability challenge.

It means the first actors governing an emerging issue do not start from scratch. They have the same legislation, their professional norms and identities, their shared ideas and predispositions, their established social relations, and a set of artefacts that enable as well as restrict their activities. Although there is still some room for initial behavioural variation, even after bringing all of that to bear, studies suggest that particular patterns of activities start to emerge rather quickly and are then used repeatedly until they become the convention (Becker, 2021c; Scott, 2014).

It is not to say that the regime of practices governing an issue is static—since institutionalisation is a dynamic, continuous process (Barley & Tolbert, 1997)—only that any updates appear to follow the same process proposed by van de Ven and Garud (1994). However, the same fragmentation, lack of coordination and reliance on the market usually remains regardless of what updates in the regime of practices that materialise (e.g., Becker, 2021c). To explain this, it is helpful to revisit the three mechanisms of institutionalisation introduced above—increasing returns, increasing commitments and increasing objectification—and to remember that their combined force is most formidable when all three align.

The governmentalisation of sustainability

It is now time to elaborate on how complex sustainability challenges become something governable on the societal level in the first place. This section is divided into two parts. First, it examines four constituent processes of governmentalisation before relating them to the impact of lurking neoliberalisation.

Four constituent processes of governmentalisation

Approaching the governmentalisation of complex sustainability challenges from the perspectives of governmentality and new institutionalism requires a systemic consideration of how regimes of practices become institutionalised. It is about studying processes rather than the processed (cf. Desmond, 2014). Such governmentalisation is never simple nor homogenous but composite and multifaceted. It has at least four constituent processes: (1) reductivisation, (2) projectification, (3) responsibilisation and (4) commodification (Figure 10.11).

Governmentalisation as reductivisation

The most fundamental constituent process of governmentalisation concerns an inherent reductivism applied to the problem. Whilst a problem can be complex in both its constitution and potential solutions, the regime of practices governing it often entails intricate fragmentation (Angst et al., 2018; Becker, 2021b; Lubell & Morrison, 2021; Rivera et al., 2015)—not only in the ways of doing (technologies) but also in the ways of thinking about it (rationalities). This reductivism concerns the latter, even if it is essential to remember that rationalities and technologies are intrinsically linked and both necessary to grasp the governing of anything (Dean, 2010; Miller & Rose, 2008).

Much of the fragmenting rationalities generally visible in such regimes of practices are reducing the complexity of the problem by partitioning it into disconnected parts that are easier to grasp in isolation (e.g., Becker, 2021b; Bond & Morrison-Saunders, 2011; Hulme, 2011; Young & Stokke, 2020). This pervasive reduction of complexity is evidently undermining the governing of risk. However, reducing the complexity of the problem is fundamental for it to become governable in the first place, as it renders reality understandable in a way that makes it amenable to analysis and action (cf. Miller & Rose, 2008, pp. 15–16). I refer to this process as reductivisation to stress the active

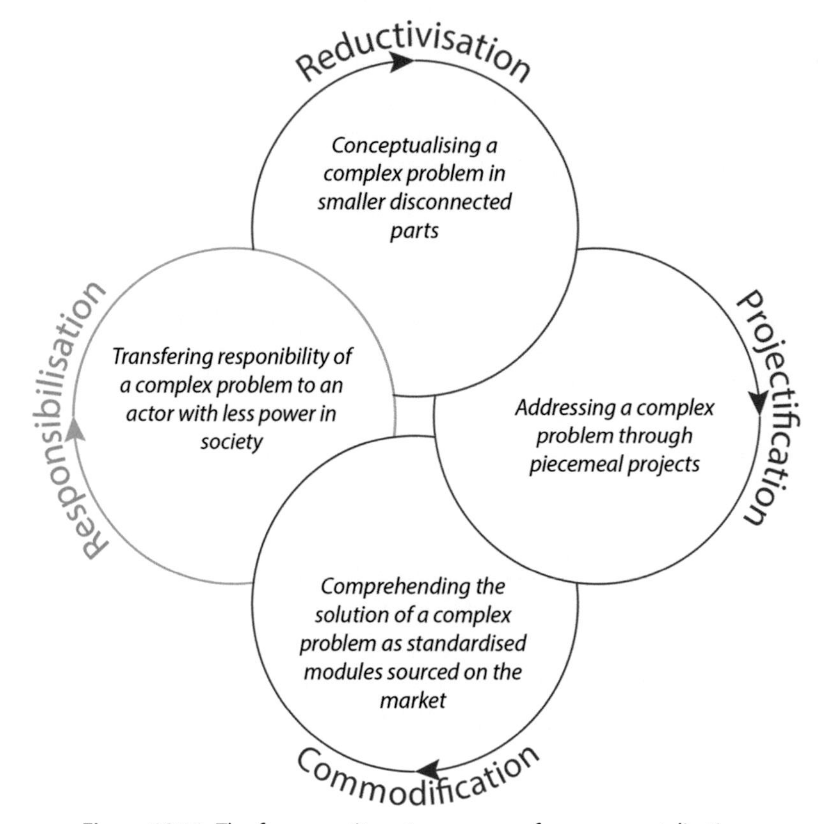

Figure 10.11 The four constituent processes of governmentalisation.

process of conceptualising a complex problem as many smaller, disconnected parts (Figure 10.11).

Luhmann (1979) suggests that reducing complexity is an existential need of all human beings. Much is becoming increasingly complex (Nowotny, 2005a), especially in the social sphere (e.g., Durkheim, 1984; Luhmann, 1995, p. 403). However, this sense of increasing complexity is also epistemological and driven by increasing knowledge of how things work. When faced with unfathomable complexity, human beings struggle to gain a sense of control in an otherwise unbearable situation (Nowotny, 2005a, pp. 19–20). We do that by reducing the complexity through a broad range of social, cultural or technological means (Luhmann, 1995). The reductivisation of identified sustainability challenges is, then, a typical human response to the otherwise seemingly overwhelming complexity of the problem.

Other scholars suggest that simplifying complex problems and the available solutions can create an image of actors having them under control (van Bommel et al., 2009) and silence alternative viewpoints (Boyd et al., 2014, pp. 148–149). Power over the framing of the problem is, thus, crucial (see Boström et al., 2017). Such framing is, for instance, visible in the dominant mode of thinking about flood risk as a local problem that requires local solutions (Becker, 2021b). If a more relational, catchment-oriented mode of thinking prevailed, flood risk mitigation would demand more attention and coordination, and many current solutions would become infeasible without significant additional investments (Becker, 2020). However, these studies of the governing of complex sustainability challenges provide little evidence of direct strategic calculation but allude to a prevalent routinisation of the fragmented rationalities. Power is exercised in the rationalisation of the current regime of practices (Flyvbjerg, 1998), and fragmentation is taken for granted—hidden in plain sight. The reductivisation can, as such, be institutionalised through the mechanism of increased objectification introduced earlier.

The reduction of complexity is not only institutionalised in the governing of complex sustainability challenges but is the foundation for the dominant Newtonian paradigm in science (Heylighen et al., 2007). Remember from Chapter 7 that this paradigm holds that 'to understand any complex phenomenon, you need to take it apart, that is, reduce it to its individual components. If these are still complex, you need to take your analysis one step further, and look at their components' (Heylighen et al., 2007, p. 118). Although there is a growing critique of such reductivism (e.g., Bond & Morrison-Saunders, 2011; Byrne & Callaghan, 2014; Castellani & Hafferty, 2009; Liverman, 2009; Nowotny, 2005b; Urry, 2005; Walby, 2007), it still dominates both in research and education. Reductivisation, therefore, appears to be profoundly institutionalised in science—the primary source of legitimisation in the modern world. Although all theories require a certain degree of complexity reduction (Cilliers, 2005), a theory that completely undermines our thinking and doing is not particularly useful.

To summarise, reductivisation is one of the constituent processes of the governmentalisation of complex sustainability challenges. It refers to the conceptualisation of a complex problem as smaller, disconnected parts (Figure 10.11).

Governmentalisation as projectification

The second constituent process of governmentalisation is tightly linked to the first and concerns how the issue is addressed. Whilst a problem is clearly complex, as described above, it is usually divided into more or less arbitrary pieces addressed in detached projects. Each action (e.g., a detailed development plan, the dyke protecting from coastal floods, or the restoration of agricultural drainage) is turned into a project with its own objectives, activities, actors and budget. I refer to this process as projectification and define it as the process of addressing a complex problem through many piecemeal projects. It comprises, in a sense, an essential part of the technology-side of the conspicuous fragmentation in the governing of complex sustainability challenges, with the rationality-side being covered by the process of reductivisation elaborated on above.

It is important to note that concepts of projectification have been used in a range of scientific disciplines since Midler (1995) first coined the term and Maylor (2006) relaunched it in the social scientific literature. Most notably in the domain of business administration and management (e.g., Bergman et al., 2013; Maylor et al., 2006), but also in the context of public administration (e.g., Fred, 2015; Godenhjelm et al., 2015). Notwithstanding substantial variation in definitions and approaches, it generally refers to repackaging activities into projects and the associated adaptation of practices. Examples include creating project teams from different formal organisational structures (Fred, 2020) and shifting from vertical to horizontal communication and power from hierarchical line managers to project managers (Maylor et al., 2006). It has been claimed that the combination of relatively stable organisational structures and the flexible mobilisation of actors and resources brings controllability and creativity to bear on the challenges organisations face (Fred, 2015, pp. 49–50). Whilst these contributions help explain change within involved organisations, I draw upon a broader sociological critique to further grasp the governmentalisation of complex sustainability challenges.

Boltanski and Chiapello (2005) show how the project has become not only the dominant form of managing formal issues in both corporate and public spheres but also a vehicle for personal fulfilment (Figure 10.12). Individual actors engage in projects with particular objectives and are rewarded when the project ends, and the stated objectives are met. Then, they move on to the next, perhaps even more interesting project, creating a sense of meaning and accomplishment (Boltanski & Chiapello, 2005, pp. 90–96). This phenomenon is pervasive in both professional and private life (Jensen, Thuesen, & Geraldi, 2016). Meeting the project's objectives is, thus, the principal goal of actors.

Figure 10.12 Projects are ubiquitous in both professional and private life.

However, these objectives are, in themselves, disjointed as the projects are designed in a piecemeal fashion over time and result in an array of projects that are related along various crucial dimensions but lack coordination and a common direction (cf. Jennings, 1994, p. 53). Consequently, a strong link exists between a fragmented public sector and the increasing importance of projects (e.g., Fred, 2015; Jensen & Trägårdh, 2012).

Whilst projects can be an efficient way to organise activities that generate expected results and meet particular objectives, it is evident that these objectives can clash when considering the many projects in the governing of even a single complex sustainability challenge. Whilst one project can meet its objectives—making it a success to be celebrated—its results can undermine the objectives of other projects. They may even aggravate the problem somewhere else or at a later point in time (Chapter 9). Such unanticipated consequences of purposive action have attracted attention for centuries (e.g., Smith, 1869, pp. 84–85) and are a fundamental aspect of projectification.

The unanticipated consequences of projects can be partly explained by the paramount concern of actors to meet their objectives. Merton (1936, p. 901) calls it the 'imperious immediacy of interest', which overshadows any considerations of other consequences of the same activities. This focus is institutionalised through the mechanism of increasing returns, introduced earlier. Merton (1936) also asserts the role of specific fundamental values in directing the involved actors towards a particular action, effectively blinding them of its further consequences. Although he is correct in pointing out significant differences between these two causes of unanticipated consequences, these values are institutionalised through the mechanism of increasing commitments to normative expectations (Selznick, 1949, pp. 256–257) that, in turn, can grow out of the repeated practices of focusing on generating the expected results. The two are closely linked.

There is an additional cause of unanticipated consequences of purposive action, in addition to Merton's (1936) framework. It is visible in the routinised and taken-for-granted ways many actors engage in their projects. Here, unanticipated consequences are not caused by actors pursuing an immediate interest or fulfilling a normative expectation but by their routinised action without reflecting on other interests and expectations. This routinisation is driven by the mechanism of increasing objectification (Berger & Luckmann, 1966), but can grow out of the repeated practices of focusing on generating the expected results or fulfilling normative expectations (Scott, 2014, pp. 147–150).

The projectification of efforts to govern complex sustainability challenges opens Pandora's box of unanticipated consequences that threaten to exacerbate risk. On the other hand, it also provides involved actors with projects that can be implemented. Projectification denotes, therefore, more than a qualitative transformation in the organisation of activities into projects and associated practices to meet some predefined objectives, and a quantitative increase in the number of such projects, regardless if driven by ideas of increased efficiency or merely as a fashionable way of organising activities (Godenhjelm et al., 2015, pp. 326–327). It is also a process of reducing complexity (cf. Fred, 2015, p. 51; Godenhjelm et al., 2015, p. 327) and plays a fundamental role in making sustainability challenges governable.

To summarise, projectification is a constituent process in the governmentalisation of sustainability challenges, in which the complex problem is addressed through piecemeal projects (Figure 10.11).

Governmentalisation as responsibilisation

The third constituent process concerns the allocation of responsibility amongst actors. The responsibility for governing many different sustainability challenges has emerged relatively recently. As mentioned, this responsibility is not evenly distributed between various actors but largely concentrated to certain actors depending on challenge and context. I refer to this as responsibilisation and define it as the process of transferring responsibility for a complex problem to an actor with less power and without appropriate resources to assume it.

There is a rich literature using concepts of responsibilisation to describe and explain how subjects become responsible for aspects of social life that were previously the duty of another—usually a state actor (Figure 10.13)—or not recognised as a responsibility at all (e.g., Dean, 1997; Joseph & Juncos, 2019; Rose, 1999; Shamir, 2008). It is most commonly considered in relation to individual subjects but can also refer to collectives of private citizens, such as families, households and communities (Dean, 2010, p. 194), as well as to organisations (Rose, 1999, pp. 236–237).

For instance, whilst private citizens are increasingly responsibilised for preparing and responding to actual floods (Rådestad & Larsson, 2018), and there are attempts to responsibilising them also for mitigating flood risk (Becker, 2021b), studies demonstrate that also

Figure 10.13 Household preparedness campaigns have been critiqued for responsibilising citizens.

municipal administrations are subject to the responsibilisation of flood risk mitigation (Becker, 2021b). I acknowledge that this is a somewhat unconventional application of the concept of responsibilisation since municipal administrations are state actors. However, the process is the same and requires asymmetry of power. This intrinsic connection between power and responsibility has been suggested for governing flood risk (Butler & Pidgeon, 2011), and I argue that this broader conceptualisation of responsibilisation may be constructive when used in relation to governmentality. Whilst the distinction between the private and the state is crucial in most theoretical perspectives, it is the power asymmetry that is the decisive feature in any responsibilisation (cf. Hannah-Moffat, 2000). Lemke (2002, p. 53) alludes to this when asserting that it is the responsibilisation of subjects that is 'forcing them to "free" decision making in fields of action', and Shamir (2008, p. 7) explains that this power 'relies on predisposing social actors to assume responsibility for their actions'.

To understand responsibilisation, it is essential to distinguish between following rules and assuming responsibility. Selznick (2002) suggests that the former concerns an actor engaging in activities under the threat of sanctions if not abiding by the rules, whilst the latter presupposes that the actor is motivated to act without coercion. Responsibilisation can, thus, work by predisposing actors to do the right thing through the cultivation of specific values (Shamir, 2008). This focus on normative values is found throughout the governmentality literature (e.g., Lemke, 2001; O'Malley, 1996; Schweber, 2014). However, it is only related to one of the three principal mechanisms of institutionalisation incorporated into new institutionalism (Scott, 2014): increasing commitments to normative values. The institutionalisation of responsibility can also be driven by increasing returns concerning incentives—like earmarked national funding for specific activities (e.g., Becker, 2021c)—and by an increasing objectification of ideas and predispositions—limiting actors to simply do the conceivable thing. Whilst governmentality perspectives pay less explicit attention to this latter process, objectification is at its core (Dean, 2010; Foucault, 1991b). Responsibilisation is, then, driven by the routinisation of ideas and predispositions that are acquired through imitation and learning.

Applying the concept of responsibilisation to lower-level state actors—such as municipal administrations—highlights the importance of considering their available resources to assume the responsibility. Otherwise, it is easily conflated with processes of decentralisation. Whilst both processes distribute responsibility, decentralisation is associated with a commensurate allocation of resources, at least in an ideal sense (Saito, 2011). Conversely, responsibilisation connotes a transfer of responsibility without appropriate resources (Joseph, 2013). The conceptual relationship between responsibilisation and empowerment is identical (Bergström, 2018), which supports my suggestion to explicitly consider the resources available also to private citizens to assume any transferred responsibility.

To summarise, responsibilisation is a constituent process in the governmentalisation of complex sustainability challenges. It refers to transferring responsibility for the complex problem to an actor with less power and no appropriate resources to assume it (Figure 10.11).

Governmentalisation as commodification

The fourth constituent process concerns how many activities for governing risk in relation to sustainability challenges are sourced as standardised modules on the market. This process refers not only to the reliance on private companies to execute particular decisions but also to their input to these decisions. I refer to this process as commodification—fully aware of the legacy of this concept—and define it as the process of comprehending the solution of a complex problem as the aggregated effect of standardised modules sourced on the market. I am indebted to Almklov and Antonsen's (2010) seminal paper on what they refer to as 'commoditisation', which is equally conceptually entangled but in business and marketing literature and not used here to reduce potential Babylonian confusion (cf. Hagelsteen & Becker, 2014a).

The theoretical underpinnings of commodification come from Marx and refer to a process through which human activities become commodities with monetary value that can be traded on markets (Abercrombie et al., 2006, p. 68). Marxists consider such commodification as the engine that drives the continuous expansion of capitalism (Friedland & Alford, 1991, p. 263) and many influential scholars—Marxist and non-Marxist—agree that there is a tendency towards the commodification of every aspect of social life (Callinicos, 2007, p. 258)—from identities (Rose, 1996, p. 344) to ecosystems (Figure 10.14) (Pellizzoni, 2016, p. 319). It should, therefore, be unsurprising to

Figure 10.14 Everything is becoming a commodity.

find significant commodification, in this traditional sense, in the governing of sustainability challenges (e.g., Becker, 2021b).

Many of the technologies of governing, or parts of such technologies, are procured on the market by various actors. For instance, consultants provide risk assessments, environmental impact assessments, landscaping plans, hydrological or hydraulic modelling, biodiversity assessments, and many more reports and services. However, not only is each procured part turned into a commodity but also the expected aggregated result (Almklov & Antonsen, 2010). Miller (2003) puts forward similar observations concerning the commodification of academic education. It means that the overall processes addressing various sustainability challenges—such as disaster risk management or climate change adaptation—have become the subject of commodification in an expanded sense since they essentially take the form of ritualised procurement of increasingly standardised modules (Becker, 2021b). Often merely to be able to exhibit them to demonstrate that the problem at hand has been seriously considered.

The process of commodification is closely linked to the other three constituent processes of governmentalisation. It depends on and facilitates the reductivisation of the problem, which aligns with the intrinsic reductivism of commodification that Radin (1996) demonstrates so eloquently. It is interdependent with projectification, where different projects and sets of associated standardised modules emerge and become increasingly institutionalised over time. Finally, commodification is largely driven by responsibilisation, where actors turn to the market when lacking the capacity to assume their responsibilities that are operationalised in a way so they can fulfil their obligations by simply procuring standardised modules.

To summarise, the governmentalisation of a complex sustainability challenge involves the constituent process of commodification, in which the solution to the complex problem is seen as the aggregation of standardised modules that can be sourced on the market (Figure 10.11).

Lurking neoliberalisation

The four constituent processes of governmentalisation are all related to the underlying process of neoliberalisation. It is important to note that neoliberalism can be approached as a policy, an ideology, or governmentality (Larner, 2000), out of which the latter is most helpful in this context. Here, it is neither 'a concrete economic doctrine' nor 'a definite set of political projects' but rather 'a complex, often incoherent, unstable and even contradictory set of practices' (Shamir, 2008, p. 3). Neoliberalisation denotes the processes that produce these practices, out of which 'marketisation and commodification have a long pedigree during the geohistory of capitalism' (Brenner et al., 2010, p. 184). I argue, however, that reductivisation, projectification and responsibilisation are also inherent in neoliberalisation and thus general to the governmentalisation of complex sustainability challenges, albeit to various degrees and in different ways depending on the penetration and diffusion of neoliberalism in each context.

Bevir (2011, p. 459) asserts that neoliberalism is less characterised by the emergence of effectively performing markets 'than by the proliferation of networks, the fragmentation of the public sector and the erosion of central control'. Fragmentation has also been pointed out as inherent to the New Public Management (Cejudo & Michel, 2017) so tightly connected to neoliberalism (Bevir, 2011, p. 464; Rose et al., 2006, p. 95). These accounts suggest that the fragmentation of the governing of sustainability challenges is likely to be a pervasive effect of neoliberalism. However, the generality of this claim can be further substantiated if the fragmentation of the regime of practices is separated into the fragmentation of doing (projectification) and the fragmentation of thinking (reductivisation).

Boltanski and Chiapello (2005) dedicate a significant part of their eloquent critique of the new spirit of capitalism to demonstrating how projectification is an immanent effect of neoliberalism. The market is not the only institutional logic in advanced liberal democracies, but it is in the intersection and contradiction of multiple institutional logics that such transformation may occur (Friedland & Alford, 1991). Consequently, the projectification of governing can be understood as the enactment of multiple institutional logics, where neoliberalism introduces and emphasises market and project logics in relation to bureaucratic and political logics (Fred, 2020). Projectification is a general feature in efforts to promote sustainable development in this context (Cerne & Jansson, 2019), and neoliberalism is widely suggested as fundamentally reductivist (e.g., Gidley et al., 2010; McAfee, 2003; Nikolakaki, 2012; Wang, 2020). Reducing the complexity of a sustainability challenge is fundamental for it to become governable, as this renders it amenable to analysis and action (cf. Miller & Rose, 2008, pp. 15–16). It is, therefore, fair to assume that the reductivisation and projectification suggested above are active to various degrees and in multiple ways in the governmentalisation of complex sustainability challenges in general.

It is also fair to assume that responsibilisation is a general feature of such governmentalisation due to the pervasiveness of neoliberalisation in this context. Governmentality is defined by the notion of free individuals who control, determine, and delimit the liberty of others and themselves (Dean, 2010; Rose & Miller, 1992). Neoliberalism promotes such freedom, understood as a personal choice that transfers responsibility to the actors themselves (Bevir, 2011, pp. 465–466). Governing is, thus, largely about making other actors assume responsibility for their situation and actions. Whilst the concept of responsibilisation is widespread in the governmentality literature, its generalisability as a constituent process of governmentalisation increases when its conventional focus on private citizens is expanded to all less-powerful actors lacking appropriate resources to assume the assigned responsibility. In this context, I argue that responsibilisation is general to the governmentalisation of complex sustainability challenges in advanced liberal

Figure 10.15 Asymmetry of power and insufficiency of resources are the fundamental aspects of responsibilisation. *(© Andrii Zastrozhnov | Dreamstime.com.)*

democracies. However, I would also argue that this conceptualisation is more useful also when maintaining the conventional focus on private citizens. An explicit focus on the asymmetry of power and insufficiency of resources (Figure 10.15), rather than the private sphere per se, may help to overcome increasing ambiguities in distinctions between the main institutional spheres of society (see Mol, 2010b, p. 32).

Whilst commodification has been affirmed as a central feature of neoliberalisation (Brenner et al., 2010), it is worth noting the particular importance of agency in this commodification. It is best exemplified by consultants who partly drive an increasing need for their services. However, other studies suggest that consultants play an essential role in pushing for neoliberalisation itself (Jupe & Funnell, 2015; Martin, 1993, p. 6).

To summarise, the constituent processes of governmentalisation are likely to be active—to various degrees and in different ways—regardless of the sustainability challenge and the context. That would confirm the neoliberalisation of sustainability itself, suggested by others (e.g., Hanna et al., 2018; McKenzie et al., 2015), since the ways through which we conceptualise and confront sustainability are laden with processes that are inherent to neoliberalism.

Conclusion

Governing risk in relation to complex sustainability challenges is fundamental for the long-term survival of humankind and Earth as we know it. Effective governing of such risk requires not only a holistic understanding of the complex sustainability challenge as such. It also requires a sufficient fit between the problem and the institutions governing it. This requirement entails coordination and collaboration between numerous actors across borders, societal spheres, and levels (Bodin, 2017). Failure to effectively govern risk would lead to irreversible damage to the planet and the well-being of future generations. Yet, the current governing of most sustainability challenges is fraught with problems.

The most conspicuous problem is the overwhelming spatial, temporal and functional fragmentation in the regime of practices governing the sustainability challenges, which undermines the desired effects. A second problem concerns the distribution of responsibility, which often is pushed towards certain actors. A third potential problem concerns the role of the market, which has expanded both qualitatively and quantitatively. It is increasingly the case that market actors not only execute specific decisions but also provide the basis for the decisions themselves. Hence, it is clear that there is a significant escalation in both the penetration and diffusion of the market into the public sphere.

These problems largely stem from the overall process through which sustainability challenges become something governable on the societal level. There are four constituent processes of this governmentalisation elaborated on in this chapter. It involves reductivisation, in which the complex problem is conceptualised as smaller, disconnected parts. The second is projectification, in which the complex problem is addressed through piecemeal projects. Whilst these two processes are intrinsically linked and combine to undermine many efforts, they are also fundamental for the sustainability challenges to become governable in the first place. The third process is responsibilisation, in which the responsibility for the complex problem is transferred to an actor that has less power and lacks the appropriate resources to assume it. Here, responsibilisation goes beyond the more conventional focus on private citizens and extends to any actor, based on the argument that it is the asymmetry of power and insufficiency of resources that are its defining features. Finally, the fourth constituent process is commodification, in which the solution is seen as the aggregation of standardised modules that can be sourced on the market. This conceptualisation encompasses both the commodification of a technology or part of a technology and of the governing itself, which increasingly takes the form of ritualised procurement of standardised modules. Commodification materialises in a vacuum of responsibilisation, as obligations are imposed on actors without commensurate resources. These four constituent processes are interdependent and form a nexus of governmentalisation.

It is easy to become disillusioned when considering the importance of governing various pressing sustainability challenges, the inherent challenges posed by their complexity, and the regimes of practices currently applied to govern them (cf. Duit et al., 2010, p. 367). Most governing ignores the requisite relationality for approaching any complex system, underestimates the potential for surprise, and disregards any intrinsic volatility, cascading effects or tipping points of such a system (Chapter 9). For more effective governing of complex sustainability challenges, we need a fundamental transformation of how we approach complexity, and then also uncertainty, ambiguity and dynamic change. One way to at least endeavour to grasp the complexity of a particular phenomenon or problem and its consequences on governing is to approach the world as human-environment systems (Chapter 11).

References

Abercrombie, N., Hill, S., & Turner, B. S. (2006). *The Penguin dictionary of sociology*. Penguin Books.

Acreman, M. C., Fisher, J., Stratford, C. J., Mould, D. J., & Mountford, J. O. (2007). Hydrological science and wetland restoration: Some case studies from Europe. *Hydrology and Earth System Sciences, 11*(1), 158–169.

Ahrne, G. (1994). *Social organizations: Interaction inside, outside and between organizations*. Sage.

Alford, J., & Head, B. W. (2017). Wicked and less wicked problems: A typology and a contingency framework. *Policy and Society, 36*(3), 397–413. https://doi.org/10.1080/14494035.2017.1361634

Almklov, P. G., & Antonsen, S. (2010). The commoditization of societal safety. *Journal of Contingencies and Crisis Management, 18*, 132–144.

Andersson, K. P., & Ostrom, E. (2008). Analyzing decentralized resource regimes from a polycentric perspective. *Policy Sciences, 41*(1), 71–93. https://doi.org/10.1007/s11077-007-9055-6

Angst, M., Widmer, A., Fischer, M., & Ingold, K. (2018). Connectors and coordinators in natural resource governance: Insights from Swiss water supply. *Ecology and Society, 23*(2), 1–15. https://doi.org/10.5751/es-10030-230201

Armaş, I. (2006). Earthquake risk perception in Bucharest, Romania. *Risk Analysis, 26*, 1223–1234.

Ashby, W. R. (1957). *An introduction to cybernetics*. Chapman & Hall Ltd.

Ashby, W. R. (1962). Principles of the self-organizing system. In H. von Foerster, & G. W. J. Zopf (Eds.), *Principles of self-organization* (pp. 255–278). Pergamon Press.

Aven, T., & Renn, O. (2010). *Risk management and governance: Concepts, guidelines and applications*. Springer.

Bäckstrand, K. (2003). Civic science for sustainability: Reframing the role of experts, policy-makers and citizens in environmental governance. *Global Environmental Politics, 3*, 24–41.

Bara, C., & Doktor, C. (2010). *Risk analysis – cooperation in civil protection*. Spain, and the UK: EU.

Barley, S. R., & Tolbert, P. S. (1997). Institutionalization and structuration: Studying the links between action and institution. *Organization Studies, 18*(1), 93–117. https://doi.org/10.1177/017084069701800106

Barley, S. R. (1986). Technology as an occasion for structuring: Evidence from observations of CT scanners and the social order of radiology departments. *Administrative Science Quarterly, 31*(1), 78–108. https://doi.org/10.2307/2392767

Becker, P., & Bodin, Ö. (2022). Brokerage activity, exclusivity and role diversity: A three-dimensional approach to brokerage in networks. *Social Networks, 70*, 267–283. https://doi.org/10.1016/j.socnet.2022.02.014

Becker, P., & Payo, A. (2013). Changing the paradigm: A requisite for safe and sustainable coastal societies. *Journal of Geography & Natural Disasters, S1*.

Becker, P., & Tehler, H. (2013). Constructing a common holistic description of what is valuable and important to protect: A possible requisite for disaster risk management. *International Journal of Disaster Risk Reduction, 6*, 18–27. https://doi.org/10.1016/j.ijdrr.2013.03.005

Becker, P., Hagelsteen, M., & Abrahamsson, M. (2021). 'Too many mice make no lining for their nest' – reasons and effects of parallel governmental structures for disaster risk reduction and climate change adaptation in Southern Africa. *Jàmbá Journal of Disaster Risk Studies, 13*(1). https://doi.org/10.4102/jamba.v13i1.1041

Becker, P. (2009). Grasping the hydra: The need for a holistic and systematic approach to disaster risk reduction. *Jàmbá: Journal of Disaster Risk Studies, 2*, 12–24.

Becker, P. (2011). Whose risks? -Gender and the ranking of hazards. *Disaster Prevention and Management, 20*, 423–433.

Becker, P. (2012). The importance of integrating multiple administrative levels in capacity assessment for disaster risk reduction and climate change adaptation. *Disaster Prevention and Management, 21*, 226–233. https://doi.org/10.1108/09653561211220016

Becker, P. (2018). Dependence, trust, and influence of external actors on municipal urban flood risk mitigation: The case of Lomma Municipality, Sweden. *International Journal of Disaster Risk Reduction, 31*(1004), 1004–1012. https://doi.org/10.1016/j.ijdrr.2018.09.005

Becker, P. (2020). The problem of fit in flood risk governance: Regulative, normative, and cultural-cognitive deliberations. *Politics and Governance, 8*(4), 281–293. https://doi.org/10.17645/pag.v8i4.3059

Becker, P. (2021a). Advancing resilience for sustainable development: A capacity development approach. In W. Leal Filho, R. Pretorius, & L. O. de Sousa (Eds.), *Sustainable development in Africa: Fostering sustainability in one of the world's most promising continents* (pp. 525–540). Springer. https://doi.org/10.1007/978-3-030-74693-3_29

Becker, P. (2021b). Fragmentation, commodification and responsibilisation in the governing of flood risk mitigation in Sweden. *Environment and Planning C: Politics and Space, 39*(2), 393–413. https://doi.org/10.1177/2399654420940727

Becker, P. (2021c). Tightly coupled policies and loosely coupled networks in the governing of flood risk mitigation in municipal administrations. *Ecology and Society, 26*(2), 34. https://doi.org/10.5751/es-12441-260234

Béland, D. (2009). Ideas, institutions, and policy change. *Journal of European Public Policy, 16*(5), 701–718. https://doi.org/10.1080/13501760902983382

Benford, R. D., & Snow, D. A. (2000). Framing processes and social movements: An overview and assessment. *Annual Review of Sociology, 26*, 611–639. https://doi.org/10.1146/annurev.soc.26.1.611

Berardo, R., & Lubell, M. (2019). The ecology of games as a theory of polycentricity: Recent advances and future challenges. *Policy Studies Journal, 47*(1), 6–26. https://doi.org/10.1111/psj.12313

Berger, P. L., & Luckmann, T. (1966). *The social construction of reality*. Penguin Books.

Bergman, I., Gunnarson, S., & Räisänen, C. (2013). Decoupling and standardization in the projectification of a company. *International Journal of Managing Projects in Business, 6*(1), 106–128. https://doi.org/10.1108/17538371311291053

Bergsten, A., Galafassi, D., & Bodin, Ö. (2014). The problem of spatial fit in social-ecological systems: Detecting mismatches between ecological connectivity and land management in an urban region. *Ecology and Society, 19*(4). https://doi.org/10.5751/ES-06931-190406. art6.

Bergsten, A., Jiren, T. S., Leventon, J., Dorresteijn, I., Schultner, J., & Fischer, J. (2019). Identifying governance gaps among interlinked sustainability challenges. *Environmental Science & Policy, 91*, 27–38. https://doi.org/10.1016/j.envsci.2018.10.007

Bergström, J. (2018). An archaeology of societal resilience. *Safety Science, 110*, 31–38. https://doi.org/10.1016/j.ssci.2017.09.013

Bevir, M. (2011). Governance and governmentality after neoliberalism. *Policy & Politics, 39*(4), 457–471.

Bevir, M. (2012). *Governance: A very short introduction*. Oxford University Press. https://doi.org/10.1093/actrade/9780199606412.001.0001

Billiet, J., Meeusen, C., & Abts, K. (2021). The relationship between (sub)national identity, citizenship conceptions, and perceived ethnic threat in Flanders and Wallonia for the period 1995–2020: A measurement invariance testing strategy. *Frontiers in Political Science, 3*, 676551. https://doi.org/10.3389/fpos.2021.676551

Billig, M. (1995). *Banal nationalism*. Sage.

Blatter, J. (2003). Beyond hierarchies and networks: Institutional logics and change in transboundary spaces. *Governance, 16*(4), 503–526. https://doi.org/10.1111/1468-0491.00226

Bodin, Ö., & Tengö, M. (2012). Disentangling intangible social-ecological systems. *Global Environmental Change, 22*, 430–439. https://doi.org/10.1016/j.gloenvcha.2012.01.005

Bodin, Ö. (2017). Collaborative environmental governance: Achieving collective action in social-ecological systems. *Science, 357*(6352). https://doi.org/10.1126/science.aan1114. eaan1114.

Boltanski, L., & Chiapello, È. (2005). *The new spirit of capitalism*. Verso.

Bond, A. J., & Morrison-Saunders, A. (2011). Re-evaluating sustainability assessment: Aligning the vision and the practice. *Environmental Impact Assessment Review, 31*, 1–7. https://doi.org/10.1016/j.eiar.2010.01.007

Bond, A. J., Morrison-Saunders, A., & Pope, J. (2012). Sustainability assessment: The state of the art. *Impact Assessment and Project Appraisal, 30*, 53–62. https://doi.org/10.1080/14615517.2012.661974

Boström, M., Lidskog, R., & Uggla, Y. (2017). A reflexive look at reflexivity in environmental sociology. *Environmental Sociology, 3*(1), 6–16. https://doi.org/10.1080/23251042.2016.1237336

Bouckaert, G., Peters, B. G., & Verhoest, K. (2010). *The coordination of public sector organizations: Shifting patterns of public management*. Palgrave Macmillan.

Bourdieu, P. (1977). *Outline of a theory of practice*. Cambridge University Press.

Boyd, E., Street, R., Gawith, M., Lonsdale, K., Newton, L., Johnstone, K., & Metcalf, G. (2011). Leading the UK adaptation agenda: A landscape of stakeholders and networked organizations for adaptation to climate change. In J. D. Ford, & L. Berrang-Ford (Eds.), *Climate change adaptation in developed nations* (pp. 85–102). Springer Netherlands.

Boyd, E., Ensor, J., Broto, V. C., & Juhola, S. (2014). Environmentalities of urban climate governance in Maputo, Mozambique. *Global Environmental Change, 26*, 140–151. https://doi.org/10.1016/j.gloenvcha.2014.03.012

Brass, P. R. (2000). Foucault steals political science. *Annual Review of Political Science, 3*, 305–330. https://doi.org/10.1146/annurev.polisci.3.1.305

Brenner, N., Peck, J., & Theodore, N. (2010). Variegated neoliberalization: Geographies, modalities, pathways. *Global Networks, 10*(2), 182–222. https://doi.org/10.1111/j.1471-0374.2009.00277.x

Butler, C., & Pidgeon, N. (2011). From 'flood defence' to 'flood risk management': Exploring governance, responsibility, and blame. *Environment and Planning C, 29*(3), 533–547. https://doi.org/10.1068/c09181j

Byrne, D., & Callaghan, G. (2014). *Complexity theory and the social sciences: The state of the art*. Routledge.

Callinicos, A. (2007). *Social theory: A historical introduction*. Polity Press.

Callon, M., & Latour, B. (1981). Unscrewing the big Leviathan: How actors macro-structure reality and how sociologists help them to do so. In K. D. Knorr-Cetina, & A. V. Cicourel (Eds.), *Advances in social theory and methodology: Toward an integration of micro- and macro-sociologies* (pp. 277–303). Routledge.

Carley, K. (1992). Organizational learning and personnel turnover. *Organization Science, 3*(1), 20–46. https://doi.org/10.1287/orsc.3.1.20

Carlisle, K., & Gruby, R. L. (2019). Polycentric systems of governance: A theoretical model for the commons. *Policy Studies Journal, 47*(4), 927–952. https://doi.org/10.1111/psj.12212

Carter, T. R., Benzie, M., Campiglio, E., Carlsen, H., Fronzek, S., Hildén, M., Reyer, C. P. O., & West, C. (2021). A conceptual framework for cross-border impacts of climate change. *Global Environmental Change, 69*, 102307. https://doi.org/10.1016/j.gloenvcha.2021.102307

Castellani, B., & Hafferty, F. W. (2009). *Sociology and complexity science: A new field of inquiry*. Springer.

CBD. (2020). *Global biodiversity outlook 5*. Secretariat of the Convention on Biological Diversity.

Cedergren, A., & Tehler, H. (2014). Studying risk governance using a design perspective. *Safety Science, 68*, 89–98. https://doi.org/10.1016/j.ssci.2014.03.006

Cejudo, G. M., & Michel, C. L. (2017). Addressing fragmented government action: Coordination, coherence, and integration. *Policy Sciences, 50*(4), 745–767. https://doi.org/10.1007/s11077-017-9281-5

Cerne, A., & Jansson, J. (2019). Projectification of sustainable development: Implications from a critical review. *International Journal of Managing Projects in Business, 12*(2), 356–376. https://doi.org/10.1108/IJMPB-04-2018-0079

Chappell, M. J., & LaValle, L. A. (2011). Food security and biodiversity: Can we have both? An agroecological analysis. *Agriculture and Human Values, 28*(1), 3–26. https://doi.org/10.1007/s10460-009-9251-4

Cilliers, P. (2005). Complexity, deconstruction and relativism. *Theory, Culture & Society, 22*(5), 255–267. https://doi.org/10.1177/0263276405058052

Cooper, D. J., Ezzamel, M., & Willmott, H. (2008). Examining "institutionalization": A critical theoretic perspective. In R. Greenwood, C. Oliver, R. Suddaby, & K. Sahlin (Eds.), *The SAGE handbook of organizational institutionalism* (pp. 673–701). Sage London.

Courchamp, F., Chapuis, J.-L., & Pascal, M. (2003). Mammal invaders on islands: Impact, control and control impact. *Biological Reviews, 78*, 347–383. https://doi.org/10.1017/S1464793102006061

Dean, M. (1996). Foucault, government and the enfolding of authority. In A. Barry, T. Osborne, & N. Rose (Eds.), *Foucault and political reason* (pp. 209–229). UCL Press.

Dean, M. (1997). Sociology after society. In D. Owen (Ed.), *Sociology after postmodernism* (pp. 205–228). SAGE Publications.

Dean, M. (2010). *Governmentality: Power and rule in modern society* (2 ed.). Sage.

Dean, M. (2017). Governmentality. In B. S. Turner, C. Kyung-Sup, C. F. Epstein, P. Kivisto, W. Outhwaite, & M. Ryan (Eds.), *The wiley-blackwell encyclopedia of social theory* (pp. 1–2). Wiley-Blackwell. https://onlinelibrary.wiley.com/doi/full/10.1002/9781118430873.est0657.

DeFries, R., & Nagendra, H. (2017). Ecosystem management as a wicked problem. *Science, 356*(6335), 265–270. https://doi.org/10.1126/science.aal1950

Desmond, M. (2014). Relational ethnography. *Theory and Society, 43*, 547–579.

DiMaggio, P. J., & Powell, W. W. (1983). The iron cage revisited: Institutional isomorphism and collective rationality in organizational fields. *American Sociological Review, 48*(2), 147–160.

DiMaggio, P. J., & Powell, W. W. (1991). Introduction. In W. W. Powell, & P. J. DiMaggio (Eds.), *The new institutionalism in organizational analysis* (pp. 1–38). University of Chicago Press.

DiMaggio, P. J. (1988). Interest and agency in institutional theory. In L. G. Zucker (Ed.), *Institutional patterns and organizations: Culture and environment* (pp. 3–21). Ballinger Publishing.

DiMaggio, P. J. (1991). Constructing an organizational field as a prodessional project: U.S. Art museums, 1920-1940. In W. W. Powell, & P. J. DiMaggio (Eds.), *The new institutionalism in organizational analysis* (pp. 267–292). University of Chicago Press.

Donzelot, J. (1979). *The policing of families*. Pantheon Books.

Drori, G. S., Meyer, J. W., & Hwang, H. (Eds.). (2006). *Globalization and organization: World society and organizational change*. Oxford University Press.

Duit, A., Galaz, V., Eckerberg, K., & Ebbesson, J. (2010). Governance, complexity, and resilience. *Global Environmental Change, 20*, 363–368. https://doi.org/10.1016/j.gloenvcha.2010.04.006

Durkheim, E. (1984). *The division of labour*. The Macmillan Press.

Elias, N. (1978). *What is sociology*. Columbia University Press.

Emirbayer, M. (1997). Manifesto for a relational sociology. *American Journal of Sociology, 103*, 281–317.

Epstein, G., Pittman, J., Alexander, S. M., Berdej, S., Dyck, T., Kreitmair, U., Rathwell, K. J., Villamayor-Tomas, S., Vogt, J., & Armitage, D. (2015). Institutional fit and the sustainability of social–ecological systems. *Current Opinion in Environmental Sustainability, 14*, 34–40. https://doi.org/10.1016/j.cosust.2015.03.005

Ewald, F. (1991). Insurance and risk. In G. Burchell, C. Gordon, & P. Miller (Eds.), *The Foucault effect: Studies in governmentality, with two lectures by and an interview with Michel Foucault* (pp. 197–210). University of Chicago Press.

Flynn, J., Slovic, P., & Mertz, C. K. (1994). Gender, race, and perception of environmental health risks. *Risk Analysis, 14*, 1101–1108.

Flyvbjerg, B. (1998). *Rationality and power: Democracy in practice*. University of Chicago Press.

Folke, C., Hahn, T., Olsson, P., & Norberg, J. (2005). Adaptive governance of social-ecological systems. *Annual Review of Environment and Resources, 30*, 441–473. https://doi.org/10.1146/annurev.energy.30.050504.144511

Folke, C., Lowell Pritchard, J., Berkes, F., Colding, J., & Svedin, U. (2007). The problem of fit between ecosystems and institutions: Ten years later. *Ecology and Society, 12*(1). https://doi.org/10.5751/ES-02064-120130. art30.

Foucault, M., Fornet-Betancourt, R., Becker, H., & Gomez-Müller, A. (1987). The ethic of care for the self as a practice of freedom – an interview with Michel Foucault on January 20, 1984. *Philosophy & Social Criticism, 12*(2–3), 112–131. https://doi.org/10.1177/019145378701200202

Foucault, M. (1982). The subject and power. *Critical Inquiry, 8*(4), 777–795. https://www.jstor.org/stable/1343197.

Foucault, M. (1991a). Governmentality. In G. Burchell, C. Gordon, & P. Miller (Eds.), *The Foucault effect: Studies in governmentality, with two lectures by and an interview with Michel Foucault* (pp. 87–104). University of Chicago Press.

Foucault, M. (1991b). Question of method. In G. Burchell, C. Gordon, & P. Miller (Eds.), *The Foucault effect: Studies in governmentality, with two lectures by and an interview with Michel Foucault* (pp. 73–86). Harvester Wheatsheaf.

Foucault, M. (1995). *Discipline and punish: The birth of the prison*. Vintage Books.

Foucault, M. (2007). *Security, territory, population: Lectures at the College de France, 1977-78*. Palgrave Macmillan.

Fred, M. (2015). Projectification in Swedish municipalities: A case of porous organizations. *Scandinavian Journal of Public Administration, 19*(2), 49–68.

Fred, M. (2020). Local government projectification in practice – a multiple institutional logic perspective. *Local Government Studies, 46*(3), 351–370. https://doi.org/10.1080/03003930.2019.1606799

Friedland, R., & Alford, R. R. (1991). Bringing society back in: Symbols, practices, and institutional contradictions. In W. W. Powell, & P. J. DiMaggio (Eds.), *The new institutionalism in organizational analysis* (pp. 232–263). University of Chicago Press.

Giddens, A. (1984). *The constitution of society: Outline of the theory of structuration*. University of California Press.

Gidley, J., Hampson, G., Wheeler, L., & Bereded-Samuel, E. (2010). Social inclusion: Context, theory and practice. *The Australasian Journal of University-Community Engagement, 5*(1), 6–36.

GNCSODR. (2009). *Clouds but little rain*. Global Network of Civil Society Organisations for Disaster Reduction.

Godenhjelm, S., Lundin, R. A., & Sjöblom, S. (2015). Projectification in the public sector – the case of the European Union. *International Journal of Managing Projects in Business, 8*(2), 324–348. https://doi.org/10.1108/IJMPB-05-2014-0049

Guerrin, J., Bouleau, G., & Grelot, F. (2014). "Functional fit" versus "politics of scale" in the governance of floodplain retention capacity. *Journal of Hydrology, 519*, 2405–2414. https://doi.org/10.1016/j.jhydrol.2014.08.024

Hagelsteen, M., & Becker, P. (2014a). A great babylonian confusion: Terminological ambiguity in capacity development for disaster risk reduction in the international community. In *Proceedings of the 5th international disaster risk conference* (pp. 298–300). Switzerland: Davos, 24-28/08/2014.

Hagelsteen, M., & Becker, P. (2014b). Forwarding a challenging task: Seven elements for capacity development for disaster risk reduction. *The Planetary Report, 2*, 94–97.

Haghighatafshar, S., Becker, P., Moddemeyer, S., Persson, A., Sörensen, J., Aspegren, H., & Jönsson, K. (2020). *Paradigm shift in engineering of pluvial floods: From historical recurrence intervals to risk-based design for an uncertain future* (p. 102317). Sustainable Cities and Society. https://doi.org/10.1016/j.scs.2020.102317

Handmer, J. W., & Dovers, S. (2007). *The handbook of disaster and emergency policies and institutions*. Earthscan.

Hanna, P., Kantenbacher, J., Cohen, S., & Gössling, S. (2018). Role model advocacy for sustainable transport. *Transportation Research Part D: Transport and Environment, 61*, 373–382. https://doi.org/10.1016/j.trd.2017.07.028

Hannah-Moffat, K. (2000). Prisons that empower. *British Journal of Criminology, 40*(3), 510–531. https://doi.org/10.1093/bjc/40.3.510

Harjanne, A., Pagneux, E., Flindt Jørgensen, L., Perrels, A., van der Keur, P., Nadim, F., Rød, J. K., & Raats, E. (2016). *Resilience to natural hazards: An overview of institutional arrangements and practices in the Nordic countries*. NORDRESS.

Hedlund, J., Bodin, Ö., & Nohrstedt, D. (2021). Policy issue interdependency and the formation of collaborative networks. *People and Nature, 3*(1), 236–250. https://doi.org/10.1002/pan3.10170

Hegger, D. L. T., Driessen, P. P. J., Wiering, M., van Rijswick, H. F. M. W., Kundzewicz, Z. W., Matczak, P., Crabbé, A., Raadgever, G. T., Bakker, M. H. N., Priest, S. J., Larrue, C., & Ek, K. (2016). Toward more flood resilience: Is a diversification of flood risk management strategies the way forward. *Ecology and Society, 21*(4), 1. https://doi.org/10.5751/ES-08854-210452. Art-52.

Heylighen, F., Cilliers, P., & Gershenson, C. (2007). Complexity and philosophy. In J. Bogg, & R. Geyer (Eds.), *Complexity, science and society* (pp. 117–134). Radcliffe Publishing.

Hillman, A. J., Withers, M. C., & Collins, B. J. (2009). Resource dependence theory: A review. *Journal of Management, 35*(6), 1404–1427. https://doi.org/10.1177/0149206309343469

Hobsbawm, E. J. (1996). Ethnicity and nationalism in Europe today. In G. Balakrishan (Ed.), *Mapping the nation* (pp. 255–266). Verso.

Hollnagel, E. (2014). *Safety-I and safety-II: The past and future of safety management.* Ashgate.

Hood, C., & Dixon, R. (2015). What we have to show for 30 Years of new public management: Higher costs, more complaints: Commentary. *Governance, 28*(3), 265–267. https://doi.org/10.1111/gove.12150

Hood, C., Rothstein, H., & Baldwin, R. (2001). *The government of risk: Understanding risk regulation regimes.* Oxford University Press.

Hotimsky, S., Cobb, R., & Bond, A. (2006). Contracts or scripts? A critical review of the application of institutional theories to the study of environmental change. *Ecology and Society, 11*(1), 41. https://www.jstor.org/stable/26267814.

Hulme, M. (2011). Reducing the future to climate: A story of climate determinism and reductionism. *Osiris, 26*, 245–266. https://doi.org/10.1086/661274

Hutter, G., Leibenath, M., & Mattissek, A. (2014). Governing through resilience? Exploring flood protection in Dresden, Germany. *Social Sciences, 3*(2), 272–287. https://doi.org/10.3390/socsci3020272

Imperial, M. T. (1999). Institutional analysis and ecosystem-based management: The institutional analysis and development framework. *Environmental Management, 24*(4), 449–465. https://doi.org/10.1007/s002679900246

Ingold, K., Balsinger, J., & Hirschi, C. (2010). Climate change in mountain regions: How local communities adapt to extreme events. *Local Environment, 15*, 651–661. https://doi.org/10.1080/13549839.2010.498811

IPCC. (2022). *Climate change 2022: Impacts, adaptation and vulnerability.* Cambridge University Press.

Jennings, E. T. (1994). Building bridges in the intergovernmental arena: Coordinating employment and training programs in the American states. *Public Administration Review, 54*(1), 52. https://doi.org/10.2307/976498

Jensen, C., & Trägårdh, L. (2012). *Tämporära organisationer för permanenta problem. Om implementering av samverkansprojekt för unga som står långt från arbetsmarknaden (1).* Ungdomsstyrelsen.

Jensen, A., Thuesen, C., & Geraldi, J. (2016). The projectification of everything: Projects as a human condition. *Project Management Journal, 47*(3), 21–34. https://doi.org/10.1177/875697281604700303

Jensen, A., Ørsted Nielsen, H., & Lilleøre Nielsen, M. (2016). *Climate adaption in local governance: Institutional barriers in Danish municipalities.* Danish Centre for Environment and Energy.

Jepperson, R. L. (1991). Institutions, institutional effects, and institutionalism. In W. W. Powell, & P. J. DiMaggio (Eds.), *The new institutionalism in organizational analysis* (pp. 143–163). University of Chicago Press.

Johannessen, Å., & Granit, J. J. (2015). Integrating flood risk, river basin management and adaptive management: Gaps, barriers and opportunities, illustrated by a case study from Kristianstad, Sweden. *International Journal of Water Governance, 3*, 5–24. https://doi.org/10.7564/13-IJWG30

Johannessen, Å., Gerger Swartling, Å., Wamsler, C., Andersson, K., Arran, J. T., Hernández Vivas, D. I., & Stenström, T. A. (2019). Transforming urban water governance through social (triple-loop) learning. *Environmental Policy and Governance, 18*(11), 11–86. https://doi.org/10.1002/eet.1843

Johansson, R. (2009). Vid den institutionella analysens gränser: Institutionell organisationsteori i Sverige. *Nordiske organisasjonsstudier, 11*(3), 5–22.

Johnson, V. (2007). What is organizational imprinting? Cultural entrepreneurship in the founding of the Paris Opera. *American Journal of Sociology, 113*(1), 97–127. https://doi.org/10.1086/517899

Joseph, J., & Juncos, A. E. (2019). Resilience as an emergent European project? The EU's place in the resilience turn. *Journal of Communication and Media Studies, 57*(5), 995–1011. https://doi.org/10.1111/jcms.12881

Joseph, J. (2010). The limits of governmentality: Social theory and the international. *European Journal of International Relations, 16*(2), 223–246. https://doi.org/10.1177/1354066109346886

Joseph, J. (2013). Resilience as embedded neoliberalism: A governmentality approach. *Resilience, 1*, 38–52.

Jupe, R., & Funnell, W. (2015). Neoliberalism, consultants and the privatisation of public policy formulation: The case of Britain's rail industry. *Critical Perspectives on Accounting, 29*, 65–85. https://doi.org/10.1016/j.cpa.2015.02.001

Kelly, P. (2010). Youth at Risk: Processes of individualisation and responsibilisation in the risk society. *Discourse: Studies in the Cultural Politics of Education, 22*(1), 23–33. https://doi.org/10.1080/01596300120039731

Khaldun, I. (1969). *The muqaddimah: An introduction to history*. Princeton University Press.

Kim, Y., & Han, M. (2008). Rainwater storage tank as a remedy for a local urban flood control. *Water Science & Technology: Water Supply, 8*(1), 31–36. https://doi.org/10.2166/ws.2008.029

King, B. G., & Pearce, N. A. (2010). The contentiousness of markets: Politics, social movements, and institutional change in markets. *Annual Review of Sociology, 36*(1), 249–267. https://doi.org/10.1146/annurev.soc.012809.102606

Larner, W. (2000). Neo-liberalism: Policy, ideology, governmentality. *Studies in Political Economy, 63*(1), 5–25. https://doi.org/10.1080/19187033.2000.11675231

Latour, B. (2005). *Reassembling the social: An introduction to actor-network theory*. Oxford University Press.

Lebel, L., Nikitina, E., Pahl-Wostl, C., & Knieper, C. (2013). Institutional fit and river basin governance. *Ecology and Society, 18*(1), 1.

Leitão, J. P., Almeida, M. D. C., Simões, N. E., & Martins, A. (2013). Methodology for qualitative urban flooding risk assessment. *Water Science and Technology, 68*(4), 829–838. https://doi.org/10.2166/wst.2013.310

Lemke, T. (2001). 'The birth of bio-politics': Michel Foucault's lecture at the Collège de France on neo-liberal governmentality. *Economy and Society, 30*(2), 190–207. https://doi.org/10.1080/03085140120042271

Lemke, T. (2002). Foucault, governmentality, and critique. *Rethinking Marxism, 14*(3), 49–64. https://doi.org/10.1080/089356902101242288

Li, M., Wiedmann, T., Fang, K., & Hadjikakou, M. (2021). The role of planetary boundaries in assessing absolute environmental sustainability across scales. *Environment International, 152*, 106475. https://doi.org/10.1016/j.envint.2021.106475

Lim, W. K. (2011). Understanding risk governance: Introducing sociological neoinstitutionalism and foucauldian governmentality for further theorizing. *International Journal of Disaster Risk Science, 2*(3), 11–20. https://doi.org/10.1007/s13753-011-0012-9

Liverman, D. M. (2009). Conventions of climate change: Constructions of danger and the dispossession of the atmosphere. *Journal of Historical Geography, 35*, 279–296. https://doi.org/10.1016/j.jhg.2008.08.008

Lövbrand, E., Stripple, J., & Wiman, B. (2009). Earth system governmentality. *Global Environmental Change, 19*(1), 7–13. https://doi.org/10.1016/j.gloenvcha.2008.10.002

Lubell, M., & Morrison, T. H. (2021). Institutional navigation for polycentric sustainability governance. *Nature Sustainability, 4*(8), 664–671. https://doi.org/10.1038/s41893-021-00707-5

Luhmann, N. (1979). *Trust and power: Two works by Niklas Luhmann*. John Wiley & Sons.

Luhmann, N. (1993). *Risk: A sociological theory*. Walter de Gruyter.

Luhmann, N. (1995). *Social systems*. Stanford University Press.

Lupton, D. (2014). *Digital sociology*. Routledge.

Machiavelli, N. (2014). *The prince (T. Parks, trans.)*. Penguin Books.

Mancilla García, M., Hileman, J., Bodin, Ö., Nilsson, A., & Jacobi, P. R. (2019). The unique role of municipalities in integrated watershed governance arrangements: A new research frontier. *Ecology and Society, 24*(1), art28. https://doi.org/10.5751/ES-10793-240128

Marks, D. (2019). Assembling the 2011 Thailand floods: Protecting farmers and inundating high-value industrial estates in a fragmented hydro-social territory. *Political Geography, 68*, 66–76. https://doi.org/10.1016/j.polgeo.2018.10.002

Martin, B. (1993). *In the public interest? Privatization and public sector reform. London, U.K.*

Maylor, H., Brady, T., Cooke-Davies, T., & Hodgson, D. (2006). From projectification to programmification. *International Journal of Project Management, 24*(8), 663–674. https://doi.org/10.1016/j.ijproman.2006.09.014

Maylor, H. (2006). Special Issue on rethinking project management (EPSRC network 2004–2006). *International Journal of Project Management, 24*(8), 635–637. https://doi.org/10.1016/j.ijproman.2006.09.013

McAfee, K. (2003). Neoliberalism on the molecular scale. Economic and genetic reductionism in biotechnology battles. *Geoforum, 34*(2), 203–219. https://doi.org/10.1016/S0016-7185(02)00089-1

McKenzie, M., Bieler, A., & McNeil, R. (2015). Education policy mobility: Reimagining sustainability in neoliberal times. *Environmental Education Research, 21*(3), 319–337. https://doi.org/10.1080/13504622.2014.993934

Merton, R. K. (1936). The unanticipated consequences of purposive social action. *American Sociological Review, 1*(6), 894–904. https://doi.org/10.2307/2084615

Metz, F., Angst, M., & Fischer, M. (2020). Policy integration: Do laws or actors integrate issues relevant to flood risk management in Switzerland. *Global Environmental Change, 61*, 101945. https://doi.org/10.1016/j.gloenvcha.2019.101945

Meyer, J. W., & Rowan, B. (1977). Institutionalized organizations: Formal structure as myth and ceremony. *American Journal of Sociology, 83*(2), 340–363.

Meyer, J. W., Boli, J., & Thomas, G. M. (1987). Ontology and rationalization in western cultural account. In G. M. Thomas, J. W. Meyer, F. O. Ramirez, & J. Boli (Eds.), *Institutional structure: Constituting state, society, and the individual* (pp. 12–37). Sage Publications.

Meyer, J. W. (2010). World society, institutional theories, and the actor. *Annual Review of Sociology, 36*(1), 1–20. https://doi.org/10.1146/annurev.soc.012809.102506

Midler, C. (1995). Projectification" of the firm: The Renault case. *Scandinavian Journal of Management, 11*(4), 363–375. https://doi.org/10.1016/0956-5221(95)00035-T

Migdal-Picker, M., & Zilber, T. B. (2019). The claim for actorhood in institutional work. In H. Hwang, J. A. Colyvas, & G. S. Drori (Eds.), *Research in the sociology of organizations* (pp. 251–272). Emerald Publishing.

Miller, P., & Rose, N. (2008). *Governing the present: Administering economic, social and personal life. Polity.*

Miller, T. (2003). Governmentality or commodification? US higher education. *Cultural Studies, 17*(6), 897–904. https://doi.org/10.1080/0950238032000150084

Milly, P. C. D., Betancourt, J., Falkenmark, M., Hirsch, R. M., Kundzewicz, Z. W., Lettenmaier, D. P., & Stouffer, R. J. (2008). Stationarity is dead: Whither water management. *Science, 319*(5863), 573–574. https://doi.org/10.1126/science.1151915

Milward, H. B., & Provan, K. G. (2000). Governing the Hollow state. *Journal of Public Administration Research and Theory, 10*(2), 359–380. https://doi.org/10.1093/oxfordjournals.jpart.a024273

Mobini, S., Becker, P., Larsson, R., & Berndtsson, R. (2020). Systemic inequity in urban flood exposure and damage compensation. *Water, 12*(11), 3142. https://doi.org/10.3390/w12113152

Moisio, S., & Luukkonen, J. (2015). European spatial planning as governmentality: An inquiry into rationalities, techniques, and manifestations. *Environment and Planning C: Government and Policy, 33*(4), 828–845. https://doi.org/10.1068/c13158

Mol, A. P. J. (2010a). Ecological modernization as a social theory of environmental reform. In M. R. Redclift, & G. Woodgate (Eds.), *The international handbook of environmental sociology* (pp. 63–76). Edward Elgar Publishing.

Mol, A. P. J. (2010b). Social theories of environmental reform: Towards a third generation. In M. Gross, & H. Heinrichs (Eds.), *Environmental sociology* (pp. 19–38). Springer Netherlands. http://link.springer.com/10.1007/978-90-481-8730-0.

Murphy, J. R. (2019). Homelessness and public space offences in Australia – a human rights case for narrow interpretation. *Griffith Journal of Law & Human Dignity, 7*(1), 103–127.

Nikolakaki, M. (2012). Critical pedagogy in the new dark ages: Challenges and possibilities: An introduction. *Counterpoint, 422*, 3–31.

Nilsson, J., & Becker, P. (2009). What's important? Making what is valuable and worth protecting explicit when performing risk and vulnerability analyses. *International Journal of Risk Assessment and Management, 13*, 345–363.

Nilsson, M., Zamparutti, T., Petersen, J. E., Nykvist, B., Rudberg, P., & McGuinn, J. (2012). Understanding policy coherence: Analytical framework and examples of sector–environment policy interactions in the EU. *Environmental Policy and Governance, 22*(6), 395–423.

Nilsson, M., Griggs, D., & Visbeck, M. (2016). Policy: Map the interactions between sustainable development goals. *Nature, 534*(7607), 320–322. https://doi.org/10.1038/534320a
North, D. C. (1990). *Institutions, institutional change and economic performance*. Cambridge University Press.
Nowotny, H. (2005a). The increase of complexity and its reduction: Emergent interfaces between the natural sciences, humanities and social sciences. *Theory, Culture & Society, 22*(5), 15–31. https://doi.org/10.1177/0263276405057189
Nowotny, H. (2005b). The rise of complexity theory in the social sciences. *Theory, Culture & Society, 22*, 15–31. https://doi.org/10.1177/0263276405057189
O'Malley, P. (1996). Risk and responsibility. In A. Barry, T. Osborne, & N. Rose (Eds.), *Foucault and political reason* (pp. 189–207). UCL Press.
O'Malley, P. (2008). Governmentality and risk. In J. O. Zinn (Ed.), *Social theories of risk and uncertainty: An introduction* (pp. 52–75). Blackwell Publishing.
Öjehag-Pettersson, A., & Granberg, M. (2019). Public procurement as marketisation: Impacts on civil servants and public administration in Sweden. *Scandinavian Journal of Public Administration, 23*(3–4), 43–59.
Oliver, C. (1991). Strategic responses to institutional processes. *Academy of Management Review, 16*, 145–179. https://doi.org/10.5465/AMR.1991.4279002
Orderud, G. I., & Naustdalslid, J. (2020). Climate change adaptation in Norway: Learning–knowledge processes and the demand for transformative adaptation. *The International Journal of Sustainable Development and World Ecology, 27*(1), 15–27. https://doi.org/10.1080/13504509.2019.1673500
Orlikowski, W. J. (1992). The duality of technology: Rethinking the concept of technology in organizations. *Organization Science, 3*, 398–427. https://doi.org/10.1287/orsc.3.3.398
Ostrom, V., Tiebout, C. M., & Warren, R. (1961). The organization of government in metropolitan areas: A theoretical inquiry. *American Political Science Review, 55*(4), 831–842. https://doi.org/10.2307/1952530
Ostrom, E. (1990). *Governing the commons: The evolution of institutions for collective action*. Cambridge University Press.
Ostrom, E. (2005). *Understanding institutional diversity*. Princeton University Press.
Ostrom, E. (2010). Polycentric systems for coping with collective action and global environmental change. *Global Environmental Change, 20*(4), 550–557. https://doi.org/10.1016/j.gloenvcha.2010.07.004
Oxford English Dictionary. (2020). *Oxford dictionary of English*. Oxford University Press.
Paasi, A. (2012). Regional identities. In M. Juergensmeyer, & H. K. Anheimer (Eds.), *Encyclopedia of global studies*. Sage. https://doi.org/10.4135/9781452218557
Pahl-Wostl, C., & Knieper, C. (2014). The capacity of water governance to deal with the climate change adaptation challenge: Using fuzzy set Qualitative Comparative Analysis to distinguish between polycentric, fragmented and centralized regimes. *Global Environmental Change, 29*, 139–154. https://doi.org/10.1016/j.gloenvcha.2014.09.003
Pahl-Wostl, C., Lebel, L., Knieper, C., & Nikitina, E. (2012). From applying panaceas to mastering complexity: Toward adaptive water governance in river basins. *Environmental Science & Policy, 23*, 24–34. https://doi.org/10.1016/j.envsci.2012.07.014
Payo, A., Becker, P., Otto, A., Vervoort, J. M., & Kingsborough, A. (2015). Experiential lock-in: Characterizing avoidable maladaptation in infrastructure systems. *Journal of Infrastructure Systems, 22*, 2515001. https://doi.org/10.1061/(asce)is.1943-555x.0000268
Pedersen, S. B. (2004). Place branding: Giving the region of Øresund a competitive edge. *Journal of Urban Technology, 11*(1), 77–95.
Pellizzoni, L. (2016). Catching up with things? Environmental sociology and the material turn in social theory. *Environmental Sociology, 2*, 312–321. https://doi.org/10.1080/23251042.2016.1190490
Polanyi, M. (1951). *The logic of liberty: Reflections and rejoinders*. Routledge and Kegan Paul.
Power, M. (2011). Foucault and sociology. *Annual Review of Sociology, 37*(1), 35–56. https://doi.org/10.1146/annurev-soc-081309-150133
Rådestad, C., & Larsson, O. (2018). Responsibilization in contemporary Swedish crisis management: Expanding 'bare life' biopolitics through exceptionalism and neoliberal governmentality. *Critical Policy Studies, 0*, 1–20. https://doi.org/10.1080/19460171.2018.1530604
Radin, M. J. (1996). *Contested commodities*. Harvard University Press.

Raju, E., & Becker, P. (2013). Multi-organisational coordination for disaster recovery: The story of post-tsunami Tamil Nadu, India. *International Journal of Disaster Risk Reduction*, 82–91. https://doi.org/10.1016/j.ijdrr.2013.02.004

Renn, O. (2008). *Risk governance: Coping with uncertainty in a complex world.* Earthscan.

Rhodes, R. A. W. (2007). Understanding governance: Ten years on. *Organization Studies, 28*(8), 1243–1264. https://doi.org/10.1177/0170840607076586

Rivera, C., Tehler, H., & Wamsler, C. (2015). Fragmentation in disaster risk management systems: A barrier for integrated planning. *International Journal of Disaster Risk Reduction, 14*, 445–456. https://doi.org/10.1016/j.ijdrr.2015.09.009

Rose, N., & Miller, P. (1992). Political-power beyond the state: Problematics of government. *The British Journal of Sociology, 43*(2), 173–205. https://doi.org/10.2307/591464

Rose, N., O'Malley, P., & Valverde, M. (2006). Governmentality. *Annual Review of Law and Social Science, 2*, 83–104. https://doi.org/10.1146/annurev.lawsocsci.2.081805.105900

Rose, N. (1996). The death of the social? Re-figuring the territory of government. *Economy and Society, 25*, 327–356.

Rose, N. (1999). *Powers of freedom: Reframing political thought.* Cambridge University Press.

Rosenau, J. N. (1992). Governance, order, and change in world politics. In J. N. Rosenau, & E.-O. Czempiel (Eds.), *Governance without government: Order and change in world politics* (pp. 1–29). Cambridge University Press. https://www.cambridge.org/core/product/identifier/CBO9780511521775A008/type/book_part.

Rutherford, S. (2016). Green governmentality: Insights and opportunities in the study of nature's rule. *Progress in Human Geography, 31*(3), 291–307. https://doi.org/10.1177/0309132507077080

Saito, F. (2011). Decentralization. In *The SAGE handbook of governance* (pp. 484–500). SAGE Publications.

Sawyer, S. W. (2015). Foucault and the state. *The Tocqueville Review, 36*(1), 135–164. https://doi.org/10.1353/toc.2015.0000

Sayer, A. (1992). *Method in social science: A realist approach.* Routledge.

Sayles, J. S., & Baggio, J. A. (2017). Social–ecological network analysis of scale mismatches in estuary watershed restoration. *Proceedings of the National Academy of Sciences, 114*(10), E1776–E1785. https://doi.org/10.1073/pnas.1604405114

Scanlan, M. K. (2019). Droughts, floods, and scarcity on a climate-disrupted planet: Understanding the legal challenges and opportunities for groundwater sustainability. *Virginia Environmental Law Journal, 37*, 52–88.

Schneiberg, M. (2007). What's on the path? Path dependence, organizational diversity and the problem of institutional change in the US economy, 1900–1950. *Socio-Economic Review, 5*(1), 47–80. https://doi.org/10.1093/ser/mwl006

Schneider, V. (2012). Governance and complexity. In D. Levi-Faur (Ed.), *The Oxford handbook of governance.* Oxford University Press.

Schweber, L. (2014). The cultural role of science in policy implementation: Voluntary self-regulation in the UK building sector. In S. Frickel, & D. J. Hess (Eds.), *27Fields of knowledge: Science, politics and publics in the neoliberal age* (pp. 157–191). Emerald http://www.emeraldinsight.com/doi/10.1108/S0198-871920140000027014.

Scott, W. R. (2014). *Institutions and organizations: Ideas, interests, and identities.* SAGE Publications.

Selznick, P. (1949). *TVA and the grass roots: A study in the sociology of formal organization.* University of California Press.

Selznick, P. (1992). *The moral commonwealth: Social theory and the promise of community.* University of California Press.

Selznick, P. (2002). *The communitarian persuasion.* Johns Hopkins University Press.

Shamir, R. (2008). The age of responsibilization: On market-embedded morality. *Economy and Society, 37*(1), 1–19. https://doi.org/10.1080/03085140701760833

Shen, Y., Morsy, M. M., Huxley, C., Tahvildari, N., & Goodall, J. L. (2019). Flood risk assessment and increased resilience for coastal urban watersheds under the combined impact of storm tide and heavy rainfall. *Journal of Hydrology, 579*, 124159. https://doi.org/10.1016/j.jhydrol.2019.124159

Smith, A. (1869). *Essays by Adam smith.* Alex Murray & Son.

Soneryd, L., & Uggla, Y. (2015). Green governmentality and responsibilization: New forms of governance and responses to 'consumer responsibility'. *Environmental Politics, 24*(6), 913–931. https://doi.org/10.1080/09644016.2015.1055885

Spence, A., & Pidgeon, N. (2010). Framing and communicating climate change: The effects of distance and outcome frame manipulations. *Global Environmental Change, 20*(4), 656–667. https://doi.org/10.1016/j.gloenvcha.2010.07.002

Stinchcombe, A. L. (2000). Social structure and organizations. In *Advances in strategic management* (pp. 229–259). Emerald. https://www.emeraldinsight.com/10.1016/S0742-3322(00)17019-6.

Strang, D., & Meyer, J. W. (1993). Institutional conditions for diffusion. *Theory and Society, 22*(4), 487–511.

Suchman, M. C. (2003). The contract as social artifact. *Law & Society Review, 37*(1), 91–142.

Taylor, F. W. (1919). *The principles of scientific management*. Harper & Brothers Publishers.

Thaler, T. (2019). Commentary: Voluntary agreement in multi-use climate adaptation in the Oekense Beek from a politic-economic perspective. In T. Hartmann, L. Slavíková, & S. McCarthy (Eds.), *Nature-based flood risk management on private land* (pp. 213–218). Springer.

Thompson, J. D. (1967). *Organizations in action: Social science bases of administrative theory*. McGraw-Hill.

Tolbert, P. S., & Zucker, L. G. (1996). The institutionalization of institutional theory. In S. R. Clegg, C. Hardy, & W. R. Nord (Eds.), *Handbook of organization studies* (pp. 175–190). SAGE Publications.

Tscharntke, T., Clough, Y., Wanger, T. C., Jackson, L., Motzke, I., Perfecto, I., Vandermeer, J., & Whitbread, A. (2012). Global food security, biodiversity conservation and the future of agricultural intensification. *Biological Conservation, 151*(1), 53–59. https://doi.org/10.1016/j.biocon.2012.01.068

Urry, J. (2005). The complexity turn. *Theory, Culture & Society, 22*, 1–14.

van Asselt, M. B. A., & Renn, O. (2011). Risk governance. *Journal of Risk Research, 14*(4), 431–449. https://doi.org/10.1080/13669877.2011.553730

van Bommel, S., Röling, N., Aarts, N., & Turnhout, E. (2009). Social learning for solving complex problems: A promising solution or wishful thinking? A case study of multi-actor negotiation for the integrated management and sustainable use of the Drentsche Aa area in The Netherlands. *Environmental Policy and Governance, 19*(6), 400–412. https://doi.org/10.1002/eet.526

van de Ven, A. H., & Garud, R. (1994). The coevolution of technical and institutional events in the development of an innovation. In J. A. C. Baum, & J. V. Singh (Eds.), *Evolutionary dynamics of organizations* (pp. 425–443). Oxford University Press.

van Niekerk, D. (2015). Disaster risk governance in Africa. *Disaster Prevention and Management, 24*(3), 397–416. https://doi.org/10.1108/DPM-08-2014-0168

Walby, S. (2007). Complexity theory, systems theory, and multiple intersecting social inequalities. *Philosophy of the Social Sciences, 37*, 449–470. https://doi.org/10.1177/0048393107307663

Wang, K. C. (2020). The price of salt: The capable self in the face of heteronormative marriage pressure in the discourses of the "post-90s" Chinese lesbians. *Chinese Journal of Communication, 13*(2), 205–220. https://doi.org/10.1080/17544750.2019.1624269

Weber, M. (1978). *Economy and society: An outline of interpretative sociology* (Vol. 1). University of California Press.

Weick, K. E. (1976). Educational organizations as loosely coupled systems. *Administrative Science Quarterly, 21*(1), 1–19. https://doi.org/10.2307/2391875

Widmer, A., Herzog, L., Moser, A., & Ingold, K. (2019). Multilevel water quality management in the international rhine catchment area: How to establish social-ecological fit through collaborative governance. *Ecology and Society, 24*(3). https://doi.org/10.5751/ES-11087-240327

Wilby, R. L., Beven, K. J., & Reynard, N. S. (2008). Climate change and fluvial flood risk in the UK: More of the same. *Hydrological Processes, 22*(14), 2511–2523. https://doi.org/10.1002/hyp.6847

Wimmer, A., & Glick Schiller, N. (2002). Methodological nationalism and beyond: Nation-state building, migration and the social sciences. *Global Networks, 2*(4), 301–334. https://doi.org/10.1111/1471-0374.00043

Wynne, B. (1982). *Rationality and ritual: The windscale inquiry and nuclear decisions in britain*. British Society for the History of Science.

Young, O. R., & Stokke, O. S. (2020). Why is it hard to solve environmental problems? The perils of institutional reductionism and institutional overload. *International Environmental Agreements: Politics, Law and Economics, 20*(1), 5–19. https://doi.org/10.1007/s10784-020-09468-6

Young, O. R., & Underdal, A. (1997). *Institutional dimensions of global change (IHDP Scoping Report)*. International Human Dimensions Programme.

Zucker, L. G. (1977). The role of institutionalization in cultural persistence. *American Sociological Review, 42*(5), 726–743. https://doi.org/10.2307/2094862

CHAPTER 11

The world as human-environment systems

Introduction

The previous chapters of the second part of this book all emphasise that risk, resilience and sustainability are complex issues. However, this is not the only area where complexity constitutes a daunting challenge for scientific inquiry and practical action. Living organisms, the brain, society, climate and ecosystems are just a few other examples with something in common. Living organisms are made up of the complex interaction of myriads of cells. The brain is a vast network of neurons transmitting signals. Society is made up of interconnected individuals and organisations. And so on. In short, they can all be approached as wholes made up of complex sets of interrelated parts (Chapter 9). In attempts to grasp and learn from this complexity, some scholars find it helpful to look at the entity under study as a system, as von Bertalanffy (1960) did regarding the living organism, Ashby (1960) in regards to the brain and Buckley (1968) when studying society. These so-called systems approaches span various disciplines, all having 'a particular set of ideas, systems ideas, in trying to understand the world's complexity' (Checkland, 1981).

I argue in Chapter 7 that sustainable development requires us to (1) analyse the current situation, (2) define our preferred expected scenario, (3) analyse potential deviations from our preferred expected scenario (risk scenarios), and (4) design and implement sets of activities that maintain our preferred expected development trajectory. To do that in our complex world, we need to construct models of it (Conant & Ashby, 1970; Dörner, 1996). These models represent the parts of the world we are interested in, be it an organisation, a community, a municipality, etc. The models can rarely represent reality entirely but may still be enough for our purposes. 'Essentially, all models are wrong, but some are useful' (Box & Draper, 2007, p. 414).

This final chapter of the second part of this book provides a foundation for approaching our often overwhelmingly intricate world. It builds on the philosophical framework presented in Chapter 7 and argues why it is helpful to construct explicit human-environment systems to grasp this complexity. The chapter presents central systems approaches, concepts and principles, which form the basis for such endeavour. It then introduces important categories of tools for constructing human-environment systems in practice.

Sustainability Science, Second Edition
ISBN 978-0-323-95640-6, https://doi.org/10.1016/B978-0-323-95640-6.00016-6

Why human-environment systems?

Deviations away from the preferred expected future scenario of an organisation, community or society are not the results of linear chains of events, like dominos falling on each other (cf. Hollnagel, 2006, pp. 10–12). They are nonlinear phenomena that emerge within such complex systems themselves (Bergström & Dekker, 2019, p. 410; Hollnagel, 2006; Perrow, 1999). These systems comprise human beings, the tools they use, the buildings they are in, the air they breathe, the grass they walk on, etc. In other words, risk emerges in the intersection between the social and the environmental. Renn (2008, p. 5) takes this relationship further when stating that risk is essentially a by-product of how humans transform the natural environment into a cultural environment to serve human needs and wants (Figure 11.1). This transformation has brought about immense changes in the world, especially over the last 300 years (Turner et al., 1990) and continues to do so at an ever-increasing pace.

In this social and environmental nexus, Sustainability Science has risen to address the core challenges of humankind (Clarke & Dickson, 2003; Kates et al., 2001; Longo et al., 2021). It contributes by increasing our understanding of our world's complex and dynamic character and supporting society's capacity to guide its development by avoiding or minimising deviations from its preferred and sustainable future (Kates et al., 2001). One way of managing this complexity and dynamic character is to approach our world as complex human-environment systems (An et al., 2005; Haque & Etkin, 2007; Metzger et al., 2008; Reenberg et al., 2008; Sarkar et al., 2021; Turner et al., 2003; Yin et al., 2020; Yurui et al., 2019) and to view both the risks of deviations away from the preferred expected scenario, as well as actual disturbances, disruptions and disasters, as rooted in the same complex human-environment system that supplies human beings with opportunities (Haque & Etkin, 2007).

This type of system is at times referred to as NTS (Natural/Technical/Social) systems (Ingelstam, 2012), or more commonly as socio-ecological systems (Gallopín, 2006; Li et al., 2021). The important thing is that they include both the human and the environmental when analysing or addressing risk, resilience and sustainability issues. To highlight this, I refer to such systems as human-environment systems throughout the book. I do so to emphasise the intentional capacity of human beings, which I show in the first part of this book to have been increasingly central to the development of our world over time.

Human agency is, thus, central when explaining, understanding or improving human-environment systems. It is important to note that this approach to our complex and dynamic world only attributes agency, in the conventional sense, to individuals or groups, organisations or other collective actors comprising human beings (and potential others with the same capacity for intentional, purposeful action). However, even if the agency is situated in each actor, it is inherently relational. It arises in the interactions

Figure 11.1 The transformation of our natural environment to our cultural environment.

with other actors, the artefacts at their disposal and the natural entities in their environment. For example, a senior officer at the fire department has a fundamentally different agency when leading a substantial fire and rescue operation involving the resources of three fire stations than when having dinner at home with her wife and two children later the same day.

This approach to human-environment systems is different from actor-network theory (ANT), which views agency as an aspect of networks of human and nonhuman 'actants' per se (Hearn, 2012, p. 93). ANT has made many seminal contributions to science and technology studies (STSs) and social science in general (Blok et al., 2020; Crawford,

2020), and it has been used concerning sustainability challenges (e.g., Callon, 1990; Jenkins, 2000; Newton, 2002; Penteado et al., 2019). However, assuming that also a door and automated door-closer, a seatbelt warning signal, a computer or other artefacts or natural entities have agency since they enable or constrain our agency (see Latour, 1992) is to redefine the concept to the degree that it loses its meaning in relation to other important social scientific contributions. This radical redefinition is, therefore, also the most fiercely critiqued feature of ANT (e.g., Amsterdamska, 1990; Bloor, 1999a,b; Collins & Yearley, 1992; Hearn, 2012).

Nevertheless, the approach to human-environment systems presented here is deeply indebted to ANT in most other aspects. For instance, by removing the previously impenetrable division between the social and the natural—together with environmental sociology (e.g., Catton & Dunlap, 1978; Dunlap & Catton, 1979)—and establishing agency and power as inherently relational—together with various other relational thinkers (e.g., Emirbayer, 1997; Young, 1990). It also shares the same core endeavour of 'learning what the world is made of' (Crawford, 2020, p. 13) and seeks to include what matters without prejudging either the human or the environmental (Sayes, 2014). Frankly, the only difference is whether agency is assumed to be situated in the human or collective actors interacting amongst themselves and with artefacts and natural entities or in the networks of interacting humans, artefacts and natural entities as such.

Systems approaches and concepts

'Hard' and 'soft' systems approaches

There is variability concerning details between different systems approaches, but the objective is always to explain, understand and improve systems. A system can be defined as 'a group of interacting, interrelated, or interdependent elements forming a complex whole' (American Heritage Dictionary, 2000). However, depending on which part of the real world we want to approach as a system and for what purpose, different systems approaches may be appropriate.

Starting in the 1940s and 1950s, cybernetics emerged as 'the science of control and communication in the animal and the machine' (Weiner, 2019) and general systems theory as a transdisciplinary approach to understanding and organising complex systems (von Bertalanffy, 1950). These were the early days of thinking about systems—the dawn of 'the systems age' (Skyttner, 2005, pp. 36–47). These early systems approaches were conceived in a range of disciplines and not always called cybernetics or general systems theory, but made up a relatively cohesive whole through more or less shared theoretical assumptions and frameworks (cf. Ingelstam, 2012, pp. 36–108). Other notable contributors are, for example, Churchman and Ackoff (1950), Parsons (1951, 1971) and Boulding (1956).

Cybernetics viewed any kind of system as a machine, whether 'electronic, mechanical, neural, or economic' (Ashby, 1957, pp. 2–4). Its primary focus was on coordination, regulation and control of these systems, and the theories produced are still valid for many systems today. However, cybernetics was mainly constructed concerning real machines and organisms and was not designed to be directly applied to systems that include social aspects (Ingelstam, 2012). Similarly, general systems theory was initially designed for biological systems but was rapidly applied to other systems, including social systems. There is no doubt that general systems theory has aided many scientific endeavours and informed many policies over the past half-century. It still does. However, there are limits to the applicability of these approaches.

Common to cybernetics, general systems theory and their closest relatives—such as systems analysis and system dynamics (Umpleby & Dent, 1999)—is their view of the systems themselves. A system is, here, seen as exact and well-defined, with precise internal structure, boundaries and purpose. It is often seen as quantifiable, opening up for powerful problem-solving techniques (Checkland, 1985, p. 765). These 'hard' systems approaches also assume that the solution to real-world problems can be found by designing a system to reach objectives taken as given at the beginning of the process (Checkland, 1981, pp. 128–146). This process follows a linear sequential flow, is oriented towards goal-seeking and ends when the problem is solved (Skyttner, 2005, pp. 481–482). An observer using such approaches perceives the world as systems she can engineer to improve (Checkland, 1981).

Such 'hard' systems approaches have undoubtedly been successful in understanding and addressing various challenges (cf. Ingelstam, 2012; Skyttner, 2005). However, they also have limitations when addressing more ill-structured problems involving social interactions (Checkland, 1981, pp. 143–146). When applied in such contexts, there is a risk that their goal orientation makes them lose touch with aspects beyond the logic of the problem (Checkland, 1985, p. 765). There is also a real risk of losing the complexity of human agency if human beings are reduced, consciously or not, to mere components of 'machines' (Ingelstam, 2012). In these contexts, 'soft' systems approaches have more to offer.

'Soft' systems approaches constitute a much younger group of ideas and methodologies for explaining and understanding systems. They take a step away from the view that the world in itself is systemic to viewing the process of inquiry as systemic (Checkland, 1981). The goal-seeking orientation of 'hard' systems approaches is replaced with an orientation towards learning (Checkland, 1985, pp. 760–766; Skyttner, 2005, p. 481). The precise definitions of 'hard' systems are replaced by the view that the system is an abstract idea that depends on the observer's perspective, which is appropriate when considering complex, open systems (Chapter 9). In other words, the system does not really exist but is a construction to aid our understanding of something complex. In short, an observer using 'soft' systems approaches perceives the world as confusing and complex, not directly as systems, but believes she can investigate it as systems (Checkland, 1981).

'Soft' systems approaches are tailored for situations with ill-structured problems and often unclear objectives (Checkland, 1981, pp. 189–191; Skyttner, 2005, p. 481)—features that are common to most human-environment systems focused on in this book. The advantages of these approaches are that they are available to laypeople and professional systems thinkers alike and designed to keep in touch with the intricacies of human agency (Checkland, 1981). Because 'soft' systems approaches aim for learning and mutual understanding to guide intervention or change in systems, they encourage inclusive dialogue and debate (Skyttner, 2005, p. 481). On the other hand, the disadvantages are that these approaches rarely produce final answers and that the inquiry often appears to be never-ending (Checkland, 1985, p. 765).

When looking into different systems approaches—here categorised as 'hard' and 'soft'—it becomes clear that it is not an either/or relationship between them concerning risk, resilience and sustainability. 'Hard' systems approaches are appropriate when problems, objectives, and constraints are well defined, and the problem can be formulated as the search for means to achieve a defined end (Checkland, 1981, p. 316). They are appropriate for simple- and complicated problems (Chapter 9). In contrast, 'soft' systems approaches are appropriate when the situation is messier and more ambiguous. Perhaps the relationship between 'hard' and 'soft' is not like that between apples and pears but more like that between apples and fruit, in which the well-defined problems in need of solutions are the apples within the general group of more ill-defined issues calling for accommodations (Checkland, 1985, p. 765).

Two typologies of elements

As stated above, a system can be defined as interacting, interrelated or interdependent elements forming a whole. The structure of human-environment systems is, in other words, made up of elements and relationships between elements. There are many ways to categorise elements. I present one typology in Chapter 8, which is helpful when analysing what constitutes the capacity of a human-environment system to perform functions that contribute to its resilience. This typology is oriented towards grasping the capacity of a system to achieve set objectives (Box 11.1). However, to describe the system in the first place, it may be helpful first to consider the composition of elements more generally. For that, we may need a composition-oriented typology (Box 11.1).

Axelrod and Cohen (2000, pp. 4–6) suggest that systems' elements can be divided into agents and artefacts. An agent is, here, a person or a family, group, organisation, government or any collective entity that 'has the ability to interact with its environment, including other agents' (Axelrod & Cohen, 2000, p. 4). An agent has objectives and

BOX 11.1 Two typologies of elements

Capacity-oriented typology:	Composition-oriented typology:
1. Knowledge, skills, tools and other resources	a. Agents
2. Organisation on all levels	b. Artefacts
3. Rules, regulations and other formal institutions	c. Natural entities
4. Norms, values and other informal institutions	

can take action, proactively or reactively, with or without an intended purpose. The main properties of an agent are where it operates (location), how it can affect the system (capabilities) and what impressions it can bring forward from past experiences (memory) (Axelrod & Cohen, 2000, p. 4). Hence, agents in a human-environment system span everything from the national government to a farmer in a remote area, all depending on the purpose and level of the analysis.

An artefact, on the other hand, is 'an object made by a human being' (Oxford English Dictionary, 2020) and used for particular purposes by agents in the system (Axelrod & Cohen, 2000, p. 6). An artefact has a location and specific functions in the system but lacks its own objectives (Axelrod & Cohen, 2000, p. 6) and memory. Moreover, the functions of an artefact are not restricted to what the creating agent intended. For instance, once I drove a nail back into a doorsill with a wrench. It was not as good as a hammer, but it worked.

Although human agency has redesigned or at least affected much in the environment, there may still be elements that are essential parts of human-environment systems but not artefacts. These elements comprise everything in the system that is not designed or used by human beings, have the same main properties as artefacts and can be referred to as natural entities (cf. Johansson & Jönsson, 2007, p. 65). However, it is essential to remember that any element we consider a natural entity may be an artefact for someone else or if we would have another purpose for constructing the human-environment system. For instance, a piece of wood lying around in a forest may be a natural entity in a system constructed to understand and improve the governing of forest fire risk. Yet, the same piece of wood may be an artefact for someone wanting to make a fire or hit someone else on the head.

Presenting multiple typologies for how to categorise elements in human-environment systems may be confusing for the reader. However, remembering the philosophical assumptions presented in Chapter 7, it is essential to note that there are various ways to describe the same human-environment system—even if the purpose of the analyses is constant. The trick to grasping the complexity of a human-environment system—for a particular purpose—is to construct multiple descriptions of the system and connect them to one holistic picture. I go into further details about such a multiplicity of descriptions later in this chapter, but connecting descriptions based on the two typologies presented in this section is one way to build such a picture. First, consider the composition of who and what the human-environment system comprises to establish what is valuable and important to protect, which events can have a negative impact on these valuable elements and how susceptible these valuable elements are to the impact of the events. Then consider what constitutes the capacity of the human-environment system to anticipate, recognise, adapt to and learn from these events.

A typology of dependencies

In addition to elements, the structure of human-environment systems is fundamentally made up of relationships between elements. These relationships may be of various kinds, but for them to have any significance for human-environment systems, they must convey changes of some sort between elements. In other words, one element's state must depend on the state of connected elements. Such relationships are often referred to as unidirectional dependencies, and when two elements are dependent on each other, they are considered interdependent (Rinaldi et al., 2001, pp. 13–14).

In many instances, it is relatively straightforward to grasp more tangible dependencies, such as a child being dependent on his parent for food or a train being dependent on a communication system. However, other dependencies are no less real even if not as visible (Little, 2002, p. 110). To facilitate a more comprehensive understanding of dependencies, it helps to divide them into four categories: (1) physical dependencies, (2) information dependencies,[1] (3) geographical dependencies and (4) logical dependencies (see Rinaldi et al., 2001, pp. 14–16).

In this typology, physical dependencies comprise the most straightforward category, so I start with that first. A physical dependency is a dependency in which the state of one element in the human-environment system depends on another element's material output (Rinaldi et al., 2001, pp. 14–15). For example, an electrified rail network depends, amongst other things, on electric power (Figure 11.2). A car manufacturer depends, amongst other things, on the supply of parts from specialised companies. Without a steady electricity supply in the first example and car parts in the second, neither the rail network nor the car manufacturer would function appropriately. Although these dependencies are the most straightforward to identify and analyse in themselves, they are often combined in such a way that it is difficult to grasp the full complexity of human-environment systems.

[1] Called cyber dependencies in original.

Figure 11.2 A Russian train being physically dependent on electricity. *(Photo by Sergey Rodovnichenko, shared on the Creative Commons.)*

The next category of dependencies is information dependencies. An information dependency is a dependency in which the state of one element in the human-environment system depends on information transmitted from another element. For example, an ATM is not only dependent on electricity to function (physical dependency) but also on a steady flow of information concerning bank and account details. It is suggested that this type of dependency is a result of the comparatively recent processes of automation and computerisation (Rinaldi et al., 2001, p. 15), which may hold for the critical infrastructures for which the typology was initially constructed. However, information dependencies were common in human-environment systems long before the Industrial Revolution, as vital information has been communicated since the dawn of humankind. Information dependencies are closely related to physical dependencies, as it may be possible to view the information in itself as a resource in a similar way to electricity. However, separating information dependencies from physical dependencies is beneficial since mixing them may result in the less tangible information dependencies being seen as less critical or even forgotten.

The third category of dependencies is geographical dependencies. A geographical dependency is one in which two elements of the human-environment system are located so that a local event can affect them simultaneously (Rinaldi et al., 2001, pp. 15—16). This type of dependency is, in other words, not arising from the two elements being dependent on the actual functioning of each other, as in physical or information dependencies, but instead from their spatial proximity. For example, a poor household living just next to a railroad may not in any way be dependent on the trains, but the geographical proximity makes both liable to be affected by the same disruptive event. Such an event may be independent of the elements themselves—such as an explosion of a passing truck with hazardous material—or originate from one of the two elements—for example, a derailment of a train hitting the house. The main thing is that both elements are affected due to their location.

The last and most challenging category of dependencies to grasp is logical dependencies. A logical dependency is a dependency in which the state of one element in the human-environment system depends on the state of another through a mechanism that is not physical, informational or geographical (Rinaldi et al., 2001, p. 16). This definition may initially appear as a way of covering for everything else. Still, logical dependencies have something in common, are often the most difficult to grasp and play vital roles in human-environment systems. For example, an increasing proportion of Europeans choose to use IP telephones over their traditional domestic landlines, knowing that this new technology will not function in a power cut. The common reasoning behind this decision is that they will use their mobile phone to communicate if the IP telephone stops working in an emergency. However, if suddenly a substantially greater part of the population than usual attempts to use their mobile phones simultaneously, the network will be overloaded and reduce the proportion of successful calls. Other examples of logical dependencies are gridlock traffic in São Paulo on rainy days due to more people driving, fluctuations in the stock market due to rumours of the poor health of a CEO (Figure 11.3), reduced traffic congestion as a result of increased fuel price, and riots due to the publication of a caricature of a prophet on another continent. Logical dependencies are, in other words, related to human expectations, decisions and behaviour and not to any material or information input or spatial location.

Figure 11.3 The fluctuations of apple stock in 2008–09 were logically dependent on rumours concerning the health of Steve Jobs, the late CEO of Apple. *(Photo by Matthew Yohe, shared on the Creative Commons.)*

This typology of dependencies is helpful to remember when constructing human-environment systems to grasp the complexity of our world—especially as it aids us by constantly reminding us not to forget the less tangible dependencies. These dependencies often give rise to the most surprising dynamics.

Wholeness, hierarchy and multiplicity of descriptions

The challenge when constructing human-environment systems is to include enough information to sufficiently capture the complexity of the part of the world we are interested in—referred to as the principle of wholeness—whilst limiting it to what is relevant in light of the purpose of the analysis and the resources available. Whilst Chapter 7 presents the principles behind finding such a balance, the primary concern for making boundary judgements is the relevance to what we address and want to accomplish (Simon, 1990, pp. 7—13). In short, it is the purpose of our analysis and the resources available that determine how the human-environment system is demarcated.

There is a procedure to manage the effects of having to draw system boundaries in one way or another, as promised in Chapter 9. Parts of the world that we define as not directly included in the human-environment system we are constructing but still influencing or being influenced by it to a degree deemed relevant may be referred to as belonging to the surrounding of the system (Ingelstam, 2012). What distinguishes these elements in the surrounding from the elements within the system itself is that only their transboundary relations with the human-environment system are of interest, not the dependencies amongst themselves. For instance, it may be relevant to include how changing global weather patterns may impact flood risk in a municipality, but it is not relevant to include the global causal factors of climate change in the municipal risk analysis. This procedure facilitates finding the balance between Ockham's Razor and the Law of Requisite Variety (Chapter 7) by allowing us to simultaneously construct focused human-environment systems based on our specific purposes and account for the inherent openness of such complex systems (Chapter 9).

Constructing human-environment systems in relation to risk, resilience, and sustainability is likely to result in a web of elements and dependencies that is complex and essentially impossible to grasp. It is here the hierarchical aggregation from Chapter 9 is helpful, which results in human-environment systems being hierarchical in the sense that the system of interest is part of a system on a higher level and is made up of systems on a lower level of complexity (Blanchard & Fabrycky, 2006, p. 5; Simon, 1962, p. 468; Simon, 1996, p. 184). Most systems are (Simon, 1996, pp. 186—188). This hierarchy plays a vital role in grasping complexity since it makes it possible to simplify a human-environment system by aggregating sets of interdependent elements into subsystems (Rasmussen, 1985; Simon, 1996). It enables us to describe and explain the behaviour of an element/subsystem at any particular level with no need for a detailed representation of, and with only moderate concern for, the structures and behaviour on the levels above and below (Simon, 1990, p. 12). For example, a community may depend on job opportunities at a local textile mill for the livelihood of many families. This pattern calls for including the textile mill when constructing a human-environment system to analyse flood risk. However, analysing flood risk in the community will not require detailed descriptions of the mill's various departments, administrative processes or production methods. In other words, the textile mill is, in this context, treated as an element in the more general model of the community, though being a complex system in itself when looking more closely.

In addition to the principles of wholeness and hierarchy, large complex systems often require applying the principle of multiplicity of descriptions (Blauberg et al., 1977, p. 132). This principle states that representing any large and complex part of our world requires constructing a range of descriptions, each of which only covers certain aspects of the wholeness and hierarchy of the human-environment system. The principle of multiplicity of descriptions becomes particularly important in the context of this book, as it entails a wide range of agents and often the integration of various analyses. It requires the human-environment systems to be explicit since effective collaboration depends on having a shared vision of what to accomplish together (Jackson, 2003, p. 22; Senge, 2006, pp. 187—197).

Structural models and functional models

Human-environment systems can be illustrated by focussing on structure or function. Let me demonstrate this difference using a somewhat simplified example of a community living beside a river. The people in the community are sustaining their livelihood through farming their plots and selling firewood in the nearby town. They cultivate maize along the river and get the firewood from cutting trees upstream in the catchment area. This way, they balance their income if something unfortunate happens, such as a flood that destroys their harvest. In this simplified human-environment system, floods occur when enough rain falls in the river's catchment area, finding its way to a watercourse and overflowing the river. The amount of water finding its way to a watercourse depends on the amount and type of vegetation in the catchment area. The illustration of the human-environment system in Figure 11.4 is an example of a structural model of this system, where its elements can be divided into agents, artefacts and natural entities that are connected through different types of dependencies as described earlier in this chapter.

Structural models of human-environment systems are static, much like snapshots at specific points in time. However, remembering Sztompka's (1993, p. 9) words in Chapter 8, we observe reality 'not as a rigid quasi-object, but as a continuous, unending stream of events'. In other words, human-environment systems are, by definition, never static but continuously transforming through various processes that, together with elements

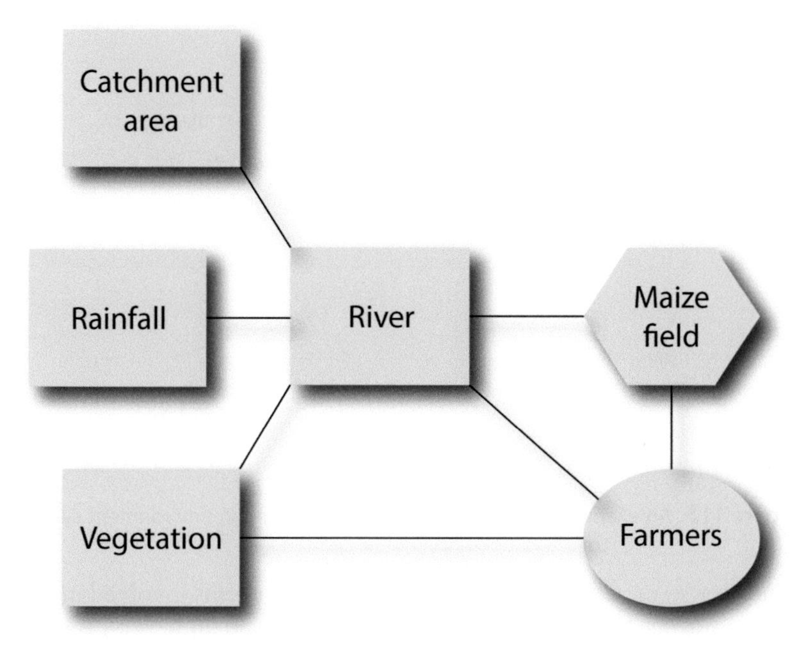

Figure 11.4 An example of a structural model of a human-environment system.

and dependencies, comprise the core building blocks of any system. To grasp this dynamic nature of human-environment systems, we need to complement structural models with functional models.

In functional models, the focus is not on how different elements cohere but on patterns of change (Senge, 2006, p. 68). Let us return to the example of the community living beside the river. Farming maize and selling firewood are the livelihood sources for the community's people. The more maize they get, the less firewood they need to sell to sustain their livelihood, which is indicated by the '—' along the arrow between 'maize harvest' and 'selling firewood' in Figure 11.5. A flood reduces the maize harvest, which in turn leads to more sold firewood that generates more cut trees, which reduces the vegetation in the catchment area and increases the flow in the river in the next rainfall, leading to more floods. If nothing is done, the community will see increasing floods that eventually may lead to a collapse of their livelihood sources. What you have is a reinforcing feedback loop strikingly similar to actual developments along the Artibonite River in Haiti.

It is important to note that structural and functional models are complementary and equally necessary to construct human-environment systems. In other words, it is neither feasible to build any functional model without at least implicitly considering the structure of the human-environment system in question nor to address issues of risk, resilience and sustainability without at least implicitly considering how changes cascade

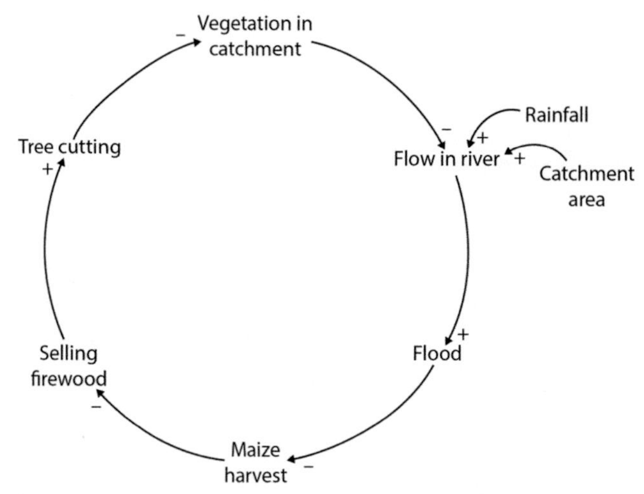

Figure 11.5 An example of a functional model of a human-environment system.

throughout the system. Considering the many different agents involved in such tasks, it is helpful to construct human-environment systems that include both types of models explicitly.

Constructing human-environment systems

With systems approaches and concepts fresh in mind, it is time to introduce important tools for constructing human-environment systems in practice. These tools have been developed for many purposes and in various disciplines but can be useful in the context of risk, resilience and sustainability. To give you an overview of these tools, I categorise them by their focus in terms of providing structural or functional models. There are other ways of modelling complexity not introduced in this book. However, using the tools provided would get you far in grasping the complexity of the human-environment systems relevant to this book's purpose.

Tools for structural models

Explicit structural models in human-environment systems can be constructed using various tools that can be categorised in many ways. For this book, I suggest the following three main categories: (1) Venn diagraming, (2) network analysis and (3) spatial mapping and analysis.

Venn diagraming

The most basic form of explicit structural models in human-environment systems is referred to as Venn diagrams—named after John Venn, who was the first to conceive graphs representing all possible logical relationships between limited collections of sets of elements (Venn, 1881). Venn diagraming is, in other words, about conceptualising groups of elements that share common properties and their relative relationships. Venn diagraming is commonly used in various disciplines spanning from pure mathematics to ecology and is useful in several ways for addressing risk, resilience and sustainability issues. In addition to the utility of Venn diagrams for grasping conditional probability, which I do not address in this book, the tool is also helpful in plotting elements (most often categories of agents in a community) and their most basic relationships (IFRC, 2007, pp. 126–132; Wisner, 2006, p. 322). Venn diagraming is, in this context, about drawing polygons that represent different elements of the human-environment system and arranging them internally to symbolise their relationships, where the size of the polygons signifies relative importance, and their position signifies how closely they are related to each other (Figure 11.6). It is important to note that for this tool to be useful, it must often allow for the participation of people involved in what is represented by the resulting human-environment system. Without such a participatory approach, the resulting Venn diagram often loses either the necessary details that only true insiders know or much of its compelling quality if it is perceived as pushed for by outsiders.

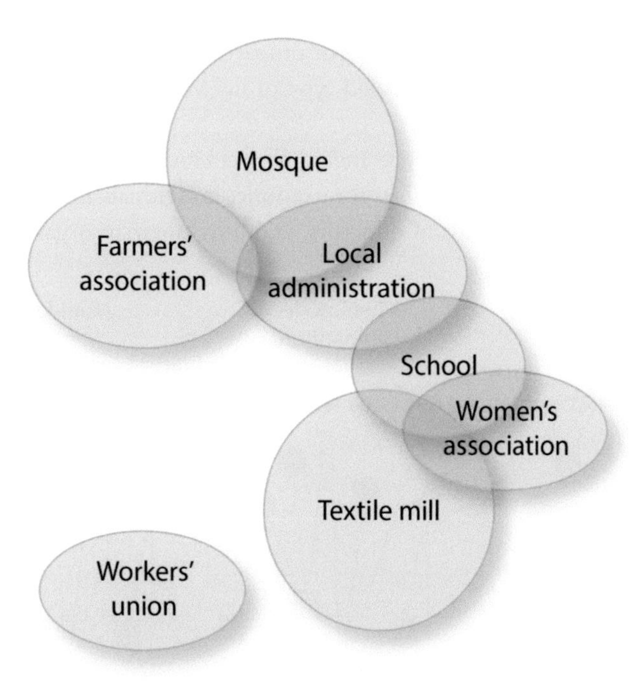

Figure 11.6 A Venn diagram of a community.

Network analysis

The next category of tools for constructing structural models is network analysis, spanning various tools used in different disciplines. What distinguishes network analysis from Venn diagraming is that it allows for more details in terms of the description of each element and focuses explicitly on the dependencies between elements. These different network analysis tools focus on representing the structure of some parts of the world as networks of elements—also referred to as nodes or vertices—and dependencies between elements—also referred to as ties, edges or arcs. I prefer nodes and ties. Examples of such tools include most forms of social network analysis (e.g., Borgatti et al., 2018; Wasserman & Faust, 1994), graph theory (e.g., Balakrishnan & Ranganathan, 2012) and network theory (e.g., Milanovic & Zhu, 2018).

The networks can be based on qualitative or quantitative data and take different forms. They can provide valuable insights and inform the construction of functional models. The networks can include qualitatively different kinds of nodes, such as the example of the structural model of the community beside the river (Figure 11.4). However, to unlock the inherent analytical powers of network analysis, the networks are organised into levels with only one kind of node per level—although each type can have many nodal attributes. Hence, networks including only one type of node—for example, individuals, organisations, policies, natural resources, computers, etc.—have only one level (Figure 11.7). In other words, the nodes are directly linked by ties between them. Such networks are called one-mode or unipartite networks (Robins, 2015, p. 21) and comprise the most frequently analysed type of network (Wasserman & Faust, 1994, p. 29).

The second common network type includes two kinds of nodes with ties only between the different types (Figure 11.7)—for example, individuals and their organisations, corporate leaders and their boards of directors, and humanitarian organisations and their

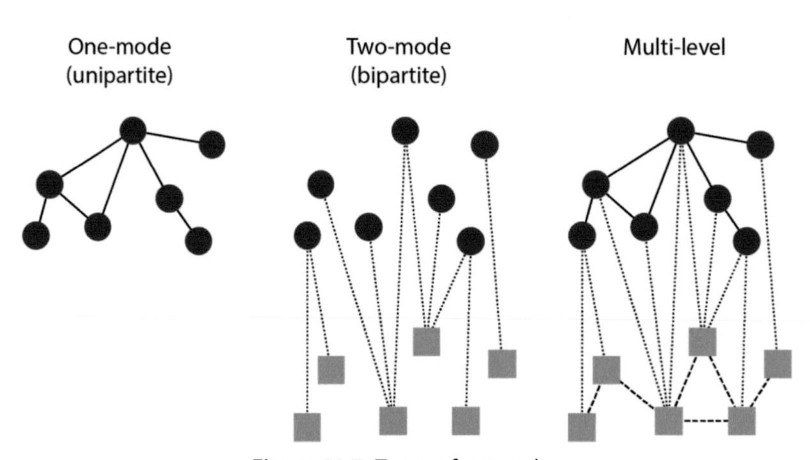

Figure 11.7 Types of networks.

clusters. These networks are called two-mode or bipartite networks (Robins, 2015, p. 21) and are often used concerning risk, resilience and sustainability (e.g., Hu et al., 2021; Malinick et al., 2013; Ongkowijoyo & Doloi, 2017). For analytical reasons, a two-mode network is sometimes transformed into a one-mode network by operationalising the indirect connection between two nodes connected to the same node of the other type as a direct tie (e.g., Malik et al., 2017). Constructing multi-mode or multipartite networks with three or more types of nodes is also possible using the same logic. However, the analytical restrictions and challenges in interpreting the results often result in the networks being transformed into sets of lower mode networks for analysis (see Coscia et al., 2013; Knoke et al., 2021).

The last generic network type also allows ties between the nodes within each category (Figure 11.7). Such networks are referred to as multi-level networks (Lazega & Snijders, 2016; Robins, 2015) and have proven tremendously useful for grasping and addressing a range of issues in human-environment systems (e.g., Bodin, 2017; Hileman & Lubell, 2018). These multi-level networks generally involve two types of nodes, such as actors and ecological components (e.g., Bodin, 2017), actors and tasks (e.g., Nohrstedt & Bodin, 2020), actors and sustainability issues (e.g., Bergsten et al., 2019) or actors and policy issues (e.g., Hedlund et al., 2021). However, there are examples of multi-level networks with more than two types of nodes, such as actors, issues and policy forums (e.g., Fried et al., 2022).

Whilst multi-level networks include different types of ties by definition, it is essential to note that all networks can consist of different kinds of ties between the same pair of nodes regardless of level. These networks are sometimes called multiplex or multivariate networks (Robins, 2015, p. 21) and are often used concerning issues of risk, resilience and sustainability (e.g., Baggio et al., 2016; Becker, 2018; Sayles et al., 2019).

Network analysis includes potent tools for identifying, analysing and visualising all kinds of dependencies in human-environment systems (Borgatti et al., 2018; Lazega & Snijders, 2016; Wasserman & Faust, 1994). There are tools for analysing personal networks around individual nodes (McCarty et al., 2019) and networks representing the parts of reality we are interested in as wholes (Borgatti et al., 2018). There are tools for understanding individual actors' influence in a network and the capacity of the network to perform as a whole. There are tools for identifying groups based on actual interaction in contrast to formal affiliation, for testing hypotheses, and much more (see Borgatti et al., 2018; Lazega & Snijders, 2016; Lusher et al., 2013; Robins, 2015).

Spatial mapping and analysis

Although the kind of network analysis just described can include geographical dependencies as long as they are defined as ties between nodes, the resulting representation is not spatial. Since many central aspects of risk, resilience and sustainability are spatial in terms of being associated with geographical locations, paths and areas, another category of tools is often needed. Many tools are available to explicitly include spatial features in constructing human-environment systems. I categorise them here as different forms of spatial mapping and analysis. However, it is essential to note that the analysis of spatial networks—such as road, river and utility networks—is sometimes also called network analysis (Curtin, 2018).

Spatial mapping is about generating visual representations of observations using spatial relationships. It is a broad term encompassing all spatial organisation of information concerning human-environment systems, using a wide range of symbolic representations from drawings to standardised map legends. It includes producing scaled representations of geographical features, such as the maps of traditional cartography, but is not limited to that. It spans from participatory mapping (e.g., Anderson & Holcombe, 2013, pp. 165–207; IFAD, 2009) to spatial visualisations based on high-tech geospatial data collection, such as LiDAR (a remote sensing technology illuminating a target with a laser and analysing the reflected light) (Heywood et al., 2006, pp. 60–61). Spatial mapping is particularly effective for identifying geographical dependencies that are often not intuitive and difficult to grasp through the generally fragmented perspectives of actors across sectorial, political, administrative, organisational or other boundaries.

Many tools have been developed over the years to facilitate spatial analysis and mapping. Ground- and sketch mapping are two examples of participatory mapping tools that plot maps from the memory of involved community members (IFAD, 2009, pp. 13–14). However, Geographical Information Systems (GIS) have opened unprecedented opportunities for analysing and visualising geospatial information (e.g., Baird et al., 2016; Huang et al., 2021; Mansourian et al., 2018; Månsson, 2019). GIS allows for combining various kinds of spatial data, from data collected through participatory processes to satellite images. It provides immense analytical power to identify different types of dependencies—also other than geographical dependencies—as well as spatial patterns. Thus, GIS is often used in relation to risk, resilience and sustainability (e.g., da Silva et al., 2020; Galappaththi et al., 2020; Yariyan et al., 2020; Wang et al., 2019; Wiek et al., 2012).

Tools for functional models

There are many ways to build functional models in human-environment systems. Here, I suggest the following six overall categories: (1) causal loop modelling, (2) system archetypes, (3) stock and flow modelling, (4) agent-based modelling, (5) network modelling, and (6) microworlds. Causal loop modelling and system archetypes are qualitative, and the others are quantitative.

Causal loop modelling

The basic building blocks for constructing functional models in human-environment systems are, again, elements and dependencies. However, this time the elements are variables that can change, and the dependencies between elements are directional and can be either positive or negative (Boardman & Sauser, 2008, p. 67; Maani & Cavana, 2000, pp. 26–27; Meadows & Wright, 2009). These dependencies cause a change in one element to spread to associated elements, creating a branching chain of causal relations through which any impact on the system could propagate to distant parts of it (Dobson et al., 2007; Hollenstein et al., 2002, pp. 56–61; OECD, 2003, pp. 44–45). The propagation of a change between each pair of elements may be immediate or delayed to various degrees, making the time for analysis critical since the timescale before the appearance of adverse effects is essential when linking risk to sustainable development (Renn, 2007, p. 15). These delays, often indicated by two parallel lines crossing the dependencies in causal loop diagrams (Figure 11.8), are also significant contributors to the complexity of systems (Maani & Cavana, 2000, p. 33; Senge, 2006, pp. 88–91). Remember from Chapter 9 that not only the number of elements determines complexity but also the dependencies between them, leading to what Senge (2006, p. 71) refers to as dynamic complexity by separating cause and effect in both space and time.

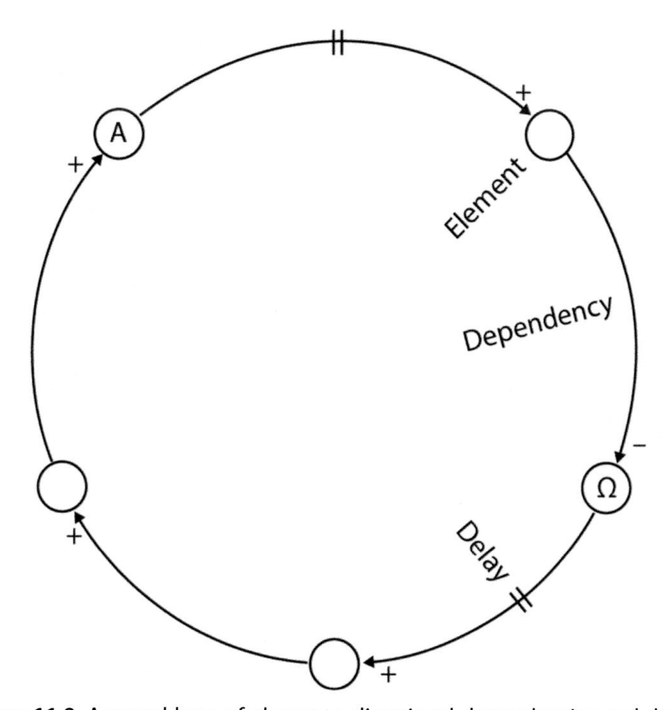

Figure 11.8 A causal loop of elements, directional dependencies and delays.

The chains of causal relations sometimes create loops—causal loops—feeding back the propagating changes to elements earlier in the chains (Figure 11.8) (Ashby, 1957, pp. 53–54; Maruyama, 1963; Senge, 2006, pp. 73–79). Such feedback loops are prevalent in our world and a fundamental manifestation of complexity (Chapter 9).

Remember from Chapter 9 that feedback can be either reinforcing or balancing. So, for causal loop modelling, reinforcing causal loops results in either continuous growth or decline in the element of interest, whilst balancing causal loops results in stability through dampening or negating changes in the element or in meeting a set target (Maani & Cavana, 2000, pp. 28–33; Senge, 2006, pp. 79–88). It is, however, essential to remember that growth, decline and stability may all be normatively positive or negative depending on values and perception.

Balancing loops that drive systems to meet set targets for specific elements are relatively easy to identify, as they attempt to reduce gaps between the actual and the desired state of the elements in question (Senge, 2006, pp. 83–88). However, balancing loops that are not goal-seeking in the sense of not having such explicit targets are not always as intuitively obvious to distinguish from reinforcing loops, as this distinction depends on the direction of the fed-back change in relation to the previous change in the element (Maani & Cavana, 2000, pp. 32–33). There are at least three ways of making this crucial distinction.

The first procedure for determining if a causal loop is reinforcing or balancing is not the easiest but has long-term benefits as it helps us develop our systems thinking. This procedure involves applying logic whilst reviewing the information on which we base the construction of the functional model. For instance, going back to the community of maize farmers along the river (Figure 11.5), it may be relatively easy for someone accustomed to thinking in systems to determine that it is a reinforcing loop without resorting to the second, more stepwise procedure.

The second procedure involves starting in any element of the loop, deciding an imaginary direction of change, and considering how the next element in the loop would change. Remember, a '+' signifies the same direction (A in Figure 11.8), and a '—' means the opposite (Ω in Figure 11.8). Then, consider how this ensuing change would influence the following element in the loop and so on until the change returns to the first element. If the returning change is in the same direction as the initial change, it is a reinforcing loop. If the direction is opposite, it is a balancing loop. This procedure is less challenging than the first but still helps us practise systems thinking as it demands deliberation.

The last procedure is by far the easiest but requires all dependencies of the loop to be specified beforehand. No deliberation is needed, so there is little opportunity for learning. With all the '+' and '—' in place, it is easy to determine if the loop is reinforcing or balancing by simply counting the number of '—' (Maani & Cavana, 2000, p. 33). If there is zero or an even number of '—' in the loop, it is a reinforcing loop. Balancing loops have an odd number of '—'. For example, the single '—' in the causal loop in Figure 11.8 reveals that it is a balancing loop.

Systems archetypes

When constructing qualitative models using causal loop modelling, for which the output is a more or less complex system of causal loops, it is sometimes possible to identify systems archetypes. A systems archetype is a pattern of behaviour of a system based on a set of elements and dependencies in a functional model, which can be found in various contexts and, thus, generalised and used to analyse systems behaviour more generally. Examples of relevant systems archetypes are 'Fixes that fail', 'Eroding goals', 'Escalation', 'Success to the successful', 'Limits to success', 'Shifting the burden', 'Growth and underinvestment', and 'Tragedy of the commons' (Figures 11.9—11.11). These systems archetypes are all conceptualised slightly differently in various literature (e.g., Jackson, 2003; Maani & Cavana, 2000; Senge, 2006).

In a 'Fixes that fail' situation, an apparent problem symptom calls for attention. A solution is implemented that alleviates the symptom but has unintended consequences that, after a delay, cause the original problem symptom to return or even worsen. In response to this, more of the same solution is applied, allowing a reinforcing loop to perpetuate or even aggravate the problem (Figure 11.9). Getting out of a 'Fixes that fail' situation usually entails acknowledging that the applied solution only relieves the symptom and committing to identifying the underlying problem. Then, combining the original solution to ease the symptom with a fundamental solution to the real problem helps ensure that we do not get stuck in an endless cycle of solving our past solutions.

'Eroding goals' involves a gap between the desired and actual state of something we value that can be closed by actively improving the actual state or lowering the desired state (Figure 11.9). Unfortunately, the gap is relatively often reduced by the latter, as the aspired improvement may be delayed or fail to materialise altogether. The drift in performance is often so gradual that we fail to notice it or its impact. Drifting performance is a common indicator of the 'Eroding goals' archetype at work, and sufficient

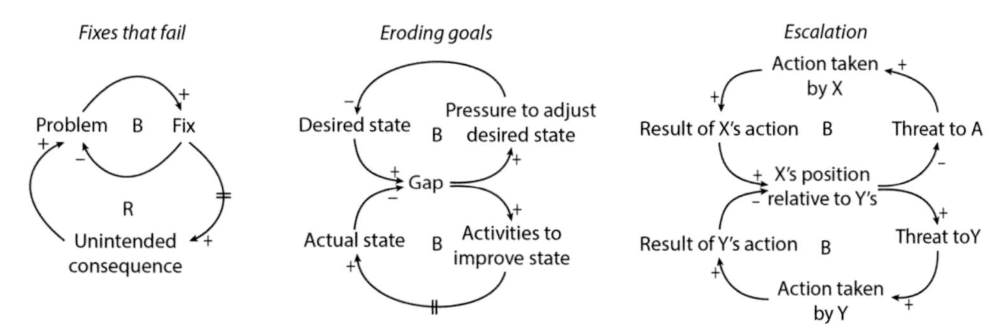

Figure 11.9 Fixes that fail, eroding goals and escalation systems archetypes.

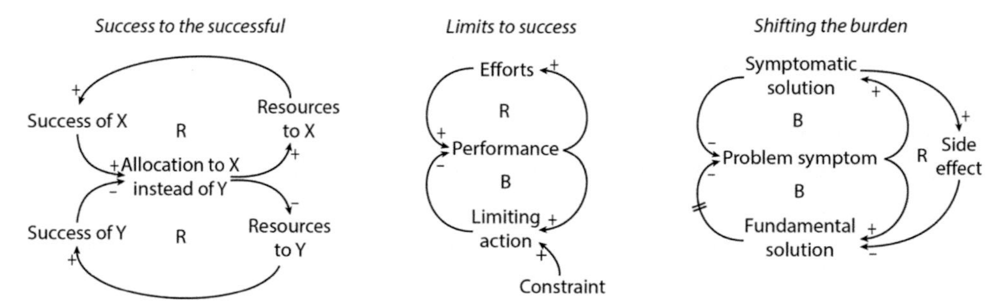

Figure 11.10 Success to the successful, limits to success and shifting the burden systems archetypes.

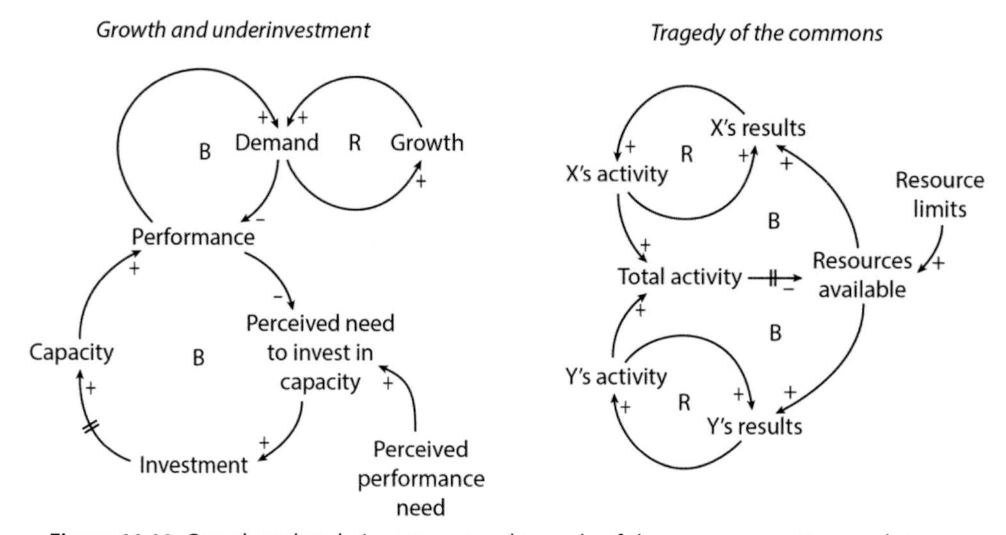

Figure 11.11 Growth and underinvestment and tragedy of the commons systems archetypes.

actions are not taken. Avoiding 'Eroding goals' is best accomplished through explicit attention to the real drivers of goal setting.

The 'Escalation' archetype involves one actor (X) perceiving a threat from another actor (Y) and taking action to counter the threat. These actions are perceived by the other actor (Y) as creating a threatening imbalance, who responds to rebalance their relationship. These subsequent actions are, in turn, perceived as threatening from X's perspective, and so on. The interplay between the two actors, each only protecting itself by balancing the perceived threat of the other, becomes a reinforcing Figure-8 pattern that escalates the situation until the pattern is broken (Figure 11.9). To halt the escalation, consider what is perceived as threatening by the actors, what we can do to address it, if any significant delays in the system may distort their perceptions, and what underlying assumptions lie beneath the actions taken.

The next systems archetype is called 'Success to the successful' and involves two or more actors, projects or initiatives depending on a shared, limited pool of resources. Suppose one of the actors is more successful than the others. In that case, it tends to attract more resources to sustain the success to the detriment of the resources allocated to the others (Figure 11.10)—regardless of whether they are superior alternatives. The initially successful is, thus, more likely to continue to succeed, whilst the others may struggle even to maintain their initial performance when their resources dwindle. Avoiding this pattern is crucial when managing most risk, resilience and sustainability issues. It requires us to understand why the system functions as a winner-take-all competition and find ways to redesign the resource allocation mechanism, turn the competitors into collaborators and define objectives and success at a higher level than them.

'Limits to success' entails a pattern of increasing efforts initially leading to improving performance. However, over time the success causes the system to encounter limits that hamper performance (Figure 11.10). When success triggers limiting action and performance declines, there is a tendency to increase the same efforts that initially worked, to no avail. It may even lead to future problems (cf. Branlat & Woods, 2010). To avoid such limits, regularly consider what constraints may exist and what kind of limiting action may be building up in the system due to the success, and find ways to counter them before the system grinds to a halt.

The 'Shifting the burden' archetype involves a problem symptom that can be addressed by applying either a symptomatic solution reducing the symptom or a more fundamental solution addressing the underlying problem. However, symptoms are usually more conspicuous than root causes, and this pattern is a real obstacle to sustainability (Ehrenfeld, 2005). With a symptomatic solution, the problem symptom is reduced or vanishes, undermining the interest in a more fundamental solution. However, not addressing the actual problem, the problem symptom reappears over time and tends to spur more of the symptomatic solution in a detrimental, reinforcing figure-8 pattern (Figure 11.10). Moreover, symptomatic solutions frequently generate side effects, further undermining the necessary attention to fundamental solutions. However, side effects can escalate and become actual problems in themselves, so determining if a solution is 'symptomatic' or 'fundamental' may depend on our perspective. Hence, overcoming this unfavourable pattern often requires exploring the system from different perspectives.

The following systems archetype is called 'Growth and underinvestment' and involves a limit to growth that could be removed or delayed with appropriate investments in developing the capacity to perform. Yet, the limitation kicks in since the initial success gives an impression of high performance, which reduces the perceived need to invest (Figure 11.11). As the performance degrades, delays in developing capacity further undermine the will to invest as investments fail to turn the decline around immediately, resulting in deteriorating performance. Avoiding this pattern requires consideration of the underlying assumptions driving decisions to invest in capacity and shifting attention from past performance towards anticipated future demands.

Finally, the 'Tragedy of the commons' archetype involves actors using shared resources to pursue their objectives without concern for their aggregated impact, leading to these commons becoming overused to the detriment of all actors (Figure 11.11). The commons may collapse at the end (Hardin, 1968). It is important to note that solutions to this pattern are rarely found on the level of individual actors but generally require proper institutions regulating resource use (Ostrom et al., 1999). Consider the actors' incentives to continue using the shared resources and find ways to communicate the potential long-term impact if nothing is done and the benefit of institutions regulating resource use to balance short-term objectives and long-term sustainability.

Stock and flow modelling

If constructing human-environment systems requires a more precise understanding and specification of the states of particular elements, quantitative functional models may be necessary. When constructing quantitative functional models, parts of the system can be transformed into what is referred to as stocks and flows but are still parts of the causal loops (Forrester, 1969, 1994; Meadows & Wright, 2009)—thus, the name stock and flow modelling. This method for modelling systems—called system dynamics (Forrester, 1994)—provides quantification and a richer conceptual language than causal loop modelling. The elements are divided into stocks and converters, the dependencies are divided into flows and connectors, and two new building blocks are added as sources and sinks (Figure 11.12).

A stock is here conceptualised as a container with a capacity for holding a quantity of some sort (Jackson, 2003, p. 73) and represents an element of a human-environment

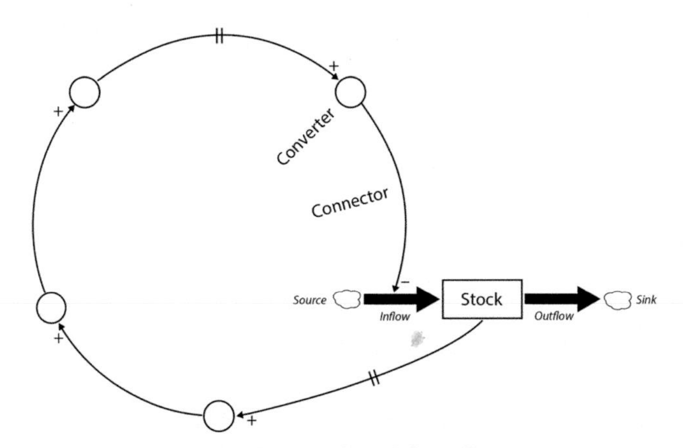

Figure 11.12 A stock and flow diagram.

system, which state at any given time depends on the past behaviour of the system. Let me illustrate this with a simplified example of cholera prevention in a refugee camp. At any given time, there are a number of cholera cases within the camp's population represented as a stock (Figure 11.13). The state of this stock can then only be determined by knowing the number of cholera cases at a previous point in time, as well as the rate of infection and the rates of cure and death since then.

Rates of change in a stock are represented as flows that cause it to either increase or decrease. Here, a flow is conceptualised as a pipe through which the quantity of the stock flows in or out. Going back to the example of the refugee camp, the infection rate determines the rate of new cholera cases, which influences the total number of cholera cases in the camp. The total number affects, in turn, the rate of infection through a whole range of causal loops, out of which I only present one to illustrate the other building blocks of stock and flow modelling.

The number of cholera cases in the refugee camp influences the attention concerning cholera of the people managing the camp. Although likely to be associated with some delay, a high number of cholera cases increase their attention to cholera, which increases the resources allocated for cholera control, which in turn intensifies the hygiene promotion campaign in the camp (Figure 11.13). An intensified hygiene promotion campaign then improves hygiene practices amongst the refugee population, reducing the infection rate. However, such change is likely to take some time. The elements contributing to this causal loop are here conceptualised as converters, that is, elements holding the information that influence the rate of flows either through direct relationships or indirectly by affecting another converter. The dependencies between these elements are in stock

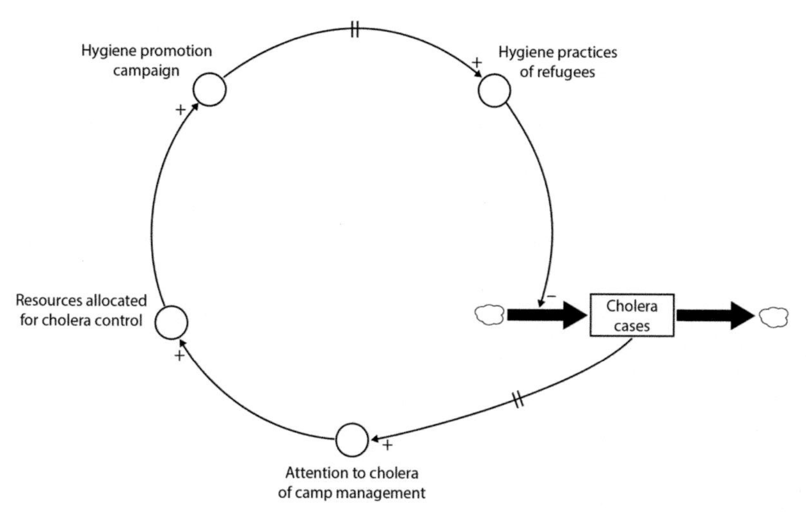

Figure 11.13 A simplified example of a stock and flow diagram of cholera control in a refugee camp.

and flow language conceptualised as connectors, with the same directional indication as in causal loop modelling. The last two building blocks of stock and flow modelling are sources and sinks, which are stocks that lie outside the boundary of the system, that is, the total population of the refugee camp.

Building a functional model using stock and flow modelling also entails some ground rules concerning how the building blocks can be combined. Most fundamentally, stocks can only be directly influenced by flows. In other words, connectors can never bypass a flow and link directly to a stock. On the other hand, flows and converters can be influenced by stocks, flows and converters—all through the connectors binding the system together. However, it is essential to note that converters also can be independent in the sense of not being influenced by any other element but acting as starting points for causal chains.

Finally, stock and flow modelling facilitates understanding and explanation of what we observe in reality. It also allows us to experiment by introducing changes to an empirically verified model to investigate what may happen if such changes were actually implemented. Such experiments are particularly valuable in situations where it may be impossible, unfeasible or unethical to test the changes in the real world.

Agent-based modelling

Instead of constructing human-environment systems with stock and flow modelling to grasp the complexity and dynamic change, it is also possible to model many systems from the bottom up. Agent-based modelling is a particularly powerful way of doing so, allowing us to seek understanding and explanation by modelling the parts of the world we are interested in as agents and their interactions (Holland, 1998; Macy & Willer, 2002; Wilensky & Rand, 2015). These models 'show how simple and predictable local interactions can generate familiar but enigmatic global patterns' (Macy & Willer, 2002, p. 143).

Agent-based modelling is instrumental when these global patterns of interest are more than the simple aggregation of local action—when the emergence behind Coleman's link 4 in Figure 9.7 challenges us the most (Hedström, 2005, pp. 115–117). Here, agent-based modelling provides methodological muscles to actually model emergent properties (Farmer & Foley, 2009; Wilensky & Rand, 2015). It exploits the fact that it is generally much more challenging to grasp and describe the global dynamics of a system than the local dynamics of the interacting agents constituting it. For instance, consider the immense challenge of capturing the complexity of a city's road traffic infrastructure system, compared to the much more straightforward task of describing the behaviour of individual vehicles in relation to the street layout, traffic rules and other vehicles and pedestrians in their immediate surroundings. Or the resilience of an ant colony as a whole compared to the minimal behavioural portfolios of ants (Chapter 9). Agent-based modelling is, in other words, about translating such micro-level knowledge into agents with attributes describing their current states, an environment with attributes in which they can interact, and rules for how agent-agent and agent-environment interactions update their attributes and future behaviour (Wilensky & Rand, 2015, p. 32). It is about designing, running and evaluating models to help us understand, explain or improve complex phenomena or problems in human-environment systems. However, these models are always simplifications of reality and will never perfectly correspond to the parts of the real world we focus on. They are not for simulating reality but for stimulating learning about that reality (Wilensky & Rand, 2015, p. 334).

Running agent-based models is done in steps, in which the attribute states of the agents and their environment are updated when interacting according to the defined rules. These steps usually represent time but can also represent a stepwise process of some other kind. The length of each step depends on what we focus on. The road traffic infrastructure system and the ant colony must be modelled in seconds to be helpful. In contrast, an ecosystem approaching desertification must be modelled in much longer steps to be feasible (cf. Reynolds et al., 2011). The results of the agent-based models include time series of data, or other stepwise series, for variables we have defined. A classic example is a predator/prey model in which the number of predators and prey coevolve over time in substantially different ways even with minuscule changes in the conditions of the model (Wilensky & Rand, 2015, pp. 163–197). The results can also include visualisations of the model's environment to identify emergent patterns. Such visualisations can look very different. A commonly used example is Schelling's groundbreaking agent-based model for racial segregation (Schelling, 1978). This model visualises increasing segregation even with agents accepting living in neighbourhoods where they are in a stark minority (25%), severe segregation even if accepting being a strong minority (33%), staggering segregation if accepting balanced neighbourhoods (50%) and complete segregation with buffer zones if demanding being in the majority (60%) (Figure 11.14). Agent-based models can also be spatial in the sense of running in an actual geography (e.g., An et al., 2005; Crooks & Heppenstall, 2012) and their results integrated with spatial mapping and analysis introduced above.

Evaluating agent-based models is crucial and entails verification, validation and replication (Wilensky & Rand, 2015, pp. 311–346). Verification is internal and checks so the model carries out what it intends to do. Validation is, instead, external in the sense of checking to what extent the model corresponds with what we can observe in reality. It involves ensuring that the agents' local behaviours, attributes and rules in the model match their real-world analogues (micro-validation) whilst ensuring that the global resultant and emergent properties in the model match observable properties in reality (macro-validation). Validation can be done by demonstrating that the properties and mechanisms of the model look like properties and mechanisms in reality (face validation) or by checking that the data generated by the model correspond to real-world data (empirical validation). Finally, replication is crucial for agent-based models that may be complex and

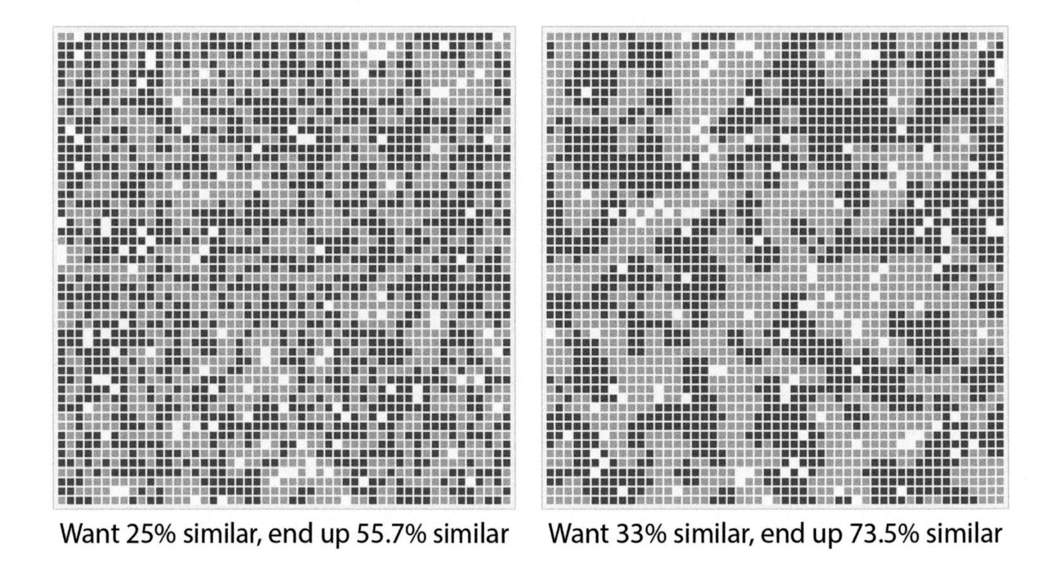

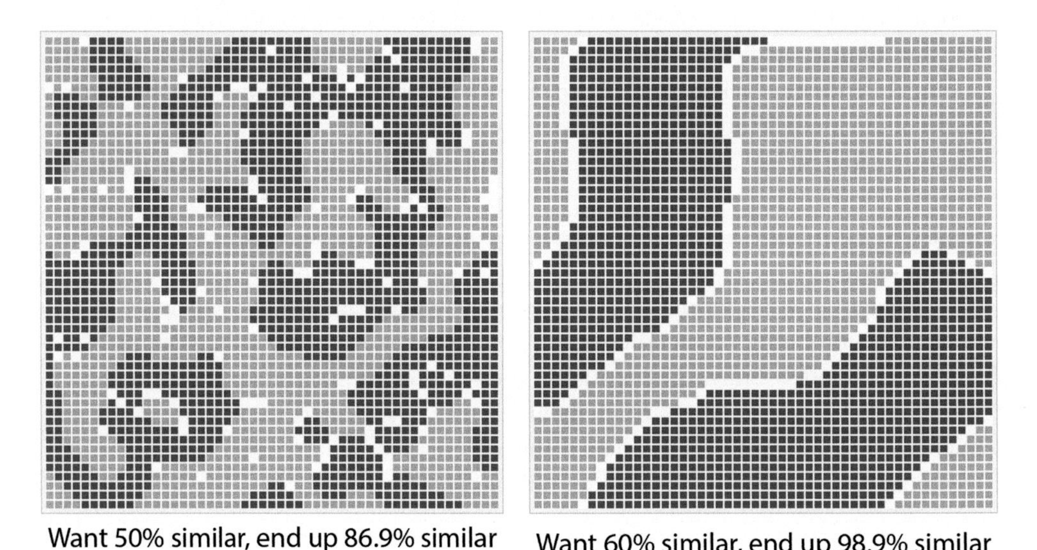

Figure 11.14 Visualisations of Schelling's agent-based segregation model using NetLogo. The model starts with around 50% similar agents and different attitudes concerning the acceptable proportion of similar neighbouring agents (i.e., wanting 25%, 33%, 50% or 60% of their neighbours to be similar).

thus sensitive to initial conditions (Chapter 9). By running the model many times whilst adjusting it, it is possible to evaluate to what extent it provides stable enough results to be useful. However, Wilensky and Rand (2015) take replication further and suggest entire models be replicated to see if they provide sufficiently similar results, including at a different point in time, on another computer, in another computer language, using different tools or algorithms, and done by other people. Nonetheless, a well-designed and successfully evaluated agent-based model is a potent tool for constructing useful human-environment systems.

Network modelling

Network modelling comprises several powerful tools for constructing functional models of human-environment systems. These tools are closely linked with network analysis designed to analyse and visualise networks (structural models) but are instead intended to model dynamic change (functional models). These tools seek to unravel the underlying processes behind observed networks by applying statistics designed for relational data that violate the conventional assumption of independence that underpins most statistical analyses (Lazega & Snijders, 2016; Lusher et al., 2013; Robins et al., 2012). Some network modelling tools can be applied to cross-sectional data, meaning the generated knowledge about the underlying processes is inferred rather than explicitly proven. Other tools use longitudinal data that allow observed change to inform the models (Butts, 2017). Regardless, network modelling facilitates the analysis of complex phenomena and problems resulting from both regularities and randomness (Lusher et al., 2013).

In brief, network modelling tools assume that different underlying processes generate different local configurations of nodes and ties through different network effects. For instance, *homophily* increases the probability of a tie between two nodes sharing an attribute, *contagion* increases the probability of two nodes sharing an attribute if connected, *reciprocity* increases the probability of a directional tie between two nodes to be reciprocated with a tie in the other direction, and *transitivity* increases the probability of a tie between two nodes already having ties to another shared node (forming a closed triangle). All these different processes occur simultaneously, to different degrees and with varying influence, and the network is the overall manifestation of them (Lusher et al., 2013).

The network modelling tools attempt to fit models based on defined sets of configurations to observed networks using maximum likelihood estimations. Conceptually, we can think about each configuration as an independent variable in an ordinary regression model, with the observed network constituting the dependent variable and the estimated parameter for each configuration capturing how strongly it contributes to explaining it. However, the parameters are all estimated simultaneously, helping the models to overcome the challenges of dependence in the data and structural entanglement—that is, a configuration encapsulated in another more complex configuration (Bodin et al., 2016). Moreover, these tools are particularly useful to disentangle and quantify different network effects and uncertainties and distinguish between different processes that may generate similar network effects (Lusher et al., 2013). Suppose a model produces a network (converge) that is sufficiently similar to the observed network (goodness-of-fit) and close enough in its descriptions of the underlying processes for the model to be adequately estimated (non-degenerate, see Goodreau et al., 2008, p. 12). Then, the parameter estimations are informative for understanding and explaining the underlying processes behind the observed network.

Two commonly used network modelling tools for cross-sectional data are Exponential random graph models (ERGMs) and Autologistic actor attribute models (ALAAM). ERGMs are tie-based statistical models for investigating and anticipating how and why ties form to create an overall network structure, permitting inferences about how ties are patterned whilst assuming nodal attributes to be fixed (Lusher et al., 2013). ALAAMs stem from the same family of models but are instead attribute-based models for investigating and anticipating how and why nodal attributes change in the network, also permitting inferences whilst assuming the network structure to be fixed (Lusher et al., 2021). These nodal attributes can include everything from an attitude to a virus infection.

Common network modelling tools for longitudinal data include Stochastic actor-oriented models (SAOM) and Relational event models (REM). There are also adaptations of ERGMs for temporal or longitudinal data, as well as SAOMs for cross-sectional data (Block et al., 2019). SAOMs are a family of models for investigating and anticipating the co-evolution of network ties and nodal attributes (Kalish, 2020). In other words, models for studying how social tie formation and actors' behaviour, performance, attitudes or health influence each other over time. These models use panel data with observations of the entire network at discrete points in time (Robins, 2015). However, longitudinal data come in other forms, demanding either the data to be restructured or the use of different network modelling tools. For data in the shape of time-stamped interactions between pairs of nodes—such as phone calls, emails and physical interaction—REMs can be used directly to model networks from the bottom up. This approach is conceptually similar to agent-based modelling seeking to capture macro-level patterns by modelling micro-level dynamics (Butts, 2008, p. 158)—even when based on hundreds of millions of pairwise interactions (Lerner & Lomi, 2020). Using the same configurations as in the other network modelling tools, REMs allow modelling relational event histories and figuring out what patterns of interaction are more or less common, taking into account past interaction patterns, nodal attributes and contextual factors (Pilny et al., 2016, p. 181).

Microworlds

Once a quantitative model has been developed using stock and flow, agent-based or network modelling, it can be extended into a microworld by providing an interactive and user-friendly interface for users to experiment with the model (Cavana & Maani, 2000; Herrera & Kopainsky, 2022; Wolstenholme, 1999). Hence, microworlds are not a distinct tool for modelling systems, as such, but for managing complex functional models—somewhat similar to the relationship between causal loop modelling and systems archetypes presented above. Their contribution can still be indispensable since they open up for much wider participation by allowing policymakers, practitioners and members of the public to join the experimentation. However, it is essential to remember that microworlds, in this sense, have nothing to do with objectively understanding how the system works and evaluating policy outcomes, like Elsawah and colleagues suggest (2015, p. 502). They are as imperfect yet useful as the models on which they are built. However, facilitating the active participation of various actors provides opportunities for including different perspectives and expertise that can enrich our understanding of the human-environment system and experiences that can assist its validation in relation to the intricate reality in which we are interested.

Conclusion

Many issues of risk, resilience and sustainability are complex and challenging to explain, understand and improve, especially as they require the integrated knowledge and effort of various actors from across functional sectors, administrative levels and geographical borders (Haimes, 2004; Renn, 2008, pp. 8–9; Renn & Schweizer, 2009). Unfortunately, efforts to manage risk and resilience for sustainable development have tended in the past to reduce problems into parts that fit functional sectors and organisational mandates (Fordham, 2007), which is a significant weakness as it continuously clouds the bigger picture (Hale & Heijer, 2006, p. 139). Twigg (2004, p. 271) may be right when stating that '[m]ulti-sectoral approaches are ideal in theory, but their complexity causes problems'. However, we have no choice but to try to grasp such complexity as the alternative has proven insufficient for decades. Managing risk and resilience for sustainable development is thus not about dividing issues into parts that fit the agenda or mandate of specific actors, but rather about grasping the dynamics and nonlinear interdependencies in complex systems (see Hollnagel, 2006, pp. 14–17).

To grasp this complexity, we need to construct human-environment systems to represent the parts of the world in which we are interested. Human-environment systems include structural and functional models, which include different elements and dependencies. These models can be either qualitative or quantitative. It may be challenging to mathematically model all dependencies between elements that appear on the surface to be involved in what the human-environment system does. It is thus important to note that it is still possible to 'determine the most important structural aspects that lie behind system viability and performance' (Jackson, 2003, p. 21). It means that qualitative methods can also elicit information on the human-environment system's structural and functional aspects, which is central to explaining, understanding and improving issues of risk, resilience and sustainability. Although it is impossible to know any system completely (Skyttner, 2005, p. 100), the goal must be an as rich and comprehensive picture as possible.

It is not only formal expertise that is vital when constructing human-environment systems to manage risk and resilience for sustainable development. The educated common sense of other actors is also important and provides a degree of moral force and political influence to the results (Ravetz, 1999, p. 651). Vickers (1968, pp. 469–470) takes this argument further when claiming that:

> *Over many decades, things which used to be regarded as 'acts of God' — war, famine, pestilence; or as part of the nature of things — crime, destitution, ignorance, have come to be regarded as controllable and are hence assumed to be somebody's responsibility. They can all be 'fixed'; it is just a matter of know-how. It is true and welcome that the degree of our control is slowly extending but the assumptions based on this extension are false and dangerous. Not everything can be fixed; and fixing is never just know-how. It is always decision, made at the cost of not fixing something else. Until both governors and governed have a common and realistic view of what can be controlled and how far and at what cost, the relations between them are bound to be disturbed; and these disturbances may be as dangerous to the system as any.*

Hence, it is not only for effectiveness that managing risk and resilience for sustainable development demands wide participation of actors, but also for pre-empting public discontent by distributing responsibility and facilitating realistic expectations—public discontent that seems to be increasing in the wake of recent examples of calamity (Renn, 2008, p. 1).

Finally, this chapter introduces a toolbox for approaching the world as human-environment systems, ranging from Venn diagraming to advanced spatial mapping and analysis for constructing structural models and between causal loop modelling and microworlds for constructing functional models. However, it is vital to realise that the more advanced tools are not necessarily better than the more elementary ones. They are simply different tools for different purposes, just like a hammer and a LiDAR. Use the right tool for what you intend and can do with the available time and resources.

References

American Heritage Dictionary. (2000). *American heritage dictionary*. American Heritage Publishing Company.

Amsterdamska, O. (1990). Surely you are joking, Monsieur Latour. *Science, Technology & Human Values, 15*(4), 495–504. http://www.jstor.org/stable/689826.

Anderson, M. G., & Holcombe, E. (2013). *Community-based landslide risk reduction: Managing disasters in small steps*. The World Bank.

An, L., Linderman, M., Qi, J., Shortridge, A., & Liu, J. (2005). Exploring complexity in a human–environment system: An agent-based spatial model for multidisciplinary and multiscale integration. *Annals of the Association of American Geographers, 95*, 54–79.

Ashby, W. R. (1957). *An introduction to cybernetics*. Chapman & Hall Ltd.

Ashby, W. R. (1960). *Design for a brain: The origin of adaptive behavior*. John Wiley & Sons.

Axelrod, R. M., & Cohen, M. D. (2000). *Harnessing complexity: Organizational implications of a scientific frontier*. Basic Books.

Baggio, J. A., BurnSilver, S. B., Arenas, A., Magdanz, J. S., Kofinas, G. P., & De Domenico, M. (2016). Multiplex social ecological network analysis reveals how social changes affect community robustness more than resource depletion. *Proceedings of the National Academy of Sciences, 113*(48), 13708–13713. https://doi.org/10.1073/pnas.1604401113

Baird, J., Jollineau, M., Plummer, R., & Valenti, J. (2016). Exploring agricultural advice networks, beneficial management practices and water quality on the landscape: A geospatial social-ecological systems analysis. *Land Use Policy, 51*, 236–243. https://doi.org/10.1016/j.landusepol.2015.11.017

Balakrishnan, R., & Ranganathan, K. (2012). *Textbook of graph theory*. Springer.

Becker, P. (2018). Dependence, trust, and influence of external actors on municipal urban flood risk mitigation: The case of Lomma Municipality, Sweden. *International Journal of Disaster Risk Reduction, 31*(1004), 1004–1012. https://doi.org/10.1016/j.ijdrr.2018.09.005

Bergsten, A., Jiren, T. S., Leventon, J., Dorresteijn, I., Schultner, J., & Fischer, J. (2019). Identifying governance gaps among interlinked sustainability challenges. *Environmental Science & Policy, 91*, 27–38. https://doi.org/10.1016/j.envsci.2018.10.007

Bergström, J., & Dekker, S. W. A. (2019). The 2010s and onward: Resilience engineering. In S. W. A. Dekker (Ed.), *Foundations of safety science* (pp. 391–429). CRC Press.

Blanchard, B. S., & Fabrycky, W. J. (2006). *Systems engineering and analysis*. Pearson/Prentice Hall.

Blauberg, I. V., Sadovsky, V. N., Yudin, E. G., Syrovatkin, S., & Germogenova, O. (1977). *Systems theory: Philosophical and methodological problems*. Progress Publishers.

Block, P., Stadtfeld, C., & Snijders, T. A. B. (2019). Forms of dependence: Comparing SAOMs and ERGMs from basic principles. *Sociological Methods & Research, 48*(1), 202–239. https://doi.org/10.1177/0049124116672680

Blok, A., Farias, I., & Roberts, C. (Eds.). (2020). *The routledge companion to actor-network theory*. Routledge.

Bloor, D. (1999a). Anti-latour. *Studies in History and Philosophy of Science, 30a*(1), 81–112. https://doi.org/10.1016/S0039-3681(98)00038-7

Bloor, D. (1999b). For David bloor and beyond: A reply to David bloor's 'anti-Latour' - reply to Bruno Latour. *Studies in History and Philosophy of Science, 30a*(1), 131–136.

Boardman, J., & Sauser, B. (2008). *Systems thinking: Coping with 21st century problems*. CRC Press.

Bodin, Ö. (2017). Collaborative environmental governance: Achieving collective action in social-ecological systems. *Science, 357*(6352). https://doi.org/10.1126/science.aan1114. eaan1114.

Bodin, Ö., Robins, G., McAllister, R. R. J., Guerrero, A. M., Crona, B., Tengö, M., & Lubell, M. (2016). Theorizing benefits and constraints in collaborative environmental governance: A transdisciplinary social-ecological network approach for empirical investigations. *Ecology and Society, 21*(1). https://doi.org/10.5751/ES-08368-210140. art40-14.

Borgatti, S. P., Everett, M. G., & Johnson, J. C. (2018). *Analysing social networks*. SAGE Publications.

Boulding, K. E. (1956). General systems theory: The skeleton of science. *Management Science, 2*, 197–208.

Box, G. E. P., & Draper, N. R. (2007). *Response surfaces, mixtures, and ridge analyses* (2nd edition). Wiley-Interscience.

Branlat, M., & Woods, D. D. (2010). How do systems manage their adaptive capacity to successfully handle disruptions? A resilience engineering perspective. In *Complex adaptive systems—resilience, robustness, and evolvability, proceedings from the association for the advancement of artificial intelligence conference, Arlington, 11-13 November 2010.*

Buckley, W. F. (1968). Society as a complex adaptive system. In W. F. Buckley (Ed.), *Modern systems research for the behavioral scientist: A sourcebook*. Aldline Publishing.

Butts, C. T. (2008). A relational event framework for social action. *Sociological Methodology, 38*(1), 155–200. https://doi.org/10.1111/j.1467-9531.2008.00203.x

Butts, C. T. (2017). Comment: Actor orientation and relational event models. *Sociological Methodology, 47*, 47–56. https://doi.org/10.2307/26429061

Callon, M. (1990). Techno-economic networks and irreversibility. *Sociological Review, 38*(1), 132–161.

Catton, W. R., Jr., & Dunlap, R. E. (1978). Environmental sociology: A new paradigm. *The American Sociologist, 13*, 41–49.

Cavana, R. Y., & Maani, K. E. (2000). Methodological framework for integrating systems thinking and system dynamics. In *Proceedings of 18th international conference of the system dynamics society* (pp. 6–10).

Checkland, P. (1981). *Systems thinking, systems practice*. John Wiley & Sons.

Checkland, P. (1985). From optimizing to learning: A development of systems thinking for the 1990s. *Journal of the Operational Research Society, 36*, 757–767.
Churchman, C. W., & Ackoff, R. L. (1950). Purposive behavior and cybernetics. *Social Forces, 29*, 32–39. https://doi.org/10.2307/2572754
Clarke, W. C., & Dickson, N. M. (2003). Sustainability science: The emerging research program. *Proceedings of the National Academy of Sciences, 100*, 8059–8061.
Collins, H. M., & Yearley, S. (1992). Epistemological chicken. In A. Pickering (Ed.), *Science as practice and culture* (pp. 301–326). University of Chicago Press.
Conant, R. C., & Ashby, W. R. (1970). Every good regulator of a system must be a model of that system. *International Journal of Systems Science, 1*, 89–97.
Coscia, M., Hausmann, R., & Hidalgo, C. A. (2013). The structure and dynamics of international development assistance. *Journal of Globalization and Development, 3*(2), 1–42. https://doi.org/10.1515/jgd-2012-0004
Crawford, T. H. (2020). Actor-network theory. In P. Rabinowitz (Ed.), *Oxford research encyclopedia of literature*. Oxford University Press. https://doi.org/10.1093/acrefore/9780190201098.013.965
Crooks, A. T., & Heppenstall, A. J. (2012). Introduction to agent-based modelling. In A. J. C. Heppenstall, T. Andrew, L. M. See, & M. Batty (Eds.), *Agent-based models of geographical systems* (pp. 85–105). Springer. https://doi.org/10.1007/978-90-481-8927-4_5
Curtin, K. M. (2018). Network analysis. In B. Huang, T. J. Cova, & M.-H. Tsou (Eds.), *Comprehensive geographic information systems: Geographic information system* (pp. 153–161). Elsevier.
da Silva, L. B. L., Humberto, J. S., Alencar, M. H., Ferreira, R. J. P., & de Almeida, A. T. (2020). GIS-based multidimensional decision model for enhancing flood risk prioritization in urban areas. *International Journal of Disaster Risk Reduction, 48*, 101582. https://doi.org/10.1016/j.ijdrr.2020.101582
Dobson, I., Carreras, B. A., Lynch, V. E., & Newman, D. E. (2007). Complex systems analysis of series of blackouts: Cascading failure, critical points, and self-organization. *Chaos, 17*, 026101–026103. https://doi.org/10.1063/1.2737822
Dörner, D. (1996). *The logic of failure: Recognizing and avoiding error in complex situations*. Metropolitan Books.
Dunlap, R. E., & Catton, W. R. J. (1979). Environmental sociology. *Annual Review of Sociology, 5*, 243–273.
Ehrenfeld, J. R. (2005). The roots of sustainability. *MIT Sloan Management Review, 46*, 23–25.
Elsawah, S., Guillaume, J. H. A., Filatova, T., Rook, J., & Jakeman, A. J. (2015). A methodology for eliciting, representing, and analysing stakeholder knowledge for decision making on complex socio-ecological systems: From cognitive maps to agent-based models. *Journal of Environmental Management, 151*, 500–516. https://doi.org/10.1016/j.jenvman.2014.11.028
Emirbayer, M. (1997). Manifesto for a relational sociology. *American Journal of Sociology, 103*, 281–317.
Farmer, J. D., & Foley, D. (2009). The economy needs agent-based modelling. *Nature, 460*(7256), 685–686. https://doi.org/10.1038/460685a
Fordham, M. H. (2007). Disaster and development research and practice: A necessary eclecticism. In H. Rodríguez, E. L. Quarantelli, & R. R. Dynes (Eds.), *Handbook of disaster research* (pp. 335–346). Springer.
Forrester, J. W. (1969). *Urban dynamics*. Productivity Press.
Forrester, J. W. (1994). System dynamics, systems thinking, and soft OR. *System Dynamics Review, 10*, 245–256.
Fried, H., Hamilton, M., & Berardo, R. (2022). Theorizing multilevel closure structures guiding forum participation. *Journal of Public Administration Research and Theory*. https://doi.org/10.1093/jopart/muac042
Galappaththi, E. K., Ichien, S. T., Hyman, A. A., Aubrac, C. J., & Ford, J. D. (2020). Climate change adaptation in aquaculture. *Reviews in Aquaculture, 12*(4), 2160–2176. https://doi.org/10.1111/raq.12427
Gallopín, G. C. (2006). Linkages between vulnerability, resilience, and adaptive capacity. *Global Environmental Change, 16*, 293–303. https://doi.org/10.1016/j.gloenvcha.2006.02.004
Goodreau, S. M., Handcock, M. S., Hunter, D. R., Butts, C. T., & Morris, M. (2008). A statnet tutorial. *Journal of Statistical Software, 24*(9). https://doi.org/10.18637/jss.v024.i09
Haimes, Y. Y. (2004). *Risk modeling, assessment, and management*. John Wiley & Sons.

Hale, A. R., & Heijer, T. (2006). Is resilience really necessary? The case of railways. In E. Hollnagel, D. D. Woods, & N. G. Leveson (Eds.), *Resilience engineering: Concepts and precepts* (pp. 125–147). Ashgate.

Haque, C. E., & Etkin, D. (2007). People and community as constituent parts of hazards: The significance of societal dimensions in hazards analysis. *Natural Hazards, 41*, 271–282. https://doi.org/10.1007/s11069-006-9035-8

Hardin, G. (1968). The tragedy of the commons. *Science, 162*, 1243–1248. https://doi.org/10.1126/science.162.3859.1243

Hearn, J. (2012). *Theorizing power*. Palgrave Macmillan.

Hedlund, J., Bodin, Ö., & Nohrstedt, D. (2021). Policy issue interdependency and the formation of collaborative networks. *People and Nature, 3*(1), 236–250. https://doi.org/10.1002/pan3.10170

Hedström, P. (2005). *Dissecting the social: On the principles of analytical sociology*. Cambridge University Press.

Herrera, H., & Kopainsky, B. (2022). Using microworlds for policymaking in the context of resilient farming systems. *Journal of Simulation*, 1–25. https://doi.org/10.1080/17477778.2022.2083990

Heywood, D. I., Cornelius, S., & Carver, S. (2006). *An introduction to geographical information systems*. Pearson Prentice Hall.

Hileman, J., & Lubell, M. (2018). The network structure of multilevel water resources governance in Central America. *Ecology and Society, 23*(2). https://doi.org/10.5751/es-10282-230248

Holland, J. H. (1998). *Emergence: From chaos to order*. Oxford University Press.

Hollenstein, K., Bieri, O., & Stückelberger, J. (2002). *Modellierung der Vulnerability vonSchadenobjekten gegenüber Naturgefahrenprozessen*.

Hollnagel, E. (2006). Resilience—the challenge of the unstable. In E. Hollnagel, D. D. Woods, & N. G. Leveson (Eds.), *Resilience engineering: Concepts and precepts* (pp. 9–17). Ashgate.

Huang, D., Wang, S., & Liu, Z. (2021). A systematic review of prediction methods for emergency management. *International Journal of Disaster Risk Reduction, 62*, 102412. https://doi.org/10.1016/j.ijdrr.2021.102412

Hu, X., Naim, K., Jia, S., & Zhengwei, Z. (2021). Disaster policy and emergency management reforms in China: From Wenchuan earthquake to Jiuzhaigou earthquake. *International Journal of Disaster Risk Reduction, 52*, 101964. https://doi.org/10.1016/j.ijdrr.2020.101964

IFAD. (2009). *Good practices in participatory mapping*.

IFRC. (2007). *VCA toolbox*.

Ingelstam, L. (2012). System: Att tänka över samhälle och teknik. *Energimyndigheten*. https://www.google.com.

Jackson, M. C. (2003). *Systems thinking: Creative holism for managers*. John Wiley & Sons.

Jenkins, T. N. (2000). Putting postmodernity into practice: Endogenous development and the role of traditional cultures in the rural development of marginal regions. *Ecological Economics, 34*, 301–314. https://doi.org/10.1016/S0921-8009(00)00191-9

Johansson, H., & Jönsson, H. (2007). *Metoder för risk- och sårbarhetsanalys ur ett systemperspektiv*.

Kalish, Y. (2020). Stochastic actor-oriented models for the Co-evolution of networks and behavior: An introduction and tutorial. *Organizational Research Methods, 23*(3), 511–534. https://doi.org/10.1177/1094428118825300

Kates, R. W., Clarke, W. C., Corell, R. W., Hall, J. M., Jaeger, C. C., Lowe, I., McCarthy, J. J., Schellnhuber, H. J., Bolin, B., Dickson, N. M., Faucheux, S., Gallopín, G. C., Huntley, B., Jodha, N. S., Kasperson, R. E., Mabogunje, A., Matson, P. A., Mooney, H., Moore, B., … Svedin, U. (2001). Sustainability science. *Science, 292*, 641–642.

Knoke, D., Diani, M., Hollway, J., & Christopoulos, D. (2021). *Multimodal political networks*. Cambridge University Press. https://doi.org/10.1017/9781108985000

Latour, B. (1992). Where are the missing masses? The sociology of a few mundane artifacts. In W. E. Bijker, & J. Law (Eds.), *Shaping technology/building society: Studies in sociotechnical change* (pp. 225–258). MIT Press.

Lazega, E., & Snijders, T. A. B. (Eds.). (2016). *Multilevel network analysis for the social sciences*. Springer. https://doi.org/10.1007/978-3-319-24520-1

Lerner, J., & Lomi, A. (2020). Reliability of relational event model estimates under sampling: How to fit a relational event model to 360 million dyadic events. *Network Science, 8*(1), 97–135. https://doi.org/10.1017/nws.2019.57

Little, R. G. (2002). Controlling cascading failure: Understanding the vulnerabilities of interconnected infrastructures. *Journal of Urban Technology, 9*, 109–123.

Li, M., Wiedmann, T., Fang, K., & Hadjikakou, M. (2021). The role of planetary boundaries in assessing absolute environmental sustainability across scales. *Environment International, 152*, 106475. https://doi.org/10.1016/j.envint.2021.106475

Longo, S. B., Isgren, E., Clark, B., Jorgenson, A. K., Jerneck, A., Olsson, L., Kelly, O. M., Harnesk, D., & York, R. (2021). Sociology for sustainability science. *Discover Sustainability, 2*(1), 1–14. https://doi.org/10.1007/s43621-021-00056-5

Lusher, D., Koskinen, J., & Robins, G. (Eds.). (2013). *Exponential random graph models for social networks: Theory, methods, and applications*. Cambridge University Press. http://ebooks.cambridge.org/ref/id/CBO9780511894701.

Lusher, D., Wang, P., Brennecke, J., Brailly, J., Faye, M., & Gallagher, C. (2021). Advances in exponential random graph models. In R. Light, & J. Moody (Eds.), *The oxford handbook of social networks* (pp. 234–253). Oxford University Press. https://doi.org/10.1093/oxfordhb/9780190251765.013.18

Maani, K. E., & Cavana, R. Y. (2000). *Systems thinking and modelling: Understanding change and complexity*. Prentice Hall.

Macy, M. W., & Willer, R. (2002). From factors to actors: Computational sociology and agent-based modeling. *Annual Review of Sociology, 28*(1), 143–166. https://doi.org/10.1146/annurev.soc.28.110601.141117

Malik, H. A. M., Mahesar, A. W., Abid, F., Waqas, A., & Wahiddin, M. R. (2017). Two-mode network modeling and analysis of dengue epidemic behavior in Gombak, Malaysia. *Applied Mathematical Modelling, 43*, 207–220. https://doi.org/10.1016/j.apm.2016.10.060

Malinick, T. E., Tindall, D. B., & Diani, M. (2013). Network centrality and social movement media coverage: A two-mode network analytic approach. *Social Networks, 35*(2), 148–158. https://doi.org/10.1016/j.socnet.2011.10.005

Mansourian, A., Pilesjö, P., Harrie, L., & van Lammeren, R. (Eds.). (2018). *Geospatial technologies for all*. Springer. https://doi.org/10.1007/978-3-319-78208-9

Månsson, P. (2019). Uncommon sense: A review of challenges and opportunities for aggregating disaster risk information. *International Journal of Disaster Risk Reduction, 40*, 101149. https://doi.org/10.1016/j.ijdrr.2019.101149

Maruyama, M. (1963). The second cybernetics: Deviation-amplifying mutual processes. *American Scientist, 5*, 164–179.

McCarty, C., Lubbers, M. J., Vacca, R., & Molina, J. L. (2019). *Conducting personal network research: A practical guide*. Guilford Publications.

Meadows, D. H., & Wright, D. (2009). *Thinking in systems: A primer*. Earthscan.

Metzger, M. J., Schröter, D., Leemans, R., & Cramer, W. (2008). A spatially explicit and quantitative vulnerability assessment of ecosystem service change in Europe. *Regional Environmental Change, 8*, 91–107. https://doi.org/10.1007/s10113-008-0044-x

Milanovic, J. V., & Zhu, W. (2018). Modeling of interconnected critical infrastructure systems using complex network theory. *IEEE Transactions on Smart Grid, 9*(5), 4637–4648. https://doi.org/10.1109/tsg.2017.2665646

Newton, T. J. (2002). Creating the new ecological order? Elias and actor-network theory. *Academy of Management Review, 27*, 523–540. https://doi.org/10.2307/4134401

Nohrstedt, D., & Bodin, Ö. (2020). Collective action problem characteristics and partner uncertainty as drivers of social tie formation in collaborative networks. *Policy Studies Journal, 48*(4), 1082–1108. https://doi.org/10.1111/psj.12309

OECD. (2003). *Emerging risks in the 21st century: An agenda for action*.

Ongkowijoyo, C., & Doloi, H. (2017). Determining critical infrastructure risks using social network analysis. *International Journal of Disaster Resilience in the Built Environment, 8*, 5–26. https://doi.org/10.1108/IJDRBE-05-2016-0016

Ostrom, E., Burger, J., Field, C. B., Norgaard, R. B., & Policansky, D. (1999). Revisiting the commons: Local lessons, global challenges. *Science, 284*, 278–282. https://doi.org/10.1126/science.284.5412.278

Oxford English Dictionary. (2020). *Oxford dictionary of English*. Oxford University Press.

Parsons, T. (1951). *The social system*. The Free Press of Glencoe.

Parsons, T. (1971). *The system of modern societies*. Prentice-Hall.

Penteado, I. M., do Nascimento, A. C. S., Corrêa, D., Moura, E. A. F., Zilles, R., Gomes, M. C. R. L., Pires, F. J., Brito, O. S., da Silva, J. F., Reis, A. V., Souza, A., & Pacífico, A. C. N. (2019). Among people and artifacts: Actor-network theory and the adoption of solar ice machines in the Brazilian Amazon. *Energy Research & Social Science, 53*, 1–9. https://doi.org/10.1016/j.erss.2019.02.013

Perrow, C. B. (1999). Organizing to reduce the vulnerabilities of complexity. *Journal of Contingencies and Crisis Management, 7*, 150–155. https://doi.org/10.1111/1468-5973.00108

Pilny, A., Schecter, A., Poole, M. S., & Contractor, N. (2016). An illustration of the relational event model to analyze group interaction processes. *Group Dynamics: Theory, Research, and Practice, 20*(3), 181–195. https://doi.org/10.1037/gdn0000042

Rasmussen, J. (1985). The role of hierarchical knowledge representation in decisionmaking and system management. *IEEE Transactions on Systems, Man and Cybernetics, 15*, 234–243.

Ravetz, J. R. (1999). What is post-normal science. *Futures, 31*, 647–653. https://doi.org/10.1016/S0016-3287(99)00024-5

Reenberg, A., Birch-Thomsen, T., Mertz, O., Fog, B., & Christiansen, S. (2008). Adaptation of human coping strategies in a small island society in the SW pacific—50 years of change in the coupled human–environment system on Bellona, Solomon Islands. *Human Ecology, 36*, 807–819. https://doi.org/10.1007/s10745-008-9199-9

Renn, O. (2007). Components of the risk governance framework. In F. Bouder, D. Slavin, & R. E. Löfstedt (Eds.), *The tolerability of risk: A new framework for risk management* (pp. 7–20). Earthscan.

Renn, O. (2008). *Risk governance: Coping with uncertainty in a complex world*. Earthscan.

Renn, O., & Schweizer, P.-J. (2009). Inclusive risk governance: Concepts and application to environmental policy making. *Environmental Policy and Governance, 19*, 174–185. https://doi.org/10.1002/eet.507

Reynolds, J. F., Grainger, A., Stafford Smith, D. M., Bastin, G., Garcia-Barrios, L., Fernández, R. J., Janssen, M. A., Jürgens, N., Scholes, R. J., Veldkamp, A., Verstraete, M. M., Von Maltitz, G., & Zdruli, P. (2011). Scientific concepts for an integrated analysis of desertification. *Land Degradation & Development, 22*(2), 166–183. https://doi.org/10.1002/ldr.1104

Rinaldi, S. M., Peerenboom, J. P., & Kelly, T. K. (2001). Identifying, understanding, and analyzing critical infrastructure interdependencies. *IEEE Control Systems Magazine, 21*, 11–25.

Robins, G. (2015). *Doing social network research: Network-based research design for social scientists*. SAGE Publications.

Robins, G., Lewis, J. M., & Wang, P. (2012). Statistical network analysis for analyzing policy networks. *Policy Studies Journal, 40*(3), 375–401. https://doi.org/10.1111/j.1541-0072.2012.00458.x

Sarkar, P., Debnath, N., & Reang, D. (2021). Coupled human-environment system amid COVID-19 crisis: A conceptual model to understand the nexus. *Science of the Total Environment, 753*, 141757. https://doi.org/10.1016/j.scitotenv.2020.141757

Sayes, E. (2014). Actor-Network Theory and methodology: Just what does it mean to say that nonhumans have agency. *Social Studies of Science, 44*(1), 134–149. https://doi.org/10.1177/0306312713511867

Sayles, J. S., Mancilla Garcia, M., Hamilton, M., Alexander, S. M., Baggio, J. A., Fischer, A. P., Ingold, K., Meredith, G. R., & Pittman, J. (2019). Social-ecological network analysis for sustainability sciences: A systematic review and innovative research agenda for the future. *Environmental Research Letters, 14*(9), 1–18. https://doi.org/10.1088/1748-9326/ab2619

Schelling, T. C. (1978). *Micromotives and macrobehavior*. W. W. Norton.

Senge, P. (2006). *The fifth discipline: The art & practise of the learning organisation*. Currency & Doubleday.

Simon, H. A. (1962). The architecture of complexity. *Proceedings of the American Philosophical Society, 106*, 467–482.

Simon, H. A. (1990). Prediction and prescription in systems modeling. *Operations Research, 38*, 7–14.

Simon, H. A. (1996). *The sciences of the artificial*. MIT Press.

Skyttner, L. (2005). *General systems theory: Problems, perspectives, practice*. World Scientific.

Sztompka, P. (1993). *The sociology of social change*. Blackwell.
Turner, B. L., II, Clarke, W. C., Kates, R. W., Richards, J. F., Mathews, J. T., & Meyer, W. B. (Eds.). (1990). *The earth as transformed by human action: Global and regional changes in the biosphere over the past 300 years*. Cambridge University Press with Clark University.
Turner, B. L., II, Matson, P. A., McCarthy, J. J., Corell, R. W., Christensen, L., Eckley, N., Hovelsrud-Broda, G. K., Kasperson, J. X., Kasperson, R. E., Luers, A., Martello, M. L., Mathiesen, S., Naylor, R., Polsky, C., Pulsipher, A., Schiller, A., Selin, H., & Tyler, N. (2003). Illustrating the coupled human-environment system for vulnerability analysis: Three case studies. *Proceedings of the National Academy of Sciences, 100*, 8080–8085. https://doi.org/10.2307/3139883
Twigg, J. (2004). *Disaster risk reduction: Mitigation and preparedness in development and emergency programming*. Overseas Development Institute.
Umpleby, S. A., & Dent, E. B. (1999). The origins and purposes of several traditions in systems theory and cybernetics. *Cybernetics & Systems, 30*, 79–103.
Venn, J. (1881). *Symbolic logic*. Macmillan and Co.
von Bertalanffy, L. (1950). An outline of general system theory. *The British Journal for the Philosophy of Science, 1*, 134–165.
von Bertalanffy, L. (1960). *Problems of life: An evaluation of modern biological thought*. Harper & Torchbook.
Vickers, G. (1968). Is adaptability enough. In W. F. Buckley (Ed.), *Modern systems research for the behavioral scientist: A sourcebook* (pp. 460–473). Aldline Publishing.
Wang, H., Pan, Y., & Luo, X. (2019). Integration of BIM and GIS in sustainable built environment: A review and bibliometric analysis. *Automation in Construction, 103*, 41–52. https://doi.org/10.1016/j.autcon.2019.03.005
Wasserman, S., & Faust, K. (1994). *Social network analysis: Methods and applications*. Cambridge University Press.
Weiner, N. (2019). *Cybernetics: Or control and communication in the animal and the machine*. MIT Press.
Wiek, A., Ness, B., Schweizer-Ries, P., Brand, F. S., & Farioli, F. (2012). From complex systems analysis to transformational change: A comparative appraisal of sustainability science projects. *Sustainability Science, 7*(S1), 5–24. https://doi.org/10.1007/s11625-011-0148-y
Wilensky, U., & Rand, W. (2015). *An introduction to agent-based modeling: Modeling natural, social, and engineered complex systems with NetLogo*. The MIT Press.
Wisner, B. (2006). Self-assessment of coping capacity: Participatory, proactive and qualitative engagement of communities in their own risk management. In J. Birkmann (Ed.), *Measuring vulnerability to natural hazards: Towards disaster resilient societies*. Teri Press.
Wolstenholme, E. F. (1999). Qualitative vs quantitative modelling: The evolving balance. *Journal of the Operational Research Society, 50*, 422–428.
Yariyan, P., Zabihi, H., Wolf, I. D., Karami, M., & Amiriyan, S. (2020). Earthquake risk assessment using an integrated fuzzy analytic hierarchy process with artificial neural networks based on GIS: A case study of Sanandaj in Iran. *International Journal of Disaster Risk Reduction, 50*, 101705. https://doi.org/10.1016/j.ijdrr.2020.101705
Yin, S., Yang, X., & Chen, J. (2020). Adaptive behavior of farmers' livelihoods in the context of human-environment system changes. *Habitat International, 100*, 102185. https://doi.org/10.1016/j.habitatint.2020.102185
Young, I. M. (1990). *Justice and the politics of difference*. Princeton University Press.
Yurui, L., Yi, L., Pengcan, F., & Hualou, L. (2019). Impacts of land consolidation on rural human–environment system in typical watershed of the Loess Plateau and implications for rural development policy. *Land Use Policy, 86*, 339–350. https://doi.org/10.1016/j.landusepol.2019.04.026

PART III

Changing the world

CHAPTER 12

Science and change

Introduction

Although the first part of this book focuses on describing the state of the world and the second part on constructing a conceptual foundation for approaching the world, it is now time to focus on changing the world. Chapter 2 suggests that science has been central to the development of the world for centuries. It indicates that the rate of change has been closely connected to the speed of scientific development, especially since the period before the Industrial Revolution. This period is often referred to as the Scientific Revolution and involved the emergence of modern science when accumulating developments in mathematics, physics, chemistry and biology overthrew previously held worldviews (Kuhn, 1970). Scientific development has continued to accelerate, and so has the world's general development. It is, therefore, a great paradox that science has not only assisted humankind in overcoming fundamental sustainability challenges in the past but also created new ones that we yet have to resolve.

Although it is a driving force behind much groundbreaking research, science is not only about satisfying curiosity. It is about explaining, understanding and improving the world we observe by developing knowledge, informing policy, addressing societal issues, developing technology, etc. The remarkable Thomas Huxley once wrote: 'The great end of life is not knowledge but action' (Huxley, 1882, p. 89). I think he is right. Knowledge is fundamental but only useful if it can be organised into action to address challenges. However, I am not saying that all scientific endeavours should be applied to particular challenges since more fundamental research is needed to provide knowledge to assist us in these endeavours and in addressing a range of challenges that we may not even know we have yet. I claim that scientific development is intrinsically linked to the development of the world, but that we have to beware of 'Hume's Guillotine' presented in Chapter 7. In other words, it is fundamental to remember the difference between descriptive statements about how something is and prescriptive statements about how it ought to be, and inferring the latter directly from the former is simply impossible.

This chapter provides a foundation for linking science and change by suggesting a way to bridge the divide between the descriptive and the prescriptive whilst maintaining scientific rigour throughout the process. It elaborates on the two forms of science introduced in Chapter 7—traditional science and design science—their purposes and limitations and how they complement each other. The chapter then details their

Sustainability Science, Second Edition
ISBN 978-0-323-95640-6, https://doi.org/10.1016/B978-0-323-95640-6.00017-8

respective processes, presents different scientific principles needed to assess scientific rigour, and discusses science's fundamental limitations for change.

The sciences of the complemental

The philosophical assumptions presented in Chapter 7 portray traditional science and design science as different in their relation to values. This difference is, however, neither unambiguous nor pitting the two against each other. Regardless of whether we embrace the influence of values in our search for knowledge or still pursue the unreachable ideal of value-neutrality, the relevance of the constructed knowledge about how the world is can only be judged in relation to normative values. This relevance usually refers to the utility of scientific knowledge for solving challenges for humankind, even if traditional science is often satisfied by simply assuming that the knowledge produced will be used at some point in the future (Lee, 2007, p. 44). However, describing the world cannot solve problems on its own, as problems entail prescriptive statements about how the world ought to be. On the other hand, design science is made for solving problems (Simon, 1996) but not for describing how the world is, which is also necessary for defining the problem and anticipating the results of potential activities (Ness et al., 2010, p. 479). Nonetheless, it is essential to note that the problem-solving of design science can also be scientifically rigorous, even if based on principles other than traditional science (Checkland & Holwell, 2007, pp. 3–4). Traditional science and design science are, thus, complementary parts of managing risk and resilience for sustainable development (Figure 12.1).

Solving problems entails changing something from the current state, which at least one human being perceives as unsatisfactory, to the desired state (Ackoff, 1962, p. 30). This definition signifies intentional and purposeful activities to change the world for human purposes (Wieringa, 2009, p. 1). However, the changes that human agency can cause in the world are limited to what our activities can influence. Purposeful products of intentional human activities are called artefacts (Hilpinen, 1993). By designing and utilising these artefacts, human beings shape both their present and future (Simon, 1996). It

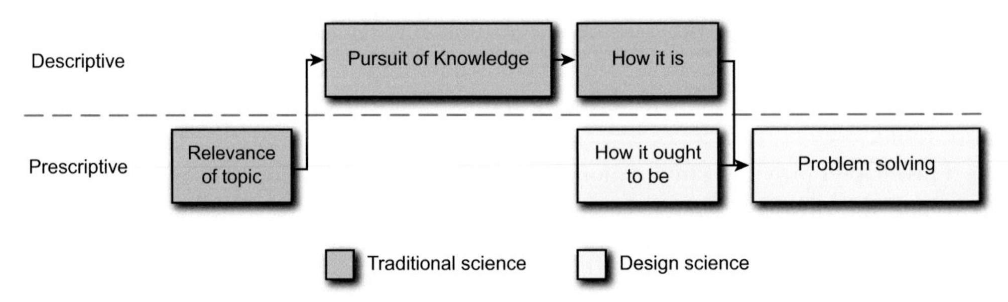

Figure 12.1 Traditional science and design science as complementary.

is, in this context, important to remember that artefacts can be either physical (tools, constructions, etc.) or conceptual (symbols, methods, etc.) (Hilpinen, 1993; Simon, 1996).

Building on Rasmussen (1985) work on functional hierarchy (Chapter 8), Brehmer (2007, pp. 212–214) suggests that every artefact has purpose, function and form. The purpose answers why the artefact exists, the function describes what it must do to meet that purpose, and the form explains how the function is fulfilled in the real world (Brehmer, 2010, p. 4; Rasmussen, 1985). The purpose is, in other words, the highest level of abstraction whilst the form is the most concrete (Rasmussen, 1985), just as presented when operationalising resilience in Chapter 8. Since an artefact is evaluated on the relationships between its purpose, intended character and its actual character (Hilpinen, 1995, p. 140), evaluation means assessing how well the form fulfils the required functions to meet the purpose when utilised.

The main activities of traditional science are to theorise and justify, whilst design science's main activities are building and evaluating (March & Smith, 1995). Managing risk and resilience for sustainable development require both sets of activities (Figure 12.2), as addressing sustainability challenges entails building and evaluating artefacts designed and evaluated against criteria for which the argumentation is informed by descriptive research.

Two scientific processes

Traditional science is about the pursuit of knowledge. It is about explaining and understanding phenomena (Keat & Urry, 1975). When looking at the disparate work in many scientific disciplines, it is clear that there are countless ways to do that. However, when focusing such investigation on the overall strategies applied for the various scientific endeavours, the diversity can be condensed into four broad logics of inquiry (Blaikie, 2010, pp. 81–92). In other words, the search for explanation and understanding can be inductive, deductive, retroductive or abductive.

The inductive research strategy aims to establish descriptions of phenomena as the basis for theory by accumulating observations, constructing generalisations and using these as patterns to explain further observations (Blaikie, 2010). The observation comes before conceptualisation and further observation. On the other hand, the deductive research strategy aims to 'test theories, to eliminate false ones and corroborate the

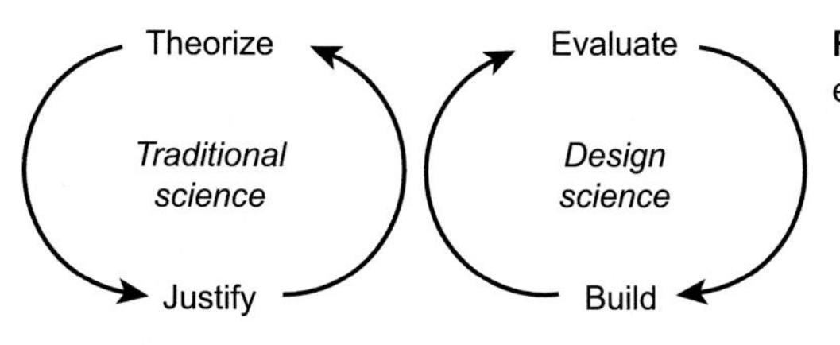

Figure 12.2 The main activities of traditional science and design science.

survivor' (Blaikie, 2010, p. 84) by forming hypotheses and testing them against observations. Conceptualisation comes before observation and further conceptualisation. The retroductive research strategy has similarities and differences with both inductive and deductive strategies. It aims to 'discover underlying mechanisms to explain observed regularities' (Blaikie, 2010, p. 84) by observing a regularity, constructing a hypothetical model of the mechanisms capable of producing the regularity and testing the model against observations (Sayer, 1992). The initial observation comes before conceptualisation and further observation. Finally, the abductive research strategy is mainly used in the social sciences (Blaikie, 2010). It aims to 'understand social life in terms of social actors' meanings and accounts' (Blaikie, 2010, p. 84) by discovering everyday concepts, meanings and motives, translating these accounts to scientific accounts and developing a theory to test iteratively (Blaikie, 2010, pp. 89–92). The observation comes before conceptualisation and further observation.

Regardless of the logic of inquiry, the process of traditional science can, in very simple terms, be illustrated as a cycle of observation and conceptualisation—or of justifying and theorising to use March and Smith's (1995) language from earlier. The main difference between the different logics of inquiry is where you start each investigation in the cycle. This illustration is a crude simplification destined to provoke scientists from various disciplines and traditions. Still, it may be helpful to illustrate further the difference between the process of traditional science and the process of design science.

If the process of traditional science can be viewed as a cycle of observation and conceptualisation—or theorising and justifying theories—then the process of design science can be viewed as a cycle of building and evaluating artefacts (March & Smith, 1995). However, to grasp the design science process, we must look more closely into what it entails and consider what makes it scientific.

Let us start with the latter. Regardless of the differences with traditional science presented above, for any process to be scientific, in any sense, it must be systematic and transparent. In other words, the process cannot be considered scientific if our activities are arbitrary and impossible for others to follow. Here, the principles of reliability and validity are commonly considered, which I elaborate on in the next section of this chapter. Nevertheless, maintaining scientific rigour in designing artefacts requires a systematic and transparent design process in which prescriptive assumptions regarding the purpose and design criteria are explicitly stated, and the choices directed by those assumptions are justified through logical reasoning (Hassel, 2010, pp. 42–47). The design process used in this book is illustrated in Figure 12.3 below.

The first step in this design process is to clearly define the purpose (or purposes) of the artefact to be designed (Cook & Ferris, 2007, pp. 173–174; Simon, 1990, p. 13; Simon, 1996, pp. 4–5, 114). This purpose is generally described in rather abstract terms and acts like an overall guiding principle for the rest of the design process (Hassel, 2010, p. 43). The second step is defining the design criteria the artefact must meet (Wieringa, 2009,

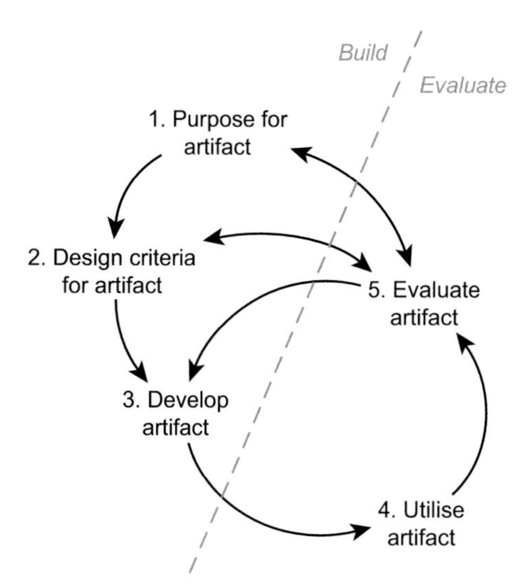

Figure 12.3 A scientific process for designing artefacts.

pp. 1–2). These design criteria are prescriptive assumptions about the artefact's required function (or functions), which must be appropriately justified through logical reasoning informed by established theory or new empirical research (Hassel, 2010, pp. 43–44). Here, the crucial link between the descriptive and the prescriptive occurs, not as an unreflective deduction over the philosophical chasm of 'Hume's Guillotine' from Chapter 7, but as conscious, transparent arguments grounded in normative values and informed by descriptive research.

The third step of the design process is to develop the actual form of the artefact based on our initial judgements regarding what is needed to meet the design criteria and purpose. The word *developed* is used here to signify that artefacts may already exist to improve or build on. The fourth step is to utilise the artefact in the intended context or in a setting designed to approximate that context (Hassel, 2010, p. 45). Utilising the artefact in the intended context is vital, as various contextual factors may influence its performance (March & Smith, 1995, p. 254; Simon, 1996, pp. 5–6). Moreover, it allows one to test theories about the context (March & Smith, 1995, p. 255; Wieringa, 2014), linking to traditional science. The application of the artefact can also cause learning that may inspire modifications in purpose and design criteria.

The fifth step of the design process is to evaluate the performance of the artefact against its design criteria and purpose (Figure 12.3). If the result of this evaluation is unsatisfactory, either the artefact must be further developed or the purpose and design criteria adjusted. Such accommodating adjustments of purpose and design criteria may be constructive if spurred by an increased understanding of the context but not if caused solely by demands to show improvement by reducing the gap between the artefact's actual and desired state. Unfortunately, this latter situation is common and constitutes the system archetype of 'Eroding goals' (Chapter 11).

Design science is not merely about designing artefacts per se but about making the design of artefacts scientific. Hence, it is about solving real-world problems scientifically. Following a systematic and transparent design process, as presented above, is a requisite. Building and evaluating artefacts to address problems can be as scientific as theorising and justifying theories. Although both scientific processes may appear similar, and some even argue that theories are artefacts in themselves (Hilpinen, 1995), they are fundamentally different in perspectives (King, 2012, p. 278) and entail differences in principles for assessing scientific quality.

Reliability, validity and workability

The quality of traditional science is commonly assessed in terms of reliability and validity (Kirk & Miller, 1986, p. 20). However, some suggest alternative definitions for qualitative research (LeCompte & Goetz, 1982), different words for the same thing (such as rigour and quality (Mason, 2002)), or new meaning and terms (e.g., validity entailing issues of plausibility and credibility (Hammersley, 1992, p. 70–71). Reliability refers to the degree of consistency of empirical results between different researchers or by the same researcher on various occasions (Figure 12.4) (Hammersley, 1992, p. 67). In other words, it refers to the degree to which these results are independent of unintentional circumstances (Kirk & Miller, 1986, p. 20). Validity, on the other hand, refers to the accuracy with which an empirical description of a particular phenomenon represents the theoretical construction intended to represent and captures the relevant features of that phenomenon (Hammersley, 1992, p. 67). In other words, it refers to the degree to which the empirical results are interpreted adequately (Kirk & Miller, 1986, p. 20).

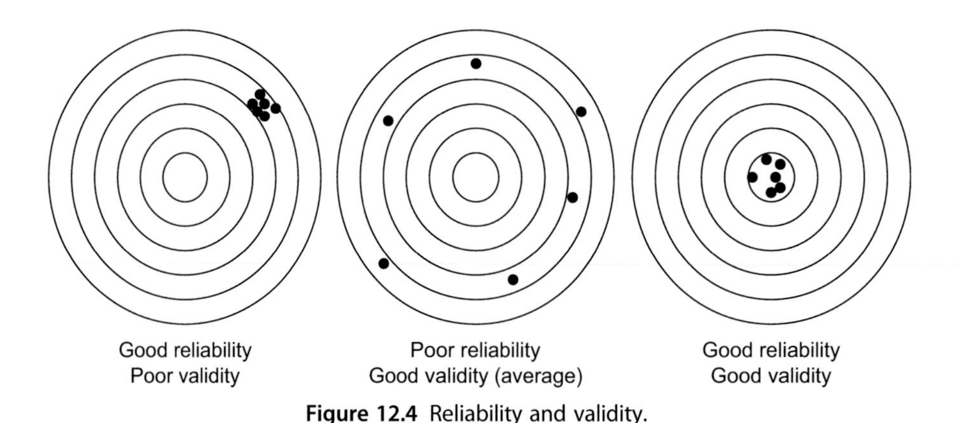

Figure 12.4 Reliability and validity.

Reliability depends essentially on transparently describing the research procedures, making it possible to distinguish at least three types of reliability (Kirk & Miller, 1986, pp. 41–42): (1) reliability as a single research method continually yielding an invariable result, (2) reliability as stability of a result over time and (3) reliability as similarity of results within the same period using different data. The first two types of reliability are not particularly useful in assuring quality when researching complex human-environment systems. The first suffers from the fact that it allows a single flawed method to continuously generate erroneous results (Kirk & Miller, 1986, p. 41) and the second from the fact that the world is constantly changing (Dewey, 1922; Keynes, 1994, p. 287) and the elements and dependencies under study are impossible to isolate (Anderson, 1999, p. 217; Checkland & Holwell, 2007, pp. 5–6). What is left is reliability in the sense of corresponding results by different research methods, which is commonly called triangulation (Buckle et al., 2003, p. 83; Denzin, 2009, pp. 297–313; Webb et al., 1966, p. 3). However, Blaikie (1991) warns us that the metaphor of triangulation connotes naïve ontological and epistemological assumptions. It suggests that it is possible to pinpoint reality by applying multiple research methods in the same way a surveyor pinpoints a geographical location. Although the metaphor is misleading, diverse data that lead to similar conclusions may still render us more confident, as different data have different biases (Atkinson & Hammersley, 2007, p. 183). It assumes, of course, that the methods used do not share the same bias (Blaikie, 2010).

With no direct access to reality, we cannot know whether, or to what extent, a theoretical construction is valid (Hammersley, 1992, p. 69) regardless of the quantity and quality of the empirical data (Atkinson & Hammersley, 2007, p. 11). However, we must still assess the validity of such a theoretical construction in relation to the adequacy of the collected empirical data (Hammersley, 1992, p. 69). Assessments of validity are based on two judgements. They are based on a judgement about the compatibility of the theoretical construction, or the empirical data supporting it, with our assumptions about the world that are presently beyond a reasonable doubt. Also, they are based on a judgement about the likelihood of error, given the conditions in which the theoretical construction is made (Hammersley, 2002, p. 73). It is important to note that these are judgements whose own validity never can be established (Hammersley, 1992, p. 78). Validity is, therefore, related to the collective judgement by the scientific community (Barnes et al., 1996; Bernard, 1995, p. 43), which Kuhn (1970) refers to as a paradigm and Said (1978, p. 275) calls ‘an academic-research consensus’.

In short, the purpose behind the ideas of reliability and validity is to provide grounds for someone to trust the research results. It makes Ramsey (1931, p. 258) idea of a reliable process central, as it becomes vital to be transparent in how data are collected, analysed and presented for others to assess the reliability and validity of the results. It also becomes crucial to be transparent in what judgements and assumptions about the world are included in the research itself and in the assessment of the quality of the study.

Although design science has a similar need for people to trust the research results, and reliability and validity play roles here as well, solving problems involves an additional way of assessing the quality of the work. When the purpose of the research is, instead, to address a problem. Then, the quality of the results can be assessed regarding the workability of the proposed solution (Olsen & Lindøe, 2004, p. 372). Workability can be assessed by whether or not the proposed solution resolves the identified problem (Greenwood & Levin, 2007, pp. 63–64). Or, in design science terminology, whether or not the form of the artefact generates a result that fulfils the required functions, as specified in the design criteria, to meet the artefact's purpose when utilised in the intended context. Some scholars even argue that workability is the highest form of validity (Greenwood & Levin, 2007, pp. 66–68). However, I would not go as far since I consider traditional science and design science complementary.

Limitations of science for change

Many people across the planet are getting increasingly worried about the state of our world. At the same time, we have never been so well-off on a global scale (cf. Gapminder, 2022). This situation may seem like a paradox, but it is not. It is quite the opposite. Although accelerating scientific developments since the outset of the scientific revolution has affected most aspects of human life—although to different degrees for different people—it has also spurred an equally growing hunger for resources. A combination of this resource hunger and the generally reductionist approach of science over the centuries—investigating specific phenomena or addressing specific problems in isolation—have created blind spots that continue to elude scientists and policymakers alike. Addressing one problem often results in creating another. The quite remarkable developments—in economic purchasing power, life expectancy, infant mortality, adult literacy, etc.—have, unfortunately, primarily been associated with unsustainable practices.

In addition to the fundamental limitation of reductionist approaches, ignoring 'Hume's Guillotine' also undermines science's potential to drive positive change. This second limitation is grounded in the common habit amongst scientists and other experts of suggesting changes in policy and practice based on the results of a particular set of scientific studies without explicit consideration of the multiple different values that may be important in the specific context. Ignoring the fundamental difference between descriptive statements about how things are and prescriptive statements about how things ought to be, is not just poor science. It constitutes a real problem. The world is full of examples where such approaches have resulted in failed policies. For instance, the rapid restrictions in land use along the tsunami-affected coastline of Sri Lanka and the state of Tamil Nadu in India were based on hasty assessments of public safety without considering the fundamental importance of living close to the primary source of livelihood for the vast majority of people residing in that zone and the lack of available land for resettlement close enough (Figure 12.5).

Although disciplinary research effectively addresses many research questions, it is fundamentally insufficient to address contemporary sustainability challenges. For science not only to be a driver for development but sustainable development, it must attempt to approach our challenges more holistically. Again, I am not claiming that more fundamental and disciplinary science is unimportant since there is a great need for such input. However, such science cannot inform effective changes in policy and practice on its own but needs to be integrated into more holistic and transdisciplinary frameworks. Sustainability Science constitutes such a framework and might be one of the more critical developments for science to address the core challenges of humankind (Clarke, 2007; Clarke & Dickson, 2003; Kates, 2011; Kates et al., 2001). It is heartening to witness these developments and humbling to attempt to contribute to them. Although many scientists are engaged and meaningful progress is made, academic structures are still overwhelmingly disciplinary. Getting recognition and funding for transdisciplinary research is still an uphill struggle.

For science to drive sustainable development, it is also vital that the scientific community is mindful of the fundamental difference between descriptive and prescriptive statements and explicitly acknowledges that difference when suggesting changes in policy and practice. Here, design science can complement traditional science, as it provides a scientific way of bridging the gap between descriptions of the world and prescriptions of how it ought to be.

Figure 12.5 Fishermen, boats and nets on a beach in Tamil Nadu, India, just a few months after the tsunami. *(Photo by Kavaiyan, shared on the Creative Commons.)*

Conclusion

History demonstrates that it is clear that science has great potential for driving change, both in addressing and inadvertently causing challenges. There may be many reasons for these inherently dual outcomes. One of the more fundamental ones concerns the reductionist approach of what Heylighen et al. (2007) refer to as Newtonian Science, which underpinned all scientific thinking for centuries and still does to a large extent. This approach has been incredibly successful in our inquisitive excavation deeper and deeper into a wide range of scientific endeavours. Still, it has also resulted in fragmented thinking and doing and a naïve confidence in our ability to control our world—a world whose complexity and dynamic character are invisible through such reductionist lenses. Another somewhat related reason for the more negative outcomes is the all too common habit amongst scientists of making prescriptive suggestions based solely on descriptive studies without taking proper care. The great Scottish Enlightenment philosopher David Hume must rotate at high speed in his grave, especially as there is a way to tackle this divide.

There are no silver bullets against failed policies and ineffective practices. Nonetheless, if applying a more holistic approach to our sustainability challenges whilst complementing traditional science with the perspectives of design science, I trust we can get further. We need traditional science to explain and understand fundamental phenomena concerning risk, resilience and sustainability in complex and dynamic human-environment systems and to use that knowledge to inform the argumentation for the design criteria of artefacts we need to develop to address our challenges. The two processes of traditional science and design science are fundamentally different but connected and equally important. Although both are liable to various errors, the quality of design science is not only assessed in terms of reliability and validity but also workability. Change entails addressing challenges, addressing challenges entails designing artefacts, and designing artefacts entails explaining and understanding phenomena.

Without silver bullets, I think we have to put our trust in the ability of Sustainability Science to bring together 'scholarship and practice, global and local perspectives from north and south, and disciplines across' all sciences (Clarke & Dickson, 2003, p. 8060). However, for Sustainability Science to reach its full potential requires transforming the academic structures that currently restrict transdisciplinary research and problem-solving-oriented scientific endeavours. There are positive signs of transdisciplinary journals and funding opportunities emerging. However, these are still generally limited and, unfortunately, not always viable options since the individuals and boards reviewing manuscripts and proposals are still disciplinary in their attitudes and composition.

References

Ackoff, R. L. (1962). *Scientific method: Optimizing applied research decisions*. Wiley.

Anderson, M. B. (1999). *Do no harm: How aid can support peace - or war*. Lynne Rienner Publishers.

Atkinson, P., & Hammersley, M. (2007). *Ethnography: Principles in practice*. Routledge.

Barnes, B., Bloor, D., & Henry, J. (1996). *Scientific knowledge: A sociological analysis*. Athlone.

Bernard, H. R. (1995). *Research methods in anthropology: Qualitative and quantitative approaches*. AltaMira Press.

Blaikie, N. W. H. (1991). A critique of the use of triangulation in social research. *Quality and Quantity, 25*, 115–136. https://doi.org/10.1007/BF00145701

Blaikie, N. W. H. (2010). *Designing social research: The logic of anticipation*. Polity Press.

Brehmer, B. (2007). Understanding the functions of C2 is the key to progress. *The International C2 Journal, 1*, 211–232. http://www.dodccrp.org/html4/journal_v1n1_07.html.

Brehmer, B. (2010). Command and control as design. In *CCRP*.

Buckle, P., Marsh, G., & Smale, S. (2003). Reframing risk, hazards, disasters, and daily life: A report of research into local appreciation of risks and threats. *Australian Journal of Emergency Management, 18*, 81–87.

Checkland, P., & Holwell, S. (2007). Action research: Its nature and validity. In N. Kock (Ed.), *Information systems action research: An applied view of emerging concepts and methods* (pp. 3–16). Springer Science.

Clarke, W. C. (2007). Sustainability science: A room of its own. *Proceedings of the National Academy of Sciences, 104*, 1737–1738.

Clarke, W. C., & Dickson, N. M. (2003). Sustainability science: The emerging research program. *Proceedings of the National Academy of Sciences, 100*, 8059–8061.

Cook, S. C., & Ferris, T. L. J. (2007). Re-evaluating systems engineering as a framework for tackling systems issues. *Systems Research and Behavioral Science, 24*, 169–181. https://doi.org/10.1002/sres.822

Denzin, N. K. (2009). *The research act: A theoretical introduction to sociological methods*. Transaction Publishers.

Dewey, J. (1922). *Human nature and conduct: An introduction to social psychology*. Henry Holt and Co.

Gapminder. (2022). *Gapminder*. www.gapminder.org/data.

Greenwood, D., & Levin, M. (2007). *Introduction to action research: Social research for social change*. Sage Publications.

Hammersley, M. (1992). *What's wrong with ethnography?: Methodological explorations*. Routledge.

Hammersley, M. (2002). Ethnography and realism. In A. M. Huberman, & M. B. Miles (Eds.), *The qualitative researcher's companion* (pp. 65–80). Sage Publications.

Hassel, H. (2010). *Risk and vulnerability analysis in society's proactive emergency management: Developing methods and improving practices*. Lund University.

Heylighen, F., Cilliers, P., & Gershenson, C. (2007). Complexity and philosophy. In J. Bogg, & R. Geyer (Eds.), *Complexity, science and society* (pp. 117–134). Radcliffe Publishing.

Hilpinen, R. (1993). Authors and artifacts. *Proceedings of the Aristotelian Society, 93*, 155–178.

Hilpinen, R. (1995). Belief systems as artifacts. *The Monist: An International Quarterly Journal of General Philosophical Inquiry, 78*, 136–155.

Huxley, T. H. (1882). Technical education. In *Science and culture, and other essays* (pp. 73–93). D Appleton and Company.

Kates, R. W. (2011). What kind of a science is sustainability science. *Proceedings of the National Academy of Sciences, 108*, 19449–19450.

Kates, R. W., Clarke, W. C., Corell, R. W., Hall, J. M., Jaeger, C. C., Lowe, I., McCarthy, J. J., Schellnhuber, H. J., Bolin, B., Dickson, N. M., Faucheux, S., Gallopín, G. C., Huntley, B., Jodha, N. S., Kasperson, R. E., Mabogunje, A., Matson, P. A., Mooney, H., Moore, B., … Svedin, U. (2001). Sustainability science. *Science, 292*, 641–642.

Keat, R., & Urry, J. (1975). *Social theory as science*. Routledge.

Keynes, J. M. (1994). Economic model construction and econometrics. In D. M. Hausman (Ed.), *The philosophy of economics: An anthology*. Cambridge University Press.

King, R. P. (2012). The science of design. *American Journal of Agricultural Economics, 94*, 275–284. https://doi.org/10.1093/ajae/aar128

Kirk, J., & Miller, M. L. (1986). *Reliability and validity in qualitative research*. Sage Publications.
Kuhn, T. (1970). *The structure of scientific revolutions*. The University of Chicago Press.
LeCompte, M. D., & Goetz, J. P. (1982). Problems of reliability and validity in ethnographic research. *Review of Educational Research, 52*(1), 31–60. https://doi.org/10.3102/00346543052001031
Lee, A. S. (2007). Action is an artifact: What action research and design science offer to each other. In N. Kock (Ed.), *Information systems action research: An applied view of emerging concepts and methods* (pp. 43–60). Springer Science.
March, S. T., & Smith, G. F. (1995). Design and natural science research on information technology. *Decision Support Systems, 15*, 251–266. https://doi.org/10.1016/0167-9236(94)00041-2
Mason, J. (2002). *Qualitative researching* (2 ed.). SAGE Publications Limited.
Ness, B., Anderberg, S., & Olsson, L. (2010). Structuring problems in sustainability science: The multi-level DPSIR framework. *Geoforum*, 479–488.
Olsen, O.-E., & Lindøe, P. (2004). Trailing research based evaluation; phases and roles. *Evaluation and Program Planning, 27*, 371–380.
Ramsey, F. P. (1931). *The Foundations of Mathematics, and other logical essays*. Kegan Paul, Trench, Trubner and Co.
Rasmussen, J. (1985). The role of hierarchical knowledge representation in decisionmaking and system management. *IEEE Transactions on Systems, Man and Cybernetics, 15*, 234–243.
Said, E. (1978). *Orientalism*. Penguin Books.
Sayer, A. (1992). *Method in social science: A realist approach*. Routledge.
Simon, H. A. (1990). Prediction and prescription in systems modeling. *Operations Research, 38*, 7–14.
Simon, H. A. (1996). *The sciences of the artificial*. MIT Press.
Webb, E. J., Campbell, D. T., Schwartz, R. D., & Sechrest, L. (1966). *Unobtrusive measures: Nonreactive research in the social sciences*. Rand McNally.
Wieringa, R. J. (2009). Design science as nested problem solving. In *Proceedings of the 4th international conference on design science research in information systems and technology* (pp. 1–12). ACM.
Wieringa, R. J. (2014). *Design science methodology for information systems and software engineering*. Springer.

CHAPTER 13

Understanding resistance to knowledge and change

Introduction

Despite the unprecedented accumulation of scientific evidence of staggering sustainability challenges, some of which may even threaten the survival of humankind, there has been so far minimal concrete action taken by policymakers and the general public concerning most of them. How is that even possible? According to UN Secretary-General António Guterres (2022a,b), we are 'on a highway to climate hell' and gradually 'committing suicide by proxy' in relation to biodiversity loss. Yet, we are still hardly doing anything to save ourselves. At least not on a scale that matters. Similar assessments are made concerning many other sustainability challenges—not only regarding global challenges but also local ones (e.g., Becker, 2021).

Understanding why people fail to pay attention and take concrete action against even the most blatant problem is an essential and difficult task. Lack of information is often considered a main reason since people may not be aware of the extent of the problem, its potential consequences if not addressed or the available solutions. People may, of course, not have access to accurate and reliable information about the problem, or they may not be able to understand it if it is not presented in an accessible way and they have adequate education. However, there cannot be a single policymaker in the world today, at least not on the national level, without access to sufficient information about climate change, biodiversity loss and several other pressing sustainability challenges to grasp their acuteness. Similarly, it is impossible to live anywhere with access to basic education and reasonably free mass media without having access to abundant information about many sustainability challenges and how our leisure travel, local transportation habits, food preferences and consumption choices exacerbate them. Whilst some individuals do change their behaviour when presented with information about their environmental footprint, the assumption that people, in general, are motivated by such ecological knowledge is flawed at best (Klintman, 2013). There are, in other words, other factors than merely a lack of information that make people susceptible to resisting knowledge and change (Figure 13.1).

This chapter focuses on presenting a variety of such factors, including factors related to the problem itself, cognitive and psychological factors, social and cultural factors, political, economic and administrative factors, and technological factors. However, it is essential first to establish what knowledge resistance and resistance to change mean in this

Sustainability Science, Second Edition
ISBN 978-0-323-95640-6, https://doi.org/10.1016/B978-0-323-95640-6.00009-9

Figure 13.1 Knowledge resistance explained by Banksy. *(Shared by Matt Brown on the Creative Commons.)*

context and, in the end, elaborate on what can be done to overcome them. The chapter ends with some concluding remarks.

Resistance to knowledge and change

Notions of knowledge resistance, or fact resistance, are becoming increasingly common in both academic and public discourse. Although there is significant variation in how these 'resistance phenomena' are described, the overall impression is that they are becoming more widespread and severe (Glüer & Wikforss, 2022, p. 29). Some scholars argue for a need to distinguish between fact resistance and knowledge resistance, with the former implying irrational resistance to clear and indisputable facts whilst the latter denotes a more complex, profound phenomenon that can have its own rationality (Klintman, 2019). This chapter is about knowledge resistance in a broad sense, spanning a wide range of phenomena in which information and knowledge plausibly inferred from empirical observation are resisted due to various factors. Such knowledge resistance is often intrinsically linked to resistance to change, which I will return to in a moment.

Knowledge resistance

Knowledge resistance entails ignorance, but not all ignorance stems from knowledge resistance (Glüer & Wikforss, 2022, p. 29). Ordinary ignorance can indeed arise from a *genuine* lack of information, with the emphasis on genuine being central and informative. If someone truly lacks the information necessary as evidence for a particular knowledge claim, we cannot assert that they resist the knowledge. Resistance is an active process. Although it can be either conscious or unconscious, for knowledge resistance to be present, the information must be available to the individual who actively resists the knowledge claim it entails (Glüer & Wikforss, 2022, p. 29).

However, actively looking the other way when presented with the evidence is also knowledge resistance since the lack of information is not genuine but purposeful. Whilst ignorance is often portrayed as something altogether negative that actors have an evident interest in overcoming or eliminating (e.g., Ungar, 2008), actors may also seek to preserve their ignorance for various reasons (McGoey, 2012). Such ideas are visible in everyday sayings, such as 'what you don't know can't hurt you' and 'ignorance is bliss'. When an individual decides not to acquire available information for whatever reason, the resulting ignorance can be referred to as 'strategic ignorance' (Carrillo & Mariotti, 2000, p. 530)—the 'practices of obfuscation and deliberate insulation from unsettling information' (McGoey, 2012, p. 555). There are many reasons for such strategic ignorance, which I will return to throughout the chapter. Information may hurt (Hirshleifer, 1971), ignorance can be a valuable tool for governing and social control (Gross, 2010; Luhmann, 1998; McGoey, 2012; Merton, 1987; Schneider, 1962), and cultivating ignorance may be more advantageous than cultivating knowledge (McGoey, 2012, p. 555). Taussig (1999) even argues that knowing what not to know is the most potent form of knowledge. However, strategic ignorance can be both consciously and unconsciously nurtured, and the most fundamental source of strategic ignorance is avoiding looking for information and knowledge in the first place (Klintman, 2019).

Knowledge resistance may also allow erroneous beliefs to prevail regardless of the amount of evidence to the contrary available to the resister. Here, knowledge resistance is not about actively avoiding the information disagreeing with the held belief but resisting the support it provides for a more plausible knowledge claim (Glüer & Wikforss, 2022, p. 30). In other words, such resistance involves engaging with the information yet resisting the knowledge inferred from it by interpreting the information or the underlying empirical observations differently or undermining their validity. It is essential to note that interpreting the available evidence differently is not knowledge resistance as long as the interpretation is plausible and consistent with empirical observations. Then, it is simply an alternative knowledge claim. Knowledge resistance arises when the erroneous belief is maintained with the backing of implausible interpretations (Klintman, 2019).

Knowledge is also resisted by actively undermining the validity of the evidence for it, doubting its quality, questioning the motives of the actors producing it, and drawing attention to ambiguities, uncertainties and alternative data. Although scrutinising the quality of the evidence, actors' motives and the need for more data are all traits of sound scepticism that is fundamental for science, it is vital to distinguish it from knowledge resistance dressed up as scepticism. The two may appear similar at first, but scepticism is the antithesis of knowledge resistance (Klintman, 2019). Scepticism is about seeking evidence before changing opinion, which is the essence of science, whereas knowledge resistance is the refusal to do so irrespectively of the quality and conclusiveness of the available evidence.

Glüer and Wikforss (2022, p. 30) propose approaching knowledge resistance 'as a form of irrational resistance to the total available empirical evidence'. It means that for a knowledge claim to be plausible, it must be consistent with the totality of all empirical evidence. That is not to say that all evidence must point towards the exact same conclusion. They rarely do in our complex world. It requires, nonetheless, consideration of all empirical evidence when making the knowledge claim and moderating how it is formulated depending on the evidential foundation. Remember from Chapter 7 that knowledge comprises beliefs in whose validity we are reasonably confident (Hammersley, 1992, p. 50). A knowledge claim requires, in other words, the backing of adequate evidence, even if some evidence may deviate, whereas an erroneous belief can be held regardless of the available evidence (Figure 13.2). In short, knowledge requires justification (Carlill, 1906, pp. 89–94), disqualifying even the most accurate guess from ever becoming knowledge without sufficient evidence.

Another aspect of Glüer and Wikforss's (2022, p. 30) approach to knowledge resistance concerns irrationality. It is important to bear in mind that they refer to a particular form of irrationality in relation to how evidence is ignored or interpreted. Looking at photos from the inaugurations of Presidents Obama and Trump and still claiming that the latter had more people attending is indeed irrational since it is simply not a plausible interpretation of the empirical evidence (Figure 13.3). However, whilst such irrationality is fundamental for knowledge resistance, it is essential to note that resisting knowledge can be rational for a resister valuing membership and status in a social group over the validity of a particular knowledge claim regardless of its overwhelming evidence (Klintman, 2019). More on that later.

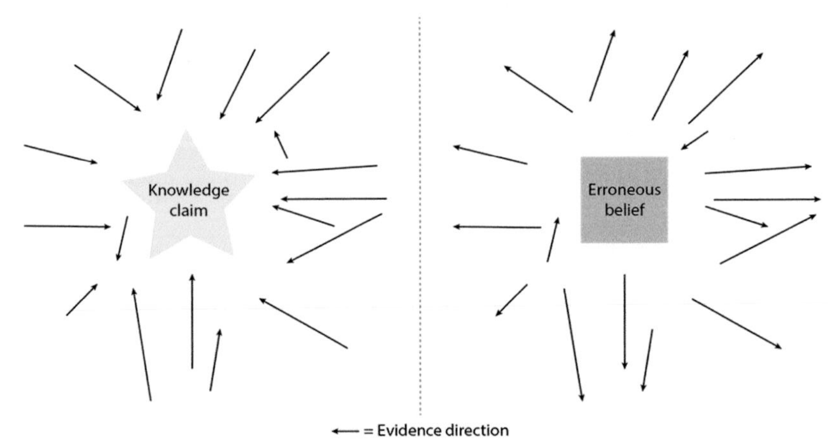

Figure 13.2 Knowledge claim versus erroneous belief.

Figure 13.3 Photos of the inauguration ceremonies of Presidents Obama (left) and Trump (right) taken from the top of the Washington Monument.

Finally, it is essential to understand that knowledge resistance only concerns knowledge claims—descriptive statements, arguments, conclusions or theories about how something is. It is irrelevant concerning value judgements (Glüer & Wikforss, 2022, pp. 31—32)—normative statements about how something ought to be. Unlike a knowledge claim based on evidence and reason, a value judgement is based on norms, values and attitudes about what is good or bad, right or wrong, important or unimportant, or otherwise desirable. Evidence and reason are, in other words, irrelevant. Although others may disagree or even question the sanity of the makers of certain value judgements, 'it is perfectly possible to accept all the relevant known empirical facts and yet reject a proposed policy without being irrational' (Glüer & Wikforss, 2022, p. 32). No knowledge is resisted, but the value judgement underpinning the framing of the problem or the formulation of the solution is rejected.

Resistance to change

Tightly connected to knowledge resistance, yet an important characteristic in itself, is resistance to change. Like knowledge resistance, being elaborated on at least since

Kierkegaard (1995, p. 5), the idea of resistance to change has a much longer pedigree than the seminal writings of Lewin (1943, 1947, 1951) who often is credited for naming it (Harich, 2010, pp. 37–38). Resistance to change was already elaborated on more than 500 years ago by Machiavelli (2021, p. 22) who writes:

> *And it ought to be remembered that there is nothing more difficult to take in hand, more perilous to conduct, or more uncertain in its success, than to take the lead in the introduction of a new order of things, because the innovator has for enemies all those who have done well under the old conditions, and lukewarm defenders in those who may do well under the new*
>
> ***Machiavelli (2021), p. 22.***

Although there are many more factors driving resistance to change than Machiavelli's shrewd examination suggests—many of which are presented in the coming sections—they all result in a tendency for an individual, organisation, society or other systems to continue its current behaviour despite measures to change it (Harich, 2010, p. 37). Such resistance to change can result from knowledge resistance, but it regularly arises regardless of what actors know. Change is resisted for various reasons, irrespective of means and ends.

This kind of resistance to change is neither uniform nor solely negative (Watson, 1971, p. 745). It operates on the level of individual actors, along the immediate relations between actors, and on different system levels (groups, communities, organisations, societies, etc.). As demonstrated throughout this chapter, resistance to change is a complex phenomenon that manifests differently both within and between these levels. Some individuals may resist change more than others, and even if many individuals aspire change, the system they constitute may resist it (Pardo del Val & Martínez Fuentes, 2003). Moreover, whilst the ambitious change agent views the resistance as an undesirable obstruction, others desiring stability may consider the same resistance as a virtue (Watson, 1971). Resistance to change can indeed play a role in maintaining stability in social systems. Change can disrupt established norms and values, and too much change too quickly can lead to chaos and confusion. However, resistance to change can also prevent development and improvement, as well as thwart the transformation necessary for sustainability. In short, resistance to change is not uniform and is influenced by a multitude of factors at the individual (micro), organisational (meso), and systemic (macro) levels.

Factors driving resistance to knowledge and change

There are many different kinds of factors that may be driving resistance to knowledge and change. In this chapter, for the sake of structure, they are presented as factors related to the problem itself, cognitive and psychological factors, social and cultural factors, political, economic and administrative factors, and technological factors.

Factors related to the problem itself

Resistance to knowledge and change is exacerbated by increasing complexity, uncertainty, ambiguity and dynamic change of the issue itself and the systems in which it exists. Without the certainty and clarity of simple issues and systems, it is more difficult to differentiate between valid knowledge claims and mere beliefs or guesses. These four fundamental challenges also combine to undermine the ability to act of even the most enlightened environmentalist. They make it rarely possible to address any one issue without affecting something else in the system (Dörner, 1996), and small changes in one place and at one point in time may have substantial impacts somewhere else, on another scale or much later (Kates et al., 2001). Knowing this, and that complex systems are liable to behaviour that is locally adaptive but globally maladaptive in both spatial and temporal terms (Branlat & Woods, 2010, pp. 27–28), can easily lead to bewilderment and hesitance. It feels like you are damned if you do and damned if you do not.

Moreover, complex issues and systems may change without human beings doing anything, rendering what worked yesterday useless today (Dörner, 1996). This dynamic feature is particularly troublesome since we are generally prone to get stuck in strategies and solutions that were successful in the past (Branlat & Woods, 2010, p. 28; Dörner, 1996) and exhaust our adaptive capacity when disturbances escalate and cascade faster than the system can adapt (Branlat & Woods, 2010, p. 27).

Cognitive and psychological factors

There are numerous cognitive and psychological factors driving resistance to knowledge and change. Although several of them are related, the most central factors are here presented under five subheadings: (1) Believing our senses, (2) Overvaluing familiarity and recent experience, (3) Taking risk, (4) Bending information and (5) Motivated reasoning.

Believing our senses

Human beings have cognitive biases predisposing us to preserve the status quo and avoid spending resources on reducing the risk of potential deviations away from our preferred expected future outside our direct realm of personal experience (Johnson & Levin, 2009, pp. 1594–1596). It is important to note that whilst human beings are almost unique in our ability to learn from the experience of others, we are also remarkable for our apparent disinclination to do so (Adams & Carwardine, 1990, p. 114). This paradox is perhaps best summarised by the French 17th-century philosopher Blaise Pascal: 'As a rule, we persuade ourselves more by the reasons we have found for ourselves than by those which have occurred to other people' (Pascal, 1900, p. 86). This pattern is explained on the most basic level by human beings being genetically hardwired to believe and react to what we see, hear, smell, taste and touch, which evolved in our prehistoric past when responding to actual threats and opportunities in the immediate surroundings where more adaptive than reacting to potential, vague, distant or abstract threats (Johnson & Levin, 2009). It also entails that we tend to assume that a lack of immediate change means that all is well (Dörner, 1996). Although such sensory biases were indeed adaptive at a time when our risk landscape and sustainability challenges were local and simple (Chapter 2), they are central obstacles to understanding and action today when our world and most of our problems are comprehensive and complex (Chapter 9). Nevertheless, the more tangible and self-experienced the risk is, the easier it is to get a reaction. 'Seeing is believing' (Slovic, 2010, pp. 95—96).

Overvaluing familiarity and recent experience

Human beings are biased by familiarity and recent experience. The more we are exposed to something, the more familiar it gets, and the more we tend to develop a preference for it. This cognitive bias is referred to as the familiarity principle (Politz, 1960) or mere exposure effect (Zajonc, 1968). In addition to being a pillar for marketing and the strategy of repeating misinformation over and over again to convince people of its validity, it has implications for how human beings perceive risk (Slovic, 2010, p. 95). For example, a person living right next to the railroad going into Jakarta in Indonesia is obviously aware of the risk of stepping too far out of her front door and onto the tracks (Figure 13.4). However, being born and raised there and experiencing several trains thundering past every day is likely to gradually reduce her aversion to the trains. Perhaps even to the extent of preferring to stay there if presented with a less familiar but safer alternative. At least until an accident occurs in her vicinity and another fundamental cognitive bias kicks in.

Figure 13.4 Houses next to the railroad in Jakarta, Indonesia. *(Photo taken by Jonathan McIntosh, shared on the Creative Commons.)*

Human beings are fundamentally biased by recent experiences. In other words, we are inclined to view the future as an extrapolation of the present (Meyer, 2006, pp. 161–163). This cognitive bias is referred to as availability heuristics (Kahneman & Tversky, 1982) and may have worked well earlier in history when our world was less connected and fluid. It still has apparent benefits in events with quick and direct feedback, such as having a careless neighbour hit by a train. However, it is also a central obstacle to knowledge and change in our increasingly complex and dynamic world. Our future is no longer an extrapolation of our past, if ever, but full of potential scenarios that are ignored if we base our anticipation only on recent experiences (Chapter 8). We should do everything possible to learn from our experiences but be aware that such a limited sample will not likely represent all possible future events. Particularly not since human beings are inclined to overestimate the frequency and probability of what they have recently experienced and underestimate other events (Tversky & Kahneman, 1973). It is not yesterday we need to adapt to. It is tomorrow.

In contrast to the quick and direct feedback of having a neighbour accidently step out in front of a train, much feedback relevant to the risk and sustainability of human-environment systems is dim, dispersed and delayed. Hence, turning our otherwise so effective trial-and-error way of learning to walk, talk and many other things we consider mundane when mastering them into a core challenge for resilience. Especially in complex human-environment systems where the effects of a change in one place and at one point in time may cascade spatially and temporally, eluding and deluding even the keenest observer (Chapter 9).

Our tendency to learn by focusing on short-term feedback also restricts learning to what has actually happened, largely ignoring what almost happened (Meyer, 2006, pp. 154–155). Near misses and close calls may play some role in some instances, but history is full of prominent examples of the opposite. For instance, the more or less complete inactivity after Hurricane Ivan missed New Orleans in 2004, even after the mayor and other authorities pointed out several shortcomings that would have substantially aggravated the disaster. Just to be caught by surprise when Hurricane Katrina hit a year later, and none of the shortcomings had been addressed (Meyer, 2006, pp. 153–155).

However, the near misses and close calls are not only ignored. They directly undermine resilience by allowing the focus on short-term feedback for learning to signal whatever implemented precautionary measures as unnecessary. Frequent false alarms of a mid-1990s dormitory fire alarm at the University of British Columbia resulted in my friends living there ignoring it and jamming it with a towel to make it possible to stay inside without too much nuisance. Similarly, recurrent cyclone warnings and evacuations, which turn out in hindsight to be unnecessary, undermine their own effectiveness over time (Figure 13.5). This focus on short-term feedback is, in other words, creating a cyclical pattern of strict compliance to precautionary measures in events following a significant event, incrementally eroding every time the measures turn out to be unnecessary

Figure 13.5 Hurricane evacuation route sign on Tulane Avenue, midtown New Orleans, with lines left by longstanding floodwater after Hurricane Katrina. *(Photo by Infrogmation, shared on the Creative Commons.)*

until the next catastrophic event proves them vital again. Availability heuristics is truly challenging in this regard.

Our tendency to learn by focusing on short-term feedback is not only undermining compliance to immediate precautionary measures, such as evacuation. It also undermines our willingness to invest in proactive measures to reduce risk more generally (Marais et al., 2006, pp. 569–570). The more effective prevention measures are, the more infrequent the events become. That is the entire logic of prevention. Similarly, the more effective mitigation measures are, the less consequences the events cause when finally occurring. Just as intended. However, when people do not experience any events, or at least not any severe consequences, the sensory biases and availability heuristics combine and cause us to increasingly underestimate the risk of the events and, thus, erode our perceived need for proactive risk reduction measures in the first place. This unfortunate pattern could be called *the tragedy of risk reduction* (Figure 13.6) and requires purposeful action to counter.

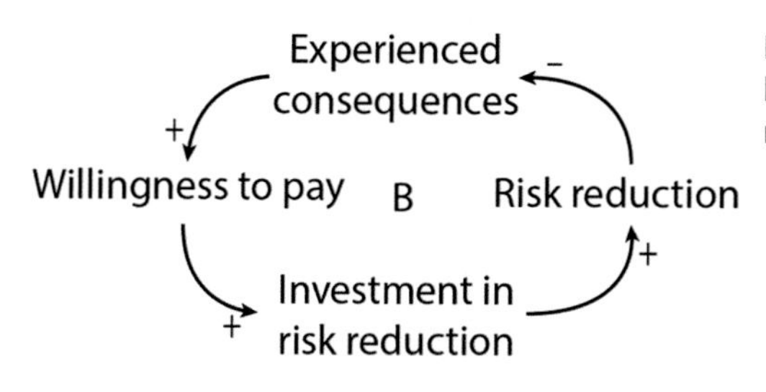

Figure 13.6 The tragedy of risk reduction as a balancing feedback loop in which successful risk reduction leads to reduced risk reduction.

Taking risk

Human beings are hardwired to take risks. Any unwillingness to invest in risk reduction can be connected to a discounting effect that makes human beings prefer immediate over long-term rewards (Johnson & Levin, 2009, p. 1596). More specifically, it is a tendency to discount the value of uncertain future benefits of proactive measures compared to the certain immediate costs of implementing them (Meyer, 2006, pp. 164–166). For example, if we know for sure that it would cost one million to reduce the risk of particular potential events that may happen and could cost one billion if occurring, we are psychologically predisposed not to make the investment even if the estimated probability of occurrence would make it good investment indeed. Just like when Nicholas Stern, the former Chief Economist of the World Bank, presented the landmark study called the Stern Review in 2006, clearly showing the immense economic benefits of immediate action in response to climate change (Stern, 2007). Yet, nothing happened. The reluctance to invest is often motivated under the guise of waiting for more certainty in how likely the events are and what the consequences would actually be—just like decision-makers did concerning the Stern Review. However, such procrastination is based on the fallacy that time will provide certainty. It will not. At least not in the complex systems of interest in this book. We may learn more about the systems we are interested in, but proactive thinking and action are, by definition, liable to uncertainty regardless of how long we wait. It is also worth noting that Nicholas Stern has admitted several times that he was wrong about climate change back in 2006: It is much worse! It means it is even easier to show that proactive investments are good investments in purely economic terms. However, most people do nothing to change their unsustainable lifestyles to contribute to the solution to our sustainability challenges in the hope that things will not be that bad after all (Figure 13.7) (Johnson & Levin, 2009, pp. 1597–1598).

Human beings also tend to have difficulty comprehending the scale of the problem. It is not merely rooted in innumeracy, although familiarity with mathematics may help people grasp the difference between large numbers (million, billion, trillion, etc.). This tendency is called scale insensitivity (Pham, 2007)—or scope insensitivity (Kahneman et al., 1999, p. 212) or scope neglect (Kahneman, 2003, p. 1464)—and results in people considering similar responses regardless of whether the problem is tens, hundreds or even thousands of times bigger. It has been demonstrated in countless studies. In the most well-

Figure 13.7 Discounting effect in a nutshell.

known study, three groups of participants were asked to state their household's willingness to contribute money to prevent the drowning of 2000, 20,000 or 200,000 migratory birds in ponds polluted with oil, respectively, which resulted in average amounts of $80, $78 and $88 and no statistically significant difference between the groups even when the consequences differed by two orders of magnitude (Desvouges et al., 1993). Other studies demonstrate similar results. For instance, residents of four states in the western United States would only pay 28% more to protect all 57 wilderness areas in those states than to protect a single site (McFadden & Leonard, 1993), or Toronto residents would pay only a little more to clean up all polluted lakes in the entire province of Ontario than the polluted lakes in only a small part of Ontario (Kahneman, 1986).

Rather than based on innumeracy, such scale insensitivity may be explained by the description of the situation evoking a mental representation of a prototypical incident—like an image of an oil-covered bird struggling for its life, a pristine wilderness, or a stinking polluted lake—and the affective value of this representation dominates the participants' attitudes to the problem, including their willingness to pay for a solution (Kahneman et al., 1999, p. 212). In other words, it is mainly the mental image of that one bird, wilderness or lake that primes us to action, whilst the varying numbers of birds, wilderness areas or lakes are difficult to visualise or otherwise imagine.

When such affective reactions are used as proxies for value, our responses are also not appropriately scaled for probability (Pham, 2007, p. 163). For instance, whilst participants in a study were willing to pay more to avoid a high probability of losing $20 than to avoid a low probability of losing the same amount, which is consistent with economic theory, they were not willing to pay much more to avoid a high probability of receiving an electric shock than to avoid a low probability of receiving the same shock (Rottenstreich & Hsee, 2001). Shocking, is not it? Scale insensitivity is regularly demonstrated in relation to risk and sustainability challenges (e.g., Remoundou et al., 2015; Sunstein, 2003; White et al., 2020) and can undermine effective action by disconnecting our willingness to invest in solutions from the scale of the risk stemming from the problem.

Such scale insensitivity has similar roots as another tendency also hampering our ability to estimate probabilities or frequencies. This tendency is called the conjunction fallacy (Tversky & Kahneman, 1983) and makes people inclined to intuitively assign a higher probability or frequency to a specific event occurring than to a broader category of events, even if the former is part of the latter. For example, people may estimate that it is more likely for a cyclone to hit Bangladesh than a natural hazard in general or that a major interstate war between China and Taiwan is more likely than a major interstate war in Asia. Regardless of how surprisingly common they are (Kahneman, 2011), such estimations are obviously logically impossible because a cyclone is a natural hazard, and a war between China and Taiwan is a war in Asia. The conjunction fallacy also makes people inclined to assign a higher probability or frequency to the conjunction of two statements than to the constituting statements themselves—hence, the name (Pinker, 2021). For instance, people may consider the statement that *a cyclone hits Bangladesh and causes floods* more likely than either of the two statements that *a cyclone hits Bangladesh* and *floods hit Bangladesh*. This reasoning is also logically flawed (Figure 13.8). Even if cyclones would always cause floods, the conjunction of a cyclone and floods can never have a higher probability than the probability of a cyclone or the probability of floods, respectively. Yet, intuitive estimations of probability or frequency do not conform to the logic of probability theory (Tversky & Kahneman, 1983).

The conjunction fallacy (Tversky & Kahneman, 1983) has essential implications for our ability to grasp and address risk and sustainability issues since it biases our probability or frequency estimations of potential events. It is, therefore, essential to consider how such an irrational tendency can be so common. Similar to scale insensitivity, the conjunction fallacy is linked to the innate propensity of human beings to prefer stories over statistics—experiential over analytical input (Chapter 8). We use the stories we have in our minds to create mental prototypes of various categories of events or objects and employ them subconsciously to process incoming information. Intuitive probability or

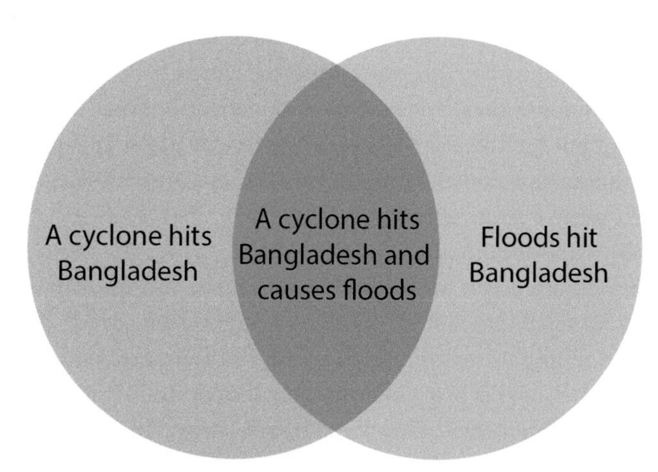

Figure 13.8 The conjunction fallacy visualised.

frequency is, thus, essentially driven by imaginability: the easier something is to visualise, the more likely it appears (Pinker, 2021). The availability heuristic from the previous section is obviously vital in this regard (Tversky & Kahneman, 1983).

However, the conjunction fallacy is also driven by a usual tendency to categorise events or objects based on how closely they resemble the typical mental prototype of that category. This tendency is referred to as the representativeness heuristic (Kahneman & Tversky, 1972) and fools us into intuiting that something more representative also is more likely. So, the statement that *a cyclone hits Bangladesh and causes floods* feels more likely because it fits our mental prototype of disaster in Bangladesh constructed by all media images of cyclones causing massive destruction and floods over the years. Similarly, a major interstate war between China and Taiwan seems more likely because the media focus on the tense cross-strait relations biases our mental prototype of interstate war in Asia, making that scenario fit. It also means that people erroneously believe that events or objects described in more detail are more probable or frequent than those described in less detail since the details hook deeper into the mental prototypes and make the events or objects more imaginable. It is the same mental processes that make scenarios such a powerful tool for mobilising people, but it is then vital to remember that 'the use of scenarios as a prime instrument for the assessment of probabilities can be highly misleading' (Tversky & Kahneman, 1983, p. 308).

In addition to the discounting effect, scale insensitivity and conjunction fallacy just considered, prospect theory is also helpful in explaining why people take risks and are so reluctant to take proactive action. This theory shows that humans tend to be more risk averse when choosing between potential gains than between potential losses (Kahneman & Tversky, 1979). It means that human beings tend to gamble more when anticipating adverse outcomes, which is, essentially, what risk is all about in the context of this book. Prospect theory is based on the idea that people's decisions are influenced not only by the possible outcomes but also by how those outcomes are framed in terms of gains or losses (Spence & Pidgeon, 2010, p. 658). It suggests that people are more prone to take risks to avoid losses than they are in their pursuit of gains because losses tend to have a more significant impact on our emotions, which have been shown to bias our judgements and decisions and are referred to as the affect heuristic (Slovic, 2010; Slovic et al., 2002). The phenomenon itself is known as loss aversion and provided an evolutionary advantage in the past since treating threats as more urgent than opportunities gives a better chance to survive and reproduce (Kahneman, 2011). This loss aversion is also consistent with the well-documented negativity bias, which refers to the tendency to give more weight and attention to negative experiences and information than positive ones (Meyerowitz & Chaiken, 1987). 'Bad is stronger than good' (Baumeister et al., 2001). However, gambling more when faced with risk is inherently problematic for sustainability. Prospect theory also proposes that people tend to be particularly risk-seeking when faced with inevitable losses. The psychological bias behind prospect theory is, in other words, unfavourable for the future of our planet since it makes us prone to gambling with it.

Human beings also tend to have positive illusions about themselves and the world. These positive illusions make people irrational optimists in three essential domains: '(1) They view themselves in unrealistically positive terms; (2) they believe they have greater control over environmental events than is actually the case; and (3) they hold views of the future that are more rosy than base-rate data can justify' (Taylor & Brown, 1994, p. 21). Whilst there are people with outright delusions concerning themselves and their abilities, positive illusions are different and 'responsive, albeit reluctantly, to reality' (Peterson, 2000, p. 46). 'Positive illusions do not bind us, they blinker us' (Johnson, 2004, p. 7).

Studies typically show that 67%—96% of people consider themselves to have better qualities than their peers, which is obviously logically impossible even if these qualities vary (Johnson, 2004, p. 7). Positive illusions make people liable to overestimate their ability to manage risk and avoid sustainability challenges, underestimate the likelihood of being personally affected, and overestimate their own ability to control and cope with events if things go south (Johnson & Levin, 2009, p. 1596). The same positive illusions also make us believe others are more at risk than ourselves (Meyer, 2006, pp. 160—161). Whilst such positive illusions can contribute to peoples' well-being (Young, 2014), they can also lead to overconfidence in their knowledge, skills or control, leading individuals to undertake tasks beyond their actual ability and take risks they cannot handle. Positive illusions can make people maintain the status quo because change appears unnecessary. Even if positive illusions can persist regardless of personal experience in actual events (Askman et al., 2018, p. 152), they are particularly prevalent in uncertain and ambiguous situations when the threat is not immanent or previously experienced (Johnson & Levin, 2009, p. 1596).

The effects of positive illusions can also be exacerbated by another cognitive bias related to the illusory superiority just mentioned. It is called the Dunning—Kruger effect, after the scholars who first described it, and suggests that people of low ability may suffer from delusory superiority and mistakenly assess their cognitive ability as greater than it is (Kruger & Dunning, 1999). Nichols (2017, p. 43) puts it more bluntly: 'The Dunning—Kruger Effect, in sum, means that the dumber you are, the more confident you are that you're not actually dumb'.

> *We've all been trapped at a party or a dinner when the least-informed person in the room holds court, never doubting his or her own intelligence and confidently lecturing the rest of us with a cascade of mistakes and misinformation. It's not your imagination: people spooling off on subjects about which they know very little and with completely unfounded confidence really happens, and science has finally figured it out*
>
> ***Nichols (2017), p. 43.***

In short, people lacking expertise or skill are more likely also to lack the necessary insight to accurately evaluate their expertise or skill, resulting in blatant overconfidence

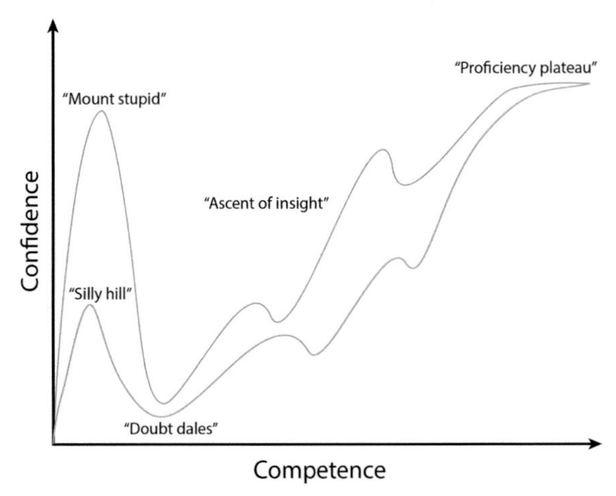

Figure 13.9 The Dunning–Kruger effect visualised.

in their competence. They suffer, as such, a dual burden: 'Not only do they reach erroneous conclusions and make unfortunate choices, but their incompetence robs them of the ability to realize it' (Kruger & Dunning, 1999, p. 1121). We are all liable to it (Dunning, 2014), but the extent of the incongruity between competence and confidence varies between people and over the course of learning. Many proficient people can recall how they gradually realised how little they actually knew as they increased their competence from the most basic level, at the same time as others get stuck on 'Mount Stupid' as their bloated self-confidence misleads them into believing that they have nothing to learn (Figure 13.9).

We all overestimate our competence, but the less competent do it more due to weaker self-reflective abilities referred to as metacognition (Nichols, 2017, pp. 44–45). However, experts within one field are still liable to the Dunning–Kruger effect irrespective of generally possessing stronger metacognition if their confidence in that field spill over into fields they do not master. Knowing a lot about a lot seems to have a tendency to fool certain people into believing that they know a lot about other things too. Regardless of Feynman (1999, p. 142) reminding us that a scientist can be 'just as dumb as the next guy' when engaging in topics removed from their discipline, scientists drawing on their epistemic authority when engaging in issues clearly outside their areas of expertise are ubiquitous (Ballantyne, 2019). Ballantyne (2019) calls this epistemic trespassing and argues that such trespassers demonstrate the same ignorance of their ignorance as anyone else exhibiting the Dunning–Kruger effect. They constitute a particularly detrimental problem since the greatest obstacle to knowledge is not ignorance, but the illusion of knowledge (Boorstin, 1983, p. 86)—something scientists are most capable of conjuring since they can talk the talk and draw on their scientific authority irrespective of whether or not it is relevant to their claims.

We must be especially humble when engaging in inter- or transdisciplinary endeavours, such as this book. Regardless of having a PhD in Sociology, another PhD in Engineering, an MRes in Politics, an MSc in Development Studies and a BSc in Engineering, several other disciplines contribute substantially to the contents of this book. To counteract epistemic trespassing, since it is impossible to be an expert in all of them, it is vital to read broadly and engage with and explicitly draw on the work of disciplinary experts instead of making our own claims outside our own expertise. How well I manage to do that in this book is up to you to assess.

Whitehead (1978) is often interpreted as arguing that it is not ignorance but the ignorance of ignorance that is the death of knowledge. However, it is not just that one may not be aware of one's ignorance, but that it is impossible to know with certainty that one does *not* suffer from ignorance (Fine, 2018, p. 4034). That demands humility, curiosity and constant reflection if we are not to drift towards taking unnecessary risk.

Bending information

Human beings cannot help bending incoming information to fit our beliefs and subconsciously updating our recollections of previously held beliefs. In situations of uncertainty and ambiguity, contradictory information causes psychological discomfort and results in human beings subconsciously trying to make the information fit existing beliefs and to avoid situations that increase such dissonance (Johnson & Levin, 2009). This phenomenon has been described for millennia (e.g., Thucydides, 2004, p. 215), but found its way into scientific accounts through Ibn Khaldun's (1969, p. 35) seminal work in the 14th century. It is now called cognitive dissonance, or confirmation bias, and causes human beings to subconsciously select, organise or distort incoming information to match their preferred and existing beliefs (Festinger, 1962). It means that two people processing the same information may interpret it entirely differently without consciously considering how they reframe it to resolve the dissonance between the information and their beliefs. Everybody is doing it. However, the more entrenched we are in our preferences, ideas, policies or ideologies, the more liable to cognitive dissonance we become (Figure 13.10). Compare this to the apparent inability of the international community to think outside the dominant paradigm of market capitalism when addressing sustainability challenges (Chapter 2) or to your friends and colleagues who cannot conceive of these challenges or the changes needed in their lifestyles to address them. Cognitive dissonance is a formidable challenge indeed. One that is skilfully used by actors wanting to preserve the status quo, or at least delay change to profit for as long as possible. For instance, by actively cultivating a sense of uncertainty and ambiguity concerning climate change, irrespective of the entire scientific community agreeing on its main causes, principles and scale of consequences.

Hindsight bias is another psychological factor contributing to knowledge resistance and resistance to change by distorting our perception of past events and reinforcing

Figure 13.10 Cognitive dissonance facilitates denial. *(Source: alphaspirit/Shutterstock.com.)*

our existing beliefs. Originally referred to as 'knew it all' attitude (Fischhoff & Beyth, 1975, p. 2), hindsight bias causes human beings to exaggerate the predictability of an event after it has actually occurred. It is as common in everyday life as in evaluations after a disaster. Most people have likely encountered hindsight bias in one way or another in the past week: A colleague telling you, 'I told you so', after the management announced a shift in policy when you cannot recollect any such unambiguous message from him before the decision was made. Hindsight bias is about recently acquired information manipulating the recollection of past information. It distorts memories, highlighting what you now know in the post-event situation was important in the pre-event situation and downplaying everything else that turned out to be less important. This unconscious curation of information regularly leads to people overestimating their ability to make accurate predictions of the future, which, in turn, can make them take greater risks, make bolder decisions and be less likely to consider alternatives or seek additional information. Hindsight bias has, over the years, helped our brains to retain and store validated and relevant information. Still, it is detrimental for learning how to act in coming situations if not explicitly considered when thinking about what happened in the past.

Hindsight bias also causes us to forget that we always have infinite potential contingent scenarios ahead of us whilst we only have one past. It makes you take the complex pre-event situation, with its many interdependent factors and countless possible pathways, and bend it into a simple linear story (Dekker, 2014, pp. 28–30). This aspect of hindsight bias combines with another bias to generate significant consequences for social learning, further elaborated on as a social factor of resistance to knowledge and change below.

Motivated reasoning

Although cognitive biases are crucial, resistance to knowledge and change can also be driven more directly by our goals, values and desires (Kunda, 1990). People are motivated by such normative frameworks to arrive at particular conclusions or decisions and attempt to construct rational justifications for their desired conclusions or decisions that will persuade others (Kunda, 1990, pp. 482–483). It means they base their justification on the desirability of a particular belief or action rather than on a plausible interpretation of all empirical evidence. They reason their way to their favoured conclusions or decisions, with their goals, values and desires influencing how evidence is collected, arguments made, and memories recalled (Epley & Gilovich, 2016, p. 133). Although related to cognitive dissonance and confirmation bias, this phenomenon is called motivated reasoning and captures 'the tendency of individuals to unconsciously conform assessment of factual information to some goal collateral to assessing its truth' (Kahan, 2016, p. 2). As such, motivated reasoning flips the conventional relationship between belief and evidence around, at the same time as people are limited to only drawing conclusions for which they can muster sufficient evidence (Kunda, 1990, p. 483). In that way, 'the reasoning process is more like a lawyer defending a client than a judge or scientist seeking truth' (Haidt, 2001, p. 820):

> *Motivated reasoning turns the relationship between ideas and facts on its head. Ideally, you base your ideas and opinions on facts. However, when using motivated reasoning, you start at the other end with a fixed idea and only accept the facts that back it up*
>
> ***Hendricks and Vestergaard (2019), p. 81.***

One of the complexities of motivated reasoning is that people have many goals, values and desires (Epley & Gilovich, 2016, p. 134; Kahan, 2016, p. 2), spanning from the basic imperatives of survival and reproduction to more proximate goals (Epley & Gilovich, 2016, p. 134). Haidt (2001, pp. 820–821) suggests two main classes called 'relatedness motives' and 'coherence motives'. Whilst coherence motives largely overlap with cognitive dissonance and comprise various defensive mechanisms triggered by such dissonance and threats to one's worldview, relatedness motives involve concerns about making impressions and smooth social interaction with others. Such relatedness motives may, of course, also help us survive and reproduce by maintaining social relationships and achieving social status (Epley & Gilovich, 2016, p. 134), although not necessarily held for those reasons. People may also be motivated by the desire to uphold a positive self-conception (Dunning, 2012) or partisan loyalty (Bolsen et al., 2014), and especially politically motivated reasoning has been suggested as a main explanation of polarisation (Glüer & Wikforss, 2022, p. 43). Moreover, motivated reasoning is commonly used to rationalise all sorts of self-serving behaviour (Hsee, 1996).

There are numerous studies of motivated reasoning in various contexts, including risk and sustainability challenges (e.g., Druckman & McGrath, 2019; Spaccatini et al., 2022), 'showing the many tricks people use to reach the conclusions they want to reach' (Haidt, 2012, p. 74). Plausible interpretations of all empirical evidence are the first casualties since people 'are not after the truth but after arguments supporting their views' (Mercier & Sperber, 2011, p. 57). Motivated reasoning is indeed a daunting challenge for safe and sustainable societies (Figure 13.11). However, after reviewing the vast scientific literature on motivated reasoning, Mercier and Sperber (2011) conclude that most of the often depressing and seemingly bizarre research findings make perfect sense once you consider motivated reasoning as having evolved not to assist us in finding truth but to help us win arguments, persuade and manipulate other people (Haidt, 2012, p. 78).

Figure 13.11 Thucydides, an ancient Greek historian and general, was amongst the first to note motivated reasoning in how human beings rationalise their beliefs and actions.

Social and cultural factors

There are also many factors behind knowledge resistance and resistance to change that manifest not within each human mind but in the interaction between them. They are here referred to as social and cultural factors. Although some of them are connected and sometimes overlapping, they are here presented under the following six subheadings: (1) Looking at each other, (2) Thinking together, (3) Blaming others, (4) Discounting global rationality, (5) Prioritising social rationality and (6) By-products of culture.

Looking at each other

The social factors closest to the cognitive and psychological factors just introduced concern how our direct relationships with people in our immediate vicinity influence what we think we know and how we act. Most fundamentally, social proof (or informational social influence) is a phenomenon first described by social psychologist Cialdini (1984) in which people tend to look to others in a group to determine what is the appropriate belief or behaviour in a particular situation. When in doubt, people assume the actions of others in an attempt to reflect the correct behaviour for that situation. This effect is particularly prominent in ambiguous social situations, where people cannot determine the appropriate mode of conduct and thus behave under the assumption that others know more about the situation and what to do (Cialdini, 2009). Looking at others is one of our most basic strategies to reduce the discomfort of ambiguity. Social proof is, as such, powerful in shaping beliefs and attitudes and can also significantly impact resistance to knowledge and change because it is likely to promote the status quo. However, it can potentially also constitute an opening for change if enough individuals begin to act confidently and resolutely in a new situation, as others may follow.

However, people are not only looking to others when in doubt. We may also conform to what they are saying and doing regardless of how sure we may be before we realise what they are saying and doing. This pattern is called social conformity and can be defined as 'a change in a person's behaviour or opinions as a result of real or imagined pressure from a person or group of people' (Aronson, 2008, p. 19). It is one of the most researched social and cultural factors relevant to knowledge resistance and resistance to change. For instance, Solomon Asch's (Asch, 1955) classic experiment in which participants were asked to compare the length of lines on cards whilst in a group of peers who had been instructed to give incorrect answers. The results showed that the participants conformed to the group's obviously incorrect answers in a significant proportion of trials (Asch, 1956). Asch's study demonstrates the powerful influence social pressure can have on an individual's behaviour and decision-making. The findings have been replicated in numerous studies since then and have helped to shape our understanding of how social influence operates in different contexts (Kundu & Cummins, 2013).

Social conformity is central in maintaining unsustainable practices (e.g., Jack & Glover, 2021, p. 293; Sturman et al., 2017) and has been suggested as an obstacle to overcome for effective risk management (Simonovic, 2010, p. 47). However, it can work in any direction (Atkinson & Jacquet, 2022, p. 623). Social conformity can also generate pressure to adopt more sustainable practices and products (e.g., Derchi et al., 2023; Ruan et al., 2022). Under certain circumstances, local policy interventions, targeted messaging or bottom-up experimentation can initiate widespread changes in attitudes and behaviour (Krueger et al., 2022, p. 5). In any case, social conformity is important for understanding resistance to knowledge and change since it shifts the main question from 'how much evidence must I have before I make this change?' to 'how unpopular must my position be before I'm willing to change it and adopt the majority view?' (MacCoun, 2012, p. 347).

Social proof and social conformity are two related mechanisms explaining how human beings tend to believe and do the same as others. However, this overall tendency is not detached from the number of people holding the opinion or practising the activity, which has different sway if 50 or 5000 people are imagined to do so. This phenomenon is called the bandwagon effect, or contagion effect, and denotes a process of public opinion impinging upon itself (Schmitt-Beck, 2015, p. 1), accelerating the diffusion of a belief or action the more people getting on board with it (Colman, 2009, p. 78). Although the term was initially coined with reference to the actual use of bandwagons in American presidential campaigns during the latter half of the 19th century (Figure 13.12), the bandwagon effect had obtained its current scientific meaning by the mid-20th-century (e.g., Cantril, 1940; Leibenstein, 1950). It suggests that the probability of an individual adopting the belief or action increases with the proportion of the group already done so

Figure 13.12 A 19th-century political cartoon of a bandwagon.

(Colman, 2009, p. 78), in a self-reinforcing process of popularity (Macy & Willer, 2002, p. 161). The bandwagon effect can drive public opinion in whatever direction and is linked to the social tipping points discussed earlier (Chapter 9). Imagine what change could ensue if the bandwagon started to roll in a more sustainable direction.

There are two more related factors that are important in this context: the false consensus effect and projection bias. The false consensus effect refers to the tendency of individuals 'to see their own behavioural choices and judgements as relatively common and appropriate to existing circumstances' (Ross et al., 1977, p. 280) and expect others to agree with them (Johnson & Levin, 2009, p. 1596). It leads people to assume that their opinions and actions are more widely accepted than they actually are. Projection bias, on the other hand, refers to the tendency of people to project their own thoughts and feelings onto others and, thus, assume that they share similar beliefs, values and motivations (Johnson & Levin, 2009, p. 1596). It regularly leads to misjudging other peoples' behaviour and inaccurate anticipation of their decisions and actions. Both of these factors can have significant impacts on issues of risk and sustainability, as they lead individuals into assuming their own beliefs and behaviour to be 'normal', whereas beliefs and behaviour that differ from their own are deemed 'deviant' (Ross et al., 1977). They play, for instance, an important role in the polarisation of the issue of climate change, in which groups on both extremes overestimate the commonness of their opinion (Marshall, 2014, p. 28).

Thinking together

Social factors influencing knowledge resistance and resistance to change concern not only our inclination to look at each other but also for thinking together. Irrespective of how closely these two are related, thinking together is here about the phenomena behind the often surprising and paradoxical sagacity and stupidity of groups referred to as the wisdom of the crowd and groupthink.

The wisdom of the crowd is a phenomenon where a group of people collectively, under certain circumstances, perform better than any individual member could alone. Although already alluded to by Aristotle (1995, pp. 4364–4365), it was first described in detail by Galton (1907) when analysing the 787 legible individual estimations of the dressed weight of an ox for a competition at the 1906 West of England Fat Stock and Poultry Exhibition in Plymouth. Whilst the individual estimations varied significantly, Galton (1907) found their distribution surprisingly informative, with the 'middlemost estimate' only deviating 1% from the correct weight. He called this outcome the 'vox populi'—the voice of the people—which, under such circumstances, is likely to approach the correct result as the individual errors of some will cancel each other out over the correct guesses of others.

Galton had unwittingly found that 'under the right circumstances, groups are remarkably intelligent, and are often smarter than the smartest people in them' (Surowiecki,

2004). However, how smart a group can be depends on four conditions (Surowiecki, 2004, p. 10): independence, diversity, decentralisation and aggregation. The first condition is somewhat paradoxical. For a group to leverage its collective wisdom, each group member must make up their own mind without being influenced by the others. Otherwise, it may become biased. For diversity, each group member must have some individual information to contribute, even if just their own eccentric interpretation (Surowiecki, 2004, p. 10). Moreover, the group should consist of individuals with diverse backgrounds, experiences and perspectives. Given that such diversity would generate diversity in individual estimations, which is likely, 'a diverse crowd always predicts more accurately than the average of the individuals' (Page, 2007, p. 209). It is simply mathematics, with the core insight that individual ability and collective diversity contribute equally to the ability of the group to make correct collective estimations (Page, 2007, pp. 205–209). Being different is as important as being good, which we will return to when considering groupthink. The third condition for the collective wisdom of groups concerns decentralisation in the sense that the group members can specialise and draw on local knowledge (Surowiecki, 2004, p. 10). The group must also be decentralised in the sense of all members being free to express and act on their opinion, with no leader or expert dominating the process (Matzler et al., 2016). Finally, there must be some mechanism for aggregating the individual estimations of the group members into the collective estimate (Surowiecki, 2004, p. 10). This process can be rather straightforward for numerical estimations, for which the average or median value is often used. A simple majority vote is often used for categorical variables. However, there are several more advanced mechanisms that weigh the votes by confidence or look for the most surprisingly popular answer (Prelec et al., 2017).

Regardless of how useful the wisdom of the crowd can be for many purposes, when considering its four conditions, it becomes clear that it also has its limitations. Whilst it has proven useful for making point estimations, eliciting accurate answers to categorical questions, and solving ordering, combinatorial and other multi-dimensional problems, the wisdom of the crowd is less useful for answering more complex, open-ended questions for which there is no unambiguously correct answer. Asking 100,000 people about the best solution to climate change would unfortunately not generate the same kind of informative result as the weight of Galton's (1907) ox. There is no practical way to aggregate the individual answers to such questions into a useful collective outcome. Moreover, it is hardly feasible to maintain the conditions of independence and decentralisation for questions embedded deeper into society itself, and there are several ways through which the social influence elaborated on above can undermine the wisdom of the crowd (Lorenz et al., 2011). That said, the conditions are helpful to keep in mind for any attempt to tap into this wisdom, however imperfectly.

There are also useful methods seeking to draw on the collective intelligence of groups whilst balancing independence and interaction to jog peoples' memory and move ideas into our focal awareness and on to explicit knowledge (Chapter 7), like different variants of the Delphi method (Linstone & Turoff, 1975). Trading off some of the necessary independence for the wisdom of the crowd, in the conventional sense, against the creative potential of interacting minds may even be necessary for the members of the group to even think about the same thing (Becker & Tehler, 2013; Vennix, 2001).

If the wisdom of the crowd requires independent, free and diverse group members, then groupthink is its all-too-common opposite. Although groupthink has most likely happened since the dawn of civilisation, it was coined by Whyte (1952) and later developed and popularised by Janis (1971, 1972). Groupthink is a phenomenon in which a group makes irrational decisions because they are more concerned with maintaining harmony and avoiding conflict within the group than with critically evaluating the problem at hand. It 'refers to a deterioration in mental efficiency, reality testing and moral judgement as a result of group pressures' (Janis, 1971, p. 43). Groupthink occurs when group members prioritise conformity to group norms over individual decision-making. However, groupthink is not based on mere instinctive conformity but a rationalised conformity in which group values are not only expedient but also considered right and good (Whyte, 1952, p. 114). Such seeking to avoid disagreements and debates can become so dominant within cohesive groups that it tends to override realistic appraisals of alternative courses of action, with 'numerous indications pointing to the development of group norms that bolster morale at the expense of critical thinking' (Janis, 1971, p. 43).

Janis (1972, pp. 197—198) lists eight symptoms of groupthink: (1) an illusion of invulnerability, creating excessive optimism and encouraging risk-taking; (2) collective rationalisation, discounting warnings before recommitting to past policy decisions; (3) unquestioned morality, ignoring the ethical or moral consequences of decisions; (4) stereotyped adversaries too evil, weak or stupid to engage with; (5) pressure for loyalty to the group and its stereotypes, illusions, or commitments; (6) self-censorship of deviants, minimising the experienced importance of any doubts and counterarguments; (7) an illusion of unanimity driven by self-censorship and augmented by the assumption that silence means consent; and (8) self-appointed mind-guards, protecting the group from adverse information. Whilst helpful to maintain morale during crises, when a group displays most or all of these symptoms, its decision-making and performance of collective tasks are ineffective, and it is likely to fail to reach its collective goals (Janis, 1972, p. 198).

Not all groups have the same susceptibility to groupthink. A group of individuals with similar backgrounds, beliefs and perspectives can result in a limited range of ideas and opinions. Such homogeneity is an important cause of groupthink (Janis, 1972, p. 192), particularly in small groups (Surowiecki, 2004, p. 36). In addition to homogeneity, groups tend to be more susceptible to groupthink the more cohesive they are since a stronger sense of belonging and loyalty to the group can cause members to feel stronger pressure to conform to group norms and values (Figure 13.13). Even amiability and camaraderie amongst group members can increase the likelihood that groupthink

Figure 13.13 Groupthink.

replaces independent critical thinking (Janis, 1972, p. 198). Moreover, groups isolated from external sources of information are more susceptible to groupthink since narrow input can provide a false sense of certainty that the group's decisions and actions are correct. Overconfidence in the group's ability to make and implement decisions and stress or time pressure are also important contributors to groupthink. Groupthink is also more likely in groups led by a dominant or charismatic leader who expresses a strong preference or opinion.

Groupthink can be an important factor undermining vital decisions and actions to govern risk and sustainability. It has played a crucial part in many flawed or at least highly questionable decisions in the past. Janis (1972) original book identifies groupthink in the nonaction before Pearl Harbour, the making of the Marshall Plan, the invasion of North Korea, the Bay of Pigs invasion, the Cuban missile crisis and the escalation of the Vietnam War. Groupthink has also been suggested as critical in the Space Shuttle Challenger disaster (Janis, 1991), the Iraq War (Badie, 2010), the Enron scandal (O'Connor, 2003), Brexit (Lees, 2021), Russia's invasion of Ukraine (Kurnyshova & Makarychev, 2022), and more. It is, therefore, vital to consider how the desire for consensus and conformity within groups also may lead to inappropriate decisions or non-decisions concerning risk and sustainability, where important ideas are silenced and change is delayed, even with increasing indications of urgency.

It is also vital to consider how to counteract groupthink. Most fundamentally, diversity mitigates groupthink, providing additional arguments for gender diversity and other pressing forms (Hofmann, 2022, p. 122). However, ensuring that the group is diverse regarding demographics, backgrounds, experiences and perspectives is not enough. This measure must be combined with activities that create an environment that encourages open communication, where all group members are comfortable expressing their ideas and constructive critique also when deviating from the rest of the group. To further facilitate diverging opinions and critique to surface within the group, allowing group members to provide feedback anonymously without fear of judgement or retribution can be a good idea. Groupthink can also be mitigated by explicitly assigning everybody the role of critical evaluator or someone the role of the devil's advocate, with the task of challenging the group's assumptions, decisions and actions, and by actively seeking input from outside the group (Janis, 1972). Please see Janis (1972, pp. 209–219) for more guidelines for counteracting groupthink.

Blaming others

The next collection of social factors concerns our inclination as human beings to blame others. This propensity is partly driven by what in social psychology is called the fundamental attribution error (Figure 13.14). The term itself was coined by Ross, 1977—although the effect had been noted earlier—and refers to the tendency of human beings

Figure 13.14 Fundamental attribution error in traffic.

to attribute their own behaviour to situational constraints whilst attributing the behaviour of others to their abilities, personalities and intentions (Ross, 1977, pp. 183–184). The fundamental attribution error leads individuals to perceive the actions of others as acting in their own interests whilst describing their own actions as the best they could do in the face of situational restrictions (Johnson & Levin, 2009, p. 1597). For instance, an environmentally minded person deciding to take the car to work is likely to motivate her car use with reference to time pressures, logistical challenges, etc., whilst attributing the frequent car use of neighbours to their laziness or indifference to their carbon footprint without considering the possibility of similar situational constraints. The fundamental attribution error can lead to misunderstandings and conflicts in personal and professional relationships, as it can prevent us from accurately understanding and interpreting the decisions and actions of others. This effect is particularly pronounced when attention is focused on what the others are doing or not doing, communication is poor, and initial distrust is high (Johnson & Levin, 2009). No wonder why international negotiations on different sustainability challenges seldom succeed.

Hindsight bias can also lead to blame gaming—often in combination with outcome bias (Dekker, 2014). Remember from the cognitive and psychological factors introduced earlier, hindsight bias refers to our tendency to view past events as more predictable than they actually were once we know the outcome. It leads us to overestimate our ability to predict an event and underestimate the importance of chance or unpredictable factors in its occurrence. When looking back at a destructive course of events, the observer has the benefit of retrospection. Hindsight bias causes us, then, to forget the infinite number of potential contingent scenarios ahead from the start of that course of events, constructing a simple linear story out of the complex pre-event situation with its many interdependent factors and countless possible pathways (Dekker, 2014, pp. 28–30). With that biased perspective, any observer can easily identify what could or should have been done to prevent the event and most often blame individuals closest in space and time for not doing so (Dekker, 2014). A complex course of events is reduced to simple human error, largely abandoning any opportunity to learn how to stop such an event from happening again.

This way of thinking is clearly visible in the all-to-common rhetorical question—'how could they not see that coming?'—from people observing an actual event when there were, in fact, myriad possible outcomes depending on numerous interdependent factors. The unfortunate blaming is, then, the result of outcome bias, which is a tendency to judge the quality of decisions based on their outcome rather than on the decision-making processes in the actual situations (Baron & Hershey, 1988). Moreover, it leads us to change our evaluation of decisions once we know the outcome, assuming that a good outcome is the result of a good decision and, conversely, that a bad outcome is the result of a bad decision (Dekker, 2014, p. 30). That is neither stringent nor fair, and only results in the people involved withholding or altering information vital for

Figure 13.15 Social biases often lead to distrust and scapegoating between groups. *(Source: Photobank gallery/ Shutterstock.com.)*

learning but attracting blame or interpreted in a way that taints their reputation or undermines their social status (Reason, 1997). To address these biases, it is vital to attempt to view decisions and events in their proper context and evaluate decisions and actions based on the information available when they were made.

The fundamental attribution error, hindsight bias and outcome bias make people prone to scapegoating others when something has gone wrong. However, scapegoating is not only happening between individuals but also between groups (Figure 13.15). Social identity theory explains this by showing that even the most randomly constructed groups tend to describe themselves and their members favourably whilst disparaging other groups and their members (Johnson & Levin, 2009; Tajfel, 1974, p. 1598). The motivation for such in-group/out-group bias is that positive stereotyping of the group members and negative stereotyping of non-members serves to maintain a positive self-image (Tajfel, 1974). However, the downside of it concerning knowledge resistance and resistance to change is significant since most sustainability challenges require cooperation between organisations, sectors, disciplines and countries. Distrust, perceived injustices and scapegoating are not at all helpful when needing to work together, whilst community, trust and social institutions have proven vital for the sustainable management of shared resources (e.g., Ostrom, 2010). Moreover, as negative input is processed more thoroughly and has a more profound impact than positive input—due to the negativity bias mentioned earlier—negative impressions and stereotypes are quicker to form and more resistant to disconfirmation than positive ones (Johnson & Levin, 2009, p. 1595).

Discounting global rationality

Another social factor relevant to understanding resistance to knowledge and change concerns a fundamental tendency of human beings to discount global rationality for our local rationality, even when undermining our own goals over time. Although connected to motivated reasoning and other cognitive and psychological factors driving selfish beliefs, decisions and actions, it calls for being treated as a distinct factor because it can explain so much of the contemporary challenges of humankind. The tragedy of the commons and game theory are two related frameworks that help elucidate and explain this tendency.

The tragedy of the commons, introduced in Chapter 11, refers to a situation where a shared resource—such as a pasture, fish stock or atmosphere—is overused, depleted or polluted due to locally rational decisions and actions. It is locally rational for each actor to use the shared resource as much as they can because that maximises the value for each of them—whatever the value may be (Acheson, 2019). The Maasai pastoralist can maximise his wealth and social status by having as many cattle as he can grazing their semi-arid lands, the Fijian villager can maximise his income by selling as much fish as possible to Chinese fish merchants, and the Danish lawyer can maximise his standard of living and social status by consuming as much as possible in terms of a large house, a holiday home, several cars, luxury products and frequent faraway holidays. However, if many actors use the shared resource as much as they can, it eventually leads to resource overuse, depletion or destruction, which undermines and finally spoils the extractable value for everyone. The Maasai lands become overgrazed to the verge of desertification, the fish stocks disappear, and climate change compromises life as we know it. Local rationality leads to global irrationality—and actually to local irrationality as well over time.

Whilst the tragedy of the commons has circumscribed human societies since the dawn of time (Pauketat, 2000), it was only named as such by Hardin (1968) a bit more than half a century ago. He initially described it as an inevitable limit of all developing societies, as locally rational individuals can only continue to pursue their aspirations as long as the impact of their aggregated actions remains well below the carrying capacity of the common resource (Hardin, 1968). This suggestion attracted a lot of attention from scholars, who found myriad examples of various common resources that had not ended up overused or depleted but were managed sustainably by different institutional arrangements (e.g., Berkes, 1985; Ostrom, 1975, 1977, 1990). The Maasai had traditional institutions to manage their land to prevent overgrazing (Blewett, 1995) and customary tabu areas where fishing is banned are commonplace in Fiji (Golden et al., 2014). Hardin (1991) then qualified his idea as 'the Tragedy of the *Unmanaged* Commons', which better highlights one of our main challenges today.

The core problem of the tragedy of the commons is that different people think and act differently concerning the use of limited shared resources in different situations. Ostrom (1999, p. 279) suggests four main categories of people:

(1) those who always behave in a narrow, self-interested way and never cooperate in dilemma situations (free-riders); (2) those who are unwilling to cooperate with others unless assured that they will not be exploited by free-riders; (3) those who are willing to initiate reciprocal cooperation in the hopes that others will return their trust; and (4) perhaps a few genuine altruists who always try to achieve higher returns for a group.

Different kinds of game theory are regularly used to explain these categories. In the classic version pioneered by von Neumann and Morgenstern (1944), the theory assumes individual rational actors and focuses on conflicting interests. Its most famous game is called the prisoner's dilemma (Figure 13.16). Although there are many versions, the game demonstrates the tension between individual self-interest and collective cooperation. Basically, two guilty suspects are brought in for questioning by the police, and each suspect is given the option to either stay silent or blame the other. If both suspects remain silent, they both receive a light sentence. If only one blames the other, the snitch goes free, whilst the silent one receives harsh punishment. Unfortunately, the high reward for blaming the other and the high risk of staying silent motivate both to betray each other, resulting in a moderate sentence for both sharing the harsh prison sentence. They both end up worse off being selfish than if they had cooperated.

The governing of all shared resources can be modelled as a prisoner's dilemma (Acheson, 2019, p. 17). If everybody restricts their resource use, everybody would benefit. However, the ones continuing to exploit the resource as much as they can whilst the rest restricts their use benefit much more, given that there are no enforceable formal or informal institutions penalising such behaviour to a corresponding extent. If not, the

Figure 13.16 The proverbial prison in game theory.

Maasai cattle owner or Fijian villager may simply continue using their shared resource as much as possible if the benefits of free-riding outweigh any penalties. With no effective institutions regulating individual greenhouse gas emissions, even if not admitting to being a free-rider, the Danish lawyer may motivate continuing his massive personal contribution to climate change with reference to his brother emitting even more greenhouse gases. In these cases, it is simply locally rational to continue maximising the benefits because of selfishness or not wanting to be exploited by other selfish people. Even people willing to initiate reciprocal cooperation may be dissuaded to continue restricting their resource use if enough others are not.

Classic game theory helps explain the tragedy of the commons but fails to explain how some groups of people keep coming up with institutional arrangements to govern some shared resources. However, Axelrod (1984) shows how cooperation can evolve over the course of many iterations of the classic prisoner's dilemma, as the short-term benefit of being selfish in a single iteration is trumped by the long-term benefit of developing mutual cooperation with the other. He also suggests that such cooperation can remain stable if the 'future is important enough relative to the present' (Axelrod, 1984, p. 118). Moreover, human beings have a more complex motivational structure than assumed by pure rational-choice theory and are more capable of solving social dilemmas than such theory posits (Ostrom, 2010, p. 664). Whilst retaining the mathematical formalism and generality across games that made classic game theory so useful, behavioural game theory incorporates cognitive, psychological and social variables to better approximate reality (Camerer, 2010). Behavioural game theory helps explain the evolution of institutional arrangements to govern common resources by highlighting the role of social preferences in shaping human action. For example, people may carry social norms for conformity, altruism, inequity aversion, trust and reciprocity that motivate them to conserve a common resource even when it is not in their immediate self-interest (Schindler, 2012). However, the emergence of such social norms requires social interaction amongst individuals who can at least imagine to form groups with commonly held values, receiving feedback when following or breaking the norm, and transmitting the norm to each other through various cultural practices. Without such social norms, game theory demonstrates how the tendency to discount global rationality prevails, and we end up in the tragedy of the commons.

There are countless examples of groups of human beings demonstrating the ability to sustainably govern the use of shared resources, often without centralised, top-down regulation or the privatisation almost instinctively promoted by proponents of capitalism (Ostrom, 1990, pp. 12—15; Partelow et al., 2019). Numerous common resources cannot be privatised practically, and many more for which cooperative institutional arrangements are more effective (Ostrom, 1990). What is left to regulate the use of common resources is often categorised as either top-down coercive schemes or bottom-up collective institutions, again with many more functioning examples leaning towards the latter

(Ostrom, 2010). It clearly shows that although people adopt a narrow and self-interested perspective in many situations, we can also use reciprocity to overcome social dilemmas (Ostrom et al., 1999, p. 279). However, whilst there are working global examples, such as the institutions addressing ozone depletion (Chapter 4), most of them concern more local governing of local resources (Acheson, 2019; Boyd et al., 2018; Feeny et al., 1990). Moreover, most of these local examples concern traditional or customary institutions (e.g., Berkes, 1985; de Schutter & Rajagopal, 2020; Kurien, 2007), many of which are being undermined or have been abolished by the coming of modernity (e.g., Becker, 2017; Sirimorok & Asfriyanto, 2020)—like the traditional Maasai institutions for land use referred to earlier (Blewett, 1995).

The rarity of global institutions and the overwhelming majority of traditional or customary institutions amongst all examples of effective institutions are not coincidences. The effectiveness of traditional or customary institutions is not only due to their generally more local scale but also their social and cultural embeddedness (Box 13.1). It means they are not only implemented through formal coercion, cooperation, exhortation or combinations of the three (Chapter 8) but also through shared norms that guide human behaviour and keep group members in line through their innate desire to belong to and have social status within that group—more on that in the coming subsection. The same social and cultural embeddedness is difficult to bring about in modern society and seems impossible globally. Yet, it may be the only way to counter the tendency of human beings to discount global rationality for our own short-term local rationality.

BOX 13.1 The Fijian tabu system

The traditional Fijian tabu system is a form of community-based system for governing the use of shared marine resources. It has been used in Fiji for centuries but has recently attracted attention as an example of an institution that circumvents the tragedy of the commons. The word 'tabu' means 'sacred' or 'forbidden' in Fijian and some Polynesian languages of the South Pacific, and is the etymological root of the English word 'taboo' that was first brought back by Captain James Cook (Oxford English Dictionary, 2020).

The system is implemented by the chief or council of clan elders and involves setting aside certain areas of a reef or lagoon as *tabu* or off-limits to fishing for different periods of time or permanently, depending on the reason and need. It is deeply embedded into the social and cultural fabric of Fijian society and is strictly enforced through a range of mechanisms, such as fines, shame, social exclusion, witchcraft and spiritual beliefs. The system prevents overfishing by allowing particularly important areas for fish stocks and other marine resources to recover and replenish. After the tabu ends, the area is reopened, but often with restrictions on resource use.

Prioritising social rationality

Knowledge resistance and resistance to change are also driven by our fundamental tendency to prioritise social rationality over all other rationalities. Even if it is entirely irrational in the conventional sense for an individual to look at photos from the inaugurations of Presidents Obama and Trump and claim the latter had more people attending, it may be completely rational from a social point of view if that individual's social status or even membership in an important group may be in jeopardy if not adhering to the mainstream view of that group (Klintman, 2019). The same human need to belong was also behind much of the irrational reluctance of certain groups to wear a face mask or get vaccinated during the Covid-19 pandemic (Baxter-King et al., 2022).

Belonging to a group and having social status in that group are primary human needs only trumped by physiological and safety needs essential for survival (Maslow, 1943). Our species is genetically hardwired to be social since group membership and status have provided significant evolutionary advantages (Trivers, 1985). We have, thus, evolved to be sensitive to being socially excluded from our group since that would have had dire consequences for our survival and reproduction (Kurzban & Leary, 2001) and to desire social status within that group (Cheng et al., 2010) since that increased such chances (Mark, 2018). This perspective explains the persistent importance of other people's opinions of us. It also explains why most people fear public speaking, which is picked up in an amusingly accurate joke by Seinfeld (1993):

> *According to most studies, people's number one fear is public speaking. Number two is death. Death is number two! Now, this means, to the average person, if you have to go to a funeral, you're better off in the casket than doing the eulogy.*

Prioritising social rationality over plausible interpretations of all available evidence or the most rational course of action does not require actually believing in the flawed belief or illogical action, although lots do. There may be group members expressing ideas they do not genuinely hold in order to signal allegiance to the group, referred to as 'cheerleading' (Glüer & Wikforss, 2022, p. 31). People claiming to be against Covid-19 vaccinations might even get vaccinated secretly, without telling friends and family, not to visibly defect from what they perceive as widespread social norms (Dow et al., 2022, p. 261). Similarly, the educated middle-class in Sweden is increasingly trying to hide that their holidays still involve flying as an alternative, low-carbon discourse is emerging that clashes with the dominant notion of flying as a status marker (Ullström et al., 2023).

However, many, if not most, members of groups exhibiting widespread resistance behaviour hold the flawed belief or partake in illogical action wholeheartedly, irrespective of overwhelming evidence to the contrary. Our urge to belong and for social status trumps our reason, and evolution has provided us with a powerful ability to help: self-deception (Trivers, 1991). It means we are genetically hardwired to be able to fool ourselves to be better at fooling others (Trivers, 2013). To signal a full and unquestionable commitment to our group's central ideas and actions, we subconsciously make ourselves believe in them. This process is connected to several cognitive and psychological factors, such as cognitive dissonance and motivated reasoning, and the resulting incongruity of beliefs and behaviours brings Orwell's (1949) notion of doublethink to mind.

Whilst we tend to think about flat-earthers, anti-vaxers and other groups clearly diverging from the mainstream, it is essential to note that we all prioritise social rationality to various degrees in different situations (Klintman, 2019). Even scientists who are supposed to follow the evidence are members of groups, which have a significant influence on the selection of research topics, the formulation of research questions, the selection of methods, the interpretation of data, and the dissemination of scientific knowledge (Barnes et al., 1996; Bloor, 1991). These groups can obviously be wrong, yet pressure their members to adhere to the flawed belief. There are several examples of scientists being ridiculed for presenting findings that clashed with the scientific consensus of that time, such as the scientists first suggesting tectonic movements, kin selection, lead in petrol, development of nervous tissue, and the infant incubator (Allen & Reedy, 2020). There is also the everyday pressure of committing to the theories and methodologies of the scientific communities we aspire to belong to and acquire status in, which not only entails the risk of status loss and ridicule for the norm breaker but also of harsh peer reviews, unsuccessful funding applications and hampered career advancement (Knorr-Cetina, 1999). Luckily, some scientists are persistent and follow the data instead of the flock, and some of their suggestions eventually amass enough empirical and social support to end up as a new scientific consensus or as part of the core of a new scientific community. The same self-correcting mechanism is unfortunately rare outside of science.

The human tendency to prioritise social rationality has severe consequences on humankind's ability to address our contemporary sustainability challenges. Not only because it currently drives knowledge resistance but also because it keeps the vast majority of the main polluters of the global population from wanting to change their environmentally disastrous behaviours. As long as social status is essentially equated with consumption—and you do not have to watch much television, read many magazines or glance at many social media accounts of so-called influencers to realise that is the case (Figure 13.17)—our current disastrous path towards doom will be maintained. You do not have to be religious to see the parallels between the consumerist lifestyles of the affluent and the worship of Mammon or Haman, warned against in Christian and Muslim traditions. As long as consumption is the primary process through which individuals gain social status amongst their peers, we are in trouble. Other norms and values need to be revitalised—not only within the minority groups of environmentally and social justice-minded people that may already be on the way (Chapter 16) but amongst the general population.

Figure 13.17 Most influencers encourage blind consumerism since the most common way to earn money is to get paid by companies to promote their products. *(Shared by Eğitim Kutusu on the Creative Commons.)*

By-products of culture

Culture influences knowledge resistance (Huber et al., 2019; Klintman, 2019) and has been restricting change to maintain the stability of societies since the dawn of civilisation (Shennan, 2018). Aside from being a central source of knowledge, skills, organisation, rules, norms, etc., that are crucial for resilience (Oliver-Smith & Hoffman, 2020; Reason, 1997), culture may, thus, also entail aspects that undermine resilience. In addition to the social factors just presented—several of which are possible to frame as cultural factors too (e.g., Brewer & Yuki, 2007; Knorr-Cetina, 1999; McCay & Acheson, 1987)—other by-products of culture can influence resistance to knowledge and change more directly.

Culture may justify and maintain social structures that put some groups at increased risk of specific events. For example, norms concerning women's clothing have been described as part of the reason why women are disproportionally affected in particular events since restrictive clothing hampers running and swimming (Neumayer & Plümper, 2007). Other cultural norms make young men prone to take risks that are visible in mortality and morbidity statistics (Creighton & Oliffe, 2010). Effective proactive measures for reducing the risk of deviations away from the preferred expected development scenario may not be culturally accepted and restrict what can be done in practice. Such by-products of culture should not be ignored, even if they are often sensitive issues to address.

Political, economic and administrative factors

The cognitive, psychological, social and cultural factors introduced so far affect people and groups across all spheres of society, including politicians and political parties, market actors, lawyers and courts, journalists and the media, and bureaucrats and bureaucracies. However, there are also factors driving resistance to knowledge and change that can be seen as political, economic or administrative in themselves. The most relevant of them are presented under the following five subheadings: (1) Lacking will or resources, (2) Policy lag, (3) Organising in silos, (4) Short-termism and (5) Manipulating messages.

Lacking will or resources

Lack of will can have significant impacts on knowledge resistance. Sufficient political will requires enough decision-makers with a common understanding of a particular problem committed to supporting a commonly perceived and potentially effective policy solution (Post et al., 2010, p. 671). Lack of political will can, therefore, revolve around insufficient agreement concerning the problem definition, the solution or both. These ambiguities provide a fertile foundation for knowledge resistance regardless of whether enough evidence is available for a particular problem definition and solution (cf. Ingre et al., 2022, p. 276). However, such situations are not only potential incubators of knowledge resistance but can also be engendered by the knowledge resistance of the politicians and bureaucrats promoting alternative perspectives that cannot be plausibly inferred from the available evidence or simply undermining the ones that can to maintain the status quo. Knowledge resistance propagates knowledge resistance. Here, strategic ignorance becomes indispensable for the knowledge resister to mobilise the unknowns to command resources, deny liability later, and assert control in the face of foreseeably unpredictable outcomes (McGoey, 2012, p. 555). Market actors are, of course, liable to the same self-reinforcing process, with ExxonMobil's orchestrated amnesia regarding their own climate change findings being a formative example (Chapter 4).

Moreover, when politicians fail to acknowledge the existence of knowledge resistance or even engage in it themselves, they may inadvertently or deliberately contribute to its proliferation. This problem is particularly critical in situations with populist politicians and marked political polarisation (Lecheler & Egelhofer, 2022, p. 69) and is further discussed in relation to manipulating messages below.

Lack of will or resources can also significantly impact resistance to change. For real change to materialise, the involved group of actors must have vision, skills, incentives, resources, and an action plan (Ambrose, 1987; Lippitt, 1987). Having skills, incentives, resources, and an action plan but no vision of what to accomplish together leads to confusion that undermines change (Figure 13.18). Having a vision, incentives, resources and an action plan but inadequate skills leads to anxiety amongst the involved

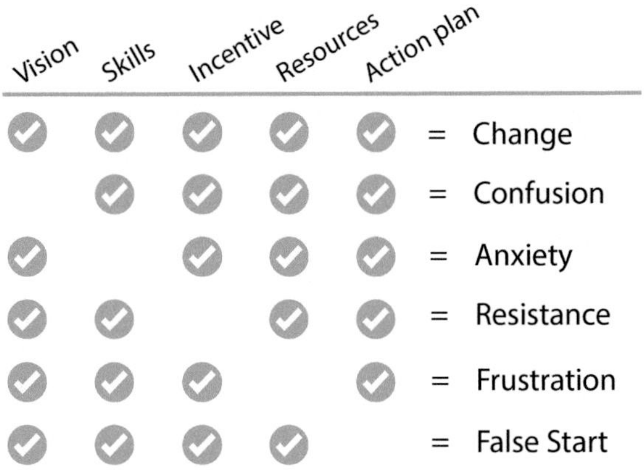

Figure 13.18 Five requisites for managing complex change. *(Based on Ambrose (1987), Lippitt (1987).)*

Vision	Skills	Incentive	Resources	Action plan		
✓	✓	✓	✓	✓	=	Change
	✓	✓	✓	✓	=	Confusion
✓		✓	✓	✓	=	Anxiety
✓	✓		✓	✓	=	Resistance
✓	✓	✓		✓	=	Frustration
✓	✓	✓	✓		=	False Start

actors that cripple change. Having everything in place but incentives—such as rewards, recognition and celebrations—leads, at best, to slow, gradual change that may take years to accomplish. Lacking the necessary resources—such as money, time and equipment—leads to frustration and the change left undone. Finally, having a vision, skills, incentives and resources but no action plan to accomplish the change, the actors are likely to experience a lot of false starts, taking off in various directions only to realise that change is not materialising due to important components being overlooked (Figure 13.18). In short, a lack of will and resources may exacerbate resistance to change by generating confusion, anxiety, lack of incentives, frustration, and false starts (Ambrose, 1987; Lippitt, 1987).

Policy lag

Even with the necessary will and resources, there may be a delay between the emergence and resolution of the problem that can prompt resistance to knowledge and change. Policy lag refers to the delay between when a problem or issue arises and when a policy has generated the impact to address it (Okun, 1970). Such delay can occur for various reasons, including bureaucratic processes or inertia in the system.

Policy lag can be divided into three parts—(1) recognition lag, (2) implementation lag, and (3) impact lag—depending on in which part of the policy process the lag materialises (Ihori, 2017, pp. 43—44). Although related to a more general mindset lag between a rapidly changing context and the pace of change in the mindsets of the actors involved (Hagelsteen & Becker, 2019, pp. 7—8), recognition lag concerns the delay between the emergence and recognition of the problem (Ihori, 2017, p. 43). For instance, a lot of people were infected by Covid-19 in Europe before the competent authorities around Europe realised the gravity of the situation, and it took a century after the discovery of the greenhouse effect (Arrhenius, 1896) for climate change to be widely recognised as a problem. Implementation lag, on the other hand, concerns the delay between recognising the problem and implementing the policy to address it (Ihori, 2017, pp. 43–44). Even when the relevant actors have realised that the problem exists, it takes time to set the agenda, consider policy options, decide on the policy to be implemented, mobilise resources and implement the policy. This is essentially where the world is stuck regarding climate change—in a protracted implementation lag that threatens the world as we know it (Bradshaw & Borchers, 2000; IPCC, 2022). Finally, impact lag concerns the delay between the implementation and the impact of the policy (Ihori, 2017, p. 44), which can be substantial in complex systems. For example, the number of Covid patients continued to escalate for 2–3 weeks after imposing a strict lockdown (Vinceti et al., 2020), and temperatures would continue to rise for decades even if we stop emitting greenhouse gases altogether, but with more than 90% of the warming realised after the first decade (Zickfeld & Herrington, 2015).

Policy lag can both drive and be driven by knowledge resistance and resistance to change. Knowledge resistance can play a vital role in generating recognition lag by delaying the identification or acknowledgement of the situation as a problem. Similarly, resistance to knowledge and change can also generate implementation lag by undermining agenda-setting, misleading policy options, obstructing decision-making, opposing resource mobilisation and hampering policy implementation. On the other hand, policy lag can also drive knowledge resistance by causing people to overlook or downplay the problem as policymakers seem to either ignore it (recognition lag) or drag out the process before doing anything (implementation lag). How important can it really be if they are not doing anything? Moreover, policy lag can also drive resistance to knowledge and change when the best possible actions are implemented without enough immediate impact on the problem (impact lag), making the policy solution or problem definition seem wrong until much later. As elaborated below, such policy lag can also be skilfully used by actors intentionally manipulating messages.

Organising in silos

Resistance to knowledge and change is not only exacerbated by the complexity of the problem itself and the systems in which it exists. It is also driven by a matching tendency of humankind to organise and work in silos. Working in silos means organising the work thematically and hierarchically, seeking to maximise vertical coordination within the silo at the expense of horizontal coordination between the silos (Scott & Gong, 2021, p. 20). It leads to interdependent or even overlapping organisations that hardly communicate. Such siloing entails turf wars, competition over limited resources and general diffusion of responsibility, resulting in inefficiency or blatant failure as essential interdependencies are ignored and important issues fall through the cracks. Kramer's (2005) seminal paper on the failure of the CIA, FBI and NSA to communicate before the 9/11 terrorist attacks provides an extreme account of what, unfortunately, is more common than not. However, these challenges are particularly likely in complex

BOX 13.2 Systemic reasons for silo-thinking in a UN agency

The senior management of one of the largest and most well-regarded UN agencies had been increasingly frustrated for years by the lack of cooperation and synergies between the different programme areas of the organisation. They disseminated message after message throughout the entire agency, prescribing cross-programme linkages and escalating the tone from gentle nudge to virtual command, but more or less nothing happened. They simply could not understand why. They had worked diligently to implement a system linking the overall goals of the organisation all the way down to staff-level outcomes, which were used as performance indicators for staff appraisals. However, irrespective of its potential merits, this system did not include any performance indicators requiring cross-programme cooperation. On the contrary, it provided a wide range of performance indicators for staff to work against to contribute to meeting the objectives of each programme area in isolation, and scoring well on each performance indicator was perceived as essential to further one's career. The entrenched silo-thinking of the organisation had, in other words, inadvertently created a system not only making it locally rational not to cooperate across programme areas but explicitly rewarding such rationality.

organisational settings rife with conflicting objectives and misunderstandings concerning priorities (Dörner, 1996). Silo-thinking is also ubiquitous within organisations, as departments are generally not rewarded for things they do that benefit other departments (Box 13.2).

Organising in silos drives resistance to knowledge and change in several ways. It tends to focus actors' attention on evidence generated within their silo, often ignoring what others have, even when they share it openly. For instance, an unusually bold World Bank study shows that about a third of all their openly published reports are never downloaded (Doemeland & Trevino, 2014, p. 4)—not even once! On the more confrontational side of organising in silos, turf wars and competition over resources drive resistance to knowledge and change by introducing political and economic incentives for ignoring or undermining the competitors' evidence or policy suggestions—exacerbating the effects of the in-group/out-group bias, fundamental attribution error and the tendency to prioritise social rationality introduced earlier.

Short-termism

Another essential factor in this overall category is pervasive short-termism across all spheres of society. Addressing issues of risk and sustainability is, by definition, a long-term endeavour that demands a corresponding long-term focus. However, decision-makers worldwide are generally focused on short-term thinking regardless of whether representing the state, market or civil society—on winning the next election, maximising this year's profit to get a large bonus, or securing funding that also funds the NGO itself through overheads. Regular budgets are almost exclusively annual, project budgets tend to extend over a few years maximum, and money not spent tends to result in budget cuts for coming years. Although such a short-term focus is not only bad, since it may facilitate responsiveness to a rapidly changing world, decision-makers must also have a long-term focus if the much-needed transformation towards sustainability is not to continue to elude us.

Short-termism can drive knowledge resistance and resistance to change in several ways. When organisations or individuals are politically, economically or administratively pushed to prioritise short-term gains over long-term change, they are less likely to actively look for or accept new information that could challenge their currently held beliefs or practices. This pressure to produce quick results exacerbates knowledge resistance by deepening several previously described cognitive, psychological and social factors. For instance, there is simply no time to seek out input other than directly (sensory bias) or recently experienced (availability heuristic), reflexively consider alternative interpretations (cognitive dissonance), or actively include free and diverse opinions and encourage open communication (groupthink). Additionally, short-termism tends to direct attention to the visible symptoms of the problem that can be addressed immediately or soon, which generally leads to a lack of investment in developing the necessary capacities to grasp and address the root causes of the problem and to adapt to changing circumstances. Finally, politicians prioritising short-term political gains over long-term solutions to complex problems often avoid tackling knowledge resistance if politically difficult or unpopular, which can perpetuate flawed beliefs and illogical practices.

Manipulating messages

Knowledge resistance and resistance to change are substantially driven by actors manipulating the masses through misinformation, disinformation and propaganda. Misinformation is here defined as unintentionally promulgated inaccurate information, disinformation as intentionally inaccurate information that is spread deliberately, and propaganda as information that may or may not be accurate and spread to rally public support (Born & Edgington, 2017, p. 4). With misinformation, neither the creation nor dissemination of false information is intentional or even conscious (Lecheler & Egelhofer, 2022, pp. 72–73). The actors involved simply do not know that the information is incorrect, such as sloppy journalists, the average retweeter or social media liker, most attendees at populist political conferences, and others actively relaying inaccurate information without practising fact-checking or source criticism. On the other hand, actors involved in creating or disseminating disinformation do know it is false and intend to deceive (Lecheler & Egelhofer, 2022, p. 73). However, when distinguishing between misinformation and disinformation, it is essential to note that incorrect information can be intentionally created yet unintentionally disseminated (Egelhofer & Lecheler, 2019). In short, the intention behind its creation can differ from the intention behind its dissemination, 'meaning that misinformation can become disinformation and vice versa' (Lecheler & Egelhofer, 2022, p. 73).

It is also important to note that the so frequently used term 'fake news' is both used to describe deliberate pseudojournalistic disinformation and to undermine the legitimacy of genuine news media (Egelhofer & Lecheler, 2019)—both to describe a particular form of disinformation and *as* disinformation to delegitimise important scrutinising and fact-checking institutions. In any case, evidence suggests that mass media have a strong influence on people's perceptions of the environment (Johnson & Levin, 2009, p. 1595) and ignorance about sustainability challenges and solutions may be the result of 'selective exposure' in the sense of 'active selection and/or avoidance of certain sources of information' (Glüer & Wikforss, 2022, p. 42).

Misinformation, disinformation and propaganda have obviously fundamental impacts on knowledge resistance and resistance to change (Figure 13.19). Politicians who actively promote such manipulating messages lend legitimacy to flawed beliefs and encourage their supporters to reject accurate information from credible sources, leading to a breakdown in trust in vital societal institutions and eroding the foundations of democracy (Lecheler & Egelhofer, 2022).

Figure 13.19 Marching for facts.

> *A particularly acute example of the danger posed by knowledge resistance can be seen by then US President Trump's attempt to overturn his electoral loss to Joe Biden. Whilst it is impossible to know what the former President truly believes, his rhetoric following the election sent an unambiguously clear (though factually incorrect) message — that Joe Biden only won the election through voter fraud. After weeks of inciting his supporters through these inaccurate and unsubstantiated claims of voter fraud, on January 6, 2021, the US witnessed an attack on its democratic institutions unprecedented in modern times*
>
> ***Szewach et al. (2022), p. 167.***

Moreover, misinformation and disinformation can undermine people's trust in expertise (Nichols, 2017), leading to knowledge resistance and resistance to change. People exposed to false or misleading information may become increasingly suspicious of experts and their recommendations. They are, thus, more likely to resist efforts to change beliefs or practices, especially since misinformation and disinformation foster a sense of uncertainty about what is correct and the right way of acting in the situation. By promoting inaccurate information or intentionally spreading falsehoods, actors engaging in disinformation campaigns are skilfully weakening the public's trust in the scientific community best equipped to describe the problems and suggest solutions. This manipulated resistance to knowledge and change is particularly problematic regarding sustainability challenges that require prompt, concerted action, as misinformation and disinformation can distort public perception and lead to apathy or misguided efforts. Ultimately, such manipulating messages make it more difficult to address our pressing sustainability challenges, as people are less likely to take the necessary steps if they do not trust the expertise of those describing the problems and suggesting the solutions (Nichols, 2017)—just as the disinformation industry is doing with climate change (McIntyre, 2022; Oreskes & Conway, 2010).

The challenge of manipulating messages is exacerbated by 'the shadowy but many-tentacled disinformation industry' (Schiller, 1989, p. 154). The disinformation industry is a nexus of organisations and individuals intentionally spreading false or misleading information for various reasons, often political or financial. It includes buying scientists who follow the money instead of the evidence, using their scientific authority and knowledge of the scientific community to cultivate uncertainty and ambiguity by misrepresenting or cherry-picking scientific evidence, and attacking the credibility of other scientists (Oreskes & Conway, 2010). It employs sophisticated technology and techniques to create and disseminate false information or fabricated data that are difficult for people to distinguish from genuine information and data. Moreover, the disinformation industry is increasingly proficient in tailoring the manipulating messages to the individual target (Risso, 2018; Wylie, 2019), which may increase their persuasiveness (Hinds et al., 2020; Hirsh et al., 2012, p. 2). Although not part of the disinformation industry as such, genuine news media is largely and unwittingly assisting its success by giving its misleading messages equal time to the knowledge and recommendations of genuine experts, even when the scientific consensus is overwhelming (Oreskes & Conway, 2010).

Technological factors

Finally, there are factors related to our ever-developing technology that also drive knowledge resistance and resistance to change. Although there may be many factors, all of which can be categorised in several different ways, the most relevant ones are here presented under the following two subheadings: (1) Tailoring technologies and (2) Technological lock-ins.

Tailoring technologies

Research shows that the ever-advancing information and communication technologies not only make the creation and dissemination of false information increasingly successful but also easier and cheaper (Lecheler & Egelhofer, 2022, p. 69). What only a few government-sponsored intelligence agencies and science laboratories could do in the past in terms of photo, sound and later video manipulation, tech-savvy teenagers are now doing on their gaming computers. The development of synthetic media, also known as deepfake, is currently very similar, with far-reaching consequences for society (Westerlund, 2019; Zegart & Morell, 2019). Being able to tailor a completely authentic-looking video of anyone you like, saying whatever you want them to say, has evidently radical consequences for knowledge resistance (Figure 13.20). If seeing is believing, as sensory biases have us do, just imagine what impact deepfake can have.

Figure 13.20 In 2017, researchers at the University of Washington shared how they used video and audio material of President Barack Obama to create a deepfake video (Suwajanakorn et al., 2017).

However, technology is not only providing the ultimate tool for the disinformation industry and foreign or domestic actors deliberately manipulating public opinion. The designs of internet search engines and social media platforms also have fundamental unintentional consequences for knowledge resistance and resistance to change by placing us in 'filter bubbles' and 'echo chambers' (Flaxman et al., 2016).

To help users find the information they consider useful amongst the mind-blowing amount of data available on the internet, search engines apply complex algorithms based not only on the search query but also location, browsing history, social media activity and other factors. These personalised algorithms are intended to save time and effort by showing the users search results they are likely interested in. However, they also create a feedback loop where added internet activity further specifies what the algorithms deem interesting for the users in future searches, resulting in so-called 'filter bubbles'—a personalised information universe—that fundamentally alter the way users encounter information (Pariser, 2011, p. 9). It means users will increasingly only see information reinforcing their existing beliefs and values whilst being insulated from alternative perspectives. Social media platforms also contribute substantially to filter bubbles through similar algorithms (Spohr, 2017).

Moreover, social media platforms are also liable to place their users in so-called 'echo chambers' (Jamieson & Cappella, 2008), where 'the opinion, political leaning, or belief of users about a topic gets reinforced due to repeated interactions with peers or sources having similar tendencies and attitudes' (Cinelli et al., 2021, p. 1). The tendency to relate to and associate with similar others is ubiquitous in human affairs and has fundamental consequences, by itself, for the interactions people experience, the information they receive, and the attitudes they form (McPherson et al., 2001). However, social media platforms amplify this tendency substantially due to their user and algorithmic design.

The algorithms of social media platforms are not only designed to present information the users are likely to be interested in, contributing to the filter bubbles described earlier. They also use search queries, location, demographics, browsing history, social media activity and other factors to tailor their suggestions of which other users to follow and groups to join. Whilst users already tend to favour information adhering to their beliefs and to join groups created around shared beliefs and values, social media platforms make it immensely easier to find and connect with them (Cinelli et al., 2021). The resulting echo chamber effect reinforces knowledge resistance and polarisation (Box 13.3) since users are exposed to a narrow range of perspectives that amplifies a wide range of the cognitive, psychological and social factors described earlier.

Whilst filter bubbles and echo chambers are distinct concepts, they often coincide to create a self-reinforcing feedback loop of ideological isolation—filter bubbles leading to echo chambers, which intensify filter bubbles, etc.

BOX 13.3 Six factors driving the polarisation of people

1. Economic inequality: As the gap between the rich and the poor continues to widen, it can lead to increased resentment and mistrust between social classes.
2. Political polarisation: As political parties become more ideologically divided, it can lead to increased polarisation amongst voters.
3. Demographic changes: As society becomes more diverse, it can lead to increased polarisation as different groups struggle to find common ground.
4. Social media and the internet: The rise of social media and the internet has made it easier for people to find and connect with like-minded individuals, leading to increased polarisation as people become more isolated in their own echo chambers.
5. Populist leaders and movements: Populist leaders and movements that appeal to people's emotions and prejudices can exacerbate polarisation by stoking fear and mistrust of outsiders and minorities.
6. Media bias: The way news is reported can also play a role in polarising the public. Studies have shown that the media can have a negative impact on public perceptions when they present news in a biased or misleading manner.
7. Education level and access to information: People with different levels of education or access to the same information can have different perspectives on the same issues, leading to different opinions and polarisation.

Technological lock-in

Resistance to change can also be driven by path dependence and lock-ins. To recapitulate, path dependence refers to the phenomenon where past decisions and actions influence present decisions and actions, and lock-in effects capture the instances where this forcing has become so strong that deviating from the path is onerous or even impossible (Chapter 9). Technological path dependence and lock-ins introduce significant limits to adaptation (Payo et al., 2015), as additional decisions and investments create an increasingly narrow path and gradually more daunting costs to deviate from it over time—a feedback loop where resistance to change becomes increasingly entrenched. Path dependence and lock-ins can, as such, be significant drivers of resistance to change, as they create a dynamic inertia that favours the established direction (Omidvar et al., 2023). This pattern is not only visible in the prevailing dominance of inferior yet successful technologies (Chapter 9). It is behind much of our persistent use of unsustainable technologies, such as contemporary western car use (Upham et al., 2013) and much of the reluctance to stop using fossil fuels in general (IPCC, 2021, pp. 1—112). It is also behind the ever-present urban processes aggravating disaster risk by generating a radial extension of urban areas outwards from a relatively safe centre into hazardous locations. Be it Oxford in the United Kingdom, Ebute Metta in Nigeria or elsewhere, the areas initially settled are generally not exposed to any common hazards. Yet, regardless of when

Figure 13.21 A disaster-prone neighbourhood of Ebute Metta, Nigeria. *(Source: Dulue Mbachu/IRIN News.)*

urbanisation kicks in, the resulting path-dependent sprawling tends to eventually result in some people living in disaster-prone locations (Figure 13.21).

Overcoming resistance to knowledge and change

A grim picture of the prospect of sustainability emerges when considering the various factors that drive knowledge resistance and resistance to change. However, we cannot afford to despair. Time is of the essence, and much needs to be done. Whilst some factors cannot be prevented, their effects can be mitigated by awareness, reflexivity and conscious countermeasures. Other factors are indeed possible to prevent, although that would generally require significant attention, will and tenacity.

Most cognitive and psychological factors are rooted in our genetic legacy and cannot be fully eliminated. However, their effects can be mitigated by increased awareness of them. By simply recognising how our thinking and the thinking of others may be biased, we can become more mindful of why we think what we think and do what we do and adjust our beliefs, actions and interactions accordingly. For instance, by simply being aware of how we are biased by our own and recent experiences, as well as to take risks and bend information to suit our preconceived understanding, we may become more inclined to take a step back, seek additional information and engage our reason before making any decisions and actions—what Kahneman calls engaging 'System 2' (Kahneman,

2011). Similarly, the effects of cognitive and psychological biases can also be mitigated by encouraging diverse perspectives, active listening and critical thinking. Exposing people to a broader range of beliefs and values—ideally, whilst they try to keep open minds—can help them challenge preconceived notions, become more aware of their own biases, and be more open to changing their perspectives. Encouraging critical thinking and reflexivity also helps mitigate the effects of cognitive and psychological biases by providing a way for people to question their assumptions and values and evaluate the veracity of any claims. Moreover, promoting curiosity, exploration, and a willingness to learn can make people more open to new ideas and less resistant to change. The education systems worldwide must endeavour to keep up with the escalating needs of critical thinking and reflexivity to thwart the proliferation of knowledge resistance.

The social and cultural factors driving resistance to knowledge and change are also somewhat genetically based, through our genetic predisposition to be social. We can, therefore, never do away with our fundamental motivation to belong to groups and have social status in those groups, nor should we try. However, the effects of the social and cultural factors elaborated on in this chapter can be mitigated using much of the same measures as for the cognitive and psychological biases, with some crucial additions. Awareness-raising and education are key for people to understand how also these social and cultural factors influence their beliefs and actions, giving them a chance to overcome them—the same with critical thinking and reflexivity. Furthermore, actively enabling wide inclusion and increasing diversity of groups help address several social factors driving knowledge resistance and resistance to change—as do encouraging open dialogue, creating safe spaces or anonymous lines of communication, and facilitating communication across group boundaries. The tragedy of the commons is most likely overcome if the institution for governing the use of the shared resource is socially and culturally embedded (Ostrom, 1990), and the social rationality behind logically irrational beliefs and actions is often best addressed with a related socially rational alternative (Klintman, 2019).

Political, economic and administrative factors driving resistance to knowledge and change are different yet often connected to cognitive, psychological, social and cultural factors. Whilst all of the measures previously mentioned can have an indirect impact also on these aspects, there is another set of measures requiring attention here. The obvious measures are also not particularly interesting to discuss, such as simply increasing will and resources or stopping the dissemination of manipulating messages. Several of the measures left have to do with fostering more holistic thinking. For instance, including externalities—the indirect or unintended costs of actions or inaction—provides decision-makers with a more comprehensive and correct basis for decisions that, in terms of most sustainability challenges, is likely to motivate increased attention (cf. Tompkins & Eakin, 2012). Including explicit, measurable targets requiring action, interaction and collaboration across borders and boundaries is another way to foster holistic thinking.

However, most fundamentally, to mitigate political, economic and administrative factors, people need to hold politicians, bureaucrats and corporate leaders accountable for their actions or inaction. We have the power as citizens and consumers not to vote for myopic or lying populists, to reject siloed administration and to boycott companies actively maintaining the status quo. Whilst genuine news media has inadvertently facilitated knowledge resistance in the past, there are also examples of it providing vital fact-checking mechanisms that could be further developed.

Similarly, whilst technology is currently exacerbating knowledge resistance substantially, with relatively little effort, it could also be mitigated by technology (Spohr, 2017, p. 157). It is just a matter of design. Technology is our only bulwark against deepfake since it is increasingly impossible for people to distinguish between real and fake without technological assistance. Search engines and social media platforms could easily be provided with adjusted algorithms to help pop filter bubbles by ensuring exposure to high-quality information from different perspectives. Social media platforms could be easily adjusted to open up echo chambers by facilitating interaction with more diverse people and groups. Such algorithmic adjustments are not only technologically feasible but simple to do and healthy for society overall (Spohr, 2017, p. 157), having the potential to mitigate several psychological and social factors too. Policymakers and tech corporations just need to get their act together.

Conclusion

It is easy to become disillusioned when going through various factors driving resistance to knowledge and change. We are working against millions of years of natural selection that left us highly adapted to hunter-gatherer life in small, tight-knit groups—an entirely different life than that of 99.9% of the global population today (cf. Burger & Fristoe, 2018, p. 1138). In addition to the evoked behaviour rooted in our genes, we are also working against the results of cultural and social selection that underlie our behaviour acquired through imitation and learning and imposed through norms and sanctions (Runciman, 1998), and against other coevolutionary processes continuously reconstituting society. Hence, there are many cognitive, psychological, social, cultural, political, economic, administrative and technological factors influencing knowledge resistance and resistance to change. A pool player would say that 'we are behind the eight ball', which is a difficult position indeed—especially with opponents skilfully undermining our game by doing everything they can to cultivate uncertainty and ambiguity to maintain the status quo.

However, it is not impossible to move from here. We have to. We have no other choice. We either overcome pervasive resistance to knowledge and change and address our sustainability challenges, or we keep acting like we have some unspoken 'Collective Suicide Pact' (Guterres, 2022b). We can never undo our genetic legacy, but we can spread awareness of the drivers of knowledge resistance and resistance to change, educate people in critical thinking and reflexivity, ensure broad inclusion and diversity of perspectives, and have technology work for us and not against us. There is also a wide range of mitigation measures to address specific factors. We must do our utmost to overcome knowledge resistance and resistance to change. However, Ruskin (1882, p. 165) is right when stating: 'What we think, or what we know, or what we believe, is in the end, of little consequence. The only thing of consequence is what we do'.

References

Acheson, J. M. (2019). The tragedy of the commons: A theoretical update. In L. R. Lozny, & T. H. McGovern (Eds.), *Global perspectives on long term community resource management* (pp. 9–22). Springer.

Adams, D., & Carwardine, M. (1990). *Last chance to see*. Ballantine Books.

Allen, D. M., & Reedy, E. A. (2020). Seven cases: Examples of how important ideas were initially attacked or ridiculed by the professions. In D. M. Allen, & J. W. Howell (Eds.), *Groupthink in science* (pp. 49–62). Springer International Publishing. https://doi.org/10.1007/978-3-030-36822-7_5

Ambrose, D. (1987). *Managing complex change*. The Enterprise Group Ltd.

Aristotle. (1995). *The complete works of Aristotle*. In J. Barnes (Ed.). Princeton: Princeton University Press.

Aronson, E. (2008). *The social animal* (10 ed.). Worth Publishers.

Arrhenius, S. P. S. (1896). On the influence of carbonic acid in the air upon the temperature of the ground. *The London, Edinburgh and Dublin Philosophical Magazine and Journal of Science, 41*, 237–276.

Asch, S. E. (1955). Opinions and social pressure. *Scientific American, 193*(5), 31–35. https://doi.org/10.1038/scientificamerican1155-31

Asch, S. E. (1956). Studies of independence and conformity: I. A minority of one against a unanimous majority. *Psychological Monographs: General and Applied, 70*(9), 1–70. https://doi.org/10.1037/h0093718

Askman, J., Nilsson, O., & Becker, P. (2018). Why people live in flood-prone areas in Akuressa, Sri Lanka. *International Journal of Disaster Risk Science, 9*, 143–156. https://doi.org/10.1007/s13753-018-0167-8

Atkinson, Q. D., & Jacquet, J. (2022). Challenging the idea that humans are not designed to solve climate change. *Perspectives on Psychological Science, 17*(3), 619–630.

Axelrod, R. (1984). *Evolution of cooperation*. Basic Books.

Badie, D. (2010). Groupthink, Iraq, and the war on terror: Explaining US policy shift toward Iraq. *Foreign Policy Analysis, 6*(4), 277–296. https://doi.org/10.1111/j.1743-8594.2010.00113.x

Ballantyne, N. (2019). Epistemic trespassing. *Mind, 128*(510), 367–395. https://doi.org/10.1093/mind/fzx042

Barnes, B., Bloor, D., & Henry, J. (1996). *Scientific knowledge: A sociological analysis. Athlone.*

Baron, J., & Hershey, J. C. (1988). Outcome bias in decision evaluation. *Journal of Personality and Social Psychology, 54*(4), 569–579. https://doi.org/10.1037//0022-3514.54.4.569

Baumeister, R. F., Bratslavsky, E., Finkenauer, C., & Vohs, K. D. (2001). Bad is stronger than good. *Review of General Psychology, 5*(4), 323–370. https://doi.org/10.1037/1089-2680.5.4.323

Baxter-King, R., Brown, J. R., Enos, R. D., Naeim, A., & Vavreck, L. (2022). How local partisan context conditions prosocial behaviors: Mask wearing during COVID-19. *Proceedings of the National Academy of Sciences, 119*(21). https://doi.org/10.1073/pnas.2116311119. e2116311119.

Becker, P., & Tehler, H. (2013). Constructing a common holistic description of what is valuable and important to protect: A possible requisite for disaster risk management. *International Journal of Disaster Risk Reduction, 6*, 18–27. https://doi.org/10.1016/j.ijdrr.2013.03.005

Becker, P. (2017). Dark side of development: Modernity, disaster risk and sustainable livelihoods in two coastal communities in Fiji. *Sustainability, 9*, 1–23. https://doi.org/10.3390/su9122315

Becker, P. (2021). Fragmentation, commodification and responsibilisation in the governing of flood risk mitigation in Sweden. *Environment and Planning C: Politics and Space, 39*(2), 393–413. https://doi.org/10.1177/2399654420940727

Berkes, F. (1985). Fishermen and 'the tragedy of the commons'. *Environmental Conservation, 12*(3), 199–206. http://www.jstor.org/stable/10.2307/44520413.

Blewett, R. A. (1995). Property rights as a cause of the tragedy of the commons: Institutional change and the pastoral Maasai of Kenya. *Eastern Economic Journal, 21*(4), 477–490. http://www.jstor.org/stable/10.2307/40325668.

Bloor, D. (1991). *Knowledge and social imagery*. University of Chicago Press.

Bolsen, T., Druckman, J. N., & Cook, F. L. (2014). The influence of partisan motivated reasoning on public opinion. *Political Behavior, 36*(2), 235–262. https://doi.org/10.1007/s11109-013-9238-0

Boorstin, D. J. (1983). *The discoverers*. Random House.

Born, K., & Edgington, N. (2017). *Analysis of philanthropic opportunities to mitigate the disinformation/propaganda problem*. Hewlet Foundation. https://doi.org/10.2139/ssrn.3144139

Boyd, R., Richerson, P. J., Meinzen-Dick, R., De Moor, T., Jackson, M. O., Gjerde, K. M., Harden-Davies, H., Frischmann, B. M., Madison, M. J., Strandburg, K. J., McLean, A. R., & Dye, C. (2018). Tragedy revisited. *Science, 362*(6420), 1236–1241. https://doi.org/10.1126/science.aaw0911

Bradshaw, G. A., & Borchers, J. G. (2000). Uncertainty as information: Narrowing the science-policy gap. *Conservation Ecology, 4*(1). http://www.jstor.org/stable/10.2307/26271749.

Branlat, M., & Woods, D. D. (2010). How do systems manage their adaptive capacity to successfully handle disruptions? A resilience engineering perspective. In *Complex adaptive systems—resilience, robustness, and evolvability, proceedings from the association for the advancement of artificial intelligence conference, Arlington, 11-13 November 2010*.

Brewer, M. B., & Yuki, M. (2007). Culture and social identity. In S. Kitayama, & C. Dov (Eds.), *Handbook of cultural psychology* (pp. 307–322). The Guilford Press.

Burger, J. R., & Fristoe, T. S. (2018). Hunter-gatherer populations inform modern ecology. *Proceedings of the National Academy of Sciences, 115*(6), 1137–1139. https://doi.org/10.1073/pnas.1721726115

Camerer, C. F. (2010). Behavioural game theory. In S. N. Durlauf, & L. E. Blume (Eds.), *Behavioural and experimental economics* (pp. 42–50). Palgrave Macmillan. https://doi.org/10.1057/9780230280786

Cantril, H. (1940). The public opinion polls: Dr. Jekyll or Mr. Hyde? *Public Opinion Quarterly, 4*(2), 212–217. http://www.jstor.org/stable/10.2307/2744630.

Carlill, H. F. (1906). *The theaetetus and philebus of plato*. Swan Sonnenschein & Co (Carlill, H. F., Trans.).

Carrillo, J. D., & Mariotti, T. (2000). Strategic ignorance as a self-disciplining device. *The Review of Economic Studies, 67*(3), 529–544. https://doi.org/10.1111/1467-937x.00142

Cheng, J. T., Tracy, J. L., & Henrich, J. (2010). Pride, personality, and the evolutionary foundations of human social status. *Evolution and Human Behavior, 31*(5), 334–347. https://doi.org/10.1016/j.evolhumbehav.2010.02.004

Cialdini, R. B. (1984). *Influence: How and why people agree to things*. William Morrow & Company.

Cialdini, R. B. (2009). *Influence: The psychology of persuasion*. Harper Collins.

Cinelli, M., De Francisci Morales, G., Galeazzi, A., Quattrociocchi, W., & Starnini, M. (2021). The echo chamber effect on social media. *Proceedings of the National Academy of Sciences, 118*(9). e2023301118 https://www.pnas.org/doi/full/10.1073/pnas.2023301118.

Colman, A. M. (2009). *Oxford dictionary of psychology* (3 ed.). Oxford University Press.

Creighton, G., & Oliffe, J. L. (2010). Theorising masculinities and men's health: A brief history with a view to practice. *Health Sociology Review, 19*(4), 409–418. https://doi.org/10.5172/hesr.2010.19.4.409

de Schutter, O., & Rajagopal, B. (2020). Conclusion: The revival of the "commons" and the redefinition of property rights. In O. de Schutter, & B. Rajagopal (Eds.), *Property rights from below: Commodification of land and the counter-movement* (pp. 203–232). Routledge.

Dekker, S. W. A. (2014). *The field guide to understanding human error* (3 ed.). Ashgate.

Derchi, G.-B., Davila, A., & Oyon, D. (2023). Green incentives for environmental goals. *Management Accounting Research*, 100830. https://doi.org/10.1016/j.mar.2022.100830

Desvouges, W. H., Johnson, F. R., Dunford, R. W., Hudson, S. P., Wilson, K. N., & Boyle, K. J. (1993). Measuring natural resource damages with contingent valuation: Tests of validity and reliability. In J. A. Hausman (Ed.), *Contingent valuation: A critical assessment* (pp. 91–164). Elsevier. https://doi.org/10.1016/b978-0-444-81469-2.50009-2

Doemeland, D., & Trevino, J. (2014). Which World Bank reports are widely read? *Policy Research Working Paper, 8651*, 1–32.
Dörner, D. (1996). *The logic of failure: Recognizing and avoiding error in complex situations*. Metropolitan Books.
Dow, B. J., Wang, C. S., Whitson, J. A., & Deng, Y. (2022). Mitigating and managing COVID-19 conspiratorial beliefs. *BMJ Lead, 6*(4), 259–262. https://doi.org/10.1136/leader-2022-000600
Druckman, J. N., & McGrath, M. C. (2019). The evidence for motivated reasoning in climate change preference formation. *Nature Climate Change, 9*(2), 111–119. https://doi.org/10.1038/s41558-018-0360-1
Dunning, D. (2012). The relation of self to social perception. In M. R. Leary, & J. P. Tangney (Eds.), *Handbook of self and identity* (2 ed., pp. 481–501). Guilford Press.
Dunning, D. (2014). *We are all confident idiots – pacific standard*. Retrieved 2023-02-28.
Egelhofer, J. L., & Lecheler, S. (2019). Fake news as a two-dimensional phenomenon: A framework and research agenda. *Annals of the International Communication Association, 43*(2), 97–116. https://doi.org/10.1080/23808985.2019.1602782
Epley, N., & Gilovich, T. (2016). The mechanics of motivated reasoning. *The Journal of Economic Perspectives, 30*(3), 133–140. https://doi.org/10.1257/jep.30.3.133
Feeny, D., Berkes, F., Mccay, B. J., & Acheson, J. M. (1990). The tragedy of the commons: Twenty-two years later. *Human Ecology, 18*(1), 1–19. https://doi.org/10.1007/BF00889070
Festinger, L. (1962). *A theory of cognitive dissonance*. Stanford University Press.
Feynman, R. P. (1999). *The pleasure of finding things out: The best short works of Richard P. Feynman*. Perseus Books.
Fine, K. (2018). Ignorance of ignorance. *Synthese, 195*(9), 4031–4045. https://doi.org/10.1007/s11229-017-1406-z
Fischhoff, B., & Beyth, R. (1975). I knew it would happen: Remembered probabilities of once—future things. *Organizational Behavior & Human Performance, 13*, 1–16. https://doi.org/10.1016/0030-5073(75)90002-1
Flaxman, S., Goel, S., & Rao, J. M. (2016). Filter bubbles, echo chambers, and online news consumption. *Public Opinion Quarterly, 80*(S1), 298–320. https://doi.org/10.1093/poq/nfw006
Galton, F. (1907). Vox populi. *Nature, 75*(1949), 450–451. https://doi.org/10.1038/075450a0
Glüer, K., & Wikforss, Å. (2022). What is knowledge resistance? In J. Strömbäck, Å. Wikforss, K. Glüer, T. Lindholm, & H. Oscarsson (Eds.), *Knowledge resistance in high-choice information environments* (pp. 29–48). Routledge. https://doi.org/10.4324/9781003111474-2
Golden, A. S., Naisilsisili, W., Ligairi, I., & Drew, J. A. (2014). Combining natural history collections with fisher knowledge for community-based conservation in Fiji. *PLoS One, 9*. https://doi.org/10.1371/journal.pone.0098036
Gross, M. (2010). *Ignorance and surprise: Science, society, and ecological design*. The MIT Press. https://doi.org/10.7551/mitpress/9780262013482.001.0001
Guterres, A. (2022a). *COP15 – UN secretary-general's remarks to the UN biodiversity conference*. United Nations.
Guterres, A. (2022b). *Secretary-general's remarks to high-level opening of COP27*. United Nations.
Hagelsteen, M., & Becker, P. (2019). Systemic problems of capacity development for disaster risk reduction in a complex, uncertain, dynamic, and ambiguous world. *International Journal of Disaster Risk Reduction, 36*, 101102. https://doi.org/10.1016/j.ijdrr.2019.101102
Haidt, J. (2001). The emotional dog and its rational tail: A social intuitionist approach to moral judgment. *Psychological Review, 108*, 814–834. https://doi.org/10.1037//0033-295X
Haidt, J. (2012). *The righteous mind: Why good people are divided by politics and religion*. Pantheon Books.
Hammersley, M. (1992). *What's wrong with ethnography?: Methodological explorations*. Routledge.
Hardin, G. (1968). The tragedy of the commons. *Science, 162*, 1243–1248. https://doi.org/10.1126/science.162.3859.1243
Hardin, G. (1991). The tragedy of the unmanaged commons: Population and the disguises of providence. In R. V. Andelson (Ed.), *Commons without tragedy: Protecting the environment from overpopulation—a new approach* (pp. 162–185). Shepheard-Walwyn.
Harich, J. (2010). Change resistance as the crux of the environmental sustainability problem. *System Dynamics Review, 26*(1), 35–72. https://doi.org/10.1002/sdr.431
Hendricks, V. F., & Vestergaard, M. (2019). *Reality lost*. Springer. https://doi.org/10.1007/978-3-030-00813-0

Hinds, J., Williams, E. J., & Joinson, A. N. (2020). "It wouldn't happen to me": Privacy concerns and perspectives following the Cambridge Analytica scandal. *International Journal of Human-Computer Studies, 143*, 102498. https://doi.org/10.1016/j.ijhcs.2020.102498

Hirsh, J. B., Kang, S. K., & Bodenhausen, G. V. (2012). Personalized persuasion: Tailoring persuasive appeals to recipients' personality traits. *Psychological Science, 23*(6), 578–581. https://doi.org/10.1177/0956797611436349

Hirshleifer, J. (1971). The private and social value of information and the reward to inventive activity. *The American Economic Review, 61*(4), 561–574.

Hofmann, A. (2022). *The ten commandments of risk leadership*. Springer. https://doi.org/10.1007/978-3-030-88797-1

Hsee, C. K. (1996). Elastic justification: How unjustifiable factors influence judgments. *Organizational Behavior and Human Decision Processes, 66*(1), 122–129. https://doi.org/10.1006/obhd.1996.0043

Huber, B., Barnidge, M., Gil de Zúñiga, H., & Liu, J. (2019). Fostering public trust in science: The role of social media. *Public Understanding of Science, 28*(7), 759–777. https://journals.sagepub.com/doi/pdf/10.1177/0963662519869097.

Ihori, T. (2017). *Principles of public finance*. Springer. https://doi.org/10.1007/978-981-10-2389-7

Ingre, M., Lindholm, T., & Strömbäck, J. (2022). Overcoming knowledge resistance: A systematic review of experimental studies. In J. Strömbäck, Å. Wikforss, K. Glüer, T. Lindholm, & H. Oscarsson (Eds.), *Knowledge resistance in high-choice information environments* (pp. 255–280). Routledge. https://doi.org/10.4324/9781003111474

IPCC. (2021). *Climate change 2021: The physical science basis*. Cambridge University Press.

IPCC. (2022). *Climate change 2022: Impacts, adaptation and vulnerability*. Cambridge University Press.

Jack, T., & Glover, A. (2021). Online conferencing in the midst of COVID-19: An "already existing experiment" in academic internationalization without air travel. *Sustainability: Science, Practice and Policy, 17*(1), 292–304. https://doi.org/10.1080/15487733.2021.1946297

Jamieson, K. H., & Cappella, J. N. (2008). *Echo chamber: Rush limbaugh and the conservative media establishment*. Oxford University Press.

Janis, I. L. (1971). Groupthink: The desperate drive for consensus at any cost that suppresses dissent among the mighty in the corridors of power. *Psychology Today, 43–44*(46), 74.

Janis, I. L. (1972). *Victims of groupthink: A psychological study of foreign-policy decisions and fiascoes*. Houghton Mifflin.

Janis, I. L. (1991). Groupthink. In E. A. Griffin (Ed.), *A first look at communication theory* (pp. 235–246). McGraw-Hill.

Johnson, D., & Levin, S. A. (2009). The tragedy of cognition: Psychological biases and environmental inaction. *Current Science, 97*, 1593–1603.

Johnson, D. D. P. (2004). *Overconfidence and war: The havoc and glory of positive illusions*. Harvard University Press.

Kahan, D. M. (2016). The politically motivated reasoning paradigm, Part 1: What politically motivated reasoning is and how to measure it. In R. Scott, & S. Kosslyn (Eds.), *Emerging trends in the social and behavioral sciences* (pp. 1–16). John Wiley & Sons. https://doi.org/10.1002/9781118900772.etrds0417

Kahneman, D., & Tversky, A. (1972). Subjective probability: A judgment of representativeness. *Cognitive Psychology, 3*, 430–454.

Kahneman, D., & Tversky, A. (1979). Prospect theory: An analysis of decision under risk. *Econometrica: Journal of the Econometric Society, 47*, 263–291.

Kahneman, D., & Tversky, A. (1982). The psychology of preferences. *Scientific American, 246*, 160–173.

Kahneman, D., Ritov, I., & Schkade, D. (1999). Economic preferences or attitude expressions?: An analysis of dollar responses to public issues. *Journal of Risk and Uncertainty, 19*(1–3), 203–235. http://www.jstor.org/stable/10.2307/41760962.

Kahneman, D. (1986). Comments by professor Daniel Kahneman. In R. G. Cummings, D. S. Brookshire, & W. D. Schulze (Eds.), *Valuing environmental goods: A state of the arts—assessment of the contingent valuation method, 1B* pp. 226–235). Rowman & Allanheld.

Kahneman, D. (2003). Maps of bounded rationality: Psychology for behavioral economics. *The American Economic Review, 93*, 1449–1475. https://doi.org/10.1257/000282803322655392

Kahneman, D. (2011). *Thinking, fast and slow*. Farrar, Straus and Giroux.
Kates, R. W., Clarke, W. C., Corell, R. W., Hall, J. M., Jaeger, C. C., Lowe, I., McCarthy, J. J., Schellnhuber, H. J., Bolin, B., Dickson, N. M., Faucheux, S., Gallopín, G. C., Huntley, B., Jodha, N. S., Kasperson, R. E., Mabogunje, A., Matson, P. A., Mooney, H., Moore, B., O'Riordan, T., & Svedin, U. (2001). Sustainability science. *Science, 292*, 641–642.
Khaldun, I. (1969). *The muqaddimah: An introduction to history*. Princeton University Press.
Kierkegaard, S. (1995). *Works of love (E. H. Hong & H. V. Hong, trans.)*. Princeton University Press.
Klintman, M. (2013). *Citizen-consumers and evolution: Reducing environmental harm through our social motivation*. Palgrave Macmillan.
Klintman, M. (2019). *Knowledge resistance: How we avoid insight from others*. Manchester University Press. https://doi.org/10.7765/9781526158703
Knorr-Cetina, K. (1999). *Epistemic cultures: How the sciences make knowledge*. Harvard University Press.
Kramer, R. M. (2005). A failure to communicate: 9/11 and the tragedy of the informational commons. *International Public Management Journal, 8*, 397–416.
Krueger, E. H., Constantino, S. M., Centeno, M. A., Elmqvist, T., Weber, E. U., & Levin, S. A. (2022). Governing sustainable transformations of urban social-ecological-technological systems. *npj Urban Sustainability, 2*(1), 1–12. https://doi.org/10.1038/s42949-022-00053-1
Kruger, J., & Dunning, D. (1999). Unskilled and unaware of it: How difficulties in recognizing one's own incompetence lead to inflated self-assessments. *Journal of Personality and Social Psychology, 77*(6), 1121–1134. https://doi.org/10.1037/0022-3514.77.6.1121
Kunda, Z. (1990). The case for motivated reasoning. *Psychological Bulletin, 108*(3), 480–498. https://doi.org/10.1037/0033-2909.108.3.480
Kundu, P., & Cummins, D. D. (2013). Morality and conformity: The Asch paradigm applied to moral decisions. *Social Influence, 8*(4), 268–279. https://doi.org/10.1080/15534510.2012.727767
Kurien, J. (2007). The blessing of the commons: Small-scale fisheries, community property rights, and coastal natural assets. In J. K. Boyce, S. Narain, & E. A. Stanton (Eds.), *Reclaiming nature: Environmental justice and ecological restoration* (pp. 23–54). Anthem Press. https://doi.org/10.7135/upo9781843313465.002
Kurnyshova, Y., & Makarychev, A. (2022). Explaining Russia's war against Ukraine: How can foreign policy analysis and political theory be helpful. *Studies in East European Thought, 74*(4), 507–519. https://doi.org/10.1007/s11212-022-09494-x
Kurzban, R., & Leary, M. R. (2001). Evolutionary origins of stigmatization: The functions of social exclusion. *Psychological Bulletin, 127*(2), 187–208. https://doi.org/10.1037/0033-2909.127.2.187
Lecheler, S., & Egelhofer, J. L. (2022). Disinformation, misinformation, and fake news: Understanding the supply side. In J. Strömbäck, Å. Wikforss, K. Glüer, T. Lindholm, & H. Oscarsson (Eds.), *Knowledge resistance in high-choice information environments* (pp. 69–87). Routledge. https://doi.org/10.4324/9781003111474
Lees, C. (2021). Brexit, the failure of the British political class, and the case for greater diversity in UK political recruitment. *British Politics, 16*(1), 36–57. https://doi.org/10.1057/s41293-020-00136-6
Leibenstein, H. (1950). Bandwagon, snob, and veblen effects in the theory of consumers' demand. *Quarterly Journal of Economics, 64*, 183–207. https://doi.org/10.2307/1882692
Lewin, K. (1943). Defining the 'field at a given time'. *Psychological Review, 50*(3), 292–310. https://doi.org/10.1037/h0062738
Lewin, K. (1947). Frontiers in group dynamics: Concept, method and reality in social science; social equilibria and social change. *Human Relations, 1*(1), 5–41. https://doi.org/10.1177/001872674700100201
Lewin, K. (1951). *Field theory in social science: Selected theoretical papers*. Harper & Brothers.
Linstone, H. A., & Turoff, M. (Eds.). (1975). *The Delphi method: Techniques and applications*. Addison-Wesley Publishing Company.
Lippitt, M. (1987). *The managing complex change model*. Enterprise Management Ltd.
Lorenz, J., Rauhut, H., Schweitzer, F., & Helbing, D. (2011). How social influence can undermine the wisdom of crowd effect. *Proceedings of the National Academy of Sciences, 108*(22), 9020–9025. https://doi.org/10.1073/pnas.1008636108
Luhmann, N. (1998). *Observations on modernity*. Stanford University Press.

MacCoun, R. J. (2012). The burden of social proof: Shared thresholds and social influence. *Psychological Review, 119*(2), 345–372. https://doi.org/10.1037/a0027121
Machiavelli, N. (2021). *The prince (W. K. Marriott, trans.). Global grey.*
Macy, M. W., & Willer, R. (2002). From factors to actors: Computational sociology and agent-based modeling. *Annual Review of Sociology, 28*(1), 143–166. https://doi.org/10.1146/annurev.soc.28.110601.141117
Marais, K., Saleh, J. H., & Leveson, N. G. (2006). Archetypes for organizational safety. *Safety Science, 44*, 565–582. https://doi.org/10.1016/j.ssci.2005.12.004
Mark, N. P. (2018). Status organizes cooperation: An evolutionary theory of status and social order. *American Journal of Sociology, 123*(6), 1601–1634. https://doi.org/10.1086/696683
Marshall, G. (2014). *Don't even think about it: Why our brains are wired to ignore climate change*. Bloomsbury.
Maslow, A. H. (1943). A theory of human motivation. *Psychological Review, 50*(4), 370–396. https://doi.org/10.1037/h0054346
Matzler, K., Strobl, A., & Bailom, F. (2016). Leadership and the wisdom of crowds: How to tap into the collective intelligence of an organization. *Strategy & Leadership, 44*(1), 30–35. https://doi.org/10.1108/sl-06-2015-0049
McCay, B. J., & Acheson, J. M. (Eds.). (1987). *The question of the commons: The culture and ecology of communal resources*. The University of Arizona Press.
McFadden, D., & Leonard, G. K. (1993). Issues in the contingent valuation of environmental goods: Methodologies for data collection and analysis. In J. A. Hausman (Ed.), *Contingent valuation—a critical assessment* (pp. 165–215). Elsevier. https://doi.org/10.1016/b978-0-444-81469-2.50010-9
McGoey, L. (2012). The logic of strategic ignorance. *British Journal of Sociology, 63*(3), 553–576. https://doi.org/10.1111/j.1468-4446.2012.01424.x
McIntyre, M. E. (2022). Climate uncertainties: A personal view. *Meteorology, 1*(2), 162–170. https://doi.org/10.3390/meteorology1020011
McPherson, M., Smith-Lovin, L., & Cook, J. M. (2001). Birds of a feather: Homophily in social networks. *Annual Review of Sociology, 27*(1), 415–444. https://doi.org/10.1146/annurev.soc.27.1.415
Mercier, H., & Sperber, D. (2011). Why do humans reason? Arguments for an argumentative theory. *Behavioral and Brain Sciences, 34*(2), 57–74. https://doi.org/10.1017/s0140525x10000968
Merton, R. K. (1987). Three fragments from a sociologist's notebooks: Establishing the phenomenon, specified ignorance, and strategic research materials. *Annual Review of Sociology, 13*(1), 1–29. https://doi.org/10.1146/annurev.so.13.080187.000245
Meyer, R. (2006). Why we under-prepare for hazards. In R. J. Daniels, & D. F. Kettl (Eds.), *On risk and disaster: Lessons from Hurricane Katrina* (pp. 153–174). University of Pennsylvania Press.
Meyerowitz, B. E., & Chaiken, S. (1987). The effect of message framing on breast self-examination attitudes, intentions, and behavior. *Journal of Personality and Social Psychology, 52*(3), 500–510. https://doi.org/10.1037//0022-3514.52.3.500
Neumayer, E., & Plümper, T. (2007). The gendered nature of natural disasters: The impact of catastrophic events on the gender gap in life expectancy, 1981-2002. *Annals of the Association of American Geographers, 97*, 551–566.
Nichols, T. M. (2017). *The death of expertise: The campaign against established knowledge and why it matters*. Oxford University Press.
O'Connor, M. A. (2003). *The Enron board: The perils of groupthink* (p. 71). University of Cincinnati Law Review.
Okun, A. M. (1970). *The political economy of prosperity*. W. W. Norton.
Oliver-Smith, A., & Hoffman, S. M. (2020). *The angry earth: Disaster in anthropological perspective* (2 ed.). Routledge.
Omidvar, O., Safavi, M., & Glaser, V. L. (2023). Algorithmic routines and dynamic inertia: How organizations avoid adapting to changes in the environment. *Journal of Management Studies, 60*(2), 313–345. https://doi.org/10.1111/joms.12819
Oreskes, N., & Conway, E. M. (2010). *Merchants of doubt: How a handful of scientists obscured the truth on issues from tobacco smoke to global warming*. Bloomsbury Press.

Orwell, G. (1949). *Nineteen eighty-four.* Secker & Warburg. https://gutenberg.ca/ebooks/orwellg-nineteeneightyfour/orwellg-nineteeneightyfour-00-h.html.
Ostrom, E., Burger, J., Field, C. B., Norgaard, R. B., & Policansky, D. (1999). Revisiting the commons: Local lessons, global challenges. *Science, 284*, 278–282. https://doi.org/10.1126/science.284.5412.278
Ostrom, V. (1975). Alternative approaches to the organization of public proprietary interests. *Natural Resources Journal, 15*(4), 765–789. http://www.jstor.org/stable/10.2307/24880909.
Ostrom, E. (1977). Collective action and the tragedy of the commons. In G. J. Hardin, & J. Baden (Eds.), *Managing the commons* (pp. 173–181). W.H. Freeman.
Ostrom, E. (1990). *Governing the commons: The evolution of institutions for collective action.* Cambridge University Press.
Ostrom, E. (2010). Beyond markets and states: Polycentric governance of complex economic systems. *The American Economic Review, 100*, 641–672.
Oxford English Dictionary. (2020). *Oxford dictionary of English.* Oxford University Press.
Page, S. E. (2007). *The difference: How the power of diversity creates better groups, firms, schools, and societies.* Princeton University Press.
Pardo del Val, M., & Martínez Fuentes, C. (2003). Resistance to change: A literature review and empirical study. *Management Decision, 41*(2), 148–155. https://doi.org/10.1108/00251740310457597
Pariser, E. (2011). *The filter bubble: What the internet is hiding from you.* Penguin Press.
Partelow, S., Abson, D. J., Schlüter, A., Fernández-Giménez, M., Von Wehrden, H., & Collier, N. (2019). Privatizing the commons: New approaches need broader evaluative criteria for sustainability. *International Journal of the Commons, 13*(1), 747. https://doi.org/10.18352/ijc.938
Pascal, B. (1900). *Pascal's pensées; or, thoughts on religion.* Peter Pauper Press.
Pauketat, T. R. (2000). The tragedy of the commoners. In M.-A. Dobres, & J. Robb (Eds.), *Agency in archaeology* (pp. 113–129). Routledge. https://doi.org/10.4324/9781315866000
Payo, A., Becker, P., Otto, A., Vervoort, J. M., & Kingsborough, A. (2015). Experiential lock-in: Characterizing avoidable maladaptation in infrastructure systems. *Journal of Infrastructure Systems, 22*, 2515001. https://doi.org/10.1061/(asce)is.1943-555x.0000268
Peterson, C. (2000). The future of optimism. *American Psychologist, 55*(1), 44–55. https://doi.org/10.1037//0003-066x.55.1.44
Pham, M. T. (2007). Emotion and rationality: A critical review and interpretation of empirical evidence. *Review of General Psychology, 11*(2), 155–178. https://doi.org/10.1037/1089-2680.11.2.155
Pinker, S. (2021). *Rationality: What it is, why it seems scarce, why it matters. Viking.*
Politz, A. (1960). The dilemma of creative advertising. *Journal of Marketing, 25*, 1–6. https://doi.org/10.2307/1248603
Post, L. A., Raile, A. N. W., & Raile, E. D. (2010). Defining political will. *Politics & Policy, 38*(4), 653–676. https://doi.org/10.1111/j.1747-1346.2010.00253.x
Prelec, D., Seung, H. S., & McCoy, J. (2017). A solution to the single-question crowd wisdom problem. *Nature, 541*(7638), 532–535. https://doi.org/10.1038/nature21054
Reason, J. (1997). *Managing the risks of organizational accidents.* Routledge.
Remoundou, K., Diaz-Simal, P., Koundouri, P., & Rulleau, B. (2015). Valuing climate change mitigation: A choice experiment on a coastal and marine ecosystem. *Ecosystem Services, 11*, 87–94. https://doi.org/10.1016/j.ecoser.2014.11.003
Risso, L. (2018). Harvesting your soul? Cambridge analytica and Brexit. In C. Jansohn (Ed.), *Brexit means Brexit?—the selected proceedings of the symposium.* Akademie der Wissenschaften und der Literatur.
Ross, L., Greene, D., & House, P. (1977). The "false consensus effect": An egocentric bias in social perception and attribution processes. *Journal of Experimental Social Psychology, 13*, 279–301. https://doi.org/10.1016/0022-1031(77)90049-X
Ross, L. (1977). The intuitive psychologist and his shortcomings: Distortions in the attribution process. *Advances in Experimental Social Psychology, 10*, 173–220.
Rottenstreich, Y., & Hsee, C. K. (2001). Money, kisses, and electric shocks: On the affective psychology of risk. *Psychological Science, 12*(3), 185–190. https://doi.org/10.1111/1467-9280.00334
Ruan, W. J., Wong, I. A., & Lan, J. (2022). Uniting ecological belief and social conformity in green events. *Journal of Hospitality and Tourism Management, 53*, 61–69. https://doi.org/10.1016/j.jhtm.2022.09.001

Runciman, W. G. (1998). *The social animal.* HarperCollins.
Ruskin, J. (1882). *The crown of wild olive. George Allen.*
Schiller, H. I. (1989). *Culture, Inc: The corporate takeover of public expression.* Oxford University Press.
Schindler, J. (2012). Rethinking the tragedy of the commons: The integration of socio-psychological dispositions. *The Journal of Artificial Societies and Social Simulation, 15*(1). https://doi.org/10.18564/jasss.1822
Schmitt-Beck, R. (2015). Bandwagon effect. In G. Mazzoleni (Ed.), *The international encyclopedia of political communication* (pp. 1–5). John Wiley & Sons. https://doi.org/10.1002/9781118541555.wbiepc015
Schneider, L. (1962). The role of the category of ignorance in sociological theory: An exploratory statement. *American Sociological Review, 27*(4), 492–508. https://doi.org/10.2307/2090030
Scott, I., & Gong, T. (2021). Coordinating government silos: Challenges and opportunities. *Global Public Policy and Governance, 1*(1), 20–38. https://doi.org/10.1007/s43508-021-00004-z
Seinfeld, J. (1993). *Seinfeld (TV series) episode: The Pilot.* New York: NBC. Retrieved 2023-03-16.
Shennan, S. (2018). *The first farmers of Europe: An evolutionary perspective.* Cambridge University Press. https://doi.org/10.1017/9781108386029
Simonovic, S. P. (2010). *Systems approach to management of disasters: Methods and applications.* John Wiley & Sons.
Sirimorok, N., & Asfriyanto, A. (2020). The return of the muro: Institutional bricolage, customary institutions, and protection of the commons in Lembata Island, Nusa Tenggara. *Forestry and Society, 4*(1), 61. https://doi.org/10.24259/fs.v4i1.7676
Slovic, P., Finucane, M. L., Peters, E., & MacGregor, D. G. (2002). Rational actors or rational fools: Implications of the affect heuristic for behavioral economics. *The Journal of Socio-Economics, 31*, 329–342. https://doi.org/10.1016/S1053-5357(02)00174-9
Slovic, P. (2010). *The feeling of risk: New perspectives on risk perception.* Earthscan.
Spaccatini, F., Richetin, J., Riva, P., Pancani, L., Ariccio, S., & Sacchi, S. (2022). Trust in science and solution aversion: Attitudes toward adaptation measures predict flood risk perception. *International Journal of Disaster Risk Reduction, 76*, 103024. https://doi.org/10.1016/j.ijdrr.2022.103024
Spence, A., & Pidgeon, N. (2010). Framing and communicating climate change: The effects of distance and outcome frame manipulations. *Global Environmental Change, 20*(4), 656–667. https://doi.org/10.1016/j.gloenvcha.2010.07.002
Spohr, D. (2017). Fake news and ideological polarization. *Business Information Review, 34*(3), 150–160. https://doi.org/10.1177/0266382117722446
Stern, N. (2007). *The economics of climate change: The stern review.* Cambridge University Press.
Sturman, E. D., Dufford, A., Bremser, J., & Chantel, C. (2017). Status striving and hypercompetitiveness as they relate to overconsumption and climate change. *Ecopsychology, 9*(1), 44–50. https://doi.org/10.1089/eco.2016.0027
Sunstein, C. R. (2003). Terrorism and probability neglect. *Journal of Risk and Uncertainty, 26*(2/3), 121–136. Special Issue on the Risks of Terrorisum) http://www.jstor.org/stable/10.2307/41755012.
Surowiecki, J. (2004). *The wisdom of crowds. Anchor books.*
Suwajanakorn, S., Seitz, S. M., & Kemelmacher-Shlizerman, I. (2017). Synthesizing Obama. *ACM Transactions on Graphics, 36*(4), 1–13. https://doi.org/10.1145/3072959.3073640
Szewach, P., Reifler, J., & Henrik, O. (2022). Is resistance futile? Citizen knowledge, motivated reasoning, and fact-checking. In J. Strömbäck, Å. Wikforss, K. Glüer, T. Lindholm, & H. Oscarsson (Eds.), *Knowledge resistance in high-choice information environments* (pp. 166–186). Routledge. https://doi.org/10.4324/9781003111474
Tajfel, H. (1974). Social identity and intergroup behaviour. *Social Science Information, 13*, 65–93. https://doi.org/10.1177/053901847401300204
Taussig, M. T. (1999). *Defacement: Public secrecy and the labor of the negative.* Stanford University Press.
Taylor, S. E., & Brown, J. D. (1994). Positive illusions and well-being revisited: Separating fact from fiction. *Psychological Bulletin, 116*(1), 21–27. https://doi.org/10.1037/0033-2909.116.1.21
Thucydides. (2004). *The peloponnesian war (C. Richard, trans.).* Dover Publications.
Tompkins, E. L., & Eakin, H. C. (2012). Managing private and public adaptation to climate change. *Global Environmental Change, 22*, 3–11.

Trivers, R. (1985). *Social evolution*. Benjamin/Cummings Publishing Company.
Trivers, R. (1991). Deceit and self-deception: The relationship between communication and consciousness. In M. H. Robinson, & L. Tiger (Eds.), *Man and beast revisited* (pp. 175–191) (Smithsonian).
Trivers, R. (2013). *The folly of fools: The logic of deceit and self-deception in human life*. Basic Books.
Tversky, A., & Kahneman, D. (1973). Availability: A heuristic for judging frequency and probability. *Cognitive Psychology, 5*, 207–232. https://doi.org/10.1016/0010-0285(73)90033-9
Tversky, A., & Kahneman, D. (1983). Extensional versus intuitive reasoning: The conjunction fallacy in probability judgment. *Psychological Review, 90*(4), 293–315. https://doi.org/10.1037/0033-295x.90.4.293
Ullström, S., Stripple, J., & Nicholas, K. A. (2023). From aspirational luxury to hypermobility to staying on the ground: Changing discourses of holiday air travel in Sweden. *Journal of Sustainable Tourism, 31*(3), 688–705. https://doi.org/10.1080/09669582.2021.1998079
Ungar, S. (2008). Ignorance as an under-identified social problem. *British Journal of Sociology, 59*(2), 301–326. https://doi.org/10.1111/j.1468-4446.2008.00195.x
Upham, P., Kivimaa, P., & Virkamäki, V. (2013). Path dependence and technological expectations in transport policy: The case of Finland and the UK. *Journal of Transport Geography, 32*, 12–22. https://doi.org/10.1016/j.jtrangeo.2013.08.004
Vennix, J. A. M. (2001). *Group model building: Facilitating team learning using system dynamics*. John Wiley.
Vinceti, M., Filippini, T., Rothman, K. J., Ferrari, F., Goffi, A., Maffeis, G., & Orsini, N. (2020). Lockdown timing and efficacy in controlling COVID-19 using mobile phone tracking. *EClinicalMedicine, 25*, 100457. https://doi.org/10.1016/j.eclinm.2020.100457
von Neumann, J., & Morgenstern, O. (1944). *The theory of games and economic behavior*. Princeton University Press.
Watson, G. (1971). Resistance to change. *American Behavioral Scientist, 14*(5), 745–766. https://doi.org/10.1177/000276427101400507
Westerlund, M. (2019). The emergence of deepfake technology: A review. *Technology Innovation Management Review, 9*(11), 39–52. https://doi.org/10.22215/timreview/1282
White, M. P., Bratman, G. N., Pahl, S., Young, G., Cracknell, D., & Elliott, L. R. (2020). Affective reactions to losses and gains in biodiversity: Testing a prospect theory approach. *Journal of Environmental Psychology, 72*, 101502. https://doi.org/10.1016/j.jenvp.2020.101502
Whitehead, A. N. (1978). *Process and reality*. The Free Press.
Whyte, W. H. J. (1952). *Groupthink* (pp. 114–117). Fortune.
Wylie, C. (2019). *Mindf*ck: Cambridge analytica and the plot to break America*. Random House. http://books.google.se/books?id=qj6REAAAQBAJ&hl=&source=gbs_api.
Young, M. C. (2014). Do positive illusions contribute to human well-being. *Philosophical Psychology, 27*(4), 536–552. https://doi.org/10.1080/09515089.2013.764860
Zajonc, R. B. (1968). Attitudinal effects of mere exposure. *Journal of Personality and Social Psychology, 9*, 1. https://doi.org/10.1037/h0025848
Zegart, A., & Morell, M. (2019). Spies, lies, and algorithms: Why U.S. Intelligence agencies must adapt or fail. *Foreign Affairs, 98*(3), 85–97.
Zickfeld, K., & Herrington, T. (2015). The time lag between a carbon dioxide emission and maximum warming increases with the size of the emission. *Environmental Research Letters, 10*(3), 031001. https://doi.org/10.1088/1748-9326/10/3/031001

CHAPTER 14

Capacity development for resilience

Introduction

We witness increasing discrepancies between our preferred and actual development trajectories resulting in local calamity and global tribulation, and all but very few scientists anticipate more trouble in the future (IPCC, 2022). The great developmental leaps we have taken since the outset of the Industrial Revolution have not only resulted in increased economic purchasing power, life expectancy, adult literacy and other improvements but also in a range of sustainability challenges introduced in the first part of this book. Most influential voices of the international community cry out the need for addressing issues of risk and unsustainable development, and many of them frame a way forward in terms of capacity development (UNCSD, 2012; United Nations, 2015a,b). In other words, whilst a community, organisation, society or any other human-environment system must improve its own capacities to be resilient, external actors can play essential roles in supporting such development.

Capacity development for resilience in the face of a change, trend or shock entails overcoming two basic challenges. First, we must grasp what constitutes resilience—the hierarchy of functions and the complex assemblages of actual, observable elements and interactions that contribute to performing them in practice (Chapter 8). All human-environment systems are resilient to varying degrees and can indeed strengthen their resilience on their own by identifying and enhancing the assemblages of elements and interactions necessary for performing particular functions and their crucial interplay. For that, you do not need to engage with the contents of this chapter. However, there are many situations in which such change processes are supported in various ways by actors that are external to the system itself, such as a consultant supporting a company's reorganisation, a national authority supporting a municipal administration to implement a new policy, a country's Red Cross Society supported by its sister society from elsewhere, or NGO volunteers supporting a disaster-prone community to be better able to reduce risk. The second basic challenge concerns the added intricacies when actors aspiring to develop their capacity to do something are supported by external actors to do so. It is in these situations the concept of capacity development is generally used (Becker, 2021; OECD, 2009), and it is for such situations that this chapter can be useful.

This chapter presents key ideas concerning capacity development for resilience, focusing on its potential for changing the world in terms of making it safer and more sustainable. It starts by defining capacity development, elaborating on the concept of

Sustainability Science, Second Edition
ISBN 978-0-323-95640-6, https://doi.org/10.1016/B978-0-323-95640-6.00015-4

capacity, different definitions of capacity development and its predecessors, and presenting its main principles. Then, it sets out to decondition the current practices of capacity development, comparing the principles with actual performance and elaborating on how certain principles clash with each other and with contemporary attitudes and viewpoints in our complex, uncertain, ambiguous and dynamic world. Finally, the chapter suggests important aspects to consider when designing capacity development, and ends with some concluding remarks.

Defining capacity development

To define capacity development requires considering what constitutes capacity, how capacity development has commonly been defined over the years, and what principles guide how capacity development should be done.

Reconsidering capacity

Before it is possible to define the concept of capacity development, we need to consider the meaning of capacity in the first place. Amongst other meanings, the Oxford English Dictionary (2020) defines capacity as 'the ability or power to do or understand something'. In other words, capacity has, in this context, to do with a specific human-environment system's ability to perform functions central to its resilience. However, one of the most influential definitions of capacity in relation to capacity development focuses on a more concrete level of abstraction when stating that capacity is '[t]he combination of all the strengths, attributes and resources available within a community, society or organization that can be used to achieve agreed goals' (UNISDR, 2009, p. 5). Remembering Rasmussen (1985) abstraction hierarchy (Chapter 8), these two definitions do not compete but complement each other. They only focus on different levels of abstraction—the former focuses on function, and the latter focuses on form. If capacity is the ability of a specific human-environment system to perform functions that are central to its resilience, then this capacity stems from the complex assemblages of elements and interactions across the different categories introduced earlier (Figure 14.1). For example, the capacity of a community to construct levees to protect it from floods involves not only many trained and equipped community members contributing with their knowledge and labour but also how they organise the work, the policies and legislation relevant for such endeavour, the norms and values guiding social action and interaction, etc.

This approach is slightly different to the conventional approach to capacity, suggesting that it resides on three interrelated levels (Figure 14.2): the individual level, the organisational level and the enabling environment (e.g., UNDP, 2009, p. 11; CADRI, 2011, pp. 9—11; OECD, 2012, pp. 28—29). The individual level of this conventional approach concerns the knowledge and skills of people that allow them to perform key tasks, whilst the organisational level concerns internal policies, procedures and processes that allow an

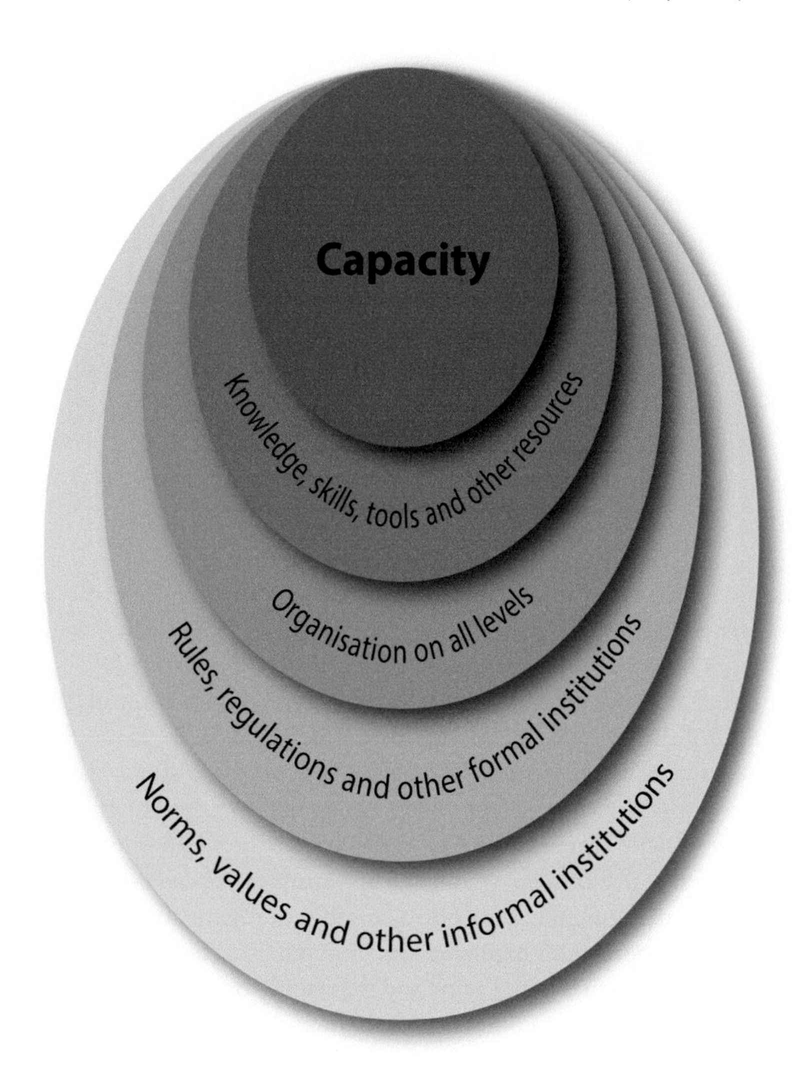

Figure 14.1 Four categories of form contributing to the capacity of a human-environment system to perform functions necessary for its resilience.

organisation to perform by enabling individual capacities to come together to achieve its goals (CADRI, 2011, p. 10). The enabling environment concerns aspects of the broader human-environment system within which individuals and organisations perform their tasks. This final level is at times referred to as the societal level (e.g., Enemark & Van der Molen, 2008; Lusthaus et al., 2000) or the network level (e.g., Chaskin, 2001, p. 298) and can either facilitate or seriously impede capacity, regardless of what these approaches refer to as individual and organisational capacities. The enabling environment is less straightforward to grasp than the other two levels. It includes the foundations for how

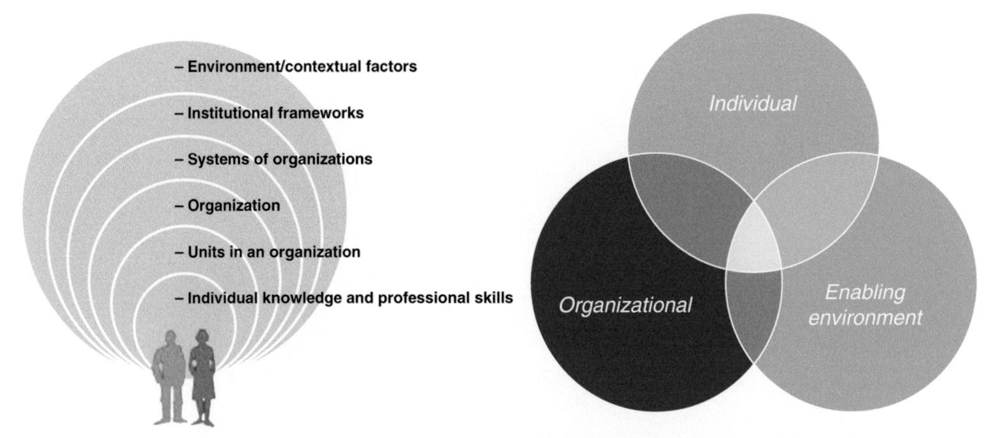

Figure 14.2 Capacity levels according to Schulz et al. (2005, p. 32) and UNDP/CADRI/OECD (CADRI, 2011, p. 11).

a society works, including the interaction between organisations, legislation, policies, procedures, power relations, norms, values, etc. (CADRI, 2011). However, it is often here that changes must be made for real transformation to be sustainable.

Another influential approach to capacity suggests it resides on six levels (Schulz et al., 2005), ranging from individual knowledge and professional skills, through units of an organisation, organisation and system of organisations, to institutional frameworks and environment/contextual factors (Figure 14.2). It is mainly a structural approach with the four first levels strictly corresponding to hierarchal levels of elements of human-environment systems—that is, a number of individuals form a unit in an organisation, a number of such units form an organisation, and a number of organisations form a system of organisations. This approach also highlights the importance of formal and informal institutions that lay down the 'rules of the game and the standards for the ways in which actors may participate and relate to each other and to their environment', as well as the context in terms 'of available resources and of obstacles and opportunities that influence operations' (Schulz et al., 2005, p. 32).

Both these approaches have their merits, as well as their similarities and differences. They describe the individual level similarly, focusing on individual knowledge and skills. Although they are right that such knowledge and skills are fundamental for the capacity of a human-environment system to perform vital functions for its resilience, they neglect the importance of other necessary resources for these individuals to be able to contribute, such as tools, equipment, vehicles and money. Indeed, Schulz et al. (2005) include some kinds of resources in their most overarching level of 'environment/contextual factors', but they do so in a very macroscopic sense of inflation rates, quality of infrastructure, access to electricity and telecommunications, etc. These are important, but I think it is more fruitful to focus on the more specific resources needed to perform particular functions for resilience.

Moreover, whilst the six-level approach helps point out different structural levels of hierarchical organisation to consider, the three-level approach focuses more on functional aspects of organisation and merges, as such, three of the four levels into one level. The last of the four hierarchical levels (i.e. 'system of organisations') is instead included in the 'enabling environment', which is a weakness since this vital aspect for grasping capacity is, then, easily lost in the complexity of everything else in the enabling environment.

Finally, the levels referred to as 'institutional frameworks' (six-level approach) and the rest of the 'enabling environment' (three-level approach) both include a mix of legislation, policy, procedures, power relations, norms, values, etc. It is nothing principally wrong with that since everything they list is vital for grasping capacity. Yet, both approaches mix formal and informal institutions into one category, which is an important reason for the general lack of focus on the latter, simply because it is much easier to address more tangible rules and regulations than underlying norms and values. However, please remember that informal institutions are at least as important as formal institutions for the capacity of human-environment systems to perform vital functions for their resilience.

Regardless of what literature you read or which approach to capacity you choose, it is fundamental to appreciate the importance of grasping and addressing capacity holistically. Disregarding any level severely undermines the effectiveness of most capacity development initiatives I can imagine. Furthermore, it is misleading to talk about 'individual capacities' and 'organisational capacities' as some kind of synonyms for the two levels (e.g., CADRI, 2011, p. 10; OECD, 2012). It is the totality of the complex assemblage of elements and interactions that constitute the capacity of the community, organisation, society or any other human-environment system to perform a particular function. It is, therefore, pointless to talk about capacity on any level in isolation. I suggest being more explicit in focusing on the functions that contribute to the resilience of a human-environment system and on the assemblages of all kinds of elements and their interactions needed to be able to perform them. This approach connects back to the operationalisation of resilience in Chapter 8 and entails restructuring the contents of the levels of the conventional approaches just presented. The capacity of a human-environment system to perform the functions contributing to its resilience is, therefore, determined by the complex assemblage of the knowledge, skills, tools and other resources available in the system; its organisation on all levels; its rules, regulations and other formal institutions; and the norms, values and other informal institutions underlying them (Figure 14.1).

Definitions of capacity development

Although capacity development is a much more specialised concept than capacity, which should cater for less terminological ambiguity, it is interesting to see the plethora of definitions from most organisations involved in practice (Box 14.1). Not that all versions

BOX 14.1 Many competing definitions of capacity development

UNDP: 'the process through which individuals, organizations and societies obtain, strengthen and maintain the capabilities to set and achieve their own development objectives over time' (UNDP, 2009, p. 5).

United Nations (in relation to DRR): 'the process by which people, organizations and society systematically stimulate and develop their capacities over time to achieve social and economic goals' (United Nations, 2016, p. 12).

UNISDR: 'The process by which people, organizations and society systematically stimulate and develop their capacities over time to achieve social and economic goals, including through improvement of knowledge, skills, systems, and institutions' (UNISDR, 2009, p. 6).

The World Bank: 'a locally driven process of learning by leaders, coalitions and other agents of change that brings about changes in sociopolitical, policy-related, and organizational factors to enhance local ownership for and the effectiveness and efficiency of efforts to achieve a development goal' (Otoo et al., 2009, p. 3).

OECD: 'the process by which people, organisations and society as a whole create, strengthen and maintain their capacity over time' (OECD, 2010, p. 1).

African Union: 'A process of enabling individuals, groups, organizations, institutions and societies to sustainably define, articulate, engage and actualize their vision or developmental goals building on their own resources and learning in the context of a pan-African paradigm' (African Union & NEPAD, 2012, p. 13).

DFID: 'the process whereby people, organisations and society as a whole unleash, strengthen, create, adapt and maintain capacity over time' (DFID, 2013, p. 2).

bring anything new to the table. On the contrary, most new definitions of capacity development I find mostly say the same but with slightly different words. First of all, I would dare to say that more or less all of them describe capacity development as a process. Secondly, almost all specify different levels of society on which capacity development can occur, out of which the three levels of individual, organisation and society are the most prevalent. Finally, most definitions describe capacity development in relation to the capacity to achieve goals or objectives. However, regardless of the definitions rarely explicitly acknowledging it, capacity development is inherently connected to situations in which actors within human-environment systems get outside support to develop capacity (Becker, 2021; Hagelsteen & Becker, 2013; OECD, 2009). Otherwise, we tend to call it something else, such as some kind of change management or an improvement in performance. In other words, capacity development entails a relationship between internal partners who aspire and attempt to develop the capacity of a particular human-environment system they are part of—be it an organisation, community or society—and external partners who are not part of that system but supporting the process (Hagelsteen & Becker, 2013, p. 4; Hagelsteen & Becker, 2019, p. 3).

It is first when an external partner gets involved that the change process turns into capacity development, with all its challenges and potential benefits. That said, capacity development is not requiring international support, although that is what most capacity development literature is about. It also includes the support of a municipal agricultural extension worker to the development of a community's capacity to reduce drought risk, a county administrative board's support to the capacity development of municipalities' preparedness planning, the support of an organisational consultant to turn an outdated governmental authority into a well-functioning agency, etc. In other words, it is not the geographical or administrative scale that is of the essence here, but the relationship between internal and external partners.

Conceptual change or continuity

Although the concept of capacity building did not come into wider use until the 1990s and capacity development roughly a decade later, similar ideas have been around since more or less the start of organised international development cooperation in the 1950s (Smillie, 2001, p. 8). The focus of these conceptual predecessors, such as institution building (Esman, 1967; Esman & Montgomery, 1969) or institutional development (Whyte, 1968), was also on supporting the development of capacities to meet various development objectives on different levels. They provided comprehensive guidelines for how external actors could facilitate such developments (e.g., Esman, 1967) but were in practice too often equated with technical assistance or cooperation in which external actors regularly assumed prominent roles as drivers of change (Moore, 1995, p. 91). This widespread failure to follow the guidelines, in turn, frequently led to activities that were designed and implemented by external actors and not generating many sustainable results (Oxenham & Chambers, 1978).

Whilst theory underwrote the ideas behind these early predecessors of capacity development, but practice failed, the solution that emerged was the reinvention of them under the name of capacity building. The focus of capacity building was again on supporting the development of capacities of actors internal to the human-environment systems of interest to this book to meet their needs, aspirations and objectives, with comprehensive guidelines for how to accomplish that (e.g., Eade, 1997). Then, after another decade of inadequate results, the ideas were again reinvented under the name of capacity development.

Although there are conceptual differences for each step of this conceptual development because scientists, policymakers and practitioners learnt something over time, these differences are much smaller than indicated by the proponents of the new concepts. It is clear that most of the arguments for the new concepts are made by describing the predecessor in terms of how it was applied in practice whilst presenting the new concept in terms of how it is described in theory. For instance, capacity building is now described as having a narrow scope, focusing on building capacities from scratch, being mainly concerned with external actors' activities, and having a short-term focus, whilst capacity development is described as having broader scope, focusing on developing existing capacities, being mainly concerned with creating local ownership, and having long-term focus (CADRI, 2011, p. 14). That is not at all a fair comparison but simply putting words into the mouths of our forerunners to make us appear as having progressed more than we actually have. Reading Eade's (1997) influential work, it is blatantly clear that capacity building never had a narrow scope, nor was it ever focused on building capacities from scratch. Even Whyte's (1968) early writings on institutional development highlight the importance of local ownership, and Gant's (1966, pp. 219–220) definition of institution building clearly prescribes a long-term focus. We have indeed learnt a lot over the years, but such unfair conceptual revisionism is simply bad form, and it is never a good idea to disregard the actual roots of much contemporary knowledge.

Principles of capacity development

To further grasp what capacity development is, it is helpful to consider the main principles for how it should be done. There are, again, several ways to present such content, but I find that the following eight principles capture the essence of it well: (1) ownership, (2) partnership, (3) contextualisation, (4) flexibility, (5) learning, (6) accountability, (7) long-term and (8) sustainability (Hagelsteen & Becker, 2019).

Ownership

The first of the cornerstones of capacity development is local ownership (Cameron & Low, 2012; Hagelsteen & Becker, 2013, p. 5; OECD, 2009, p. 3). It means that internal partners are always primarily responsible for their capacity development, whilst external partners have supporting roles (OECD, 2009, p. 3). In short, since ownership is about the formation and possession of ideas, strategies, development processes, resources and results (Schulz et al., 2005, pp. 23–26), local ownership is about internal partners driving their own capacity development (Figure 14.3) (Hagelsteen & Becker, 2019, p. 3).

Although capacity development must come from inside the community, organisation, society, or whatever human-environment system in question (Krznaric, 2007, p. 17; Hagelsteen & Becker, 2019, p. 3), it may be initiated with the support of external actors and nurtured through participatory approaches (Anderson, 1999, p. 28); always consciously making sure that the support never undermines local ownership (Lopes & Theisohn, 2003, p. 29). It has also been noted that external partners 'can play an important role in giving legitimacy to nationally led initiatives and processes' (CADRI, 2011, p. 17) as long as they base their support on internal partners' strategies, institutions and procedures (i.e. the principle of alignment) (OECD, 2009, p. 3). It is also worth noting that *all* external partners involved must ensure their respective support is 'harmonised, transparent and collectively effective' (OECD, 2009, p. 6), without undermining the ownership of the internal partners (i.e. the principle of harmonisation).

Figure 14.3 Vulnerability and capacity assessment (VCA) is an influential tool to facilitate local ownership.

Moreover, ownership entails a commitment to the particular capacity development initiative in question and the ability to translate this commitment into action (Lopes & Theisohn, 2003, pp. 29–30). Shouldering ownership is, thus, voluntary and cannot be imposed by outsiders (Anderson, 1999, p. 239; Lopes & Theisohn, 2003, p. 22). Local ownership can, in other words, not be fostered by persuasion (Lopes & Theisohn, 2003, p. 33) but instead by external partners continuously adapting their support to meet the internal partners' needs (Hagelsteen et al., 2021, p. 7).

Partnership

The second principle of capacity development is partnership. Capacity development is essentially a partnership amongst internal and external partners (Becker, 2021; Hagelsteen

& Becker, 2013; OECD, 2009). However, such a partnership entails a particular kind of relationship between the partners. It generally conveys a sense of mutual legitimisation, participation and commitment to some cooperative process (Casey & Delaney, 2022, p. 54). It entails an open dialogue between partners (Sida, 1997, p. 36) and mutual trust and commitment to achieve agreed objectives (Neuhann & Barteit, 2017, p. 10). Partnership also entails, in principle, some sense of equality between the partners (Bontenbal, 2009, p. 100–101; Hagelsteen et al., 2021, pp. 3–4). It is important to note that equality does not mean sameness since there can be immense diversity of actors involved in the partnership (Woodhill, 2010). Equality is, in this context, instead about all partners participating on an equal footing in the capacity development partnership regardless of their diversity.

For such a partnership of diverse actors to be possible, clear and mutually agreed roles and responsibilities are essential (Flaspohler et al., 2007, pp. 116–117; Hagelsteen & Becker, 2013, p. 5). It means all partners must know what is expected of them and what to expect from others whilst continuously ensuring local ownership (Becker, 2021, pp. 534). In other words, any capacity development partnership would be fruitless if the roles and responsibilities, regardless of their clarity and mutual agreeability, end up transferring ownership to external partners. It is also important to note that the roles of neither internal nor external partners should be fixed throughout the partnership but flexible and adjusted or changed to meet the changing needs and abilities of the partners (Hagelsteen et al., 2021, p. 7). More on flexibility in a moment.

Contextualisation

Capacity development must not only be driven by internal partners but integrated into and tailored for each particular local context (Sida, 1997, p. 36; Crewe & Harrison, 1998; Schulz et al., 2005, p. 63; Hagelsteen & Becker, 2013, pp. 8–9; Hagelsteen & Becker, 2019, p. 4). This requisite contextualisation of capacity development encompasses a comprehensive understanding of the community, organisation, society or whatever human-environment system, including the context it is embedded in, and its current capacity to perform the functions the partnership aspires to develop further (Figure 14.4). In short, any capacity development initiative must be based on a systemic understanding of the problem it intends to address and how the human-environment system works in general.

Whilst it may be handy for many larger external partners to focus on the national level (e.g., OECD, 2006; Sida, 1997, p. 36), it is often appropriate for capacity development for resilience also to think more locally (Becker, 2012), for instance, not only to consider that people live in dangerous locations but also why they live there in the first place (Askman et al., 2018; Buckle, 1998, p. 23). That is of utmost importance since communities, organisations, societies and most other human-environment systems are not homogeneous but comprise different groups with different vulnerabilities, capacities and needs (Anderson, 1999, p. 17; Smith & Wenger, 2007, pp. 237–238; Becker, 2017).

Figure 14.4 All contexts are different.

Flexibility

Flexibility is also a central principle of capacity development since the only permanent feature of our world is its impermanence (Plato, 2008). In other words, capacity development must not only be tailored to its particular context at its outset but continuously adjusted as the context changes (Hagelsteen & Becker, 2019)—and rest assured, the context is always changing. Communities, organisations, societies and all other kinds of human-environment systems are never static, nor do they exist in static surroundings. Although the planned activities to develop capacity may remain relevant and feasible despite specific changes, other changes or particular combinations or aggregation of such changes may require more or less substantial revision of the plan for any capacity to be developed. These changes can stem from the results of implemented activities within the capacity development initiative, arise out of activities implemented as part of other purposeful projects or programmes, or emerge in the coevolution of the human-environment system and its surroundings in general.

Learning

Capacity development is largely about learning (Hagelsteen & Becker, 2019, p. 4). It is about acquiring knowledge and understanding, competence and skills, attitudes and

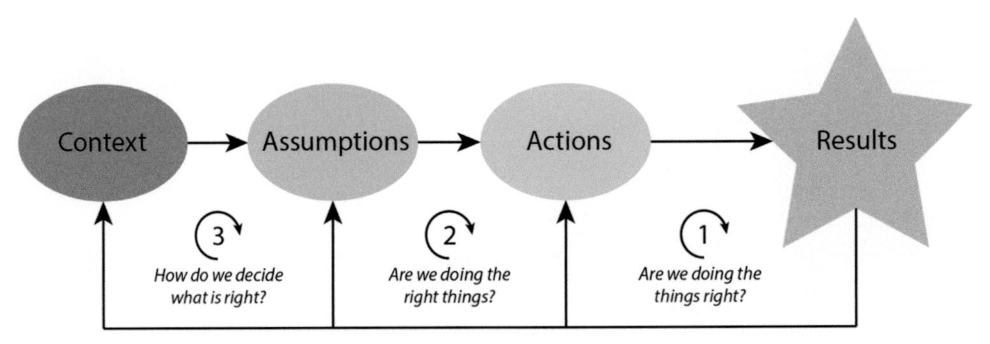

Figure 14.5 Triple-loop learning.

preferences, and values and behaviours. It is not only about learning to do things right but also if we are doing the right things and how we decide what is right (Argyris & Schön, 1996). These distinct yet interconnected aspects of learning can be conceived as three reflexive loops (Figure 14.5), and engaging in only the first can be referred to as single-loop learning, in the first and the second as double-loop learning, and in all three as triple-loop learning (Argyris & Schön, 1996, although triple-loop learning is not explicitly labelled). It is not only about developing capacities to address some identified problems but also about learning how to develop capacities better to develop those capacities, which I will come back to shortly. This link to learning fits the approach to resilience presented in this book (Chapters 7–8).

It is essential to note that this principle of capacity development is not only about the learning of internal partners but about mutual learning of both internal and external partners (Casey & Delaney, 2022; Hagelsteen et al., 2022; Taylor & Clarke, 2008), although different partners are likely to learn different things (Hagelsteen et al., 2021, pp. 5–7). Moreover, it is not only about individual learning but also social learning (Wenger-Trayner, 2012). Social learning is a process in which learning goes beyond the individuals, occurs through social interactions and processes and becomes situated within human-environment systems (Reed et al., 2010). It is learning on both the individual and collective levels at once (Wenger, 1998).

Accountability

Accountability is also a crucial principle of capacity development (Baser & Morgan, 2008, p. 69; Hagelsteen & Becker, 2019; OECD, 2009). Accountability refers to being responsible for one's decisions and actions. It involves being transparent about them and willing to take responsibility for their effects, regardless of whether positive or negative, intended or unintended. Whilst accountability is relatively straightforward in the context of individual action, it often involves explicit schemes in organisational contexts—such as policies, procedures, and reporting or whistleblowing mechanisms—to ensure that individual or organisational actors are held accountable for their decisions and actions. Accountability also helps to ensure that resources are used efficiently, that capacity development initiatives are designed to achieve specific objectives, and that progress is monitored and evaluated against established targets. In the context of capacity development, accountability promotes transparency and facilitates building trust amongst involved actors, which is essential for sustained engagement and support. Accountability is vital across several different relationships. Partners in capacity development are accountable to the people affected by the initiative, to each other and to the donors financing it (if different), whilst donors are accountable to their boards or politicians, who in turn are accountable to their back donors or taxpayers (Watson, 2006, p. 17).

Long-term

Capacity development is a long-term process (Baser & Morgan, 2008, p. 120; Becker, 2021, p. 535; Hagelsteen & Becker, 2019). Sustainable change takes time (Figure 14.6), and there is no way to rush it (Watson, 2006, p. vii). Capacity development involves investing in people and resources, facilitating for them to organise, designing and implementing policies and procedures, and changing behaviours, norms, attitudes and preferences. It is a gradual and iterative process that requires long-term engagement and patience to achieve meaningful results, although some quick and visible results can be important to bolster political support (cf. McEntire, 2007, p. 398), promote further investment, and facilitate willingness to change (Becker, 2021, p. 533).

Sustainability

Finally, for capacity development to have any real meaning, the developed capacities must be sustainable in the sense of lasting over time (Baser & Morgan, 2008; Brinkerhoff

Figure 14.6 Change takes time.

& Morgan, 2010; Hagelsteen & Becker, 2019). Although this kind of sustainability largely depends on following most of the other principles of capacity development introduced above, it demands the explicit attention of all involved partners. Otherwise, all efforts and invested resources are likely to be in vain. This principle stipulates that capacity development cannot only focus on creating additional capacities of internal partners but must also focus on them utilising and retaining their capacities (Hagelsteen et al., 2021)—individually, organisationally and systemically.

Deconditioning capacity development

Knowing what capacity development should be, it is utterly discouraging to consider what it currently involves in general. We know what to do, yet continue to do something else, or at least only detached parts. Morpheus (1999) was right when telling Neo that: 'There's a difference of knowing the path and walking the path'. This inadequacy of the practical implementation of most capacity development initiatives has been pointed out for as long as we have used that name for them. For instance:

> *Whilst the importance of capacity is widely recognised, how it emerges, how to develop and evaluate it and how to sustain it is for many less clear. There are a number of experiences, tools and resources that are now available in the field of disaster risk reduction and relate to capacity development. Lessons of past experience, for example, point to many inappropriate approaches with short-lived impacts on the part of development cooperation partners. There is however the need for many to better familiarise with the link between capacity, its development and disaster risk reduction. The evidence and knowledge available within the disaster risk reduction community on how to support the development of capacity 'in practice' is still not widely systematised and shared, although examples do exist*
>
> ***CADRI (2011, pp. 7–8).***

This statement from one of the leading specialist organisations on capacity development for resilience within the international community provides a clear-sighted and sobering view that, unfortunately, is still valid. Although some actors are up for the challenge and ready to do it right, many others must abandon much of their habitual mode of thinking about capacity development for it to have any lasting effects on our future. Instead of merely renaming capacity development again, it is time for some serious deconditioning of the entire system.

Principles versus performance

Whilst the principles for capacity development all appear reasonable and feasible, they are generally not adhered to in any comprehensive sense (Hagelsteen & Becker, 2013, 2019; Hagelsteen & Burke, 2016).

Performing the principles

There is a broad consensus that lack of local ownership is a fundamental reason for the generally poor results of many capacity development initiatives over the years (Godfrey et al., 2002, p. 365; Morgan, 2002; Lopes & Theisohn, 2003, pp. 29–31; Devereux, 2008, pp. 360–361; Hagelsteen & Becker, 2013, pp. 8–9; Hagelsteen & Becker, 2012, p. 95; UNDRR & Coppola, 2019, p. 90). Capacity development is, instead, often designed and driven by external partners and donor priorities (Hagelsteen & Becker, 2013, 2019; Hagelsteen et al., 2021).

Whilst virtually all capacity development initiatives are explicitly described as partnerships, many, if not most, of them struggle with the essential mutuality and equality between internal and external partners (Hagelsteen & Becker, 2013; Hagelsteen et al., 2021) due to differences in organisational cultures, tight time frames, lack of ability to cooperate, and power imbalances (Becker & Hagelsteen, 2016). Moreover, there is generally a lack of consensus on what ownership, partnership and other key concepts mean, resulting in misunderstandings between partners (e.g., Hagelsteen & Becker, 2013, p. 8; Månsson et al., 2015; Raju & Becker, 2013). Such Babylonian confusion is detrimental in itself to partnerships for capacity development (Lopes & Theisohn, 2003, pp. 29–31; Hagelsteen & Becker, 2014).

Tailoring the capacity development initiatives to the actual needs and context is generally not happening to any sufficient extent in practice either (Figure 14.7). Instead, external partners often have a 'fix-it' or 'know better' approach to capacity development (Hagelsteen & Becker, 2013, p. 4) and regularly come up with blueprint solutions that fit their own and their donors' agendas (Anderson et al., 2012, p. 25; Hagelsteen & Becker, 2019, pp. 7–8; Hagelsteen et al., 2021, p. 8). Thus, regularly ignoring established strategies, institutions and procedures and creating parallel structures (Twigg, 2004, p. 289; Becker et al., 2021), potentially obstructing the space for novelty and local experimentation and undermining existing capacity (Andrews et al., 2013, pp. 241–242) and local ownership (Anderson et al., 2012, p. 63).

Although there have been calls for flexibility for decades (e.g., Bolger, 2000, p. 2; Watson, 2006, p. 6), most capacity development initiatives still come with pre-set conditionalities and little or no flexibility to adjust to changing situations (Hagelsteen et al., 2021, p. 3). However, even when donors do provide flexible conditions, many external partners still use strict result frameworks and issue, in turn, rigid conditions for their internal partners (Hagelsteen et al., 2022, pp. 3–4). The desire for control trumps the need for flexibility.

Whilst learning is indeed a central aspect of almost all capacity development initiatives, they are overwhelmingly focused on training individuals (Hagelsteen & Burke, 2016, p. 48; Hagelsteen & Becker, 2019, p. 5). Actually, most of the USD 15 billion

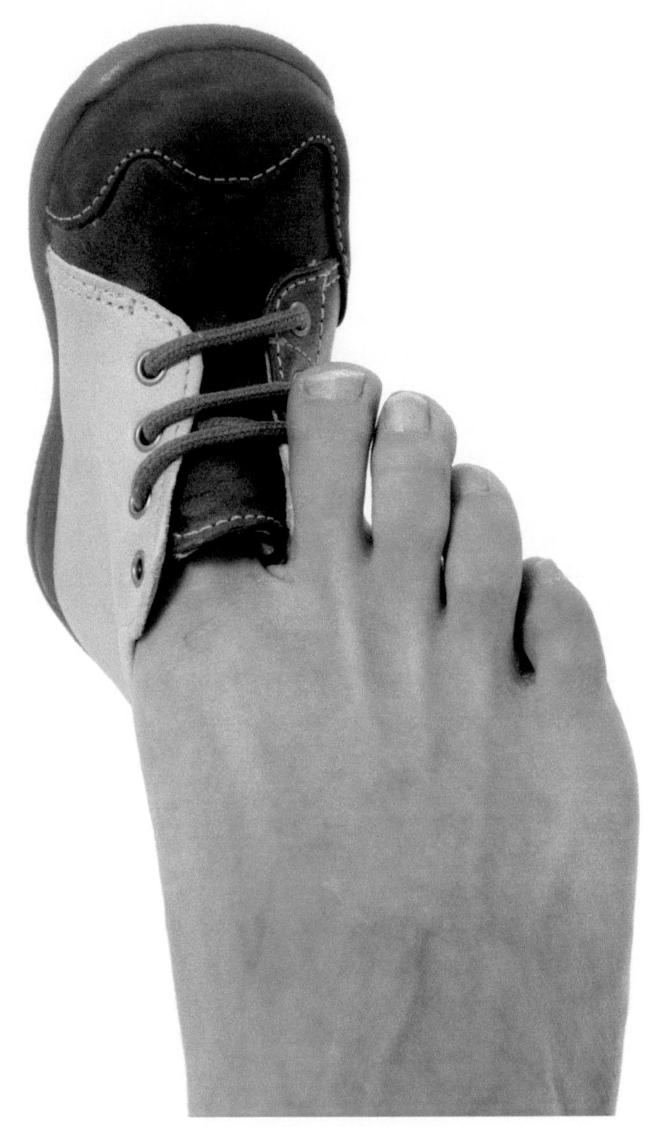

Figure 14.7 One size fit not all. *(Source: © Lars Christnsen | Dreamstime.com.)*

invested in capacity development per year is spent narrowly on training, with disappointing results (Denney, 2017). Although training of individuals may be necessary, it is rarely sufficient if their organisation is not able to utilise and maintain the acquired knowledge and skills (Eade, 1997, p. 31; Becker, 2021) and the individual learning is not complemented by social learning (Wenger, 2009). Moreover, training can never produce sustainable results if delivered ad hoc by external partners and is not institutionalised within the human-environment system in question (Becker, 2009, p. 19; Hagelsteen & Becker, 2013, p. 10).

The extent to which the principle of accountability is generally met in capacity development initiatives is a somewhat more complicated issue due to the multiplicity of accountability relations and directions. However, to grasp the relationship between principle and performance, it helps to look at these relations as forming a chain from the people supposedly benefitting from the capacity development initiative all the way to the board members or taxpayers bankrolling it. From this perspective, it is blatantly clear that most partners in capacity development are almost exclusively focusing on their accountability upward along this aid chain. It means that the partners implementing activities are primarily accountable to the partners having the budget for the activities, who are primarily accountable to donors providing the funds, who are primarily accountable to their back donors, and so on until the end of the aid chain (Hagelsteen & Becker, 2019; Hagelsteen et al., 2022)—generally glossing over their paramount accountability to the people on its other end (Anderson et al., 2012; Hagelsteen et al., 2021).

Contemporary capacity development is overwhelmingly short term (Hagelsteen & Becker, 2019). Although quick wins and early positive feedback are important for building faith and commitment in capacity development initiatives (Kotter & Cohen, 2002, pp. 127–141), short-term activities must only be the first phase of longer-term initiatives (Hagelsteen & Becker, 2013). Yet, most funding available for capacity development is short term—often for just 12, 15 or 24 months (Anderson et al., 2012, p. 72; Hagelsteen & Becker, 2013, p. 8)—whilst 5–10 years have been suggested as more appropriate timescales (Binger et al., 2002). Most partners in capacity development are, thus, stuck focusing on producing 'a stream of bite-sized and discrete projects', driven by their modus operandi 'to organise their work around designing and funding projects' whilst forgetting or ignoring other vital aspects needed to facilitate real capacity development (Tendler, 2002, pp. 2–4).

As stated earlier, not following the other principles of capacity development seriously undermine its sustainability. However, the sustainability of any results is also directly compromised by most capacity development initiatives mainly paying lip service concerning sustainability whilst creating capacities that cannot be maintained without the external support provided during the capacity development initiative, or at least not focusing on how the created capacity could be utilised and retained over time (Hagelsteen et al., 2021).

Seven types of failure

Analysing the discrepancy between principles and performance, Hagelsteen et al. (2021) suggest a typology of seven common types of failures in capacity development that ultimately undermine any potential effects (Figure 14.8). Whilst these failures are depicted as

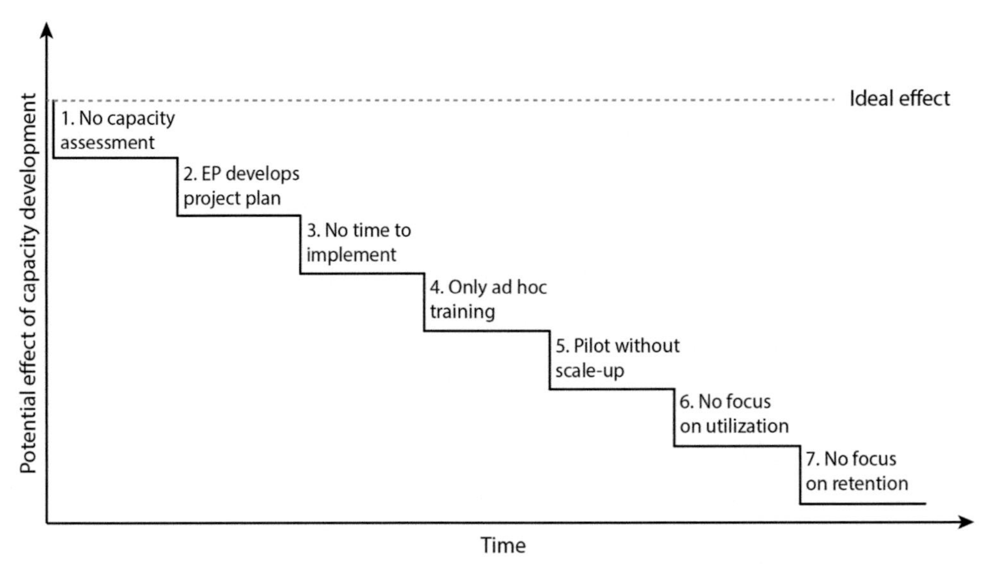

Figure 14.8 A typology of seven common failures (Hagelsteen et al., 2021).

steps downward towards complete failure, it is important to note that they can occur in various combinations with varying results.

Type 1 failure: no capacity assessment

Capacity assessment is essential for external partners to understand the local context and existing capacities and to start building relations with potential partners and other actors, and internal partners must be central to and facilitate the process (Hagelsteen et al., 2021, pp. 7–8). Such mutual commitment from the beginning has also been suggested as important to facilitate trust between prospective partners (UNDP, 2008, p. 22). However, properly understanding the local context and pre-existing capacities is challenging without sufficient time and resources, which are both generally scarce when needed due to short-term funding that is almost exclusively only available until after the capacity development initiative has been designed and approved (Hagelsteen & Becker, 2013, pp. 8–9). Moreover, external partners are also liable to 'the expert blind spot'—an excessive focus on one's own expertise and perspectives (Hagelsteen et al., 2021, p. 8)—that biases them only to see what they are interested in and focus on their own priorities and areas of expertise, whilst overlooking other issues and available capacities (Hagelsteen & Becker, 2019, p. 4).

Type 2 failure: external partners design initiative

Even with an adequate capacity assessment, external partners generally have their funding opportunities, areas of expertise, goals, conditions and blueprint solutions (Hagelsteen &

Becker, 2019, p. 5; Hagelsteen et al., 2021, p. 8). Having specific funding that demands results within a particular area and within a short timeframe generates a deep need for control, erodes flexibility and points accountability towards the donors instead of the internal partners and the people benefitting from the initiative (Hagelsteen & Becker, 2019, pp. 6–7). Capacity development initiatives are, thus, often procedurally designed and managed by external partners, undermining the influence of internal partners who end up in dependency (Anderson et al., 2012, pp. 21, 67, 79–81). Furthermore, internal partners seeking increased independence and self-efficacy are, then, often hindered by their external partners' static models and blueprint solutions of what they believe is needed (Hagelsteen et al., 2021, p. 8).

Type 3 failure: insufficient time to implement

Even with an adequate capacity assessment and well-designed plan of activities, there is often insufficient time to implement actual capacity development activities (Hagelsteen et al., 2021, p. 8). Ironically enough, there appears to exist an 'assessment dilemma', with either too much focus and time spent on assessments in relation to the actual implementation of activities or no assessments done at all (Hagelsteen et al., 2021, p. 8). Moreover, many capacity development initiatives for resilience are short term (IFRC, 2015, p. 56), with assessments and planning taking up too much of the time and budget or the funds for implementation arriving late, forcing the most important parts of the initiatives to be rushed (Hagelsteen et al., 2021, p. 8) with too little time to establish any institutional memory (Hagelsteen & Becker, 2013, pp. 4–5). It is also common that funding is secured for the initial phase, with little possibility of additional funding allocated for the implementation (Hagelsteen et al., 2021, p. 8). Finally, the donors sometimes alter funding conditions, leaving the partners no room to implement the planned activities.

Type 4 failure: only ad hoc training

Many capacity development projects focus overwhelmingly on ad hoc, short-term technical training activities, such as 1- or 2-day, or maybe up to 1-week training workshops (Lipson & Warren, 2006, p. 6; Hagelsteen & Burke, 2016, pp. 48–49; Hagelsteen & Becker, 2019, p. 5), irrespective of having a limited impact at best (Denney, 2017; Hagelsteen et al., 2021, p. 8). Capacity development is often even thought of as such training by default (Pearson, 2011b, pp. 16–19; Hagelsteen & Becker, 2013, p. 10; Hagelsteen & Becker, 2019, p. 5). Although the trained individuals may, of course, learn something useful and important, ad hoc training workshops rarely make any sustainable contributions to the capacity of the human-environment system since individuals move between roles, leave or might be the wrong person to be trained in the first place (Becker & van Niekerk, 2014; Hagelsteen et al., 2021, p. 8). These staff turnover problems or wrong participants are particularly problematic in relation to ad hoc training activities because there is no predetermined way to train additional people. Studies also suggest that such training workshops often are delivered by technical experts, who, regardless of their advanced technical expertise, often lack the pedagogical knowledge and skills to conduct effective educational activities (Hagelsteen et al., 2021, p. 8; Hagelsteen et al., 2022, p. 6).

Type 5 failure: pilot without scale-up

Another typical failure of capacity development initiatives is a common fixation on pilot projects that are often never scaled up (Hagelsteen et al., 2021, p. 8). Whilst capacity development initiatives are frequently designed to start in a couple of selected communities or districts but with the explicit rhetoric and grand plans for later expansion to many others (Figure 14.9), donors habitually finance only the pilot and pull out afterwards (Hagelsteen et al., 2021, p. 8). It means that the promised scale-up that often motivated the allocation of external funds is entirely contingent on often unrealistic economic commitments from the internal partners that are either simply impossible or not prioritised when the external support ends. This challenge for scaling up is long established in international development cooperation (Edwards & Hulme, 1992).

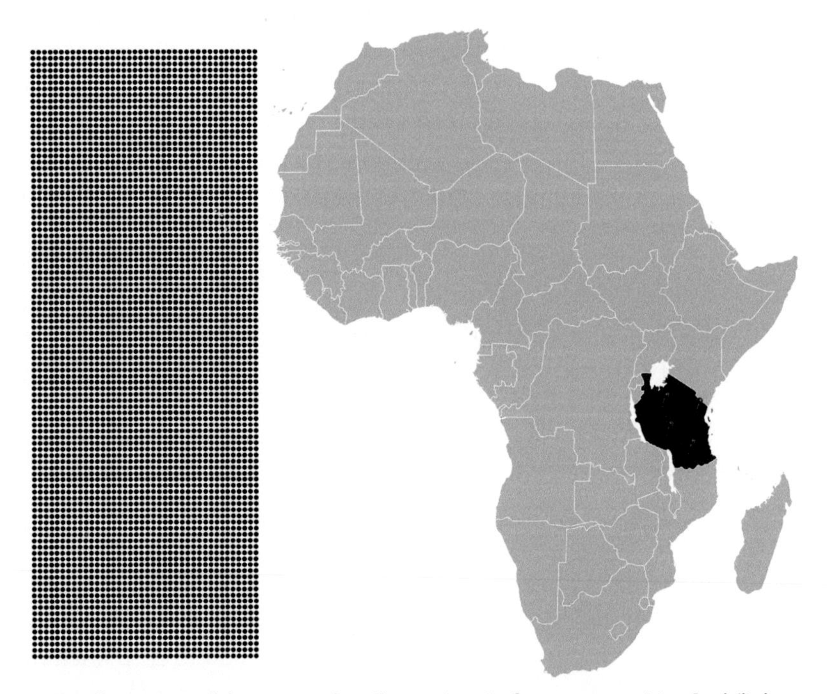

Figure 14.9 An illustration of the cover of a pilot project in four communities *[red (light grey in print version)]* out of the 4000 communities of Tanzania.

Type 6 failure: no focus on utilisation

Virtually all capacity development initiatives focus overwhelmingly, if not only, on creating capacity— for example, on developing individual skills, establishing procedures, policies and regulations—with no or marginal focus on supporting the internal partners in utilising the developed ability (Hagelsteen & Burke, 2016, p. 49; Hagelsteen et al., 2021, p. 9). This lack of capacity utilisation seriously undermines capacity development since it separates whatever is taught and promoted from the essential process of learning by doing and creates a barrier to the type of adaptive learning that actually could facilitate real change (Armstrong, 2013, p. 213). It means that it is not evident whether or not the individuals being trained and organisations getting new policies or procedures actually can utilise them, which is the obvious purpose of capacity development. This problem is aggravated further by the internal partners' full agendas, small units, different priorities, and limited and competing resources, often with many initiatives going on simultaneously without sufficient alignment and harmonisation (Hagelsteen et al., 2021, p. 9). Internal partners are, in other words, constantly being capacitated, without time and resources to reflect on and apply the skills or integrate the new processes and procedures within the organisation (Bolger, 2000, p. 5; Hagelsteen et al., 2021, p. 9).

Type 7 failure: no focus on retention

Furthermore, the almost exclusive focus on capacity creation also results in most capacity development initiatives neglecting the capacity retention necessary for any results to be sustainable (Hagelsteen et al., 2021, p. 9). For instance, staff turnover is a massive challenge in this regard (Hagelsteen & Becker, 2019, p. 5; Hagelsteen & Burke, 2016, p. 49; van Riet & van Niekerk, 2012), as it continuously undermines the vital institutional memory (cf. UNDRR & Coppola, 2019, p. 44). Capacity creation can even undermine capacity retention when the increasing knowledge and skills of internal partners' key staff increase their chances of changing positions or getting international jobs, which is further exacerbated by a general lack of job security for many of them as they are not guaranteed permanent positions when the external funding ends (Hagelsteen et al., 2021, p. 9). However, capacity retention is also generally overlooked concerning other essential aspects of capacity development, such as the vital capacity to regularly adapt policies, procedures, regulations and processes, which is most often ignored after they have first been developed (Hagelsteen et al., 2021, p. 9).

Principles vs principles

After decades of studies pointing out the continuously poor performance of capacity development (e.g., Anderson et al., 2012; Baser & Morgan, 2008; CADRI, 2011; Hagelsteen & Becker, 2013; Hagelsteen & Burke, 2016), the underlying reasons for this predicament has started to attract increasing attention (e.g., Hagelsteen & Becker, 2019; Hagelsteen et al., 2021; Hagelsteen et al., 2022). One of the primary reasons found for the generally poor performance concerns clashing principles (Hagelsteen & Becker, 2019). In other words, whilst most actors involved in capacity development are well aware of how it should be done in principle, they cannot work in accordance with all principles since some of them are contradictory, resulting, most pertinently, in misguided accountability and temporal discord.

Misguided accountability

Most partners in capacity development mainly focus their accountability upward along the aid chain (Figure 14.10), towards the donors, instead of downward, towards the people who are to benefit from the capacity development (Hagelsteen & Becker, 2019, pp. 6–7). This misguided accountability clashes, either directly or indirectly, with the other principles. Whilst the paramount accountability should be directed towards the beneficiaries (OECD, 2009), the system is rigged against that (Hagelsteen & Becker, 2019, p. 6) as it demands predictability, deliverables and reporting back to the donor

Figure 14.10 An aid chain for a capacity development programme of the Red Cross.

(Hagelsteen & Becker, 2013). The principles of ownership and flexibility clash directly with such upward accountability as most external partners are reluctant to relinquish control over the planned activities due to the perceived risk of the initiative not delivering as expected (Hagelsteen & Becker, 2019, p. 6). The principle of learning also clashes directly with this upward accountability since it usually entails an overwhelming focus on monitoring and evaluating capacity development initiatives against predefined deliverables and not on the learning and actual impact (Guijt, 2010).

There are also indirect clashes between this misguided accountability and other principles. It clashes with the principle of long-term engagement due to the latter's dependence on flexibility to adjust to changes that become increasingly impossible to fully anticipate the longer the timeframe (Hagelsteen & Becker, 2019, p. 6). It also clashes indirectly with sustainability through the latter's dependence on several other principles, such as local ownership, contextualisation, long term and learning. However, upward accountability can also clash with sustainability as a result of its tendency to bias the contents of capacity development interventions towards activities focusing on capacity creation with visible, controllable results that can be reported back to the donors (Hagelsteen & Becker, 2019, pp. 6—7).

This kind of misguided accountability is inherent to the aid system and other contexts in which external partners in capacity development rely on resources from a third-party donor, such as a Swedish County Administrative Board supporting the development of municipal administrations' capacity to manage the risk of coastal floods using obtained national funding earmarked to strengthen Swedish crisis preparedness. There is no way for the partners to resolve the underlying clash of principles by themselves. The problem is, in other words, systemic and only possible to address through deliberately transforming the system (Hagelsteen & Becker, 2019).

Temporal discord

Clashing principles also result in a temporal discord between capacity development demanding long-term engagement and the aid system requiring short-term feedback (Hagelsteen & Becker, 2019, p. 7). This temporal discord is institutionalised in the short-term funding cycles mentioned earlier, which are particularly common in capacity development for resilience, with funding often coming from humanitarian instead of development budgets (Hagelsteen & Burke, 2016). Such short-term funding cycles can, in fact, directly undermine capacity development due to their inherent inflexibility whilst pressuring the partners to show visible results without providing adequate time even to understand the local context, let alone ensure local ownership and form a functioning partnership (Hagelsteen & Becker, 2013, p. 11). Ownership and partnership entail trust, which takes time to develop but can be destroyed in an instant (Slovic, 1993, p. 677), and learning takes time because of the lag between assimilation and accommodation of information (Piaget, 1971). Contextualisation demands time during the initial period when funding is commonly unavailable or, at least, uncertain, which often results in underfunded and rushed preparatory processes that exacerbate the impact of the expert blind spot introduced earlier and confine the partners to implement rushed initial plans largely designed by disconnected and single-minded external experts (Hagelsteen & Becker, 2019, p. 7). This temporal discord frequently results in static, short-term and high-intensity initiatives with much less potential for developing capacities for resilience than flexible, long-term and low-intensity initiatives using the same resources but over a longer time (Hagelsteen & Becker, 2019, p. 7). Just as for misguided accountability, this temporal discord is systemic and impossible for the partners in capacity development to address themselves fully. For that, systemic change is needed.

Principles versus perspectives

The principles of capacity development not only clash with each other in our complex, uncertain, ambiguous and dynamic world. These four fundamental challenges also influence our underlying perspectives in the sense of our attitudes and viewpoints towards capacity development, the human-environment system in question, and the world, which can also work against the principles of capacity development.

Mindset lag

First of all, there is a clear lag between the rapidly changing world and the pace of change in the mindsets of the actors involved in capacity development. Hagelsteen and Becker (2019, pp. 7—8) call this 'mindset lag' and demonstrate how it undermines the effectiveness of capacity development as most external partners continue to work the same way they have always done regardless of the astonishing developments in many parts of the world. External partners focus their assistance on politically defined priority countries and design capacity development based on the resource they have available rather than the need of the internal partners, push for their predefined objectives with blueprint solutions and do most of the work themselves whilst blaming the disinterest of internal partners on culture, lack of capacity, corruption, etc. (Hagelsteen & Becker, 2019)—essentially a contemporary version of the White Man's Burden (Figure 14.11) (Eriksson Baaz, 2005). Although this approach to capacity development has never worked (Smillie, 2001), it is high time for external partners to realise that the contexts they engage in are diverse and hardly any places are still like they were in the 1950s or even 10 years ago (Hagelsteen & Becker, 2019, p. 7).

Look at Botswana, which has made substantial progress in terms of human development since the end of the British protectorate in 1965. The economic purchasing power per person has increased around 18 times since then, the adult literacy rate is approaching 90%, child mortality has gone from 14.4% of the children dying before the age of 5 in 1965 to 3.3% in 2022, and fertility rates have dropped from 6.7 to 2.5 children per

Figure 14.11 The White Man's Burden, depicted by Gillam (1899) with clear racist undertones and showing John Bull (UK) and Uncle Sam (USA) carrying the world's people of colour towards civilisation.

woman during the same period (Gapminder, 2023). It is only life expectancy that has not followed the same positive trend due to the high prevalence of HIV/AIDS, with rapidly declining life expectancy from the late 1980s until massive investments turned the trend around again early in the new millennium (Gapminder, 2023). More than 90% of the urban population has access to improved sanitation, there have been more mobile phone subscriptions than people since 2010, and almost 70% of the population are using the internet (Gapminder, 2023). They have people with master's degrees and even PhDs working for the government offices responsible for disaster risk management and climate change adaptation.

Many countries have similar developments, although Botswana is amongst the most impressive, and most of the least developed countries are also changing rapidly. Yet, external partners tend to be stuck in outdated and often rather patronising mindsets (Eriksson Baaz, 2005; Hagelsteen et al., 2021). Whilst some internal partners accept this treatment—at least in the face of their external partners and donors who may still offer resources in contexts of scarcity—more and more are not. That is a good sign. External partners and donors must overcome their mindset lag and develop their capacity to listen and learn. They may still have a lot to offer, but that must be tailored to the actual needs now and neither to some outdated or imaginary image nor to what they happen to have resources for. Getting over this mindset lag is complicated by many of the drivers of resistance to knowledge and change discussed in Chapter 13 (e.g., Hagelsteen & Becker, 2019, p. 8). The more entrenched people are in their preferences, ideas, policies or ideologies, the more difficult it is to overcome their mindset lag. To completely change the mindset of the entire aid industry is indeed difficult and demands patience and motivation. Still, it must be done for capacity development to work.

Quixotic control

The next underlying perspective that undermines the principles for capacity development is the immense need for control that permeates the entire system for capacity development. Hagelsteen and Becker (2019, p. 7) refer to this as 'quixotic control', with an apparent reference to Don Quijote's refusal to change his beliefs and actions regardless of the obvious need to do so.

Donors commonly have predetermined priority countries and issues and often define in advance where and for what their funding can be used. That undermines the actual potential for needs-driven assistance already from the start. Partners wanting to engage in capacity development typically have to apply for funding, relating to these conditionalities and describing what they would achieve with the money and how they would accomplish that. If granted the money, these descriptions add to the conditionalities since they are now what the donor expects from the initiative. For each step downwards along the aid chain, the partner with the funding adds conditionalities for the partner below, increasingly restricting the use of the money until little or no flexibility is left when it finally reaches the people actually implementing the activities on the ground (Hagelsteen & Becker, 2019, p. 7). These cascading conditionalities effectively undermine most of the principles for capacity development, with devastating effects on effectiveness (Hagelsteen & Becker, 2019, p. 7).

This quixotic control stems largely from actors overwhelmed by complexity, uncertainty, ambiguity and dynamic change. Failing to acknowledge complexity, most actors adopt linear thinking and approach complex problems as merely simple or complicated (e.g., Coetzee et al., 2016) using planning tools in ways that reify linearity (Davies, 2004, 2005) and addressing the illusory problems with familiar blueprint solutions.[1] Whilst rarely generating any real and sustainable impact (Becker, 2009), this approach provides projects comprising sets of activities that are possible to control (Hagelsteen & Becker, 2019, p. 7). This problem is also exacerbated by the seemingly ever-increasing rate of change, which is also essentially ignored by actors having to present detailed plans of all activities already from the start to access funding (Hagelsteen & Becker, 2019, p. 7). Moreover, complexity and dynamic change generate uncertainty that increases exponentially with an increasing timeframe since it becomes increasingly difficult to foresee what may happen. This uncertainty is uncomfortable for actors being accountable to

[1] Sometimes referred to as logic-less frames or lack-frames, with reference to the logical framework approach commonly used (Gasper, 2000).

Figure 14.12 When everybody finally is on board, nobody wants to open the discussion again.

implement what they have committed to do (Hagelsteen & Becker, 2019, p. 7), and they seek to avoid it. Whilst people involved in planning dislike uncertainty, decision-makers generally detest it and usually neglect it altogether behind 'illusions of certainty' (Boulding, 1975, p. 11). However, ignoring uncertainty and concealing it behind pretty, dressed-up plans that appear controllable, or have been so watered-down in terms of timeframe or scope that they have no real impact on capacity, is counterproductive at best and disastrous at worst and one of the main reasons behind the failure of capacity development in general.

Quixotic control can, paradoxically enough, also be a result of the level of participation. The more people participate in designing the capacity development initiative, the more difficult and time-consuming the decision-making becomes (Figure 14.12). The more energy and resources invested in finally getting all actors on board, the less willing the actors are to propose changes that require them to redo at least part of the process, binding them to the initial plan via a distributed network of control irrespective of what happens over the course of the initiative[2] (Hagelsteen & Becker, 2019, p. 7).

Finally, even when donors are increasingly aware of these problems and attempting to alleviate the constraints of conventional development cooperation by allowing more flexible conditions—like advocated by the Doing Development Differently dialogue (Honig & Gulrajani, 2017)—actors further down the aid chain remain reluctant to let go of control

[2] Sometimes referred to as 'lock-frame', with reference to the logical framework approach commonly used (Gasper, 2000).

since they still are accountable to deliver and do not want to risk failure (Hagelsteen et al., 2022, pp. 3–4). This risk aversion is understandable, considering their more or less complete dependence on external funding and the increasing competition for resources. It is also important to note that this more progressive movement amongst donors is still marginal and not influential enough to change the system, which unfortunately seems to push most donors even further towards flag-waving and short-term visible results (Hagelsteen & Becker, 2019, p. 7). However, as long as the focus is on maintaining control and ignoring uncertainty, there is no chance for capacity development to be effective. What is needed is, again, a complete change of mindset amongst all actors.

Lack of motivation for change

Several of the problems of capacity development can only be addressed by substantial changes in expectations, attitudes, relationships, practices, procedures and the system. Yet, there seems to be little real motivation to change at all (Hagelsteen & Becker, 2019, p. 8). Most people involved are aware of many of the problems, but there is a sharp disconnect between what they say and do. There are simply insufficient incentives to change since the wheels keep turning anyway (Hagelsteen & Becker, 2019, p. 8).

In addition to several drivers of resistance to change introduced in Chapter 13, internal partners may resist change as powerful local elites may reap benefits from the current system (Mansuri & Rao, 2012). Lower-level managers and staff may also benefit by getting a job or lucrative per diems when travelling or attending workshops, but the best chances of earning extra money are often seized by more powerful individuals, often leading to the wrong people getting trained or certain people receiving the same training over and over again (Hagelsteen et al., 2021, p. 6). Similarly, external partners may be resistant to change as they often get exorbitant contracts and high status, at the same time as their organisations finance immense bureaucracies through overheads added to the costs generated by the current system (Hagelsteen & Becker, 2019, p. 8).

However, the incentives do not have to be direct. There are also incentives to continue the current practices based on the mechanism of increasing returns (North, 1990), described in Chapter 10, where the investments of time, energy and resources into the practices make it increasingly difficult to change them. Remember that flawed practices remain because additional work in the same direction continues to be rewarded at the same time as the costs of changing to an alternative increase over time (Scott, 2014), through learning effects, coordination effects, and adaptive expectations that help institutionalise practices in general (North, 1990). The same mechanism is at play in capacity development.

The lack of motivation to change is also driven by the other two mechanisms of institutionalisation (Chapter 10). Repeating a particular practice within a professional group leads to increasing commitments to normative expectations (Selznick, 1949, pp. 256–257) in which the practice becomes part of the identity of the particular professional groups (Scott, 2014, pp. 145–148). Regardless of being flawed, that is simply what they do within that group—what is expected of them. With enough repetition, the practice becomes increasingly habitualised, routinised, and taken for granted (Berger & Luckmann, 1966)—hidden in plain sight.

The direct and indirect incentives for continuing the current practices of capacity development, the normative expectations on the members of different professional groups, and the habitualised and routinised character of many practices align to demotivate the urgently needed changes. The world is changing, but the system and most actors are stuck in their outdated ways (Hagelsteen & Becker, 2019, p. 8). Whilst donors and external partners may still have other roles to play (Hagelsteen & Becker, 2013; Hagelsteen et al., 2021; Stone Motes and McCartt Hess, 2007), they seem collectively hellbent on continuing down the beaten path. Without the intensive engagement of external partners, many international organisations would dwindle, and many international experts would be unemployed (Hagelsteen & Becker, 2019, p. 8). However, leaving the driver's seat to internal partners and stepping back has not only potential implications for money and status but is fundamentally about relinquishing power.

Power imbalances

Most fundamentally, the kernel of the problems in capacity development concerns the unequal distribution of power between the actors involved. This inequality is not arbitrary or by chance but systematic and intrinsic to the current order (Figure 14.13).

Figure 14.13 Who is most powerful?

All relationships between human beings involve power (Hearn, 2012, p. 217), and power is an inherent part of relationships between organisations (Brinkerhoff, 2002, p. 217). Yet, power is often overlooked in capacity development (Degnbol-Martinussen & Engberg-Pedersen, 2003, p. 5). This general obliviousness to such a fundamental feature of our human existence may be surprising but is a result of external partners almost exclusively being the more powerful (Lopes & Theisohn, 2003, pp. 41–44) and century-old colonial constructions of how such relationships 'should' be. In other words, power is invisible since it follows preconceived notions of how most people expect it to be distributed—a result of prevailing mental images of 'us' and 'them' amongst both external and internal partners, which often materialises in overly confident postures of the former and lack of self-esteem of the latter (Lopes & Theisohn, 2003). Moreover, in a partnership for capacity development, power can influence what needs others express and cause them to behave in the interests of the more powerful (Brinkerhoff, 2002, p. 217).

First, power in capacity development largely follows the money, where donors are on top and beneficiaries are at the bottom. Between is a hierarchy of actors controlling the funding downwards along the aid chain (Hagelsteen & Becker, 2019, p. 8). The actors controlling the funding are, thus, capable of forcing their will onto the actors below, which is what Lukes (2005) refers to as the first dimension of power. However, they can also use their power to set the agenda for what can be discussed and addressed within the partnership, which is the second dimension of power (Lukes, 2005). Finally, the more powerful actors are also more able even to influence the preferences, aspirations and motives of the less powerful to suit their interests—the third dimension of power (Lukes, 2005). However, it is essential to remember that power is not only about domination—exercised coercion, force and oppression—but also about the legitimate authority that is voluntarily obeyed (Hearn, 2012, p. 23).

Sometimes authority can be rooted in tradition (Weber, 1947, pp. 341–358)—for example, the authority of the British Monarchy, Roman Catholic Church, Fijian House of Chiefs—or in 'the specific and exceptional sanctity, heroism or exemplary character of an individual' (Weber, 1947, p. 328)—for example, Gandhi, King and Mandela. There are myriad less prominent examples of such traditional and charismatic authority in the many more localised traditions and informal leaders with significant power out there. However, other forms of authority are more relevant for the pronounced power imbalances in capacity development.

The power imbalances in capacity development stem not only from control over funding, which can be referred to as induced authority (Wrong, 1980) or reward power (French & Raven, 1959) when the powerful can make others obey voluntarily if that would lead to some reward. Whilst the power imbalances could also be rooted in the legal and institutional frameworks—codified or not—that determine rules, roles, responsibilities, etc. (Weber, 1947), the formal agreements concerning capacity development explicitly call for balancing power (e.g., OECD, 2009). The power imbalances are also rooted in the recognition of specialised knowledge and skills of experts, which can be called competent authority (Wrong, 1980) or expert power (French & Raven, 1959) and are currently seen as residing mainly with external partners due to narrow definitions of what constitutes expertise (Hagelsteen et al., 2021, pp. 5–6). Finally, they can stem from what French and Raven (1959) call informational power, in which one partner has access to certain information that another partner needs or wants or is important for the capacity development initiative. This kind of authority is also commonly biased in favour of external partners regardless of internal partners often having certain information external partners should need or want, which could be vital for the initiative because they simply tend to overlook such input (Hagelsteen et al., 2021). Whilst new donors are emerging and expertise is accumulating in all corners of the world, the balance of power is still heavily biased in favour of donors and external partners in its most affluent parts.

However, power is not only a matter of one actor having power over another but also of the power of the collective to do something (Morriss, 2002, p. 32)—two sides of power that are intrinsically linked (Hearn, 2012, pp. 6–7). Consider, for instance, a meeting of the Tanzania Disaster Relief Committee (TANDREC), which is a committee with decision-making powers consisting of high-level officials from the prime minister's office and relevant ministries. This committee has substantial power to allocate the resources made available by the president when declaring a disaster area, direct activities towards specific locations, etc. Between each of the members of this committee, there are relationships of power that may be balanced or more or less asymmetrical (see Wrong, 1980, pp. 10–13), granting some members power over other members in the context of the meeting. These members have greater potential for influencing the committee's decisions, but without such imbalances, it might be difficult for the committee to decide anything at all and coordinate their activities.

There are more or less egalitarian human-environment systems (Hofstede, 1983), but it is difficult to imagine a functioning society without some kind of social contract conferring authority to some actors overseeing and facilitating human transactions. In other words, the power of the collective to perform key functions for resilience depends on—like it or not—some actors having power over other actors. Each context has its specific nexus of power relations between partners, often including different forms of domination and authority. Partners may try to force their will at meetings or control what is discussed in the first place. Internal partners generally have legal authority in the context of their area of responsibility—be it a country, municipality or functional sector—whilst external partners may come with financial resources, formal expertise and vital information. If we are serious about facilitating local ownership and strong partnerships, we must do our utmost to navigate this minefield of power.

Designing capacity development

After elaborating on defining capacity development and pointing out the current practices that require radical deconditioning, it is time to focus on designing capacity development. However, designing capacity development is not a new challenge, but has baffled development professionals for more than half a century. In 1969, the US Agency for International Development (USAID) commissioned a study that found several common challenges—that is, vague planning, unclear responsibilities, and difficult evaluation (Rosenberg & Posner, 1979). It sparked the development of the logical framework approach (LFA), which has been adopted, adapted, and altered in several steps by numerous organisations and spread all over the world (e.g., AusAID, 2005; Breuer et al., 2016; CIDA, 2008; Göbel & Helming, 1997; Mayne, 2007; NORAD, 1999; Örtengren, 2003). These design frameworks have been called different names over the years, such as the LFA, Goal-Oriented Project Planning (GOPP), Results-Based Management (RBM), and Theory of Change (Figure 14.14). Although there are noticeable differences between them as the originators of each new framework tried to address perceived shortcomings in its predecessor, the common challenges identified in the USAID-study persisted and still do.

That is not to say that these frameworks are useless. Quite the contrary, there is nothing critically wrong with the frameworks themselves. It is how they are applied that is usually problematic, resulting in pre-existing blueprint solutions disguised as the results of a thorough process (Gasper, 2000), disconnected project-based interventions that simplify the actual problem (Tendler, 2002), sets of activities that omit vital parts that are required to generate the intended results (Gasper, 2000), and linear thinking and overlooked complexity (e.g., Davies, 2004, 2005). These problems largely remain regardless of what the frameworks are called.

Pick whichever framework you feel comfortable with, or use the one your organisation or prospective donor requires. It will likely provide you with several steps to guide the design process. Make sure to address all steps, but understand that completing them is not a linear process since there are many interdependencies requiring iteration (Davies, 2004, 2005). The rest of this chapter is intended to help overcome the challenges currently undermining capacity development by providing practical guidelines for addressing common problems in the design of capacity development initiatives. It is not repeating every problem identified previously, so keep them in mind when engaging in the design process. These guidelines are divided into the following seven categories: (1) the human-environment system, its problems and the intervention, (2) roles and responsibilities, (3) relationships, trust and power, (4) mutual learning, (5) requisite institutionalisation, (6) explicit risk sharing and (7) a double design process.

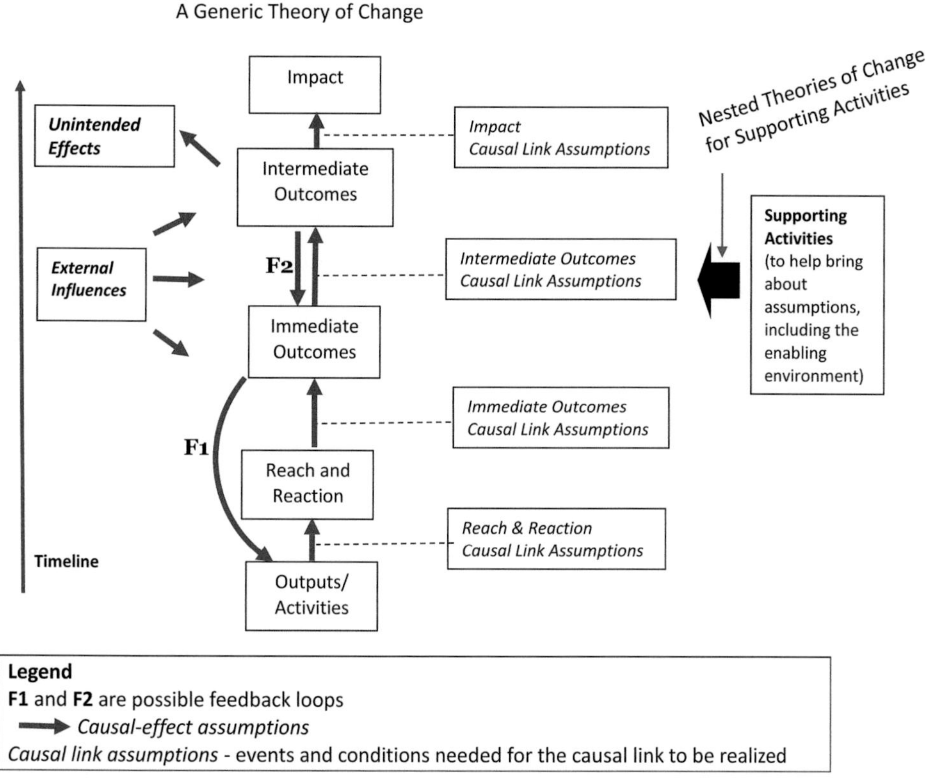

Figure 14.14 A generic theory of change.

The human-environment system, its problems and the intervention

To be able to develop whatever capacity of a human-environment system at all, the involved actors must understand its current state and what the problems are, decide its desired state and how to get there, and design purposeful activities to do it (compare with the definition of development in Chapter 7). Without any one of these parts, capacity development would fumble and most likely fail. All design frameworks mentioned spend most of their steps helping the actors accomplish precisely that. For instance, the first five steps of particular versions of the LFA include what they refer to as (1) analysis of project context, (2) stakeholder analysis, (3) situation analysis, (4) objectives analysis and (5) plan of activities (Becker & Abrahamsson, 2012; Örtengren, 2003), whilst some versions of Theory of Change are very similar but with additional emphasis on the mechanisms of change and sequencing of activities (e.g., Noble, 2019). Yet, this is where most capacity development initiatives founder, as described earlier in the chapter. Instead of blaming the design frameworks, it is helpful to consider the following when utilising them to design capacity development initiatives.

First, remember that the human-environment systems of interest within the context of this book are complex (Chapter 9) and that their resilience is an emergent property of the performance of a hierarchy of functions, which emerge in particular assemblages of perceivable elements and their interrelations—the knowledge, skills, tools and other resources, the organisation of them on all levels, the rules, regulations and other formal institutions, and the norms, values and other informal institutions (Chapter 8). It means it is rarely possible to develop capacity for resilience, or for doing anything but the simplest things, without identifying and addressing *all* the parts and links between parts necessary for that (Becker, 2021). Otherwise, the capacity development initiative will just waste resources since the aspired performance can never emerge if only addressing some arbitrary parts and links whilst other essential aspects are missing (Becker, 2009). For example, educating staff does not help if the organisation cannot coordinate them and utilise their new knowledge (Eade, 1997) and technical equipment is unusable if nobody knows how to operate or maintain it (Becker, 2009).

Regardless of whether applying some LFA or Theory of Change, the involved actors must be relentless in seeking a deeper understanding of how the human-environment system works, what really constitutes the aspired capacity in that particular context, what the problems are, who should be involved to address them, etc., and not simply settling for what worked elsewhere or comes first to mind. They must keep enquiring into these aspects and probing their interdependencies. Not to belittle the vital importance of this process, but the involved actors must be like inquisitive children, never really settling for the answer and always following up with additional who, what, where, when, why and how—for example, *What else is needed?*, *Who must do that?* and *How are they dependent?* Only armed with a deeper, systemic understanding, do they have a real chance of designing successful capacity development. Otherwise, it is simply dumb luck if the activities just happen to address what is needed (Figure 14.15).

It is important to note that having a deeper, systemic understanding does not entail trying to address everything. That could require unlimited resources. Quite the contrary, such understanding actually makes it possible to design capacity development initiatives that can meet their overall purposes with the limited resources available. Each initiative still needs to address *all* necessary parts and links to generate its objectives, but with this essential understanding, the involved actors can design a capacity development initiative that balances its objectives and activities with the resources available to still meet its overall purpose (Becker, 2021). However, it is essential also to consider if the overall purpose is meaningful in itself and not only in combination with other things that may not materialise.

Whilst most design frameworks include explicit consideration of what actors are needed to be involved and which other stakeholders could influence or be influenced by the capacity development initiative, it is essential that such considerations are exercised continuously throughout the entire process. It is not an early step of the design

Figure 14.15 Missing one piece can ruin the entire effect. *(Shared by Willi Heidelbach on the Creative Commons.)*

framework to tick off and be done with, although these considerations must commence at the beginning (Örtengren, 2003). There are commonly a relatively small number of actors initiating the design of a particular capacity development initiative, who generally identify various numbers of other potential actors or stakeholders from their perspective. However, as the involved actors develop their understanding of the human-environment system, what the problems are and what to do about them, they are likely to realise that they need to involve others—especially when realising there are vital interdependencies across boundaries and borders. It could also be that particular actors realise they have nothing relevant to contribute or no interest in the initiative and withdraw from the process. Regardless, the design process must not only be participatory and involve the right people, but the group of participants must also be open and actively adapt its membership to meet its changing needs throughout the process.

In addition to considering complexity and the importance of deep, systemic understanding, actors involved in capacity development must also remember the dynamic nature of the world. Most capacity development initiatives ignore that things change regardless of what is done or not done within them (the assumption of ceteris paribus in Chapter 9) and are often planned out in detail from the start with slight chances to adapt to changing circumstances over time (Hagelsteen & Becker, 2019). Moreover, it is almost certain that capacity development initiatives will not go as initially planned, irrespective of what goes on around them, and flexibility is indispensable regardless of how comprehensive internal risk analyses have been done. It is, therefore, crucial that the participating actors allow for flexibility in the design of the capacity development and monitor the organisation, community, or society they focus on and the environment around it to identify and adapt the activities to the changing circumstances (Becker, 2021). If continuing to ignore the dynamic change in and around the human-environment system in question, there is imminent danger of wasting resources on activities that are no longer necessary and ignoring new problems as they arise. Thus, the ability of the involved actors to strike a balance between the parts and the whole, as well as between control and flexibility, is a key component of effective capacity development for resilience (Hagelsteen & Becker, 2019).

Finally, the partners in capacity development must be ready to motivate their decisions in the design process and to dissuade funding cuts, which are commonplace with some donors. Some donors even seem to routinely push budgets down without removing any planned activities, which may partly be explained by the fact that there are partners inflating their budgets on purpose. However, these donors must realise that cutting one part of the capacity development initiative may undermine its entire chance of success, and the partners inflating their budgets must realise that this practice undermines donors' trust in general. If forced into haggling over budgets, the partners in capacity development must never accept unsolicited cuts of specific parts selected by the donor without appropriate systemic understanding, nor should they accept proportional cuts across the board, reducing their ability to implement all planned activities. Instead, when entering a budget negotiation, the partners must bring with them several plans covering different funding scenarios, ensuring that each comprises a set of well-motivated purpose, objectives and activities in relation to each anticipated level of funding. That way, the partners are not only more likely to exit the negotiations with a balanced capacity development initiative but also better equipped for an informed dialogue with their prospective donors.

Roles and responsibilities

Virtually everybody involved in capacity development is aware of the paramount importance of mutual partnership and local ownership. Yet, as described in previous sections, most capacity development initiatives struggle with both. Partnership requires clear and mutually agreed roles and responsibilities for all partners, and local ownership requires internal partners to drive their capacity development whilst external partners support. This support can take many forms, from providing technical assistance to facilitating capacity development processes (Hagelsteen & Becker, 2013, p. 5), and internal partners also have a range of roles they can assume in partnerships for capacity development. However, most design frameworks focus more on different types of stakeholders—for example, beneficiaries, decision-makers, implementers and financiers (Örtengren, 2003)—than on various roles in capacity development per se. To enable roles and responsibilities that facilitate capacity development, it is helpful to consider the following when applying whichever framework when designing such initiatives.

First, it is vital to appreciate that holding the pen when designing a capacity development initiative confers extraordinary power over the process. It grants a critical prerogative to interpret frequently ambiguous signals from diverse actors, thus facilitating for the author to bias the design of the capacity development towards her own preferences—knowingly or unknowingly. As long as external partners are exclusively holding the pen and communicating with donors and other actors, local ownership is virtually impossible. Local ownership requires, in other words, more than simply involving internal partners in the process. It requires local authorship, or at least shared authorship of capacity development proposals and reports, in which internal partners fully engage in all aspects of the process.

It is also essential to contemplate who should be considered an expert (Box 14.2). Remember that not only formal expertise is vital here, as repeatedly pointed out in the second part of this book. For example, if the members of a disaster-prone community are invited to share their understanding and experiences of the situation the initiative is intended to address, this does not only bring vital local knowledge of the issues but is also likely to spur interest, commitment and ownership when the community members see how their input influence the design of the capacity development initiative. It is also vital to remember that the internal partners have essential expertise that no external partner can match, regardless of their prowess (Hagelsteen et al., 2021, pp. 5–6)—more on that in the next section.

There are also other, perhaps less obvious, roles and responsibilities that are crucial to consider when designing capacity development initiatives. First of all, the broad participation of actors demands explicit assignment of the responsibility for assuring that the

BOX 14.2 Who is the expert?

I may be an expert in risk assessment, but I do not have the same immediate appreciation of what heavy rainfall in the Ukaguru Mountains in Tanzania will cause in a matter of days, as an illiterate subsistent farmer outside the town of Kilosa. I may have concepts and tools for estimating likelihood and consequences, but I am dependent on the farmer to tell me when the river Mkondoa has flooded the area in the past and describe to me what happened then. The narratives of the farmer and others having extensive experience of living there for decades, even centuries as a community, are fundamental input for my work and should never be diminished. We always have to ask ourselves what we are experts in and accept that what we find is likely to be rather narrow in relation to the bigger picture of the particular situation we are in. I may be an expert in risk assessment, but I am not an expert in the risk landscape that the farmer and his community live in. Quite the opposite. However, it is okay to be an expert in one thing and a moron in another. Actually, I think true expertise is founded on our aptitude to accept that and make it explicit in our communication with others. Humility has nothing to do with being humiliated, even if the two words have the same etymological root.

capacity development will result in the expected improvement (Ulrich, 2000, p. 258). In other words, all involved actors should know where they could seek some guarantee that the planned activities will achieve improvement. This responsibility needs to be assigned to one of the more influential actors involved or to a group of them (Becker & Abrahamsson, 2012). Otherwise, the overall responsibility for the capacity development initiative is blurred, and accountability is severely reduced.

Moreover, it is important for legitimacy to also include some actor who explicitly argues the case of those who cannot speak for themselves and who seeks the emancipation of those affected but not involved (Ulrich, 2000, p. 258)—e.g., marginalised groups, future generations, the environment, etc. Without such an advocate, the risk of the capacity development initiative causing collateral damage increases significantly. It is obviously impossible to consider the well-being of all possible actors or aspects of human-environment systems across space and time, but explicitly assigning an advocate for that purpose is a big step in the right direction.

The next set of roles to consider concerns the roles of external partners whilst implementing the capacity development initiative being designed. Whilst there are many roles suggested (e.g., Champion et al., 2010), Hagelsteen et al. (2021, p. 7) suggest five general roles that external partners can assume when engaging in capacity development—that is, expert, adviser, teacher, facilitator and coach—varying in balance between implementing activities and supporting the growth of the internal partner (Figure 14.16). At one end of the spectrum, the expert performs the task and solves the problem herself, and at the other, the coach lets the internal partner perform the task and solve the problem whilst observing, providing feedback and discussing the pros and cons with different alternatives (cf. Champion et al., 2010, p. 60). In short, the expert focuses more or less exclusively on the implementation of activities, whilst the coach concentrates almost entirely on the internal partner's growth (Figure 14.16). Between these opposites, you find the adviser assisting the internal partner in solving a specific problem with the adviser's knowledge and experience, the teacher explaining basic principles and the skills required to conduct a general task and solve a problem, and the facilitator facilitating brainstorms, planning

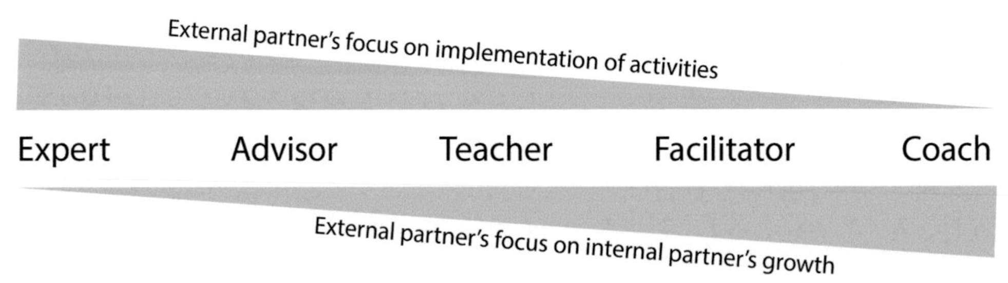

Figure 14.16 Five general roles and the focus of external partners assuming them.

and meetings (Hagelsteen et al., 2021, p. 7). For each of these three intermediate roles, the focus on the implementation of activities decreases, and the focus on the internal partner's growth increases (Figure 14.16).

It is important to note that all five roles are useful in different situations, at different times, and for different purposes (cf. Champion et al., 2010, pp. 61–62), and the best role external partners could assume depends on the needs of the internal partners and what support the external partners are capable of providing (Flaspohler et al., 2007, p. 117). A role needed at the start may not be helpful later, and a role that could become vital relatively far into a partnership may not work at all if assumed too early (Hagelsteen et al., 2021, p. 7). However, the need for roles focusing more and more on the internal partner's growth could be seen as a proxy indicator of project progression (cf. Bolger, 2000, pp. 5–6; Yachkaschi, 2010, p. 201). To meet the internal partner's changing needs, the external partner's role should be openly discussed and routinely renegotiated (Hagelsteen et al., 2021, p. 7).

Many highlight the importance of leadership for capacity development to work (e.g., Chaskin, 2001, p. 292; Lopes & Theisohn, 2003, pp. 35–40; Hagelsteen et al., 2021, p. 5). Good leadership is not only crucial for recognising needs and allocating resources in terms of time, funds, equipment and personnel (Floyd, 2007, pp. 101–106), but also for making change happen by having 'the courage to take risks, expand implementation, overcome obstacles and empower others' (Figure 14.17) (Lopes & Theisohn, 2003, p. 35). Poor leadership, on the other hand, can rapidly destroy the slow incremental results of decades of patient capacity development, for which there are countless examples (e.g., Bolger, 2000, p. 3; Godfrey et al., 2002, p. 357; Hope, 2009; Jae Moon & Wu, 2022; Lopes & Theisohn, 2003, p. 115). Whilst there is no doubt that poor leadership is a

Figure 14.17 Leadership to empower others. *(Photo by Magnus Hagelsteen.)*

significant challenge for capacity development for resilience, it is important not to blame all problems on poor leadership as soon as something goes wrong (Mitchell, 2006, p. 236).

Good leadership in capacity development is about enhancing understanding, improving relationships, generating commitment and fostering ownership (Lopes & Theisohn, 2003, p. 40). However, leadership may need to transform throughout the capacity development process. For instance, leaders in the first embryonic phase of capacity development initiatives may need to be charismatic and inspiring to mobilise others (Floyd, 2007, p. 102), but such leadership may undermine commitment and ownership later as too much of the initiative revolve around the charismatic leader. Also, whilst strong charismatic leaders may force radical transformation that is apparently successful, these gains are often fragile as they depend on the continuous presence of the leaders not to fade away (Fullan, 1992, p. 19). Then, it may be more appropriate with more facilitating and shared leadership (Floyd, 2007, pp. 102–103), as vision can be blinding if it comes from one pair of eyes only (Fullan, 1992). Moreover, good leadership is not about being an instructional leader but the coordinator of many instructional leaders (Glickman, 1991). In other words, good leadership is about making the construction of vision into a collective exercise (Fullan, 1992, p. 20), making explicit activities to develop local leadership vital for capacity development (Chaskin, 2001; Kubisch et al., 2002)—especially since leadership must be sustained over time not to lose vital momentum in the capacity development process. In short, good leadership for capacity development is adaptive, shared and sustained, which most often demands a core group of several leaders with different skills, perspectives and leadership styles (Floyd, 2007, pp. 102–103).

Finally, leadership does not necessarily rely on formal management structures in complex human-environment systems (Schneider & Somers, 2006, p. 356). However, such cases generally involve an intrinsic conflict between formal and informal leadership that is often difficult to grasp for external partners (Kubisch et al., 2002, p. 36). It is here that the role of a champion comes in (Brinkerhoff, 2002, pp. 220–223), with individuals vigorously supporting the capacity development process from the inside (Agranoff, 2005, p. 34; Schacter, 2000, pp. 7–8; Westoby & Blerk, 2012, p. 1093). Although the role of the champions may shift over time (Mahanty et al., 2009, p. 864), champions are generally 'entrepreneurial individuals who advocate on behalf of the partnership and the partnership approach within their home organizations, within the partnership as a whole, and externally' (Brinkerhoff, 2002, pp. 220–223). However, championship entails not only strong abilities to communicate, negotiate and organise various aspects of capacity development but also legitimacy amongst all partners (Brinkerhoff, 2002). It is even suggested that external partners should focus their support on internal partners with already identified champions who are willing and able to contribute to making change happen even in situations of resistance (Schacter, 2000, p. 16; Mahanty et al., 2009, p. 861).

Relationships, trust and power

As repeatedly indicated above, the relationship between partners is fundamental for the success of any capacity development initiative. Although it is possible to talk about relationships between organisations, or even international relations, it is essential to remember that these relationships are always built on personal relationships between individuals (Ahrne, 1994). However, the technocrats commonly involved in capacity development for resilience often forget this and focus exclusively on the tasks at hand, ignoring the importance of building relationships of trust between partners.

Trust is a multifaceted concept applied in various situations and scientific disciplines (Blomqvist, 1997; Gambetta, 1988). One of the more common approaches to trust links the concept to 'the willingness to be vulnerable under conditions of risk and interdependence' (Rousseau et al., 1998, p. 395). This approach to trust entails three crucial things when considering two individuals involved in capacity development. First of all, the concept of trust does not make any sense if there is no relationship between the two individuals in the first place. In other words, if both are just minding their own business and are not in any way dependent on what the other one is doing, trust has no meaning. Moreover, the concept of trust also has no meaning if none of the two risk anything in the relationship with the other. In short, if there is no likelihood that the actions of one could have any negative consequences on something that the other values, then trust is not important. Finally, when one individual is dependent on another for the state of something she values, trust entails consciously making herself vulnerable to the impact of the other person's agency.

Returning to capacity development, it is evident that partners are interdependent, and everybody takes risks by sharing information, acknowledging problems, linkup activities, etc. Without trust, partners are not willing to do any of that, which in turn undermines the effectiveness of the capacity development initiative completely. Trust is, as such, essential for capacity development (Hagelsteen & Becker, 2013, p. 10). It is, therefore, somewhat shocking that such a fundamental requisite is so often forgotten or ignored—especially since that mistake generally is repeated over and over again (Box 14.3). Building trust takes time whilst destroying trust is instantaneous. Open communication channels and continuity in personal relations support mutual trust or even friendship (Bass, 1990), which I return to in a moment. Girgis (2007, p. 357) takes this one step further and points to the value of spending non-professional time together to facilitate trust—not only meeting in the office and discussing professional matters but having a cup of tea, dinner or a run together, and talking about family, sports or whatever interests you may share. Here, most external partners have much to learn, with their usual professional fixation and fix-it attitudes.

There is a particular type of relationship that involves all the requisites for facilitating partnership for effective capacity development: friendship. Friendship is 'a state of mutual trust and support' (Oxford English Dictionary, 2020) and has been suggested as an

BOX 14.3 A steady stream of strangers

Personal relationships of mutual trust are fundamental requisites for healthy partnerships in capacity development for resilience. It is therefore particularly upsetting to find countless examples of situations in which the formation of such relationships is completely undermined by the basic design of how capacity development is managed in practice. First of all, capacity development for resilience should be a long-term process of change, but is regularly restricted by short-term funding for 12–18 months. It should be development-oriented with a substantial planning horizon for its activities, but each activity is often implemented in a rushed and at times even ad hoc manner. Most external partners have also very rapid staff turnover, with people on temporary contracts who get used to move around. These three factors result in a steady stream of strangers moving in and out of partnerships in capacity development for resilience, with no time to build trustful relationships. As long as external partners ignore this fundamental flaw, the effectiveness of their projects and programmes will be severely restricted.

explicit part of effective capacity development frameworks (Girgis, 2007). Friendship is built on trust. It entails mutual support between friends. It is generally conceived as one of the more egalitarian relationships regarding power possible between human beings. However, friendship also involves affection and loyalty, which may complicate things when the situation calls for uncomfortable decisions for the partners. Although I have not found any studies of the balance between the pros and cons of friendship in capacity development, I think the positive sides of mutual trust and support between equals outweigh the negative sides of potentially being overly lenient when dealing with problems in the partnership. Close personal friendship is not a requisite for capacity development. Still, enmity is definitely damaging, indicating at least some level of rapport as a minimum for capacity development to work at all.

Trust is not the only fundamental aspect of the relationships between partners in capacity development. Power is also paramount and, unfortunately, almost always imbalanced towards the external partner and upwards along the aid chain, as shown in earlier sections. We can never end power and should definitely not ignore it (Hearn, 2012, p. 217). Instead, we must explicitly consider power in the design, implementation and evaluation of capacity development and do our utmost to rebalance it towards the internal partners (Hagelsteen et al., 2021, p. 7; Hagelsteen et al., 2022, pp. 7–9). As long as powerful donors and external partners remain reluctant to surrender some of their power, capacity development will fail miserably. Therefore, we must be vigilant and resolute in identifying and addressing power relations that undermine capacity development, using the framework presented earlier in the chapter. We must be mindful of the immense power stemming from controlling the economic resources needed to do anything and dare to reform the financial arrangements to improve the balance of power between the partners. We must ensure that internal partners are directly involved in writing proposals and reports, as well as in the communication with donors and other key actors around the partnership. It is also crucial that internal partners, to the furthest possible extent, have direct access to vital information and not only mediated access through external partners. Another important way of rebalancing capacity development partnerships in terms of power is to acknowledge that external and internal partners are equally competent, although their abilities may differ in type (Hagelsteen et al., 2021, pp. 5–6), which I elaborate on in the next section.

Mutual learning

Mutual engagement has been suggested to facilitate trust between partners (UNDP, 2008, p. 22) and mutual learning as fundamental to capacity development (Hagelsteen & Becker, 2019). However, mutual learning is talked about so much yet happens so rarely it has become a cliché (Hagelsteen et al., 2021, p. 6). Individuals engaging in some activities always learn something, but mutual learning in the context of capacity development requires both internal and external partners to learn something significant for them. Moreover, mutual learning in this context occurs within each individual and in the relationships between them (Wenger, 1998). Hagelsteen et al. (2021) explain the general lack of mutual learning in terms of biased attention to particular kinds of competencies, which can be categorised as either technical, processual or contextual.

Technical competence concerns the ability to perform the required technical activities (CADRI, 2011, p. 11), such as a risk assessment or preparedness planning. Processual competence concerns the ability to both drive the project and organisation as a whole and to facilitate capacity development processes, with the former part comprising general organisational and project management skills—for example, ability to assess, plan, formulate, implement and evaluate visions, policies and strategies and manage resources (UNDRR & Coppola, 2019, p. 20)—and the latter comprising more social, relational, intangible and invisible abilities—for example, leadership, learning, self-reflection, conflict resolution, intercultural communication, change management, problem-solving, negotiation and relational skills (Acquaye-Baddoo, 2010, pp. 66–70; Hagelsteen & Burke, 2016, p. 50). Finally, contextual competence concerns the ability to understand the local context and the existing capacities and needs, which is not explicitly part of other influential frameworks for understanding capacity (e.g., Pearson, 2011a; UNDRR & Coppola, 2019). These three types of competencies are interdependent of each other (Figure 14.18), resulting in the lack of one undermining the utility of another (Hagelsteen et al., 2021, pp. 5–6).

The lack of mutual learning in capacity development can at least partly be explained by an overwhelming focus on technical competencies the internal partners are expected

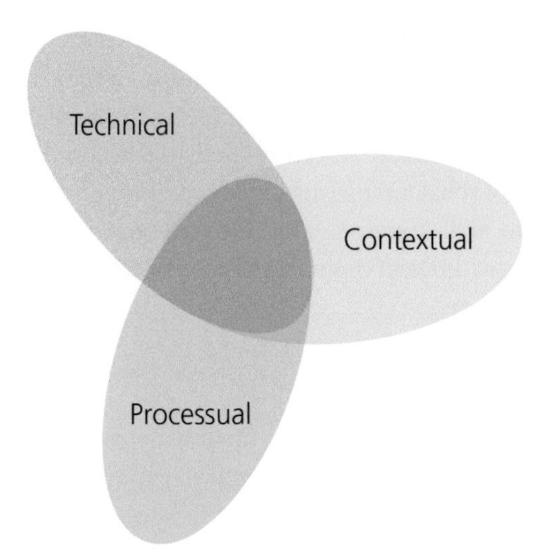

Figure 14.18 Three requisite competencies.

to need and desire and the external partners are expected to have and deliver (Hagelsteen et al., 2021, pp. 5–6). Whilst there are technical details that even the most knowledgeable external partner can learn from supporting internal partners, as well as internal partners with great technical expertise, there is an imbalance in technical competence by default as the entire raison d'être of capacity development initiatives is to focus on what the internal partners lack. Related to this raison d'être is the imbalance in competent authority discussed above. This imbalance is sometimes reversed when external partners send underqualified staff without the technical competence the internal partners expect or inexperienced staff with textbook knowledge but insufficient experience and maturity (Hagelsteen et al., 2021, p. 6). However, internal partners also often send either junior staff to participate in capacity development activities due to language issues or more experienced staff being busy (Hagelsteen & Becker, 2019, p. 5), or the wrong staff if participation is rewarded with lucrative per diems and status as mentioned earlier (Hagelsteen et al., 2021, p. 6). Although that may cater for mutual learning on the individual level, it is not helpful to invest in the technical competencies of individuals who are not in a position to contribute to change on the organisational level and is unlikely to lead to capacity development. For real mutual learning to be possible, the other two types of competencies must also count and receive explicit attention (Hagelsteen et al., 2021, p. 6).

Given that external partners have the necessary technical competence and the processual competence to facilitate capacity development processes, they still need to understand and adapt to the local context (Hagelsteen et al., 2021, p. 6). This contextual competence is regularly overlooked in the contemporary discourse of capacity development as something external partners can deliver and must be learnt over time in each context (Lavergne & Saxby, 2001, p. 10). Internal partners are clearly the experts in this regard, and external partners must learn from them (Hagelsteen et al., 2021, p. 6). Explicitly recognising external partners' need for developing contextual competence may inspire them to listen more to their partners and question entrenched assumptions about them and the context.

In short, capacity development can involve mutual learning only if both technical and contextual competence receive explicit attention, with external partners usually contributing most of the former and learning the latter and internal partners usually contributing most of the latter and learning the former (Hagelsteen et al., 2021, p. 6). Some internal partners have immense technical competence, and some external partners have substantial experience working and living in the context, but explicitly acknowledging the equal importance of both types of competence facilitates mutual learning and more equal partnerships.

Moreover, mutual learning can only happen through communicating experiences and social participation amongst partners (Wenger, 1998). This more relational notion of learning seems essentially overlooked, which may perhaps be explained by the general lack of organised ways of sharing experiences (cf. Wenger, 2009, p. 214). All learning starts with a disjuncture between what the partners know and currently experience (Jarvis, 2009, pp. 25—30), in which they realise that their entrenched ideas and habitual actions are no longer working (Elkjær, 2009, p. 83). For mutual learning to occur, the partners must not only acknowledge their respective contributions and unlearn old convictions and habits but set aside time for open dialogue about their perceptions and perspectives on the past, present, and future of their partnership (Hagelsteen et al., 2021).

Requisite institutionalisation

As stated in the section deconditioning capacity development, most initiatives fail partly because they only focus on creating capacity and not on facilitating for the internal partners to utilise and retain their acquired knowledge and new equipment, improved policies and procedures, etc. Whatever results a capacity development initiative can generate must be institutionalised to have the slightest chance of being sustainable. The training courses or workshops cannot just be designed and delivered in an ad hoc, short-term and even one-off manner simply because that is controllable and easy to report output from. Whilst a specific training course may be needed to inject the required input to initiate a change process, it must be part of a comprehensive plan of how to ensure that the acquired knowledge is put to use (capacity utilisation) and can be maintained (capacity retention) in the human-environment system in question. Similarly, an improved policy or procedure is useless in itself if not implemented and will eventually lose its meaning if not updated (Figure 14.19). In short, whatever is created by the capacity development initiative must be institutionalised to have either any immediate or lasting

Figure 14.19 None of these policies are helpful if not implemented and updated.

impact, and this requisite institutionalisation is not an autonomous process but must generally be facilitated. To ensure that, it must virtually always be facilitated by the same capacity development initiative—at least to establish the practices—since it is improbable that some other initiative will step in afterwards.

Remembering the mechanisms of institutionalisation from earlier, it is of utmost importance to facilitate iteration or repetition of the developed practices. In the same way the mechanisms of increasing returns, commitments, and objectification (Figure 10.4) can bind actors to flawed practices, they can also institutionalise the appropriate developed practices. Investing time, energy and resources into applying these practices help the actors to cement them and demonstrate how the contribution of each actor is facilitated by all actors following the same practices (cf. North, 1990). After the practices have taken hold, new actors are motivated to adopt them as they appear to be commonly accepted (Scott, 2014). Moreover, performing the practices over and over again eventually creates normative expectations of professional groups to perform the practices (Selznick, 1992), which after even more repetition may become increasingly routinised until they are taken for granted (Berger & Luckmann, 1966). Although it is rarely feasible to reach such deep institutionalisation within the duration of a capacity development initiative, allowing the partners to apply what they have developed in at least some iterations is crucial for initiating the institutionalisation in a more appropriate direction.

For capacity retention, it is essential to include activities with the explicit purpose of facilitating for the human-environment system to continue to maintain, restore and adapt the developed capacity long after the initiative has ended (Hagelsteen et al., 2021). For instance, to develop the ability to maintain necessary equipment, provide the training course without external support, or update the policy or procedure. Some of these abilities can be developed within the internal partners' organisation itself, whilst there are other organisations more suitable to help retain other abilities. For instance, all but the smallest countries in the world have universities that can play an important role with their pedagogical knowhow and in their position to facilitate institutionalisation and sustainability of knowledge and skills through their education (Becker & van Niekerk, 2014). Instead of the conventional ad hoc, short-term and even one-off training courses delivered by external partners themselves, it is far better to collaborate with universities to facilitate the integration of the necessary content into their educational curricula. There are universities in many countries already offering education in various topics relevant to risk, resilience and sustainability (Becker & van Niekerk, 2014). By collaborating with them instead of going alone, external partners can not only ensure the retention of necessary knowledge and skills when the capacity development initiative is over but also facilitate scale-up and wider change as the steady stream of graduates enters the organisations tasked to do the actual work on the national to local levels.

Explicit risk sharing

To overcome misguided accountability and quixotic control, it is of utmost importance to address their root causes. Donors must provide conditions that allow partners in capacity development to work according to the principles, but they must also address the underlying reason why external partners are still stuck in their ways even when given such conditions. Hagelsteen et al. (2022) explain how donor staff often fear losing control and not allowing flexible, adaptable, and long-term capacity development as they are also accountable to others further up the aid chain and heavily constrained by rigid and burdensome administrative systems. They also explain how external partners' risk aversion, in relation to failing to produce and report the results they promised their donors when granted the funding, constrains them even when provided appropriate conditions for successful capacity development (Hagelsteen et al., 2022). Donor staff do not want to risk reprimands or being ridiculed or even dismissed if something goes wrong, and external partners' jobs often depend more or less entirely on external funding from a few donors that may stop funding them if not satisfied. Overcoming misguided accountability and quixotic control requires explicit risk-sharing agreements along the aid chain (Hagelsteen et al., 2022, p. 10), delineating the distribution of responsibilities whilst acknowledging the vital importance of the principles of capacity development and the fundamental challenges of complexity, uncertainty, ambiguity and dynamic change.

A double design process

Finally, capacity development initiatives are generally including only objectives, activities and indicators focusing on developing the abilities of the human-environment system to perform particular functions for resilience as such—with little or no

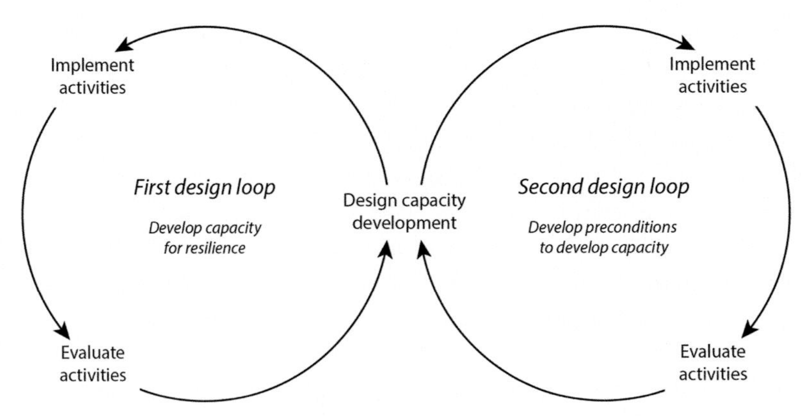

Figure 14.20 Capacity development as a double design process.

attention to the activities and results necessary for developing local ownership, mutual partnership, vital roles and responsibilities, trustful relationships and other aspects essential for effective capacity development (Hagelsteen & Burke, 2016). No wonder the overall results are generally insufficient. For capacity development for resilience to work, it is helpful to approach it as a double design problem, with the first design loop focusing on designing, implementing and evaluating sets of activities to develop the abilities of the human-environment system in question to perform particular functions for resilience, and the second design loop focusing on designing, implementing and evaluating sets of activities with the explicit purpose of engendering the preconditions necessary for capacity development to work (Figure 14.20). These preconditions require explicit attention and not just talking about them and hoping they are addressed as some fortunate by-product of the activities of the first loop. We must include explicit objectives, activities and indicators that directly address clear roles and responsibilities, allow time to build relationships of mutual trust, and develop softer skill sets. That way, the partners in capacity development must pay real attention to its principles, get a venue for dialogue about how to ensure them and must allocate time and resources for the work. Only then is there a chance that capacity development for resilience can become the key tool for sustainable development it has been intended to be since the beginning (Becker, 2021).

Conclusion

Capacity development is nothing new just because we started to call it that in the last decades or so. Its intellectual roots date back to the 1950s, and although we have learnt a lot over the decades, much of the theoretical underpinnings are still similar. The periodic name changes for what we now call capacity development have, in other words, been more reactions to poor implementation in practice than anything else. Considering the continuously poor results of much capacity development, it is interesting to speculate how long we will use this name to describe our efforts to support other actors in developing their capacities. However, I hope the concept is here to stay and that we focus our frustration and efforts on doing what we are supposed to do in theory instead of wasting more energy by continuing this seemingly perpetual renaming exercise.

Ad hoc and short-term activities do not suffice, regardless of what we call the initiatives. Instead, capacity development for resilience entails holistic and systematic action that addresses all significant parts and links in the causal chains behind particular challenges. It must address these challenges on all necessary levels, including developing knowledge, skills, tools and other resources; organisation on all levels; rules, regulations and other formal institutions; and norms, values and other informal institutions. To be able to do so, a capacity development initiative must be designed, implemented and evaluated with attention to dependencies in both the human-environment system in question and the capacity development initiative as such. There are many versions and adaptations of the original LFA that are valuable tools for accomplishing that, given that they are applied with explicit consideration of complexity, uncertainty, ambiguity and dynamic change. The guidelines suggested in this chapter are intended to help with that.

Finally, capacity development is better approached as a double design problem, in which the first design loop focus on designing, implementing and evaluating sets of activities to develop the abilities of the human-environment system to perform particular functions for resilience, and the second design loop focus on designing, implementing and evaluating sets of activities with the explicit purpose of engendering the preconditions necessary for capacity development to work. That way, capacity development has a much better chance of becoming the crucial process for the sustainability of humankind asserted in the current global frameworks (United Nations, 2015a,b).

References

Acquaye-Baddoo, N.-A. (2010). Thematic and change expertise: The balanced practitioner. In J. Ubels, N.-A. Acquaye-Baddoo, & A. Fowler (Eds.), *Capacity development in practice* (pp. 65–79). Earthscan.

African Union, NEPAD. (2012). *Africa's capacity development strategic framework*. NEPAD.

Agranoff, R. (2005). Managing collaborative performance: Changing the boundaries of the state. *Public Performance and Management Review, 29*, 18–45. https://doi.org/10.2307/20447574

Ahrne, G. (1994). *Social organizations: Interaction inside, outside and between organizations*. Sage.

Anderson, M. B. (1999). *Do no harm: How aid can support peace - or war*. Lynne Rienner Publishers.

Anderson, M. B., Brown, D., & Jean, I. (2012). *Time to listen: Hearing people on the receiving end of international aid*. CDA Collaborative Learning Projects.

Andrews, M., Pritchett, L., & Woolcock, M. (2013). Escaping capability Traps through problem driven iterative adaptation (PDIA). *World Development, 51*, 234–244. https://doi.org/10.1016/j.worlddev.2013.05.011

Argyris, C., & Schön, D. A. (1996). *Organizational learning II: Theory, method and practice.* Addison-Wesley Publishin.

Armstrong, J. (2013). *Improving international capacity development.* Palgrave Macmillan. https://doi.org/10.1057/9781137310118

Askman, J., Nilsson, O., & Becker, P. (2018). Why people live in flood-prone areas in Akuressa, Sri Lanka. *International Journal of Disaster Risk Science, 9*, 143–156. https://doi.org/10.1007/s13753-018-0167-8

AusAID. (2005). *AusGuideline 3.3 - the logical framework approach.*

Baser, H., & Morgan, P. J. (2008). *Capacity, change and performance, 59B* (pp. 1–157). ECDPM Discussion Paper.

Bass, B. M. (1990). *Bass and Stogdill's handbook of leadership: Theory, research, and managerial applications.* Free Press.

Becker, P. (2009). Grasping the hydra: The need for a holistic and systematic approach to disaster risk reduction. *Jàmbá: Journal of Disaster Risk Studies, 2*, 12–24.

Becker, P. (2012). The importance of integrating multiple administrative levels in capacity assessment for disaster risk reduction and climate change adaptation. *Disaster Prevention and Management, 21*, 226–233. https://doi.org/10.1108/09653561211220016

Becker, P. (2017). Dark side of development: Modernity, disaster risk and sustainable livelihoods in two coastal communities in Fiji. *Sustainability, 9*, 1–23. https://doi.org/10.3390/su9122315

Becker, P. (2021). Advancing resilience for sustainable development: A capacity development approach. In W. Leal Filho, R. Pretorius, & L. O. de Sousa (Eds.), *Sustainable development in Africa: Fostering sustainability in one of the world's most promising continents* (pp. 525–540). Springer. https://doi.org/10.1007/978-3-030-74693-3_29

Becker, P., & Abrahamsson, M. (2012). *Designing capacity development for disaster risk management: A logical framework approach.* MSB.

Becker, P., & Hagelsteen, M. (2016). Kapacitetsutveckling för katastrofriskreducering. In S. Baez Ullberg, & P. Becker (Eds.), *Katastrofriskreducering: Perspektiv, praktik, potential* (pp. 265–291). Studentlitteratur.

Becker, P., Hagelsteen, M., & Abrahamsson, M. (2021). 'Too many mice make no lining for their nest' — reasons and effects of parallel governmental structures for disaster risk reduction and climate change adaptation in Southern Africa. *Jàmbá Journal of Disaster Risk Studies, 13*(1). https://doi.org/10.4102/jamba.v13i1.1041

Becker, P., & van Niekerk, D. (2014). Developing sustainable capacity for disaster risk reduction in Southern Africa. In A. E. Collins, S. Jones, B. Manyena, & J. Jayawickrama (Eds.), *Hazards, risks and disasters in society* (pp. 63–78). Elsevier.

Berger, P. L., & Luckmann, T. (1966). *The social construction of reality.* Penguin Books.

Binger, A., Georgieva, K., Khosla, A., Makram-Ebeid, M., Savane, M.-A., & Umaña, A. (2002). *Capacity 21: Global evaluation 1993-2001.* UNDP.

Blomqvist, K. (1997). The many faces of trust. *Scandinavian Journal of Management, 13*, 271–286.

Bolger, J. (2000). Capacity development: Why, what and how. *Capacity Development Occasional Series, 1*, 1–8.

Bontenbal, M. (2009). Understanding North–South municipal partnership conditions for capacity development: A Dutch–Peruvian example. *Habitat International, 33*(1), 100–105. https://doi.org/10.1016/j.habitatint.2008.05.003

Boulding, K. E. (1975). Reflections on planning: The value of uncertainty. *Planning Review, 3*(2), 11–12. https://doi.org/10.1108/eb053705

Breuer, E., Lee, L., De Silva, M., & Lund, C. (2016). Using theory of change to design and evaluate public health interventions: A systematic review. *Implementation Science, 11*, 63. https://doi.org/10.1186/s13012-016-0422-6

Brinkerhoff, J. M. (2002). Assessing and improving partnership relationships and outcomes: A proposed framework. *Evaluation and Program Planning, 25*, 215–231.

Brinkerhoff, D. W., & Morgan, P. J. (2010). Capacity and capacity development: Coping with complexity. *Public Administration and Development, 30*, 2–10. https://doi.org/10.1002/pad.559

Buckle, P. (1998). Re-defining community and vulnerability in the context of emergency management. *Australian Journal of Emergency Management, 13*, 21–26.

CADRI. (2011). *Basics of capacity development for disaster risk reduction.*

Cameron, C., & Low, S. (2012). Aid-effectiveness and donor coordination from Paris to busan: A Cambodian case study. *Law and Development Review, 5*, 167–193. https://doi.org/10.1515/1943-3867.1155
Casey, C., & Delaney, H. (2022). The effort of partnership: Capacity development and moral capital in partnership for mutual gains. *Economic and Industrial Democracy, 43*(1), 52–71.
Champion, D. P., Kiel, D. H., & McLendon, J. A. (2010). Choosing a consulting role: Principles and dynamics of matching role to situation. In J. Ubels, N.-A. Acquaye-Baddoo, & A. Fowler (Eds.), *Capacity development in practice* (pp. 57–64). Earthscan.
Chaskin, R. J. (2001). Building community capacity: A definitional framework and case studies from a comprehensive community initiative. *Urban Affairs Review, 36*, 291–323. https://doi.org/10.1177/10780870122184876
CIDA. (2008). *Results-based management policy statement.*
Coetzee, C., Van Niekerk, D., & Raju, E. (2016). Disaster resilience and complex adaptive systems theory. *Disaster Prevention and Management, 25*(2), 196–211. https://doi.org/10.1108/dpm-07-2015-0153
Crewe, E., & Harrison, E. (1998). *Whose development? An ethnography of aid.* Zed Books.
Davies, R. (2004). Scale, complexity and the representation of Theories of change. *Evaluation, 10*(1), 101–121. https://doi.org/10.1177/1356389004043124
Davies, R. (2005). Scale, complexity and the representation of theories of change: Part II. *Evaluation, 11*(2), 133–149.
Degnbol-Martinussen, J., & Engberg-Pedersen, P. (2003). *Aid: Understanding international development cooperation.* Zed Books.
Denney, L. (2017). *$15bn is spent every year on training, with disappointing results. Why the aid industry needs to rethink 'capacity building'.* https://frompoverty.oxfam.org.uk/15bn-is-spent-every-year-on-aid-for-training-with-disappointing-results-why-the-aid-industry-needs-to-rethink-its-approach-to-capacity-building/.
Devereux, P. (2008). International volunteering for development and sustainability: Outdated paternalism or a radical response to globalisation. *Development in Practice, 18*, 357–370. https://doi.org/10.1080/09614520802030409
DFID. (2013). *How to note - capacity development.* DFID.
Eade, D. (1997). Capacity-building: An approach to people-centred development. Oxford American.
Edwards, M., & Hulme, D. (1992). Scaling-up the development impact of NGOs: Concepts and experiences. In M. Edwards, & D. Hulme (Eds.), *Making a difference: NGOs and development in changing world* (pp. 13–27). Earthscan.
Elkjær, B. (2009). Pragmatism: A learning theory for the future. In K. Illeris (Ed.), *Contemporary theories of learning: Learning theorists.in their own words* (pp. 74–89). Routledge.
Enemark, S., & Van der Molen, P. (2008). *Capacity assessment in land administration.* International Federation of Surveyors.
Eriksson Baaz, M. (2005). *The paternalism of partnership: A postcolonial reading of identity in development aid.* Zed Books.
Esman, M. J. (1967). *The institution building concepts: An interim appraisal.* University of Pittsburgh.
Esman, M. J., & Montgomery, J. D. (1969). Systems approaches to technical cooperation: The role of development administration. *Public Administration Review, 29*, 507–539. https://doi.org/10.2307/973472
Flaspohler, P., Ledgerwood, A., & Bowers Andrews, A. (2007). Putting it all together: Building capacity for strategic planning. In P. Stone Motes, & P. McCartt Hess (Eds.), *Collaborating with community-based organizations through consultation and technical assistance* (pp. 116–136). Columbia University Press.
Floyd, A. (2007). Collaboratives: Avenues to build community capacity. In P. Stone Motes, & P. McCartt Hess (Eds.), *Collaborating with community-based organizations through consultation and technical assistance* (pp. 81–115). Columbia University Press.
French, J. R., & Raven, B. (1959). The bases of social power. In D. Cartwright (Ed.), *Studies in social power* (pp. 150–167). University of Michigan Press.
Fullan, M. G. (1992). Visions that blind. *Educational Leadership, 49*, 19–22.
Gambetta, D. (1988). Can we trust trust. In *Trust: Making and breaking cooperative relations* (pp. 213–237). Basil Blackwell.

Gant, G. F. (1966). The institution building project. *International Review of Administrative Sciences, 32*, 219–225.
Gapminder. (2023). *Gapminder*. www.gapminder.org/data.
Gasper, D. (2000). Evaluating the 'logical framework approach' towards learning-oriented development evaluation. *Public Administration and Development, 20*(1), 17–28. https://doi.org/10.1002/1099-162x(200002)20:1<17::aid-pad89>3.0.co;2-5
Gillam, V. (1899). *The white Man's burden (apologies to Rudyard Kipling)* — Judge. Retrieved 2023-06-04.
Girgis, M. (2007). The capacity-building paradox: Using friendship to build capacity in the South. *Development in Practice, 17*, 353–366.
Glickman, C. (1991). Pretending not to know what we know. *Educational Leadership, 48*, 4–10.
Göbel, M., & Helming, S. (1997). *Ziel Orientierte Projekt Planung -ZOPP: Eine Orientierung für die Planung bei neuen und laufenden Projekten und Programmen*. GTZ.
Godfrey, M., Sophal, C., Kato, T., Piseth, L. V., Dorina, P., Saravy, T., Savora, T., & Sovannarith, S. (2002). Technical assistance and capacity development in an aid-dependent economy: The experience of Cambodia. *World Development, 30*, 355–373. https://doi.org/10.1016/S0305-750X(01)00121-8
Guijt, I. (2010). Accountability and learning. In J. Ubels, N.-A. Acquaye-Baddoo, & A. Fowler (Eds.), *Capacity development in practice* (pp. 277–291). Earthscan.
Hagelsteen, M., & Becker, P. (2012). *Seven elements for capacity development for disaster risk reduction*.
Hagelsteen, M., & Becker, P. (2013). Challenging disparities in capacity development for disaster risk reduction. *International Journal of Disaster Risk Reduction, 3*, 4–13. https://doi.org/10.1016/j.ijdrr.2012.11.001
Hagelsteen, M., & Becker, P. (2014). A great babylonian confusion: Terminological ambiguity in capacity development for disaster risk reduction in the international community. In *Proceedings of the 5th international disaster risk conference, Davos, Switzerland, 24-28/08/2014* (pp. 298–300).
Hagelsteen, M., & Becker, P. (2019). Systemic problems of capacity development for disaster risk reduction in a complex, uncertain, dynamic, and ambiguous world. *International Journal of Disaster Risk Reduction, 36*, 101102. https://doi.org/10.1016/j.ijdrr.2019.101102
Hagelsteen, M., Becker, P., & Abrahamsson, M. (2021). Troubling partnerships: Perspectives from the receiving end of capacity development. *International Journal of Disaster Risk Reduction, 59*, 102231. https://doi.org/10.1016/j.ijdrr.2021.102231
Hagelsteen, M., & Burke, J. (2016). Practical aspects of capacity development in the context of disaster risk reduction. *International Journal of Disaster Risk Reduction, 16*, 43–52. https://doi.org/10.1016/j.ijdrr.2016.01.010
Hagelsteen, M., Gutheil, J., Morales Burkle, M. D. M., & Becker, P. (2022). Caught between principles and politics: Challenges and opportunities for capacity development from governmental donors' perspectives. *International Journal of Disaster Risk Reduction, 70*, 102785. https://doi.org/10.1016/j.ijdrr.2022.102785
Hearn, J. (2012). *Theorizing power*. Palgrave Macmillan.
Hofstede, G. (1983). National cultures in four dimensions: A research-based theory of cultural differences among nations. *International Studies of Management and Organization, 13*, 46–74.
Honig, D., & Gulrajani, N. (2017). Making good on donors' desire to do development differently. *Third World Quarterly, 39*(1), 68–84. https://doi.org/10.1080/01436597.2017.1369030
Hope, K. R. S. (2009). Capacity development for good governance in developing societies: Lessons from the field. *Development in Practice, 19*, 79–86. https://doi.org/10.2307/27752012
IFRC. (2015). World Disaster Report 2015: Focus on local actors, the key to humanitarian effectiveness. *International Federation of Red Cross and Red Crescent Societies*.
IPCC. (2022). *Climate change 2022: Impacts, adaptation and vulnerability*. Cambridge University Press.
Jae Moon, M., & Wu, X. (2022). Sustaining Asia's development amidst the COVID-19 pandemic: Capacity development and governance innovation. *Journal of Asian Public Policy, 15*(2), 165–174. https://doi.org/10.1080/17516234.2021.2015850
Jarvis, P. (2009). Learning to be a person in society. In K. Illeris (Ed.), *Contemporary theories of learning: Learning theorists.in their own words* (pp. 21–34). Routledge.

Kotter, J. P., & Cohen, D. S. (2002). *The heart of change: Real-life stories of how people change their organizations.* Harvard Business Review Press. http://books.google.se/books?id=83-&hl=&source=gbs_api.
Krznaric, R. (2007). *How change happens: Interdisciplinary perspectives for human development.* Oxfam.
Kubisch, A. C., Auspos, P., Brown, P., Chaskin, R. J., Fullbright-Anderson, K., & Hamilton, R. (2002). *Voices from the field II: Reflections on comprehensive community change.* Aspen Institute.
Lavergne, R., & Saxby, J. (2001). Capacity development: Vision and implications. *Capacity Development: Occasional Series, 3*, 1–11.
Lipson, B., & Warren, H. (2006). *'Taking stock' – a snapshot of INGO engagement in civil society capacity building.* INTRAC.
Lopes, C., & Theisohn, T. (2003). *Ownership, leadership, and transformation: Can we do better for capacity development.* Earthscan.
Lukes, S. (2005). *Power: A radical view.* Palgrave Macmillan.
Lusthaus, C., Adrien, M.-H., & Morgan, P. J. (2000). Integrating capacity development into project design and evaluation. *Global Environment Facility, Monitoring and Evaluation Working Paper 5*, 1–25.
Mahanty, S., Yasmi, Y., Guernier, J., Ukkerman, R., & Nass, L. (2009). Relationships, learning, and trust: Lessons from the SNV-RECOFTC partnership. *Development in Practice, 19*, 859–872. https://doi.org/10.2307/27752140
Månsson, P., Abrahamsson, M., Hassel, H., & Tehler, H. (2015). On common terms with shared risks – Studying the communication of risk between local, regional and national authorities in Sweden. *International Journal of Disaster Risk Reduction, 13*, 441–453. https://doi.org/10.1016/j.ijdrr.2015.08.003
Mansuri, G., & Rao, V. (2012). *Localizing development: Does participation work?* The World Bank. https://doi.org/10.1596/978-0-8213-8256-1
Mayne, J. (2007). Challenges and lessons in implementing results-based management. *Evaluation, 13*(1), 87–109. https://doi.org/10.1177/1356389007073683
McEntire, D. A. (2007). *Disaster response and recovery: Strategies and tactics for resilience.* Wiley.
Mitchell, J. K. (2006). The primacy of partnership: Scoping a new national disaster recovery policy. *The Annals of the American Academy of Political and Social Science, 604*, 228–255. https://doi.org/10.2307/25097790
Moore, M. (1995). Promoting good government by supporting institutional development. *IDS Bulletin, 26*, 89–96. https://doi.org/10.1111/j.1759-5436.1995.mp26002010.x
Morgan, P. J. (2002). Technical assistance: Correcting the precedents. *UNPD Development Policy Journal, 2*, 1–22.
Morpheus. (1999). *The matrix [motion picture]: Warner Bros., village roadshow pictures, Groucho II film partnership and silver pictures.* Retrieved 2023-02-08.
Morriss, P. (2002). *Power: A philosophical analysis.* Manchester University Press.
Neuhann, F., & Barteit, S. (2017). Lessons learnt from the MAGNET Malawian-German Hospital partnership: The German perspective on contributions to patient care and capacity development. *Globalization and Health, 13*(1), 50. https://doi.org/10.1186/s12992-017-0270-4
Noble, J. (2019). *Theory of change in ten steps.* NPC.
NORAD. (1999). *The logical framework approach (LFA): Handbook for objectives-oriented planning.* NORAD.
North, D. C. (1990). *Institutions, institutional change and economic performance.* Cambridge University Press.
OECD. (2006). *The challenge of capacity development: Working towards good practice.* OECD.
OECD. (2009). *The Paris declaration on aid effectiveness (2005) and the accra agenda for action (2008).*
OECD. (2010). *Capacity development: A DAC priority.* OECD.
OECD. (2012). *Greening development: Enhancing capacity for environmental management and governance.* OECD. https://doi.org/10.1787/9789264167896-en
Örtengren, K. (2003). *Logical Framework Approach - a summary of the theory behind the LFA method.* Sida.
Otoo, S., Agapitova, N., & Behrens, J. (2009). *The capacity development results framework: A strategic and results-oriented approach to learning for capacity development.*
Oxenham, J., & Chambers, R. (1978). *Organising education and training for rural development: Problems and challenges.* International Institute for Educational Planning.
Oxford English Dictionary. (2020). *Oxford dictionary of English.* Oxford University Press.

Pearson, J. (2011a). *LenCD learning package on capacity development, part 2: How-to learning network on capacity development (LenCD).*

Pearson, J. (2011b). Training and beyond: Seeking better practices for capacity development. *OECD Development Co-operation Working Papers, 1*, 1–55.

Piaget, J. (1971). *The construction of reality in the child (M. Cook, Trans.).* Ballantine Books. https://doi.org/10.1037/11168-000

Plato. (2008). *Cratylus - 380BC.* Project Gutenberg.

Raju, E., & Becker, P. (2013). Multi-organisational coordination for disaster recovery: The story of post-tsunami Tamil Nadu, India. *International Journal of Disaster Risk Reduction*, 82–91. https://doi.org/10.1016/j.ijdrr.2013.02.004

Rasmussen, J. (1985). The role of hierarchical knowledge representation in decisionmaking and system management. *IEEE Transactions on Systems, Man and Cybernetics, 15*, 234–243.

Reed, M. S., Evely, A. C., Cundill, G., Fazey, I., Glass, J., Laing, A., Newig, J., Parrish, B., Prell, C., Raymond, C., & Stringer, L. C. (2010). What is social learning. *Ecology and Society, 15*(4). http://www.jstor.org/stable/10.2307/26268235.

Rosenberg, L. J., & Posner, L. D. (1979). *The logical framework: A manager's guide to a scientific approach to design and evaluation.* Practical Concepts Incorporated (PCI).

Rousseau, D. M., Sitkin, S. B., Burt, R. S., & Camerer, C. (1998). Not so different after all: A cross-discipline view of trust. *Academy of Management Review, 23*, 393–404.

Schacter, M. (2000). Monitoring and evaluation capacity development in Sub-Saharan Africa: Lessons from experience in supporting sound governance. *ECD Working Paper Series, 7*, 1–27.

Schneider, M., & Somers, M. (2006). Organizations as complex adaptive systems: Implications of Complexity Theory for leadership research. *The Leadership Quarterly, 17*, 351–365. https://doi.org/10.1016/j.leaqua.2006.04.006

Schulz, K., Gustafsson, I., & Illes, E. (2005). *Manual for capacity development.* Sida.

Scott, W. R. (2014). *Institutions and organizations: Ideas, interests, and identities.* SAGE Publications.

Selznick, P. (1949). *TVA and the grass roots: A study in the sociology of formal organization.* University of California Press.

Selznick, P. (1992). *The moral commonwealth: Social theory and the promise of community.* University of California Press.

Sida. (1997). *Sida looks forward: Sida's programme for global development.* Sida.

Slovic, P. (1993). Perceived risk, trust, and democracy. *Risk Analysis, 13*, 675–682. https://doi.org/10.1111/j.1539-6924.1993.tb01329.x

Smillie, I. (2001). Capacity building and the humanitarian enterprise. In I. Smillie (Ed.), *Patronage or partnership: Local capacity building in humanitarian crises* (pp. 7–23). Kumarian Press.

Smith, G. P., & Wenger, D. (2007). Sustainable disaster recovery: Operationalizing an existing agenda. In H. Rodríguez, E. L. Quarantelli, & R. R. Dynes (Eds.), *Handbook of disaster research* (pp. 234–257). Springer.

Stone Motes, P., & McCartt Hess, P. (Eds.). (2007). *Collaborating with community-based organizations through consultation and technical assistance.* Columbia University Press.

Taylor, P., & Clarke, P. (2008). *Capacity for a change.*

Tendler, J. (2002). *Why social policy is condemned to a residual category of safety nets and what to do about it.*

Twigg, J. (2004). *Disaster risk reduction: Mitigation and preparedness in development and emergency programming.* Overseas Development Institute.

Ulrich, W. (2000). Reflective practice in the civil society: The contribution of critical systems thinking. *Reflective Practice, 1*, 247–268.

UNCSD. (2012). *The future we want.*

UNDP. (2008). *Capacity Assessment - practice note.*

UNDP. (2009). *Capacity development: A UNDP primer.*

UNDRR, & Coppola, D. P. (2019). *Strategic approach to capacity development for implementation of the sendai framework for disaster risk reduction: A vision of risk-informed sustainable development by 2013.* UNDRR.

UNISDR. (2009). *UNISDR terminology on disaster risk reduction.*

United Nations. (2015a). *Sendai framework for disaster risk reduction 2015-2030. United Nations.*

United Nations. (2015b). *Transforming our world: The 2030 agenda for sustainable development (A/RES/70/1). United Nations.*

United Nations. (2016). *Report of the open-ended intergovernmental expert working group on indicators and terminology relating to disaster risk reduction (A/71/644). United Nations.*

van Riet, G., & van Niekerk, D. (2012). Capacity development for participatory disaster risk assessment. *Environmental Hazards, 11*(3), 213–225. https://doi.org/10.1080/17477891.2012.688793

Watson, D. (2006). *Monitoring and evaluation of capacity and capacity development, 58B* (pp. 1–31). ECDPM Discussion Paper.

Weber, M. (1947). *The theory of social and economic organization.* The Free Press.

Wenger, E. (1998). *Communities of practice: Learning, meaning, and identity.* Cambridge University Press.

Wenger, E. (2009). A social theory of learning. In K. Illeris (Ed.), *Contemporary theories of learning: Learning theorists.in their own words* (pp. 209–218). Routledge.

Wenger-Trayner, E. (2012). Developing complex capabilities: The case of disaster risk reduction. *Natural Hazards Informer, 5*, 59–67.

Westoby, P., & Blerk, R. V. (2012). An investigation into the training of community development workers within South Africa. *Development in Practice, 22*, 1082–1096. https://doi.org/10.1080/09614524.2012.714354

Whyte, W. F. (1968). Imitation or innovation: Reflections on the institutional development of Peru. *Administrative Science Quarterly, 13*, 370–385. https://doi.org/10.2307/2391048

Woodhill, J. (2010). Multiple actors: Capacity lives between multiple stakeholders. In J. Ubels, N.-A. Acquaye-Baddoo, & A. Fowler (Eds.), *Capacity development in practice* (pp. 25–41). Earthscan.

Wrong, D. H. (1980). *Power: Its forms, bases, and uses.* Harper Colophon Books.

Yachkaschi, S. (2010). Lessons from below: Capacity development and communities. In J. Ubels, N.-A. Acquaye-Baddoo, & A. Fowler (Eds.), *Capacity development in practice* (pp. 194–207). Earthscan.

CHAPTER 15

Social change for a resilient society

Introduction

If I could venture outside my philosophical assumptions and use the term truth, I think there is a lot of it in the ancient Chinese proverb: 'Unless we change direction, we are likely to end up where we are going'. The first part of this book paints a rather bleak picture of the state of the world, although we have to remember that we have been through rough times in the past and that the relative death toll in diseases and disasters is lower today than a century ago (CRED, 2022). However, the complex combination of trends is distressing, obscuring what potential risk scenarios we may have in front of us and limiting the effectiveness of our traditional ways of managing them.

A safe and sustainable society requires not only capacity development for resilience but more fundamental transformation. Especially since it is the general population that is ultimately affected by, as well as largely affecting, the variations, trends, disturbances, disruptions or disasters that threaten our preferred development. In other words, a safe and sustainable society requires behaviour, norms, values and social organisation that are favourable for the resilience of society as a whole. These are not constant issues in space and time but are highly contextual and continuously reconstructed. Chapter 2 presents past fundamental social change, indicating that an equally significant transformation as the Neolithic Revolution and the Industrial Revolution is possible. That is heartening, considering our contemporary sustainability challenges. Yet, Chapter 2 also shows us that such fundamental social change is difficult to engineer as it requires a complex mix of economic, ideological and political factors, as well as a result that provides some competitive advantage over societies that have not transformed. However, recent history has seen more rapid changes in the transactions of humankind than ever before. These changes have been linked to the intensification of social networks that can 'accelerate behaviour change, improve organisational efficiency, enhance social change, and improve dissemination and diffusion of innovations' (Valente, 2012, p. 49).

I feel genuinely inadequate to present any comprehensive account of the fundamental social change required for our world's safety and sustainability. Nonetheless, I give it a humble go since it should be an essential part of any Sustainability Science approach to issues of risk and resilience for sustainable development. I ask for your forbearance when describing what social change can entail and prescribing what social change may be necessary for a safer and more sustainable future.

Sustainability Science, Second Edition
ISBN 978-0-323-95640-6, https://doi.org/10.1016/B978-0-323-95640-6.00010-5

Describing social change

There are many definitions of the concept of social change, but it generally includes enduring observable changes in a social system's structure and/or functioning that affect its future significantly (Oliver-Smith, 1979; Rocher, 2004, p. 341; Sztompka, 1993). Social change is, in other words, emphasising changes *of* society rather than merely changes *in* society, although the latter may accumulate and cause the former (Sztompka, 1993, p. 6). For instance, the legislation granting universal suffrage in Sweden was just one of the many changes in society that eventually resulted in substantial changes from a traditionally patriarchal society into one of the most gender-equal countries in the world (Figure 15.1). However, I want to emphasise that there is still much to be done, and there are worrying regressive signs. Social change is, then, 'the difference between various states of the same system succeeding each other in time' (Sztompka, 1993, p. 4).

Approaching social change in a system stipulates that it can stem from five main interrelated types of change. On the most basic level, there can be changes in the elements of the system, either by altering the composition by adding or removing elements or the variables of existing elements. For instance, new people may arrive, ideas may change, infrastructure may be constructed, etc. Second, there can be changes in dependencies between the elements of the system, such as new relationships with the newcomers that also affect already established dependencies, altered dependencies between previously discriminated and privileged groups, reduced dependence on firewood for cooking or heating with electrification, etc. Third, there can be changes in the system's boundary, which according to the philosophical assumptions presented in Chapter 5, is based on inherently subjective judgements. However, to talk about social change at all, it is essential to maintain consistency in the definition of what is considered part of the system or not and limit redefinitions to changes with significant empirical foundation. These changes can have physical or

Figure 15.1 Stockholm 1924 and 2009. *(Photos by Gustaf W:son Cronquist and Holger Ellgaard, shared on the Creative Commons.)*

environmental roots, like the rapid extension of mobile phone networks in Africa that completely alters how communities are imagined. They can also have social or cultural origins, such as the gradual extension of suffrage in many European countries from including only male landowners, to all men, to also include women. Finally, significant changes in the boundary of a system can also have political or economic roots, epitomised by the reunification of Germany. However, there are many less grand examples, for example, municipal reform, the annexation of territory, etc.

Changes in elements, dependencies and boundaries are all structural changes in the system. However, there can also be changes in the functions of specific system parts. For instance, newspapers may slide away from keeping the checks and balances of powerful actors to supply celebrity gossip. A beautiful beach may go from being an open toilet to becoming a primary source of income for a community. Women may expand their functions in society from childbearing and homemaking to, hopefully soon, all functions possible. In short, the fourth type comprises changes in function. The fifth and final type of change entails changes in the surrounding of the system. That is to say, changes outside the boundary of how the system is defined may still impact the system through transboundary relations, briefly introduced in Chapter 11. For example, the intensification of burning fossil fuel in industrialising Europe causes climate change in Africa's poorest region, and the US supreme mortgage crisis had repercussions for the Australian health system.

When looking at these five main types of change, it is essential to note that they are connected in such a way that a change in one most often causes other changes. Consider, for instance, a generic example of a part of the Sahel region in Africa. The emission of greenhouse gases by affluent countries is virtually certain to impact the global climate. It will likely affect Atlantic Ocean circulation (IPCC, 2021). Such changes in ocean circulation have caused long-lasting dry and wet episodes in the Sahel throughout history but will undoubtedly be exacerbated by rising temperatures that may even cause the climate system to switch into a century-long drought mode (Shanahan et al., 2009). Although there has been a recent greening of the Sahel, frequent and prolonged droughts since the 1960s have caused massive migration (Olsson et al., 2005). Livelihood patterns are being adapted, including diversification into small-scale business and trade (Mortimore & Adams, 2001). The urbanising population have better access to health care, education and diverse income opportunities than in their rural past (Njoh, 2003). At the same time, infectious diseases spread along the transit routes of increasingly mobile populations (Rebaudet et al., 2013, p. 49) and the risk of communal conflicts increase in the interface between marginalised communities (e.g., Raleigh, 2010). The list of interrelated changes can be made long, but the main argument here is that they may all interact to cause a significant overall change of society. In other words, there is not one master process of social change, but many processes of change of different spatial and temporal scales, different complexity and different direction (Tilly, 1984, pp. 33—40). Social change is the abstract concept encapsulating their aggregated and accumulated result (Sztompka, 1993, p. 187).

This view on social change fits the approach to our world as human-environment systems presented in this book. Especially since the tight couplings between natural and social systems make a complete divide unfeasible (e.g., Fordham, 2007; Hewitt, 1983; Oliver-Smith, 1999). In other words, the environment must also be included in the system when grasping and addressing social change in relation to resilience—both in the sense of the natural environment and the ideological environment (Sztompka, 1993, pp. 219—223), but not in the sense of the surrounding of the system, which unfortunately also is referred to as the environment in some systems literature (e.g., Anderson, 1999, p. 216; Boardman & Sauser, 2008; Buckley, 1967). The natural environment 'is an inescapable 'container' in which social life is flowing' (Sztompka, 1993, p. 219). It is not difficult to imagine a whole range of ways in which our 'natural environment may appear as negative constraints (barriers, blockades) or positive enablements (facilitations, resources)' (Sztompka, 1993, p. 220).

> *Think of migration and trade routes, communication networks or settlement patterns in mountainous areas as opposed to the plains, emerging in valleys or along rivers, established on the coast or islands. Or think about the hierarchies of inequality of wealth or power typical for areas poor in resources, as opposed to those emerging in conditions of natural abundance*
>
> ***Sztompka (1993, p. 220).***

However, remember from the first part of this book that the relationship between the human and the environmental roots goes both ways. Nature has always provided conditions for human agency, but the actualisation of this agency is also increasingly impacting nature, for good and bad (Sztompka, 1993, pp. 220—221).

In addition to the natural environment, Sztompka (1993) also stresses the importance of including the ideological environment when grasping and addressing social change. What people in a given society might do is significantly influenced by what they think and believe, individually or collectively, and what the dominant ideological structures make them think and believe (Sztompka, 1993, p. 222). We are all constrained in what ends we conceive as feasible, what means are available and what activities are possible. Boulding (1964, p. 7) takes this further when stating that knowledge of the system itself contributes significantly to its dynamics, meaning that changes in what we know about the system change the system itself. Consider, for instance, de Tocqueville (1856) classic observation that exploited and deprived people only rebelled when more egalitarian ideology emerged. Or the relatively recent awakening of gender equality movements after millennia of utter patriarchy when feminist ideology emerged. This connection gives hope concerning our contemporary sustainability challenges, as there seems to be a growing awareness of our dire situation. Let us just hope that the dissemination of knowledge is quick enough and motivates people to take action.

Society is not a rigid object or 'hard' system but rather a 'soft' field of relationships between individuals and other social entities in continuous movement, motion and change (Sztompka, 1993, pp. 9—10). No static structures direct human activity, but there is a process of structuration in which human actions produce and reproduce social structures that guide and restrict what actions human beings may take (Giddens, 1984, pp. 25—26). Whilst intrinsically linked, both the structure and function of such a system are continuously transforming, making it difficult to grasp any of them without studying events (Sztompka, 1993, p. 10). An event is 'a thing that happens' (Oxford English Dictionary, 2020)—something we can observe in the world. Or as Abrams (1982, p. 192) so eloquently puts it: 'An event is a moment of becoming at which action and structure meet'. In other words, events are discrete manifestations of what agents do or do not do in particular contexts. We can learn something about the structure or function of human-environment systems by looking at events. Agency is, in this view, the potential of a particular agent to act in relation to specific structures (Sztompka, 1993, pp. 217—219). It means that agency can manifest itself in actual activity for a more sustainable world or remain latent. Hence, motivation is critical for fundamental social change to happen (Klintman, 2013) as it 'concerns those processes that give behaviour its energy and direction' (Reeve, 2005, p. 6).

Motivation concerning decisions and actions for sustainability can be understood through three main approaches: the ecological motivation approach, the material motivation approach and the social motivation approach (Klintman, 2013). The ecological motivation approach rests on the assumption that a reduced ecological footprint is a primary value of individuals who do not fully understand our dire situation but would be motivated to change their behaviour if they were better informed. On the other hand, the material motivation approach rests on the assumption that incessant material accumulation is a primary value of individuals, who thus would be motivated to change behaviour if the market adjusts the prices in relation to the ecological footprint of the goods and services they buy. Finally, the social motivation approach assumes that the overriding driving force behind individual motivation resides in our evolutionary propensity for group convergence and distinction. In other words, it entails our fundamental need to belong to groups—whilst distinguishing them from other groups—and to seek and signal social status in them (Klintman, 2013).

Although events are central to grasping social change, we must remember Heraclitus's alleged saying that you cannot go into the same river twice (Plato, 2008). Our world is dynamic, and any present instance 'constitutes just one small momentary phase within the vast stream of humanity's development, which, coming from the past, debouches into the present and thrusts ahead towards possible futures' (Elias, 1987, p. 224). It is, in other words, only for practical reasons we conceptually freeze some moments in time and treat them as single events (Figure 15.2) when they are pieces of unending processes

Figure 15.2 A moment in an endless flow. *(Christian Mueller/Shutterstock.com.)*

(Sztompka, 1993, p. 12). One family moving from the countryside into a city is not particularly significant, but the urbanisation process is.

Social change can be linear, cyclical or stepwise (Sztompka, 1993, pp. 13–17), spontaneous or planned (Sztompka, 1993, pp. 20–22). Social change can be slow and evolutionary or rapid and revolutionary (Gellner, 1989; Huy, 2001). It can be partial or complete (Gellner, 1989; Sztompka, 1993). Not all social change amounts to the immense transformation of the Neolithic Revolution, nor does it always require centuries or even millennia. Especially not when it occurs as a reaction to changes that have already transformed neighbouring societies, which include many in our globalised world. For example, the rapid social change of contemporary China would simply not be possible without intense contact with other powerful societies over time, like the United Kingdom, Japan, the United States and now India. However, perhaps the most extreme example of rapid social change in relation to intrusive neighbours is Japan, which in barely more than a generation after US warships sailed into Tokyo Bay, went from being a withdrawn feudal society to an early industrial society with the capacity of winning wars against China and Russia (Beasley, 2000), and later controlling most of the western side of the Pacific. This 'neighbourhood factor' must never be underestimated as both a driver and obstacle for social change towards sustainability. A driver as soon as the new and sustainable society becomes more competitive than its neighbours in the current world system, but an obstacle as long as it is deemed more beneficial to continue on the beaten track.

Social change is thus not only about individuals' attitudes, behaviour and choices (ABCs), contained in the dominant ABC paradigm of social change (Shove, 2010). It occurs on the macro-, meso- and micro-level (Sztompka, 1993, pp. 22–23). Although individuals' ABCs are central to social change, we must not ignore the importance of the physical, environmental, social, cultural, political and economic context in which individuals exist. For instance, Reeve (2005, p. 6) argues that individuals' motivation emanates from internal experiences (needs, cognitions and emotions) of external environmental events that supply incentives for them to engage or not to engage in particular activities. Think about all the time the maintenance of the levees of New Orleans was overlooked, even after the close call of Hurricane Ivan in 2004, and then the massive investments in levee improvement after Hurricane Katrina in 2005. Klintman (2013) argues convincingly for the importance of the social when explaining, understanding and improving unsustainable consumer behaviour, whilst (Shove, 2010, p. 1274) points towards 'the extent to which governments sustain unsustainable economic institutions and ways of life, and the extent to which they have a hand in structuring options and possibilities'.

To summarise, social change entails enduring changes of a human-environment system's structure and/or function that significantly affect its future. These changes may stem from changes in that system's elements, relations, boundaries, functions or surroundings. It is the aggregated and accumulated result of various processes of different spatial and temporal scales, complexity and direction that is labelled social change. The physical, environmental, social, cultural, political and economic context set the frame for human agency, influencing these spheres through manifest human activity. However, what we have the potential to do is not the same as what we are doing, making motivation essential for fundamental social change towards sustainability.

Prescribing social change

The following section presents some of my humble ideas of social change that would favour sustainability. To give this account some structure, I organise my thoughts according to the typical levels of analysis of macro, meso and micro. However, they are closely related, as they are simply different hierarchical levels of the same human-environment systems. Since it is inherently challenging to imagine a society as radically different as after the previous two fundamental transformations without coming across as a science fiction novelist, I limit myself to suggest changes that are substantial and difficult but imaginable.

At the macro-level of countries and international systems, we must transcend the contemporary paradigm of seeing the world as divided into competing states. Such political realism has roots as far back as Thucydides (1956) in ancient Greece and is still the dominant paradigm. The invention of the UN system, the World Bank, the World Trade Organization and other international institutions merely gives structure to the still overwhelmingly competitive interaction of states. Alliances are forged if they benefit our cause, and agreements are only honoured if they do not undermine our position in the global order. There are countless examples of this. The US reluctance to sign or ratify the Kyoto Protocol to the United Nations Framework Convention on Climate Change

(UNFCCC) is a relevant example. Another more recent example is the withdrawal of the United States from the Paris Agreement on climate change under the Trump Administration. It would be highly beneficial for the world if all states ratified and followed such crucial agreements for sustainability. Still, considering the number and diversity of states worldwide, there is little hope for that as long as not doing so benefits freeloaders. When there are many freeloaders or when they include some of the wealthiest and most significant contributors to the problem, fingers start to point at each other and arguments for dropping out are heard: 'If they are not doing it, why should we?'

Another problem with the paradigm of seeing the world as divided into states is that it makes us lose perspective on our sustainability challenges. For instance, it should be evident that the per capita emissions of greenhouse gases matter more than the total national emissions. The opposite involves a rather ridiculous idea entertained by many unreflective people that we do not need to do anything in the West and should instead focus on Asia because our national emissions are minuscule compared to countries like China and India. China's total emissions are indeed double that of the United States, but the Chinese per capita emissions are less than half that of the US population (2016 data). The comparison becomes even more absurd when comparing India and Sweden, with the former emitting almost 57 times more carbon dioxide than the latter, whilst the Indian per capita emissions are considerably less than half that of the Swedish population. If only focussing on the total emissions, the solution for bringing the emissions down below some other geopolitical entity would simply be to report emissions on a lower administrative level. However, the actual emissions would stay the same.

Moreover, these comparisons hide our actual contribution to climate change by only looking at the emissions produced in each country and ignoring who consumes the produced goods and services. This issue received surprisingly little attention in scientific and policymaking circles for decades (Peters & Hertwich, 2008) and effectively obscures the actual situation from being visible and addressed. In other words, we need a consumption-based perspective on greenhouse gas emissions to grasp the root of the problem (Davis & Caldeira, 2010). It should be evident that the total consumption-based emissions matter since the solution for countries to reduce their greenhouse gas emissions would otherwise simply be to import all goods and services from elsewhere. However, global emissions would stay the same.

So, we need to focus on the consumption-based per capita emissions of greenhouse gases to grasp where the problem lies. When applying this perspective, it turns out that the Chinese and Swedish have roughly the same per capita emissions, whilst the emissions of the US population are, on average, more than 2.5 times bigger per person, and the emissions of the Indians amount to just a quarter (2019 data). Only a few populations emit more on average than the Americans when accounting for consumption, but it is important to note that the Luxembourgers emit twice as much as them. The other populations above the United States on the list include, for instance, the more affluent

Persian Gulf States, Brunei and Singapore. The populations just below the United States are Canada, Australia, Belgium, Hong Kong and South Korea. These are all affluent countries, but it is important to note that there are equally rich countries with populations enjoying a similar or even better quality of life but with significantly lower consumption-based per capita emissions. These countries include, for instance, Norway, Ireland, Iceland, Germany, Sweden, Denmark and New Zealand. Yet, there is contradictory evidence concerning a possible decoupling between development and greenhouse gas emissions in both high- and low-income countries (e.g., Hubacek et al., 2021; Knight & Schor, 2014; Tenaw & Hawitibo, 2021). The only thing we know for sure is that all affluent countries are way over sustainable levels of greenhouse gas emissions.

Finally, although it is evident that people in affluent countries contribute most to climate change, it is vital to understand that not everybody within a country contributes equally. Regardless of the Indian average consumption-based per capita emissions comprising just a quarter of the Swedish emissions per person, there are Indian households with multiple cars, big houses, self-indulgent lifestyles and frequent air travel. There are also Swedish households that do not have a car, eat vegetarian, refrain from unnecessary consumption and do not fly privately. These households are not only made up of devoted environmentalists but also of people who cannot afford such lifestyles. The same goes for every country. The differences in average emissions are then primarily explained, together with energy production and specific industrial processes, by the proportion of people within each country enjoying such carbon-intensive lifestyles. There are environmentally minded people in all income groups. However, it is staggering to note that the richest 10% of the world's population, regardless of their nationality, were responsible for 52% of the cumulative carbon emissions between 1990 and 2015, which is 7.5 times more than the poorest half of the global population (Oxfam, 2020, p. 3). It becomes even more appalling when looking at the personal carbon footprints of the really rich today, which include emissions from domestic consumption, public and private investments and imports and exports of carbon embedded in traded goods and services. Then, the richest 10% emit, on average, 31 tonnes of carbon dioxide per year—or 4.7 times the global average and 19.4 times the personal emissions of the poorest half (Chancel et al., 2022, p. 123). The sustainable level of such annual personal emissions compatible with the 1.5 °C warming limit is 1.1 tonnes of carbon dioxide, whilst a 2 °C warming allows around 3.4 tonnes (Chancel et al., 2022, p. 118). However, the richest 1% emit, on average, 100 tonnes, the richest 0.1% emit 467 tonnes, and the richest 0.01% emit, on average, 2531 tonnes of carbon dioxide annually (Chancel et al., 2022, p. 123).

To summarise, it is absurd to claim that India is the problem because its total national carbon dioxide emissions are 250 times greater than Luxembourg's when the consumption-based per capita emissions of the Luxembourgers are more than 20 times that of the Indians. However, it is equally absurd to hold every citizen equally accountable for the average emissions. The ultrarich elite of any country cannot be allowed to hide their extreme personal emissions behind the much lower national average, regardless of country. Furthermore, India's growing middle-class should not be held less accountable for their environmental footprints than the middle-class of Luxembourg if their footprints are similar. Nor should the poor in Luxembourg be held responsible for an environmental footprint to which they are not contributing. To have any chance at sustainable development, we must overcome the paradigm of only perceiving the world as divided into states and hold each other personally accountable for our contributions to the problems.

Similarly, western-based companies export toxic waste from states with implemented environmental legislation to states with more lenient approaches to waste management or dump it there when nobody is watching or turning a blind eye in exchange for money under the table. The Dutch multinational company Trafigura is one example of such a company, selling chemical waste, either knowingly or unknowingly, to dump in developing countries. For instance, in Côte d'Ivoire in 2006. Although prizewinning investigating journalists reported proof against Trafigura, and it is fair to assume that the company at least must have wondered how the newly established Ivorian company was going to process the waste for a prize everybody else declined, Trafigura paid a hefty compensation to Côte d'Ivoire and was thus exonerated from further legal processes—even when tens of people have died, and tens of thousands have turned ill as a direct consequence of their waste.

It would be naïve not to assume that criminal activity will always occur. However, if we cannot transcend this current world order and start seeing and treating our sustainability challenges holistically, I am afraid we will continue to fight a losing battle. People often refer to the need for a significant threat to unite otherwise quarrelling parties, but what can be more threatening than the challenges described in the first part of this book (Figure 15.3)?

The example above has already brought us onto the meso-level of cities, regions, governmental authorities, private companies, civil society organisations, etc., where we find similar challenges of a lack of holistic approaches. First, we need to acknowledge that the context people exist in is an indivisible whole and not sliced up into functional sectors, administrative levels and geographical areas. These are just tools to grasp the complexity of our world that, unfortunately, are often applied so that they obscure more than they illuminate (e.g., Becker, 2021). Different sectors or actors depend on each other in various ways. Sometimes to more significant degrees than within the sector or organisation itself. Similarly, there may be more significant dependencies between cities or local communities across a border than within the geopolitical areas containing them. The people of a community, city or any human-environment system perceive their context as a whole, but as soon as we organise ourselves to grasp or address this context, we focus on particular parts. That is okay, but only if we remember that the

Figure 15.3 What can be more threatening than the apocalypse? *(© Stokkete | Dreamstime.com.)*

part we focus on is intrinsically connected to the rest and that many sustainability challenges cut across these artificial boundaries. In other words, we need to change current institutional frameworks to facilitate collaboration and holistic approaches better.

Another problem with the lack of holism could be addressed by including externalities in our economic systems. If we keep market exchange as the dominant form of exchange in the future, we must include all costs when setting prices. And yes, it was equally unthinkable with market exchange before it was invented as it is to imagine something else today. Flying flowers from Kenya to Sweden is not without environmental costs, even if there seems to be a global consensus to make transportation as cheap as possible to boost economic growth. Similarly, closing a factory to move to another place with lower salaries is not without social costs regarding social security and poverty-induced crime, depression, etc. These costs (or benefits) that affect actors who did not choose to incur that cost is referred to as externalities (Buchanan & Stubblebine, 1962) and including them in the actual price of goods and services is difficult (Crocker & Tschirhart, 1992) but vital for sustainability—also in times of economic downturn (Bowen & Stern, 2010).

Although it should be rather common sense and has been repeatedly pointed out by scientists for at least 40 years, perpetual growth is impossible in a finite world (Jones et al., 2013; Meadows et al., 1972, 2004)—at least in the form of growth in material output, which has been the actual modus operandi of most economies for centuries. This paradigm has to be questioned and altered for sustainability to be possible in the first place. However, this does not mean that we cannot have an increasing quality of life for all people if we balance the use of natural resources and focus innovation on minimising our ecological footprint instead of on maximising consumption. Anything else is short-sighted in economic terms and potentially cataclysmic in social and environmental terms.

Or in the words of Jeremy Grantham (2012, p. 5), a co-founder and member of the board of directors of a global investment management firm:

> *Of all the technical weaknesses in capitalism, though, probably the most immediately dangerous is its absolute inability to process the finiteness of resources and the mathematical impossibility of maintaining rapid growth in physical output. You can have steady increases in the quality of goods and services and, I hope, the quality of life, but you can't have sustainable growth in physical output. You can have 'growth' — for now — or you can have 'sustainable' forever, but not both. This is a message brought to you by the laws of compound interest and the laws of nature*
>
> ***Grantham (2012, p. 5).***

In addition, we must halt and reverse the spread of the economism of the market into the spheres of the state and civil society. I am aware of the central role of the economy, in a broad sense, for all human societies throughout history. I acknowledge that market exchange has been most effective in driving development. However, I think it is fundamental to maintain that other societal values cannot and should never be measured in economic terms. We must never conflate 'standard of living' with 'quality of life' and never forget the fundamental difference between 'being well off' and 'well-being'. It is therefore worrying to hear governmental authorities, even in the Scandinavian bastion of the welfare state, talk about citizens as 'their customers' or to hear humanitarian civil society organisations talk about 'their competitive advantage' in relation to other organisations working in the same functional or geographical sector of a disaster area.

At the micro-level of individuals, families, groups and communities, it is of utmost importance that they put pressure on the state and market actors on the meso- and macro-levels, as well as shouldering their individual responsibility for sustainable development. They can do so by mobilising and demanding political changes, adapting their consumption patterns, etc. However, for people to do so requires motivation in any form. Although the three approaches to motivation briefly introduced above differ in fundamental assumptions and focus, I believe they all carry some wisdom that can assist us in instigating social change for a more sustainable world. We need to educate everybody concerning our sustainability challenges and what we all can do to address them, even if we as individuals are unlikely to be turned into autonomous change agents driven by such ecological motivation alone (Klintman, 2013, pp. 127—129). Similarly, we need economic incentives not to buy goods and services with large environmental footprints. However, it may be even more critical to change the social structures that determine social status (Klintman, 2013). Imagine a society where we talk about and actively address sustainability challenges with our families, friends and colleagues, where sustainable lifestyles grant more status than our current consumerist ways of life. I share this dream with many. For instance, actor Keanu Reeves said: 'I dream of a day where I walk down the street and hear people talk about morality, sustainability and philosophy instead of the Kardashians'. It is not impossible. It is necessary.

Conclusion

It is evident that fundamental social change is needed to deal with most of the core challenges of humankind (Figure 15.4). However, fundamental social change only occurs when a substantial part of a society's population changes, and then hopefully in the right direction in relation to their sustainability challenges. 'Unless we subscribe to a fatalist worldview, a major part of these issues—and the potential for changing the trends—can be boiled down to human motivation, whether that of politicians, policymakers, industrial actors, NGOs, or citizen-consumers' (Klintman, 2013, p. 3). Unfortunately, I think the words of Chesterton's famous character, Father Brown, could have been about most of these actors. When he laid down his cigar and said: 'It isn't that they can't see the solution. It is that they can't see the problem' (Chesterton, 2000, p. 141). Or perhaps more accurately, that we cannot see the extent of our problems. If we would, I am confident that we would see unprecedented transformations towards a more sustainable world in only decades.

Consider, for instance, the estimated half per cent probability that the current concentration of CO_2 in the atmosphere would cause an apocalyptic scenario with six degrees warmer average global temperature, which would completely transmute the entire world as we know it. People may be worried, and many know what they could do to contribute to a solution, but the absolute majority is simply not doing anything. The half per cent probability does not seem that persuasive after all. However, when that same probability is considered in relation to fatal flight accidents, it corresponds to 17 crashes per hour or 150,000 per year (Global Challenges Foundation, 2013). It is evident that no mentally sane person would fly under such circumstances, and commercial aviation would most likely be banned. And that is with the CO_2 concentration a decade ago, which unfortunately has increased significantly since then and continues

Figure 15.4 Change is our only chance. *(© Andrii Zastrozhnov | Dreamstime.com.)*

to do so faster than ever found in geological records. Moreover, climate change is only one sustainability challenge we face.

However, the necessary transformations for sustainability would neither be caused by autonomous well-informed individuals in isolation nor solely by economic incentives to steer consumption. It is only when the social dimension is included that substantial social change is likely to ensue. Most people can see the logical flaw of the idea of perpetual growth in a finite world and that something is fundamentally wrong with Norwegians eating fish caught in Norway but processed and packed in China or Thailand. However, we seem incapable of scrutinising the prevailing paradigm, which we must do for any substantial social change to materialise. Today, we have a greater capacity for grasping and addressing our sustainability challenges than ever in human history. Still, transformations for sustainable development are continuously undermined by actors utilising the inherent uncertainty and ambiguity to sow ambivalence and hesitation.

We will always have a group of devoted people motivated to make informed decisions and change behaviour to do their best to alleviate social and environmental problems. We will also always have a group of hardcore cynics questioning every statement concerning sustainability, regardless of their scientific foundation. However, these two groups are relatively small, and between them, a majority of people who are more or less ambivalent and hesitant. This majority is the key to social change, but to shoulder this momentous obligation, they must make up their mind and stop being impartial in this tug-of-war between idealists and cynics. Remember 'that impartiality [...] is a pompous name for indifference, which is an elegant name for ignorance' (Chesterton, 1900). Many understand that the current situation is unsustainable and know what they can do to contribute to a more sustainable world but choose for various reasons not to act. Often rationalising it by notions like 'what change can I make in the bigger scheme of things?' Hence, I think the great British adventurer and polar explorer Robert Swan is correct when stating that the 'greatest threat to our planet is the belief that someone else will save it.'

References

Abrams, P. (1982). *Historical sociology*. Cornell University Press.

Anderson, P. W. (1999). Complexity theory and Organization science. *Organization Science, 10*, 216–232. https://doi.org/10.2307/2640328

Beasley, W. G. (2000). *The rise of modern Japan*. St Martins Press.

Becker, P. (2021). Fragmentation, commodification and responsibilisation in the governing of flood risk mitigation in Sweden. *Environment and Planning C: Politics and Space, 39*(2), 393–413. https://doi.org/10.1177/2399654420940727

Boardman, J., & Sauser, B. (2008). *Systems thinking: Coping with 21st century problems*. CRC Press.

Boulding, K. E. (1964). The place of the image in the dynamics of society. In G. K. Zollschan, & W. Hirsch (Eds.), *Explorations in social change* (pp. 5–16). Houghton Mifflin.

Bowen, A., & Stern, N. (2010). Environmental policy and the economic downturn. *Oxford Review of Economic Policy, 26*, 137–163. https://doi.org/10.1093/oxrep/grq007

Buchanan, J. M., & Stubblebine, W. C. (1962). Externality. *Economica, 29*, 371. https://doi.org/10.2307/2551386
Buckley, W. F. (1967). *Sociology and modern systems theory*. Prentice-Hall.
Chancel, L., Piketty, T., Saez, E., Zucman, G., et al. (2022). *World inequality report 2022*. World Inequality Lab.
Chesterton, G. K. (1900). *Review of Puritan and Anglican by Kegan Paul. The speaker*. http://www.gkc.org.uk/gkc/books/Articles_for_Speaker.html#s00.
Chesterton, G. K. (2000). *The scandal of Father Brown*. House of Stratus.
CRED. (2022). *EM-DAT: The international disaster database*. Centre for Research on the Epidemiology of Disasters.
Crocker, T. D., & Tschirhart, J. (1992). Ecosystems, externalities, and economies. *Environmental and Resource Economics, 2*, 551–567.
Davis, S. J., & Caldeira, K. (2010). Consumption-based accounting of CO_2 emissions. *Proceedings of the National Academy of Sciences, 107*(12), 5687–5692. https://doi.org/10.1073/pnas.0906974107
de Tocqueville, A. (1856). *The old regime and the revolution*. Harper and Brothers.
Elias, N. (1987). The retreat of sociologists into the present. *Theory, Culture and Society, 4*, 223–247. https://doi.org/10.1177/026327687004002003
Fordham, M. H. (2007). Disaster and development research and practice: A necessary eclecticism. In H. Rodríguez, E. L. Quarantelli, & R. R. Dynes (Eds.), *Handbook of disaster research* (pp. 335–346). Springer.
Gellner, E. (1989). *Plough, sword and book: The structure of human history*. University of Chicago Press.
Giddens, A. (1984). *The constitution of society: Outline of the theory of structuration*. University of California Press.
Global Challenges Foundation. (2013). *Global risk and opportunity indicator (GROI)*.
Grantham, J. (2012). *Your grandchildren have No value (and other deficiencies of capitalism)*. GMO Quarterly Letter, February 2.
Hewitt, K. (1983). The idea of calamity in a technocratic age. In K. Hewitt (Ed.), *Interpretations of calamity* (pp. 3–32). Allen and Unwin.
Hubacek, K., Chen, X., Feng, K., Wiedmann, T., & Shan, Y. (2021). Evidence of decoupling consumption-based CO_2 emissions from economic growth. *Advances in Applied Energy, 4*, 100074. https://doi.org/10.1016/j.adapen.2021.100074
Huy, Q. N. (2001). Time, temporal capability, and planned change. *Academy of Management Review, 26*, 601–623. https://doi.org/10.2307/3560244
IPCC. (2021). *Climate change 2021: The physical science basis*. Cambridge University Press.
Jones, A., Allen, I., Silver, N., Cameron, C., Howarth, C., & Caldecott, B. (2013). *Resource constraints: Sharing a finite world*.
Klintman, M. (2013). *Citizen-consumers and evolution: Reducing environmental harm through our social motivation*. Palgrave Macmillan.
Knight, K., & Schor, J. (2014). Economic growth and climate change: A cross-national analysis of territorial and consumption-based carbon emissions in high-income countries. *Sustainability, 6*(6), 3722–3731. https://doi.org/10.3390/su6063722
Meadows, D., Meadows, D. H., & Randers, J. (2004). *Limits to growth: The 30-year update*. Chelsea Green Publishing.
Meadows, D. H., Randers, J., & Behrens, W. W., III (1972). *The limits to growth: A report to the club of Rome*. Universe Books.
Mortimore, M. J., & Adams, W. M. (2001). Farmer adaptation, change and 'crisis' in the Sahel. *Global Environmental Change, 11*, 49–57. https://doi.org/10.1016/S0959-3780(00)00044-3
Njoh, A. J. (2003). Urbanization and development in sub-Saharan Africa. *Cities, 20*, 167–174. https://doi.org/10.1016/S0264-2751(03)00010-6
Oliver-Smith, A. (1979). The Yungay Avalanche of 1970: Anthropological perspectives on disaster and social change. *Disasters, 3*, 95–101. https://doi.org/10.1111/j.1467-7717.1979.tb00205.x
Oliver-Smith, A. (1999). Peru's five-hundred-year earthquake: Vulnerability in historical context. In A. Oliver-Smith, & S. M. Hoffman (Eds.), *The angry earth: Disaster in anthropological perspective* (pp. 74–88). Routledge.

Olsson, L., Eklundh, L., & Ardö, J. (2005). A recent greening of the Sahel—trends, patterns and potential causes. *Journal of Arid Environments, 63*, 556–566. https://doi.org/10.1016/j.jaridenv.2005.03.008
Oxfam. (2020). Confronting carbon inequality: Putting climate justice at the heart of the COVID-19 recovery. *Oxfam*.
Oxford English Dictionary. (2020). *Oxford dictionary of English*. Oxford University Press.
Peters, G. P., & Hertwich, E. G. (2008). Post-kyoto greenhouse gas inventories: Production versus consumption. *Climatic Change, 86*, 51–66. https://doi.org/10.1007/s10584-007-9280-1
Plato. (2008). *Cratylus - 380BC*. Project Gutenberg.
Raleigh, C. (2010). Political marginalization, climate change, and conflict in African Sahel states. *International Studies Review, 12*, 69–86.
Rebaudet, S., Sudre, B., Faucher, B., & Piarroux, R. (2013). Environmental determinants of cholera outbreaks in inland Africa: A systematic review of main transmission foci and propagation routes. *Journal of Infectious Diseases, 208*, S46–S54. https://doi.org/10.1093/infdis/jit195
Reeve, J. (2005). *Understanding motivation and emotion*. Wiley.
Rocher, G. (2004). *A general introduction to sociology: A theoretical perspective*. B.K. Dhur Academic Publishers.
Shanahan, T. M., Overpeck, J. T., Anchukaitis, K. J., Beck, J. W., Cole, J. E., Dettman, D. L., Peck, J. A., Scholz, C. A., & King, J. W. (2009). Atlantic forcing of persistent drought in West Africa. *Science, 324*, 377–380.
Shove, E. (2010). Beyond the ABC: Climate change policy and theories of social change. *Environment and Planning, 42*, 1273–1285. https://doi.org/10.1068/a42282
Sztompka, P. (1993). *The sociology of social change*. Blackwell.
Tenaw, D., & Hawitibo, A. L. (2021). Carbon decoupling and economic growth in Africa: Evidence from production and consumption-based carbon emissions. *Resources, Environment and Sustainability, 6*, 100040. https://doi.org/10.1016/j.resenv.2021.100040
Thucydides. (1956). *History of the peloponnesian war (C. Forster Smith, Trans.)*. William Heinemann.
Tilly, C. (1984). *Big structures, large processes, huge comparisons*. Russell Sage Foundation.
Valente, T. W. (2012). Network interventions. *Science, 337*, 49–53. https://doi.org/10.1126/science.1217330

CHAPTER 16

On a bumpy road from *Industria* to *Sustainia*?

Introduction

Geologists divide time according to marked shifts in the state and functioning of the Earth (Lewis & Maslin, 2015). The catastrophic release of carbon dioxide from roughly two million years of massive volcanic eruptions marked the shift between the last epoch of the Permian period and the Early Triassic epoch, resulting in a mass extinction of more than 81% of marine and 89% of terrestrial species (Zhang et al., 2021)—the Great Dying. The Chicxulub impact delineated the shift from the dinosaur-dominated Late Cretaceous to the burgeoning dominion of mammals and birds of the Paleocene (Chapter 4)—again through the mass extinction of 75% of marine and terrestrial species by altering Earth's climate (Neubauer et al., 2021, p. 1). Now, humankind has had such a significant and escalating impact on the Earth that scientists have proposed an end to the Holocene—the relatively stable geological epoch necessary for the astonishing human developments since the Stone Age (Chapter 2)—and the start of the Anthropocene (Crutzen & Stoermer, 2000). In this new geological epoch, human activity exerts a dominant influence on Earth, its environment, and its climate. Hence, the Anthropocene: the new epoch of human beings.[1]

The start of the Anthropocene has been heavily debated (e.g., Certini & Scalenghe, 2011; Lewis & Maslin, 2015; Whitaker, 2020, p. 845). Yet, there is a growing scientific consensus that it started in the mid-20th century (Steffen et al., 2015; Waters et al., 2016; Zalasiewicz et al., 2015). Whilst human beings have indeed had increasing impacts on the environment since at least 12,000 years ago (Certini & Scalenghe, 2011; Ellis, 2021), only after the mid-20th century can we find strong evidence for fundamental shifts in the state and functioning of Earth beyond the variability of the Holocene (Steffen et al., 2015). Only after the mid-20th century can we identify a globally existing geological layer created by human activity—the fallout from the enormous number of nuclear bombs detonated from 1945 to 1988 (Zalasiewicz et al., 2015). Although the concentration of atmospheric carbon dioxide started to depart from Holocene levels a century earlier (Figure 6.2), the deviation had accelerated by the mid-20th century and began to result in observable warming beyond late Holocene variability (Waters et al., 2016). Similarly, human activity has resulted in biodiversity loss for millennia (Ellis, 2019). Still, the

[1]Derived from the Greek words *ánthropo* (man) and *cene* (new).

Sustainability Science, Second Edition
ISBN 978-0-323-95640-6, https://doi.org/10.1016/B978-0-323-95640-6.00001-4

compounded effects of what we are currently doing—and failing to do—generate a much faster extinction rate than during the last great mass extinction 66 million years ago (Barnosky et al., 2011). All the signs of a new geological epoch are here (Figure 16.1).

Whilst natural scientists have been studying the extent of the impacts of human activity on our planet, social scientists have been studying the transformations of the human societies that have generated the impact. Often with specific temporal—(e.g., Castells, 2010) or thematic focuses (e.g., Polanyi, 2001) but sometimes focussing on the transformations of entire societies (Gellner, 1989; Hall, 1986; Mann, 1986). Such more comprehensive perspectives become essential for understanding our past and anticipating future social change when considering the profound transformations of societies over time (Chapter 2). Amongst the most notable scholars providing holistic perspectives on past social change are John A. Hall, Michael Mann and Ernest Gellner. Hall (1986) focuses on shifts in economic, political, and ideological power when describing and explaining the disposition and development of different societies. Mann (1986, 1993, 2012, 2013) suggests that complex combinations of ideological, economic, military, and political sources of social power direct societal development. Gellner (1989), on the other hand, provides a more materialist framework that arguably gives precedence to changes in production, which generates the environmental impacts studied by natural scientists, whilst also including changes in cognition and coercion. He proposes that human history

Figure 16.1 The human footprint is now so significant it has ushered in a new geological epoch.

Figure 16.2 From a shrewdness of apes to a sustainable society? *(Based on Leremy, Shutterstock.com.)*

entails great revolutions, each resulting in distinct types of societies (Figure 16.2). Remember from Chapter 2 how the Neolithic Revolution transformed hunter-gatherer society into agrarian society—or 'Agraria' in Gellner's (1989) vocabulary—and the Industrial Revolution transforms agrarian society into industrial society—or 'Industria': the culprit behind the alarming current state of the Earth.

Although Industria emerged in Great Britain in the second half of the 18th century and spread across the Western world and to Japan and isolated pockets across the globe from the middle of the 19th century, it did not become a global feature until the mid-20th century. That was when industrialisation took off in large nonwestern countries, such as Brazil and China (Figure 2.8), and elsewhere. It was also a period of intensifying industrial developments in the already established industrialised societies. These parallel developments resulted in the Great Acceleration—a central argument for dating the start of the Anthropocene—visible in various social and economic indicators, such as population, urbanisation, GDP, energy use, fertiliser use, water use and transportation (Steffen et al., 2015). This period of acceleration is not only associated with an exponentially increasing pace of technological development (Heylighen, 2008) and extraordinary global economic growth (Maddison, 2001). It is also a period of unprecedentedly rapid and accelerating social change (Heylighen, 2008)—enough to wonder if we are in a new great revolution.

This final chapter, before concluding the book, is an attempt at the impossible. We know our past is punctuated by great revolutions that can only be discerned with hindsight. We also know that it is impossible to predict what society will look like after each transformation. Yet, this chapter attempts to anticipate social change or, at least, suggest what to look for when anticipating social change, which hopefully results in a more sustainable society.

Unfolding Sustainia

The proposed start of the Anthropocene in the mid-20th century coincides with the establishment of environmental issues on the global policy agenda—epitomised by the 1972 United Nations Conference on the Human Environment in Stockholm and the following string of Earth Summits that ended up in Stockholm again in 2022 (Chapter 3). Sustainability is now a central connecting aspect in several previously separate policy areas, most clearly visible in the Sustainable Development Goals for 2015–30.

The outcome documents from these global processes request a balance between production and conservation to safeguard sustainable development for both present and future generations. When contrasting that with contemporary society, it becomes clear that the envisioned society is so fundamentally different it may be considered a distinct kind of society. If that vision cannot be considered *Industria* anymore, let us call it *Sustainia.*

For Gellner (1989), the central theme of life in Agraria is *predation*, with elaborate coercive schemes to redistribute wealth. In Industria, on the other hand, *production* has replaced predation, and power shifted from feudal masters to a more homogeneous population of relatively flexible functional specialists (Gellner, 1989). Production is still hegemonic in contemporary society, but it is increasingly challenged by *preservation*—the central theme of life in the vision of Sustainia. It is, at best, a gradual change, and perhaps preservation never surpasses production—at least if we are to believe the more cynical observers (e.g., Žižek, 2010). However, the shift from predation to production was also gradual, opposed by established elites and full of turmoil and setbacks (Gellner, 1989).

Humankind is in dire need of a new great revolution as Industria has led us into a developmental cul-de-sac. Although preservation appears to become increasingly central in the scientific and policy dialogs of the Anthropocene, it is impossible to ascertain if we are in a new great revolution. Nobody could anticipate our previous great transformations beforehand, whilst they appear obvious in hindsight. However, remember Gellner's warning in Chapter 2 that people tend to take their institutions and assumptions as absolute, self-evident, and beyond question, actively fortifying them against any potential challenge (Gellner, 1989, p. 11). If we are in a new great revolution, it is also impossible to ascertain if it would take us to Sustainia or somewhere else. However, it is essential to engage in critical dialogue concerning contemporary institutions and assumptions and inquire resolutely into current changes and their potential resultant direction. Not that we can engineer social change with any precision, but we must still push for what we think can help the sustainability of humankind and against what we see as a continuation towards doom.

Investigating changes in production, cognition and coercion

Gellner (1989) focuses on the three central features of production, cognition and coercion when elaborating on the fundamental differences between hunter-gatherer, agrarian, and industrial societies. Perhaps we can also find hints of a new great revolution in these features of social life.

Numerous social scientists study changes in production in the Anthropocene. Schnaiberg's (1980) Treadmill of Production and Mol and Spaargaren's (2000) work on

Ecological Modernisation Theory are compelling examples of changes in perspectives on production. There are also several new forms of production being suggested and advocated, such as not-for-profit (e.g., Nussbaum, 2010), social enterprises (e.g., Short et al., 2009), ethical organisations (e.g., O'Neil, 2015), open source (e.g., Budhathoki & Haythornthwaite, 2012), crowdsourcing (e.g., Brabham, 2010), fair trade (e.g., Ponte & Gibbon, 2005) and microfinance (e.g., Sanyal, 2009). More and more examples of radically different institutions and organisations have become central and influential (Wright, 2010), such as Wikipedia, Linux, OpenStreetMap and Patagonia. Although these alternative forms of production are all still dwarfed by conventional capitalism, they may constitute hints of a potential transformation towards Sustainia.

It is important to note that all modern states have traded economic growth against environmental degradation, regardless of regime type (Figure 16.3). Our sustainability challenges would, in other words, have been much the same if state socialism had become the dominant mode of production instead of capitalism (Mann, 2013, p. 366). However, production is arguably the central element of Gellner's (1989) explanations for our previous great transformations. Therefore, it is fair to assume that production changes may also be central in the next—especially since it is Industria's kind of production that underlies our current sustainability challenges. Latour even argues that we need not only to modify the current production system but to get out of it altogether, reminding us that framing everything in terms of the economy is a new idea in human history (Watts, 2020).

Other social scientists focus on changes in what Gellner refers to as cognition, suggesting a range of significant changes since the mid-20th century. In addition to various fundamental changes in ontological and epistemological assumptions, many scholars advance ideas concerning the interconnectedness (e.g., Castells, 2010), complexity (e.g., Urry, 2005) and fluid character (Bauman, 2000) of society and the world. Knowledge is constructed at an exponentially increasing pace at the same time as scientists seem increasingly explicit about the limits of our knowledge (e.g., IPCC, 2021). The Internet makes disseminating information more immediate, far-reaching and open to

Figure 16.3 The shrinking of the Aral Sea happened during both the communist regime of the Soviet Union and the capitalist regimes of most countries in its catchment area. *(© Meiram Nurtazin | Dreamstime.com.)*

various actors and source criticism more difficult (e.g., Mintz, 2002). Whilst science is increasingly able to describe the challenges of the Anthropocene, its role in legitimising the institutions of industrial society, as Gellner (1989) suggests, is increasingly undermined by growing resistance to scientific knowledge (Hall, 2019; Klintman, 2019). These changes in cognition may have diverging effects on the potential and direction of a new great revolution. The increasing extensity, intensity, and velocity of communication enhance society's potential for mobilising and coordinating social actions towards sustainability. And yet, the sheer complexity and ambiguity in both problem descriptions and policy advice undermine decisions and actions and effectively preserve the status quo.

Finally, many social scientists study different aspects of coercion, such as deviance and crime, policing, law, military and power. Some of these investigations suggest notable changes since the mid-20th century. For instance, Conrad and Schneider (1992) show how society's perception of deviance changes from viewing it as badness to seeing it as sickness, and Bayley and Shearing's (1996) analysis of the social changes behind the restructuring of policing in developed countries. Also, there are more quantitative changes in coercion, such as mounting incarceration rates (e.g., Pager, 2003) and the growing strength of law enforcement agencies (e.g., Jacobs & Helms, 1997). The state has allowed increased individual freedom in certain matters in some parts of the world (e.g., Tremblay et al., 2011). Nonetheless, the period since the mid-20th century entails essentially a continuation of the increasing coercive capacity of the state that has resulted in an unprecedented ability to penetrate society (Mann, 2013). Not so much through the sovereign power to control and punish or pastoral power to protect and nurture, which the state exercises itself, but through the disciplinary power we exercise over ourselves to fit into society (Foucault, 1984).

This steadily increasing capacity of coercion might be fundamental for the transformation towards Sustainia, as states are still the main actors regulating action and organisation in all spheres of society (Mann, 1997, 2013). They are the signatories of vital agreements—such as the UNFCCC and the Paris Agreement, regarding climate change—but rely on the compliance of the market, civil society and the public for these instruments to have any effect. Although politics and power play peripheral roles in Gellner's (1989) account, which has been pointed out as a weakness (Haugaard, 2007), the coercive capacity of the state may be fundamental to steer transformation towards Sustainia. Not with a sword at the throat, as in Agraria, but with legislation, taxes, rewards, encouragements and cultivation of norms and values.

Although it is vital to investigate changes in production, cognition and coercion separately, analysing them together may be necessary to hint at the potential and direction of a new great revolution. We must diligently study what is going on in society and formulate bold syntheses of what our recent changes in the central features of social life may mean for our future.

Conclusion

We truly live in an age of anxiety again (Hall, 2019, p. 45). Humankind can now influence Earth, its environment, and its climate. Yet, so far, we seem collectively incapacitated to steer away from certain doom. Earth recovered from previous mass extinction events, but it took millions of years each time, and the dominant species were most often gone. All of us alive today will love somebody who will be alive in the year 2100. What we do today and in the coming decades determines if they will live in a sustainable society, enjoying at least the same quality of life as we do today, or if they will live in a human-environmental apocalypse (Figure 16.4). We are in the Anthropocene. The question is if we can use our unprecedented knowledge and resources to usher in a new and much-needed great transformation away from Industria and towards the vision of Sustainia. Perhaps we are in it already.

'There are two very silly doctrines about knowledge and the world: that we can do whatever we wish, and that everything is completely determined' (Hall, 1986, p. 5). History shows us that only hindsight can tell if we are in a new great revolution and that it is impossible to determine what society would look like on the other side. However, we must inquire resolutely into recent changes and engage in dialogue concerning any hints we may get of the potential and direction of a new transformation. We must study our past to understand our options for our future without jumping to conclusions about our potential choices (Gellner, 1989). We can use the work of Gellner (1989) and other influential historical sociologists (e.g., Hall, 1986; Mann, 2013) when investigating and engaging in the transformations needed for sustainability today.

Figure 16.4 What legacy do we want to leave? *(© Lovelyday12 | Dreamstime.com.)*

Gellner (1989) teaches us that human history is marked by fundamental transformation and that a future society may differ equally much from ours as Industria from Agraria. That is comforting, considering the immense sustainability challenges of the Anthropocene. However, not all recent changes in production, cognition and coercion identified by social scientists correspond cogently to the vision of Sustainia seemingly emerging in global policy. That begs obvious questions about the actual direction of contemporary social change but may also be a symptom of the same turmoil and setbacks found in previous revolutions. It is impossible to know if we are on the bumpy road to Sustainia, but let us sincerely hope so.

References

Barnosky, A. D., Matzke, N., Tomiya, S., Wogan, G. O. U., Swartz, B., Quental, T. B., Marshall, C., McGuire, J. L., Lindsey, E. L., Maguire, K. C., Mersey, B., & Ferrer, E. A. (2011). Has the Earth's sixth mass extinction already arrived. *Nature, 471*, 51–57. https://doi.org/10.1038/nature09678

Bauman, Z. (2000). *Liquid Modernity*. Cambridge: Polity.

Bayley, D. H., & Shearing, C. D. (1996). The future of policing. *Law & Society Review, 30*, 585–606.

Brabham, D. C. (2010). Moving the crowd at threadless: Motivations for participation in a crowdsourcing application. *Information, Communication & Society, 13*, 1122–1145. https://doi.org/10.1080/136911810 03624090

Budhathoki, N. R., & Haythornthwaite, C. (2012). Motivation for open collaboration: Crowd and community models and the case of OpenStreetMap. *American Behavioral Scientist, 57*, 548–575. https://doi.org/10.1177/0002764212469364

Castells, M. (2010). *The rise of the network society*. Wiley-Blackwell.

Certini, G., & Scalenghe, R. (2011). Anthropogenic soils are the golden spikes for the Anthropocene. *The Holocene, 21*, 1269–1274. https://doi.org/10.1177/0959683611408454

Conrad, P., & Schneider, J. W. (1992). *Deviance and medicalization: From badness to sickness*. Philadelphia: Temple University Press.

Crutzen, P. J., & Stoermer, E. F. (2000). The Anthropocene. *Global Change Newsletter, 41*, 17–18.

Ellis, E. C. (2019). Evolution: Biodiversity in the Anthropocene. *Current Biology, 29*(17), R831–R833. https://doi.org/10.1016/j.cub.2019.07.073

Ellis, E. C. (2021). Land use and ecological change: A 12,000-year history. *Annual Review of Environment and Resources, 46*(1), 1–33. https://doi.org/10.1146/annurev-environ-012220-010822

Foucault, M. (1984). *The Foucault reader*. Pantheon Books.

Gellner, E. (1989). *Plough, sword and book: The structure of human history*. University of Chicago Press.

Hall, J. A. (1986). *Powers and liberties: The causes and consequences of the rise of the west*. University of California Press.

Hall, J. A. (2019). Our current sense of anxiety or after Gellner. *Nations and Nationalism, 25*(1), 45–57.

Haugaard, M. (2007). Power, modernity and liberal democracy. In S. Malešević, & M. Haugaard (Eds.), *Ernest Gellner and contemporary social thought* (pp. 75–102). Cambridge University Press. https://doi.org/10.1017/cbo9780511488795.004

Heylighen, F. (2008). Accelerating socio-technological evolution: From ephemeralization and stigmergy to the global brain. In G. Modelski, T. Devezas, & W. R. Thompson (Eds.), *Globalization as evolutionary process: Modeling global change* (pp. 284–309). Routledge.

IPCC. (2021). *Climate change 2021: The physical science basis*. Cambridge University Press.

Jacobs, D., & Helms, R. E. (1997). Testing coercive explanations for order: The determinants of law enforcement strength over time. *Social Forces, 75*, 1361–1392. https://doi.org/10.2307/2580675

Klintman, M. (2019). *Knowledge resistance: How we avoid insight from others*. Manchester University Press. https://doi.org/10.7765/9781526158703

Lewis, S. L., & Maslin, M. A. (2015). Defining the Anthropocene. *Nature, 519*, 171–180. https://doi.org/10.1038/nature14258
Maddison, A. (2001). *The world economy: A millennial perspective*. Organisation for Economic Cooperation and Development.
Mann, M. (1986). *The sources of social power: Volume 1, A history of power from the beginning to AD 1760*. Cambridge University Press.
Mann, M. (1993). *The sources of social power: Volume 2, the rise of classes and nation states 1760–1914*. Cambridge University Press.
Mann, M. (1997). Has globalization ended the rise and rise of the nation-state. *Review of International Political Economy, 4*, 472–496. https://doi.org/10.1080/096922997347715
Mann, M. (2012). *The sources of social power: Volume 3, global empires and revolution, 1890–1945*. Cambridge University Press.
Mann, M. (2013). *The sources of social power: Volume 4, Globalizations 1945-2011*. Cambridge University Press.
Mintz, A. P. (2002). *Web of deception: Misinformation on the internet*. CyberAge Books.
Mol, A. P. J., & Spaargaren, G. (2000). Ecological modernisation theory in debate: A review. *Environmental Politics, 9*, 17–49.
Neubauer, T. A., Hauffe, T., Silvestro, D., Schauer, J., Kadolsky, D., Wesselingh, F. P., Harzhauser, M., & Wilke, T. (2021). Current extinction rate in European freshwater gastropods greatly exceeds that of the late Cretaceous mass extinction. *Communications Earth & Environment, 2*(1), 1–7. https://doi.org/10.1038/s43247-021-00167-x
Nussbaum, M. C. (2010). *Not for profit*. Princeton University Press.
O'Neil, M. (2015). Labour out of control: The political economy of capitalist and ethical organizations. *Organization Studies, 36*, 1627–1647. https://doi.org/10.1177/0170840615585339
Pager, D. (2003). The mark of a criminal record. *American Journal of Sociology Review, 108*, 937–975. https://doi.org/10.1086/374403
Polanyi, K. (2001). *The great transformation*. Beacon Press.
Ponte, S., & Gibbon, P. (2005). Quality standards, conventions and the governance of global value chains. *Economy and Society, 34*, 1–31. https://doi.org/10.1080/0308514042000329315
Sanyal, P. (2009). From credit to collective action: The role of microfinance in promoting women's social capital and normative influence. *American Sociological Review, 74*, 529–550. https://doi.org/10.1177/000312240907400402
Schnaiberg, A. (1980). *The environment: From surplus to scarcity*. Oxford University Press.
Short, J. C., Moss, T. W., & Lumpkin, G. T. (2009). What makes a process a capability? Heuristics, strategy and effective capture of opportunities. *Strategic Entrepreneurship Journal, 3*, 161–194. https://doi.org/10.1002/sej
Steffen, W., Broadgate, W., Deutsch, L., Gaffney, O., & Ludwig, C. (2015). The trajectory of the Anthropocene: The great acceleration. *The Anthropocene Review, 2*, 81–98. https://doi.org/10.1177/2053019614564785
Tremblay, M., Paternotte, D., & Johnson, C. (Eds.). (2011). *The Lesbian and gay movement and the state: Comparative insights into a transformed relationship*. Routledge.
Urry, J. (2005). The complexity turn. *Theory, Culture & Society, 22*, 1–14.
Waters, C. N., Zalasiewicz, J., Summerhayes, C., Barnosky, A. D., Poirier, C., Gałuszka, A., Cearreta, A., Edgeworth, M., Ellis, E. C., Ellis, M. A., Jeandel, C., Leinfelder, R., McNeill, J. R., Richter, D., Steffen, W., Syvitski, J. P. M., Vidas, D., Wagreich, M., Williams, M., ... Wolfe, A. P. (2016). The Anthropocene is functionally and stratigraphically distinct from the Holocene. *Science, 351*, aad2622. https://doi.org/10.1126/science.aad2622
Watts, J. (2020). *Bruno Latour: 'This is a global catastrophe that has come from within'* – the Guardian. Retrieved 2022-10-09.
Whitaker, J. A. (2020). Climatic and ontological change in the Anthropocene among the makushi in Guyana. *Ethnos, 85*(5), 843–860. https://doi.org/10.1080/00141844.2019.1626466
Wright, E. O. (2010). *Envisioning real Utopias*. Verso.
Zalasiewicz, J., Waters, C. N., Williams, M., Barnosky, A. D., Cearreta, A., Crutzen, P. J., Ellis, E. C., Ellis, M. A., Fairchild, I. J., Grinevald, J., Haff, P. K., Hajdas, I., Leinfelder, R., McNeill, J. R.,

Odada, E. O., Poirier, C., Richter, D., Steffen, W., Summerhayes, C., ... Oreskes, N. (2015). When did the Anthropocene begin? A mid-twentieth century boundary level is stratigraphically optimal. *Quaternary International, 383,* 196–203. https://doi.org/10.1016/j.quaint.2014.11.045
Zhang, H., Zhang, F., Chen, J. B., Erwin, D. H., Syverson, D. D., Ni, P., Rampino, M., Chi, Z., Cai, Y. F., Xiang, L., Li, W. Q., Liu, S. A., Wang, R. C., Wang, X. D., Feng, Z., Li, H. M., Zhang, T., Cai, H. M., Zheng, W., ... Shen, S. Z. (2021). Felsic volcanism as a factor driving the end-Permian mass extinction. *Science Advances,* 7(47), eabh1390. https://doi.org/10.1126/sciadv.abh1390
Žižek, S. (2010). *Living in the end times.* Verso.

CHAPTER 17

Concluding remarks

Introduction

There is no doubt that our world is in a grave state. Centuries of unprecedented developmental leaps in energy output, economic development, increase in life expectancy and decrease in adult illiteracy and child mortality have unfortunately had side effects that have accumulated into a range of sustainability challenges and their symptomatic events. We have pushed on to an edge of a cliff and are increasingly realising that our contemporary practices and ways of life, at least amongst the affluent, are jeopardising the entire world as we know it. However, '[h]ope is as stubborn as ever. Despite daily catastrophes in the media—effects of human-made and natural environmental disaster, social inequality, and unfair economic priorities—loud exclamations of hope are constantly heard' (Klintman, 2013, p. 2). The future is not predestined or predetermined but instead contingent on choice, decision and action. Concern should drive us into action and not depression,[1] and there are infinite possible futures ahead of us. To get the future we want, we must think about what to do and not to do today in order to bring that future about. The choice is ours.

As I demonstrate in this book, sustainable development demands conceptual and practical approaches to safety and sustainability that help us grasp or address uncertainty, complexity, ambiguity and dynamic change. This concluding chapter attempts to close the rhetorical loop between the descriptive, conceptual and transformative perspectives of the three parts of this book. It is thus focussing on drawing out the key messages from each of the three parts—*The state of the world*, *Approaching the world* and *Changing the world*—and summarising them into a Sustainability Science approach to managing risk and resilience for sustainable development.

The state of the world

Our world has a long history of change, from a burning inferno to the amazing planet we live on today. Although human beings have looked the same for around 200,000 years and had the same intellectual capacity for about 60,000 years, it was not until the onset of the Neolithic Revolution, some 10,000 years ago, that we started to more significantly alter our environment to suit our purposes. The local sustainability challenge of

[1] Pythagoras allegedly wrote something similar.

Sustainability Science, Second Edition
ISBN 978-0-323-95640-6, https://doi.org/10.1016/B978-0-323-95640-6.00003-8

overwhelming the carrying capacity of the immediate environment of hunter-gatherer society, leading to starvation or migration, turned into regional sustainability challenges in terms of depleted agricultural lands on increasing distances around the emerging cities of early agrarian society. As we learnt how to manage one sustainability challenge, we created others that, at times, even contributed to the downfall of entire societies. Then the second great transformation, the Industrial Revolution, helped us overcome the perennial agrarian limitation of relying on biomass as more or less the sole energy source, causing unprecedented developments in all spheres of society. We had yet again overcome one sustainability challenge but created new ones that, over the last three centuries, have materialised into the current core challenges of humankind. The local and regional sustainability challenges of our past have been joined by the global sustainability challenges of our present.

It is clear that we have become capable of both great deeds and great destruction, and our expanding human agency seems to have few limits. We invented the concept of risk during the last great transformation, as we started to realise our own agency, and it has proven an excellent tool for managing uncertainty (Figure 17.1). It is not that we in any way can predict our future, but we utilise the concept of risk to anticipate what could happen. Moreover, such an exercise can guide decisions and actions to steer our development towards a more sustainable future.

However, our world is increasingly complex and dynamic, and so are our sustainability challenges. The human impact on our environment has accelerated exponentially over the last centuries, demographic and socioeconomic processes are continuously redrawing our human landscape, and globalisation is increasing the extensity, intensity and velocity of interaction across the world. Furthermore, the increasing complexity of modern society undermines our ability to oversee and steer our development, allowing society to drift into danger. We are progressively more cognizant of our dire situation but seem, so far, to be collectively incapacitated to make any substantial changes on local,

Figure 17.1 Uncertainty is a fundamental challenge to manage. *(© Feng Yu | Dreamstime.com.)*

national or international levels, even when the symptoms of the unsustainable state we are in are many and growing.

It seems like our tremendous capacity for change is restricted by the inherent ambiguity of what the future holds for us. Science cannot supply completely lucid and unambiguous answers in our complex and dynamic world, and whatever scientific consensus is undermined by the ferocious spread of disinformation by agents wanting to maintain the status quo. The future will always be uncertain, but we must realise that the only thing we can do is act the best way we can with the best information we have at hand. Regardless of differences in details, virtually the entire scientific community is pointing out that we are rapidly moving towards a very undesirable future. We must act now.

The two great transformations in our past fundamentally altered our relationships amongst ourselves and with our environment. They were both unimaginable before they occurred but seem entirely self-evident with hindsight. Considering the dire state we increasingly realise we are in, we again need a new great transformation. We are indeed experiencing more rapid change than ever in human history, so perhaps we are already in it. However, we must do everything we can to intentionally align various economic, political and ideological factors to facilitate and steer such transformation towards sustainability, though the end result might be impossible to fully foresee and comprehend.

We must remember that we have faced severe sustainability challenges over millennia and have prevailed so far. At times with great costs or even entire societies collapsing, humanity has always moved on. The main issue at stake is, thus, not the survival of our species. Instead, it is what costs, in human and material terms, we will have to pay this time around. These costs are determined by when and how we start substantially addressing our contemporary sustainability challenges. The message is nonetheless very clear. The sooner the better, the more the merrier.

To summarise, Sustainability Science provides a descriptive framework of the core challenges of humankind—describing their causes and effects, as well as their past, present and many potential futures. This book has merely touched upon a fraction of the knowledge accumulated during Sustainability Science's relatively short but intense history of transdisciplinary research and action, much of which draws from important disciplinary efforts in anthropology, archaeology, design, engineering, geography, public administration, sociology, etc., but amalgamated to explain and understand our precarious situation as a whole. 'There is no such thing as a single-issue struggle because we do not live single-issue lives' (Lorde, 1982).

Approaching the world

Our future is uncertain, and our world is complex, ambiguous and dynamic. Sustainable development thus demands philosophical, theoretical and practical approaches that are not only assisting us in explaining, understanding and improving our sustainability challenges and their symptomatic events but also fit together to link between the conceptual and the actual. That is the core of Sustainability Science.

Assuming that the world exists independently of any observer but that any knowledge we can have about it is constructed in particular social contexts, development becomes a highly subjective and normative concept. Development is thus viewed as a preferred expected scenario of change in a set of variables over time, from a current to a desired state, and includes purposeful activities to drive or steer this change. Sustainable development is then a development that can be maintained over time and safeguarded from the impact of adverse events and their underlying processes. However, at any point in time, there are infinite possible alternative future scenarios ahead of us.

As soon as we can conceive a more positive and still realistic development scenario, we tend to reevaluate our situation and update our preferred expected scenario accordingly. There are thus no positive risks in Sustainability Science, only opportunities that most often cause modifications in our expectations of our future. However, to manage the uncertainty of what actual development will happen, we utilise the concept of risk and structure various risk scenarios that could divert us away from our preferred expected future. Assessing and addressing risk scenarios require first explicitly establishing what is valuable to us and important to protect—now and for the future—which in the context of sustainable development includes multiple and often interdependent aspects. Then, we analyse what hazards that could initiate events that have the potential to disturb, disrupt or destroy these valuable aspects, how vulnerable they are to the impact of each specific event and what capacities we have to address the consequences. It is important to note that human agency is not only fundamental for determining vulnerability and capacity but also for influencing most of the different types of hazards known to us. In short, risk assists us in managing uncertainty and guiding decisions and actions to maintain development along our preferred expected trajectory.

Many sustainability issues are complex and difficult to explain, understand and improve. Especially as they often involve various factors from across functional sectors, administrative levels and geographical borders. We need to understand what constitutes complexity, be mindful of its manifestations and learn to cope with its consequences—especially in our attempts to govern our challenges. To assist us in this endeavour, we can construct models of the parts of our world we are interested in whilst including relevant transboundary dependencies. Throughout this book, these models are called human-environment systems to highlight the intrinsic links between nature and society on all levels. They are tools for grasping the dynamics and nonlinear interdependencies so prevalent in our world. Human-environment systems include both structural and functional models, which in turn have different types of elements and dependencies and can be either qualitative or quantitative. Although it is impossible to know any system completely, the goal must be as a holistic picture as possible. In short, a system approach assists us in grasping complexity.

No matter how much we think we know about our world, there is always ambiguity in terms of multiple ways of explaining and understanding observed phenomena and improving experienced challenges. To cope with this ambiguity, it is not only vital to include formal experts when analysing and managing risk or constructing human-environment systems but to involve a wide range of other actors, such as policymakers and the public—not only to find mutually agreeable standpoints but also to provide a degree of moral force and political influence behind these standpoints. In short, participatory approaches assist us in coping with ambiguity.

Finally, the dynamic character of our world makes static approaches to safety and sustainability futile. Human-environment systems must, in other words, be resilient to be sustainable, meaning they must have the capacity to continuously develop along a preferred expected trajectory whilst remaining within human and environmental boundaries. Resilience is an emergent property determined by the ability of the human-environment system to anticipate, recognise, adapt to and learn from variations, changes, disturbances, disruptions and disasters that may cause harm to what human beings value. It is relative and context-dependent, just as the concepts of development and risk, and assessing and comparing resilience requires proper care concerning keeping variables and preferences constant over space and time. The functions of anticipation, recognition, adaptation and learning, as well as their constituent functions on lower hierarchical levels of abstraction, are in turn performed by a complex combination of knowledge, skills, tools and other resources; organisation on all levels; rules, regulations and other formal institutions; and norms, values and other informal institutions. In short, resilience assists us in managing the dynamic character of our world.

To summarise, Sustainability Science provides coherent conceptual approaches and tools for making sense of uncertainty, complexity, ambiguity and dynamic change necessary for grasping and addressing present and future safety and sustainability issues. I am thus truly standing on the shoulders of giants when presenting my contribution to this endeavour—giants who have not only actively contributed to establishing Sustainability Science as such but to a wide range of important disciplinary work that underpins the transdisciplinary efforts for a sustainable future.

Changing the world

There is a lot of wisdom in Mark Twain's immortal words: 'If you always do what you've always done, you'll always get what you've always had'—at least in the sense of what you had coming. We are in trouble, and everything points towards alarming consequences if we continue on the beaten track. We must, ironically enough, change direction if we are

Figure 17.2 We have a choice. *(Source: kwest/Shutterstock.com.)*

to maintain any substantial part of what we have gained in human and material terms so far. In other words, we must change a lot if we want a sustainable future (Figure 17.2).

Science has great potential for driving change. However, not directly through traditional science, with its emphasis on explaining and understanding observed phenomena, but by complementing it with design science, with its focus on improving experienced challenges. 'Our task is not to predict the future; our task is to design a future for a sustainable and acceptable world, and then to devote our efforts to bringing that future about' (Simon, 2002, p. 601). In other words, it is fundamental that scientists stop making prescriptive suggestions based solely on descriptive studies without taking proper care. We will always need traditional science to explain and understand key phenomena concerning risk, resilience and sustainability and to use that knowledge to inform the argumentation for the design criteria of artefacts we need to develop to address our challenges—vital artefacts that can be both material and conceptual. Traditional science and design science are fundamentally different but connected and equally important in Sustainability Science. Change entails addressing challenges, addressing challenges entails designing artefacts and designing artefacts entails explaining and understanding phenomena. However, for Sustainability Science to grasp and address humankind's core challenges requires transforming the academic structures that currently restrict transdisciplinary research and problem-solving-oriented scientific endeavours. Whilst disciplinary research is effective for addressing a wide range of research questions, it is fundamentally insufficient to address our contemporary sustainability challenges.

Notwithstanding the astonishing accumulation of scientific evidence of staggering sustainability challenges, humankind is still hardly doing anything to save itself. Many people simply do not believe in the problems or the recommended solutions, regardless of the available evidence. Others are reluctant to make the necessary changes, regardless of whether they believe or not. This resistance to knowledge and change is driven by a range of cognitive, psychological, social, cultural, political, economic, administrative, and technological factors, and addressing it is a crucial task for the survival of humankind.

> *Knowledge, for any conscious organism, is the means of survival; to a living consciousness, every 'is' implies an 'ought.' Man is free to choose not to be conscious, but not free to escape the penalty of unconsciousness: destruction. Man is the only living species that has the power to act as his own destroyer—and that is the way he has acted through most of his history*
>
> ***(Rand, 1961, p. 15).***

One practical way of addressing sustainability challenges is to develop a capacity for resilience, which entails holistic and systematic action that addresses all significant links in the causal chains behind particular capacity gaps. It addresses these gaps on all necessary levels, including developing knowledge, skills, tools and other resources; organisation on all levels; rules, regulations and other formal institutions; and norms, values and other informal institutions. It is, thus, vital to acknowledge dependencies in both the human-environment system in question and the capacity development initiative as such, and the various versions and adaptations of the Logical Framework Approach can be useful tools for that, given that they are applied with explicit consideration of complexity, uncertainty, ambiguity and dynamic change. Moreover, it is essential to not only focus on technical objectives and activities but also to explicitly consider roles, responsibilities, relationships and softer skill sets needed for any capacity development initiative to generate sustainable results. Hence, capacity development for resilience is a double design problem, with equal importance of objectives, activities, and indicators focused on the capacities of the human-environment system to perform key functions for its resilience and on facilitating local ownership, effective partnership, trustful relationships, etc. That is how capacity development for resilience can become the awaited key tool for sustainable development.

Finally, in addition to the changes *in* society promoted by capacity development initiatives, we urgently need more fundamental changes *of* society. This is referred to as social change and entails not only the attitudes, behaviour, motivation and choices of individuals but the physical, environmental, social, cultural, political and economic context these individuals exist in. Sustainable development requires significant changes on the macro-, meso- and micro-levels of society, and we do not have the centuries or even millennia of earlier transformations to accomplish them. We need rapid, or perhaps even revolutionary social change towards sustainability. Maybe we are in another great revolution from *Industria* to *Sustainia* already. It is impossible to tell. Although it is difficult for us to foresee or even imagine what the resulting society would look like, we must motivate policymakers and practitioners from the state, market and civil society, as well as the public, to make a myriad of small changes that accumulate into fundamental social change. Everybody must grasp the precarious situation we are in, economic incentives must steer us towards sustainable consumption, and sustainable lifestyles must engender real social status. Einstein (1946, p. 11) once said about nuclear weapons that 'a new type of thinking is essential if mankind is to survive and move towards higher levels'. I think that applies to our current sustainability challenges too (Figure 17.2).

To summarise, Sustainability Science provides a comprehensive, transformative framework for addressing the core challenges of humankind. It bridges between science and practice and bases its prescriptive transformative power on solid descriptive research and a coherent conceptual framework. Sustainability Science promotes changes *in* society, as well as changes *of* society, with the overall objective of fostering resilient communities, organisations, cities and societies for a safe and sustainable world.

Conclusion

It is easy to succumb to pessimism, like Žižek's (2010) apocalyptic view that we are closer to the end of the world than to fundamental social change towards sustainability. However, I choose to be optimistic and hope that Simon (2002, p. 605) is proven right when stating:

> *Perhaps our very salvation will come from the severity of the problems we will have to solve: finding an ecologically sustainable state for the Earth and all its living inhabitants, injecting far stronger criteria of fairness into the allocation of available resources and their products, and disarming the vicious competitions that now take place between every imaginable sort of 'we' and 'they'.*

Science has essential roles to play in facilitating such dramatic and critical changes in the transactions of humankind (Kates et al., 2001)—changes that are necessary for the continuation of our world as we know it (Figure 17.2). Most scientific disciplines have a role in this endeavour, but the holistic, systemic and transdisciplinary approach of Sustainability Science is imperative in this context as it brings together 'scholarship and practice, global and local perspectives from north and south, and disciplines across the natural and social sciences, engineering, and medicine' (Clarke & Dickson, 2003, p. 8060).

Sustainability Science provides a comprehensive description of our sustainability challenges and coherent conceptual and transformative frameworks to explain, understand and improve them. It maps the past, present and many potential futures of these challenges, including their symptomatic events and underlying change processes. It points out the necessity of acknowledging and addressing uncertainty, complexity, ambiguity and dynamic change for sustainable development to be possible in our contemporary world and provides conceptual tools and approaches to facilitate that. The concepts of risk and resilience are, thus, fundamental for Sustainability Science, as are systems approaches and participatory approaches.

Science is a crucial driver for change towards sustainability, but not only through the descriptive focus of traditional science. Sustainability Science must also be prescriptive and supplies a scientifically rigorous way of bridging the chasm between the descriptive and prescriptive by complementing traditional science with design science. It also highlights the importance of practical action and emphasises the significance of capacity development for resilience as a tool for sustainability. In addition to the changes *in* society caused by intentional capacity development or other practical initiatives, Sustainability Science also promotes fundamental changes *of* society, making social change towards a more sustainable world a central aim.

Finally, our world is a fantastic place, and we have proven to be an astonishingly adaptive species throughout history. We have conquered all corners of the world and flourished regardless of environmental circumstances. Although our current prospects are rather disheartening, considering our apparent incapacity to address our contemporary challenges substantially, I am sure we will prevail this time around too. The question is, therefore, not *if* we will change direction towards sustainability but *when* the anticipated or actual costs are high enough to make anything else so overwhelmingly troubling that even the ambivalent and hesitant majority starts to engage in addressing our situation. Hope is truly as stubborn as ever.

Many people look up at the stars and dream themselves away.

Finding another place to live, even terraforming Mars.

We need to look down at the ground and dream ourselves to stay.

On Earth, the only liveable planet we know, even with her scars.

References

Clarke, W. C., & Dickson, N. M. (2003). Sustainability science: The emerging research program. *Proceedings of the National Academy of Sciences, 100*, 8059–8061.

Einstein, A. (1946). *Atomic education urged by Einstein: Scientist in plea for $200,000 to promote new type of essential thinking*. The New York Times. Retrieved 2023-02-13 from https://www.nytimes.com/1946/05/25/archives/atomic-education-urged-by-einstein-scientist-in-plea-for-200000-to.html.

Kates, R. W., Clarke, W. C., Corell, R. W., Hall, J. M., Jaeger, C. C., Lowe, I., McCarthy, J. J., Schellnhuber, H. J., Bolin, B., Dickson, N. M., Faucheux, S., Gallopín, G. C., Huntley, B., Jodha, N. S., Kasperson, R. E., Mabogunje, A., Matson, P. A., Mooney, H., Moore, B., ... Svedin, U. (2001). Sustainability science. *Science, 292*, 641–642.

Klintman, M. (2013). *Citizen-consumers and evolution: Reducing environmental harm through our social motivation*. Palgrave Macmillan.

Lorde, A. (1982). *Learning from the 60s [Talk at the Malcolm X weekend at Harvard University]*. Harvard University.

Rand, A. (1961). *The virtue of sefishness*. Signet.

Simon, H. A. (2002). Forecasting the future or shaping it. *Industrial and Corporate Change, 11*, 601–605.

Žižek, S. (2010). *Living in the end times*. Verso.

Index

Note: 'Page numbers followed by "*f*" indicate figures and "*t*" indicate tables and "*b*" indicate boxes.'

E

F

G

O

P

Q

R

T

U

V

W

Y

Printed in the United States
by Baker & Taylor Publisher Services